T0189094

Passive Circuit Analysis with LTspice®

Colin May

Passive Circuit Analysis with LTspice®

An Interactive Approach

 Springer

Colin May (retired)
University of Westminster
London, UK

Simulation files can be found at: https://www.springer.com/us/book/9783030383039. In order to use the simulation files with the extension '.asc' referenced in the text it is necessary to download the free programme LTspice. The simulation files are grouped in folders for each chapter. These can be downloaded into an appropriate enclosing folder. The essential files in each are the schematic '.ASC' files which can be run by LTspice. Some schematics rely on symbols and sub-circuits created in the text. Most (if not all) can be found in the folders 'Mysym' and 'Mysubs'. These can be downloaded and included in the LTspice search paths. Additionally, some schematics, especially chapter 2, use commercial sub-circuits and symbols provided by device manufacturers. These too can be downloaded and must be included in the search paths. Some additional files are also included, but not for every schematic. The '.NET' files are netlist files that can be run by LTspice instead of the '.ASC' file but these do not show a circuit schematic. The '.PLT' files are pre-determined settings for how the traces should be drawn and also annotations to the trace. They are not executable files.

ISBN 978-3-030-38306-0   ISBN 978-3-030-38304-6   (eBook)
https://doi.org/10.1007/978-3-030-38304-6

© Springer Nature Switzerland AG 2020
This work is subject to copyright. All rights are reserved by the Publisher, whether the whole or part of the material is concerned, specifically the rights of translation, reprinting, reuse of illustrations, recitation, broadcasting, reproduction on microfilms or in any other physical way, and transmission or information storage and retrieval, electronic adaptation, computer software, or by similar or dissimilar methodology now known or hereafter developed.
The use of general descriptive names, registered names, trademarks, service marks, etc. in this publication does not imply, even in the absence of a specific statement, that such names are exempt from the relevant protective laws and regulations and therefore free for general use.
The publisher, the authors, and the editors are safe to assume that the advice and information in this book are believed to be true and accurate at the date of publication. Neither the publisher nor the authors or the editors give a warranty, expressed or implied, with respect to the material contained herein or for any errors or omissions that may have been made. The publisher remains neutral with regard to jurisdictional claims in published maps and institutional affiliations.

This Springer imprint is published by the registered company Springer Nature Switzerland AG
The registered company address is: Gewerbestrasse 11, 6330 Cham, Switzerland

In Memoriam
Ελένη Αθανασοπούλου
(1979–2019)
'Σε έναν 'Αγγελο...'

Eleni Athanasopoulou
She was taken too early

Foreword

This book is idiosyncratic in that it includes oddities not included in the general run of things, such as the Hamon voltage divider, the Murray loop test, the Elmore delay and the Thiele-Small loudspeaker model. These are applications of passive circuits which were encountered during countless trawls through the Internet and seemed both ingenious and interesting. But it is hoped that the text also encompasses the usual range of circuits.

SPICE is an analytical tool, just as manual analysis is, a very powerful tool that can return far more accurate results than are generally possible with pencil-and-paper analysis because it can include nonlinearities and temperature effects, but still it is just an analytical tool. At present, it is not possible to insert the required system performance and for SPICE to create an appropriate circuit: this, perhaps, is a project for Mike Engelhardt. So LTspice does not absolve the circuit designer from the onerous task of designing the circuit – it will only report how the circuit performs.

We must also answer the question "Why bother using SPICE?" To this are our several answers. Granted that it takes a little time to learn the basics – and a bit more to become proficient – it will be found afterwards that circuits can be simulated and tested faster than is possible with manual calculations. But does this mean that prototyping on the workbench is redundant? Far from it. LTspice will only analyse the data it is given: so stray capacitance and inductance that bedevil practical circuits will only come to light on the real physical circuit. On the other hand, the exigences of health and safety considerations, as well as the time needed to construct pro- totypes, make LTspice an attractive adjunct where component values can be changed and tested in seconds. And it is also invaluable in one other important area which is often overlooked – production yield. Components are not exact but have tolerances. It is clearly unreasonable to build a few thousand actual circuits to see which are out of specifications, but no problem with LTspice – even if it means leaving it to run all night.

In the early days, that is, pre-millenium, some versions of SPICE were prone to unfortunate errors; and although it is not something germane in this text, in one version at least, it was possible to mistakenly connect an opamp intended as a unity gain buffer, with the feedback taken to the positive input instead of the negative, and to find that the circuit still worked as a buffer. Nowadays, such errors have been fixed, but still the user must beware of false results.

These still occur and originate from two sources; one is the default settings of LTspice. These are accessed through **Tools- > Control Panel**, and there are three of importance: the first is that by default the results are compressed; in most cases, this is very useful since it greatly reduces the size of the simulation file – the *raw* file – by a factor of at least ten. However, it does mean that there is imprecision in measuring sharp resonant peaks or transitions, and then it must be turned off. The second is that by default, inductors have a parallel damping resistor and a small series resistance: it is this latter that can lead to false reductions in the output of a tuned circuit. The third area of concern is the rise and fall times of pulse input waveforms where the default of zero has significant slopes; setting each to something like *1e-15* results in vertical edges, but it is always worth checking. Otherwise, the default settings have been used throughout this book, only needing the Gear method to resolve a rather awkward example of the definition of a capacitor by its charge in Chap. 7, nor generally adjusting the settings in **Tools** $\rightarrow$ **Control Panel** $\rightarrow$ **SPICE**, although tolerance changes were tried for one or two simulations that ran slowly or stopped, but there did not seem to be any improvement. This leads to a second consideration, and that although it is very unlikely that LTspice miscalculates the performance of a circuit, it is always a good idea to start with a simple circuit, without the complications of nonlinearities or temperature effects, and check that the results agree with calculation.

The second source of errors is in models and subcircuits. Models are predefined in SPICE and it is not within our power to add extra parameters. Therefore any shortcomings, such as the failure of bipolar transistor models to include the base-emitter breakdown, can only be dealt with by creating a subcircuit. These are another matter and are essential where the performance of a device such as a thermistor cannot adequately be replicated by an existing model, and it is therefore necessary to construct a circuit to create the correct characteristics. In some cases, there are agreed subcircuit structures, but not always, and the performance of the subcircuit may fall short of the actual device which we shall explore, especially in Chap. 3.

The structure of this book, as far as possible, is to start with simple circuits and basic analytical tools and then progress to more complex circuits and advanced analytical methods on the basis that it is more productive to move from the known to the unknown rather than the other way round. Also, an exhaustive listing of, say, the *.meas* statement is apt to be rather indigestible, and so its features are introduced as needed, scattered over several chapters. On the other hand, there seemed no alternative but to list, with illustrations, the enormous capabilities of the voltage and current sources as Chap. 5. Additionally, some topics such as RCL circuits are covered twice, but in different fashions.

Perforce, the text is essentially mathematical because LTspice makes a no-compromise analysis of the circuit, returning results correct to six or seven figures, including aspects of the circuit's performance that are not always covered in text books. This sometimes results in puzzling results prompting a "What the (insert expletive) is happening here?" These can often be the most fruitful simulations bringing to light side effects that were overlooked in a simple manual analysis (or were omitted due to an inadequate grasp of the theory).

Worked examples with accompanying simulations pervade the book, but there are no end-of-chapter lists of questions, just a few suggested explorations at the end each section. The hope is that the theory and examples will stimulate more investigations, which should be seen as a challenge, not an onerous burden. Indeed, if the thought of having to simulate a circuit brings a deep sigh and a heavy heart, it is respectfully asked if electronics is the right career – life is too short to spend one's working days doing something that does not bring deep satisfaction.

LTspice is under continuous development, so some simulations that caused problems now run smoothly, and for others the comments may not now be relevant. Also, a few circuit symbols have changed. And sources now require five rather than four parameters.

Inevitably, there are mistakes, for which I crave the readers' indulgence and a gentle smile rather than *schadenfreude*. To err is human, but for a really good foul up you need a computer. I hope I have avoided the opacity variously attributed to Robert Browning at https://www.google.com/search?client=firefox-b-d&q=only +god+and+browning.

Patra, Greece   Colin May

Contents

Chapter 1
LTspice Essentials

1.1 Introduction

The aim of this book is to provide what, hopefully, is a useful and enjoyable introduction to LTspice mainly because the text is structured so that many simulations that go hand-in-hand with the analysis to reinforce and extend the topic are available from the website. For this reason, there are illustrative worked examples and suggested explorations, but no end-of-chapter questions since it is entirely possible to create a circuit, make the analysis and check the result by simulation. And this is open-ended leaving room for flights of fancy in circuit design. And because these are only simulations, a few hundred kV or a million amps or two will not bring down the wrath of the Health and Safety executive nor excite the fears of the laboratory staff. And under this later head, it must be emphasized that simulation is not a replacement for building and testing the actual circuit. Certainly there are cases where this is just not possible; the design of integrated circuits is the prime example. Otherwise, after the LTspice analysis, we need to build the circuit to find out if there were things we had forgotten in the simulation: stray capacitances, the internal resistance of a source, things like that.

And it must always be remembered that SPICE in whatever form is an analytical tool, a very powerful one, to be sure, but still only a means of analysis, and it does not absolve the circuit designer of the hard work of designing the circuit in the first place, although judicious use of 'what if?' simulations can often give helpful clues about the circuit's performance. In addition, by using SPICE, we can quickly find out more about a circuit's behaviour than would be economically possible to calculate. A very simple example is the optimum load resistor for maximum power transfer from a DC source. The calculation is not difficult, but it takes time to repeat the sums to find the seriousness of a mismatch. We have the answer in seconds with SPICE. And many circuits have start-up transients that are difficult to calculate and which are easily seen using SPICE. Many years ago, in the pre-SPICE

© Springer Nature Switzerland AG 2020
C. May, *Passive Circuit Analysis with LTspice*®,
https://doi.org/10.1007/978-3-030-38304-6_1

days, the start-up current surge of a main power supply could only be estimated using graphs. This was a serious matter because it could easily cause device failure if it was not limited.

But perhaps the greatest advantage of simulations is that of including the imperfections of practical components: inductors have resistance, capacitors also, and everything can change with temperature, not just linearly but with a quadratic or higher polynomial relationship, or even exponentially. These can give the LTspice user scope for ingenuity in finding ways to model these circuits. Many years ago, the late Bob Pease of Analog Devices created a SPICE shoot-out of a few circuits of no great complexity but extremely difficult to simulate to the required accuracy. CPUs have moved on since then, so perhaps now it is only time for a cup of coffee, but then it took an all-night run to get an answer. And this is another useful feature of SPICE; doing, for example, a statistical analysis to estimate the production yield means that we can get on with something useful and leave SPICE to crunch the numbers.

Those familiar with LTspice can safely skip this chapter, or perhaps skim the last section. Otherwise a good starting point to trace the interesting history and development of SPICE is: https://docs.easyeda.com/en/Simulation/Chapter14-Device-models/index.html.

The book, perforce, is mathematical. This is for two reasons: the first is that it is never safe blindly to accept the results of a simulation – there should at least be an approximate analysis that agrees in the main with the simulation. This is not to say that the simulation is faulty – although that is possible if the analysis is poorly set up – but rather as a check that the underlying concept is sound. The second reason is that LTspice will show second-order effects that are often glossed over: in particular, the resistance of the inductance in a tuned circuit and the effects of pulse excitation of a tuned circuit rather than a sinusoid. These will be exhaustively dealt with in an appropriate chapter.

There were two conflicting requirements in the layout of the book: one was to cover the gamut of LTspice's capabilities and the other to explore the range of passive circuits. Trying to cram, for example, all the 'Simulate' options into a chapter seemed rather indigestible and disruptive of an orderly flow of topics since some analyses are appropriate to DC and others not. Therefore, apart from the chapter on voltage and current sources, the approach has been to start with simple concepts and circuits – such as Ohm's Law in this chapter – and build to more complex circuits and analytical tools for DC, then to move to AC and capacitors and inductors and higher things. The result is that the LTspice commands, directives, call them what you will, are introduced as they are needed, and often only in part so that, for example, the *.meas* directive is scattered through four chapters.

The contents of this book are idiosyncratic. Certainly it is to be hoped that the analysis and applications of the most popular passive circuits have been covered somewhere, but there are also such diverse items as spark transmitters, loudspeakers and thermal modelling. In short, anything that looked interesting or unusual. Under this head comes the Hamon Potential Divider that creates a highly accurate division by 10 using not-so-close tolerance resistors, the Murray and Varley loop tests to find faults in cables and the Tapped Capacitor Impedance Divider to efficiently match a source to a load.

It should be mentioned that filters in one form or another appear in more than one chapter. It can be argued that simple RC circuits are filters of sorts. Therefore having discussed those, it is no great stretch to cascaded RC filters, although the 'T' and Bridged 'T' are somewhat more tricky. Likewise, having explored inductors, it is appropriate to deal with notch filters with inductors replacing capacitors. But it still seemed appropriate to create a special chapter on filters which deals with the higher-order RCL filters and more esoteric methods.

So now to the contents of this chapter. First we shall find out how to build a simple circuit using Ohm's Law as a vehicle for exploring some of the most important properties of the programme including (of course) how to select and place components, how to assign values, how to edit them by rotating and flipping and even removing them. But first it is worth saying a little about drafting conventions.

Then having built the circuit, we can establish the DC conditions and view the results, using the flexible cosmetic abilities of LTspice to change colours, line thicknesses, and so on. Finally we can save the results in various formats.

But above all, simulation should be fun: it should excite curiosity. If you do not enjoy the intellectual challenge, perhaps you are in the wrong business.

1.2 Representing the Circuit

Information about a circuit is communicated through a circuit diagram using symbols to represent the physical components of the circuit. There are, however, two distinct applications of a circuit diagram

The first is to represent an actual physical circuit; typically it has a reference letter and number for each component as well as it its value, for example, *Rs 56k 5%*. The symbols therefore represent the physical devices which may not be ideal: for example, a resistor may also have some inductance. These diagrams often contain test point waveforms or voltage or current measurements and are used by service and repair staff. In passing, it should be noted that the size of the symbols is of no significance – a 100 ohm resistor is not drawn larger than a 10 ohm one – nor is the relative positioning of the symbols on the sheet of paper any guide to the position of the actual component on a printed circuit board.

The second, and perhaps the most common use, is in circuit design and analysis. Here, the symbols are understood to represent ideal components: resistors only have resistance; likewise wires are perfect conductors and possess neither resistance nor inductance. If it is necessary to include, say, the inductance of a resistor, then it can be shown as a separate ideal inductor in series with the resistor. LTspice, however, allows these secondary effects to be incorporated into the symbol, as we shall see in a later chapter.

1.3   Drawing Conventions

Obviously, there must be an accepted convention for representing electronic components and in drawing circuits. In Europe, the International Electrotechnical Com-
mission (IEC) symbols predominate, whereas in the USA the IEEE usage is
preferred. The divergences are few and should not give rise to confusion: in
particular, the IEEE symbol for a resistor is a zigzag line. This can occasion heated
debate about whether the line should first go to the left or to the right (with the
attendant political overtones), how many zigzags should there be and what angle
between them, whereas, apart from the ratio of width to length, not much more can
be said about the IEC symbol of a simple rectangle. Other differences will be
discussed as we encounter them. The LTspice toolbar and the directory immediately
opened when picking a component use the IEEE symbols although there already are
a few IEC symbols in a separate folder and it is not difficult to make more.

The drawing convention is that as far as possible, a circuit is read from left to
right – that is, with inputs on the left and outputs on the right – and from top to
bottom. Lines representing conductors are drawn horizontally or vertically, or at 45°
in special circumstances. Connections are drawn as filled circles, and only three
wires should join at a point, not four. This is to avoid the 'cake crumb' effect where
lines crossing shown by '+' could be turned into four lines joining by an extraneous
speck of dust in the photocopier. LTspice, however, does not enforce this. If it is
important to emphasize the point that a component is connected directly to another,
for example, that resistor *Rs* is connected to point *A*, then an 'offset junction' can be
used with the lines at 45° as shown in the rather fanciful circuit of Fig. 1.1 which
illustrates some of the conventions. It also uses the IEC symbols rather than the ones
used by LTspice.

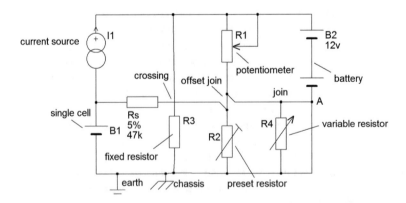

Fig. 1.1 Circuit Diagram

1.3.1   Component Symbols

The ones appropriate to this and the following chapter are as follows.

1.3.1.1   Voltage Sources

Voltage sources are either DC or AC. For our purposes, DC is a steady, unchanging voltage, either positive or negative. AC sources will be dealt with in later chapters.

So, turning to DC sources, there is a technical difference between a single cell and a battery: a single cell may either be single use (the traditional 'dry cell') or, more commonly now, rechargeable. They rely on the electrochemical potential difference between two metals. Typically this is around 1.2–1.6 V. A commercial battery consists of a number of identical cells in series and are often shown just by the single cell symbol with the voltage written beside it as *B2 12V*. Otherwise we may draw duplicated single cells, or we may draw two cells spaced apart and joined by a dashed or dotted line to indicate that there are others in between.

LTspice has symbols for both a cell and a battery in the ***misc*** folder, but – being lazy – it's easiest to use the general symbol for a voltage source consisting of a circle with '+' '−' drawn on it, and this also serves for AC sources. We should note that the cell and battery symbols are simply alternative representations of the voltage source circle and have the same attributes. So if we really want to create confusion, we can use the battery symbol and assign it an AC value, or have a cell with a pulse output.

1.3.1.2   Resistors

Simple resistors are shown as a rectangle, but notice that *Rs* also has its value and tolerance, and power rating could be added, but often, to avoid cluttering the diagram, this information is tabulated elsewhere rather than included in the diagram.

Variable resistors as *front panel controls* (i.e. those freely adjustable by the user, such as the volume control of a radio) are drawn with a diagonal arrow (R4), whereas *preset resistors,* whose value is set during calibration and not easily accessible, are shown with a 'hammer' (R2) rather than an arrow. These are two-terminal resistors, but very often, we use a *potentiometer* which is a three-terminal resistor with a sliding contact that divides the total resistance (R1). The difference is that we have a constant resistance between the ends and if we apply a signal between them, we can tap off a portion of the signal using the slider. Manufacturers generally only make potentiometers, and we make a variable resistor by leaving one end unconnected, or connect it to the slider which is how *R1* is drawn.

1.3.1.3   Current Sources

Current sources are difficult to find in practice and are often shown as two linked circles (I1), but diamonds with arrows are often used. Generally there is no confusion, and the context explains what is meant. LTspice uses a circle with an arrow pointing in the direction of current flow and is used for DC and AC sources.

1.3.1.4   Ground Connection

There is a subtle but important difference between the symbol for an earth connection which implies a low resistance conduction path to the earth itself (ideally zero ohms) and a chassis connection where the connection is made to a substantial metallic structure – traditionally the 'chassis' or framework on which the circuit was built – but which may not actually be connected to earth. LTspice uses a triangle, point down.

1.3.1.5   Connections

As was stated before, these need not be physical wires, but printed-circuit tracks or anything else offering a path with low resistance. It is implicit on the circuit diagram that these have zero resistance. LTspice uses a straight line for the wire and a small square for a join.

1.4   Drawing the Circuit and Ohm's Law

Now before we say that this is blindingly obvious, and why did it take so long to arrive at this law, it is worth remembering the state of electrical science at that point. Stable, reproducible voltage sources were not easy to come by, and the galvanometer had only recently been invented. Wikipedia has some excellent articles on this. So, let us state the law, then see how we can observe it (not prove it) using LTspice.

$$V = IR$$

1.4.1   Drawing the Circuit

The workspace is where the circuit schematic will be drawn. In addition, text comments can be added, and the schematic given a name, if so desired. It will also be found that, by default, the SPICE commands are also shown on the workspace. This can be inhibited, but in general, it is very useful to see what simulation has been

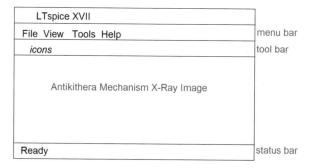

Fig. 1.2 LTspice screen

performed. In the following sections, stylized menus and edit boxes are shown with comments enclosed in brackets and some – but not all – of the captions, menu items and so on.

1.4.1.1   The Opening Screen

Download and open LTspice. A simplified opening screen is shown in Fig. 1.2. The very top row has the LTspice icon in the left corner followed by the version name, currently **LTspice XV11**. Below that is the 'menu bar' and underneath the icons of the 'tool bar' and then the workspace. The current default background is an X-ray of the Antikythera Mechanism – a truly remarkable machine showing that the ancient Greeks were not only philosophers, mathematicians and architects of the highest order but also instrument makers whose skill would not be equalled, let alone surpassed, for at least another thousand years. In this and later dialogues and windows the text presented by LTspice is shown in regular font, comments and explanations are shown in red, and typical user input is in bold. Menu selections are shown by an adjacent '<'.

Then, at the bottom of the window, the 'status bar' carries the single word – 'Ready'. Users of Ubuntu or other versions of Linux may see something different.

1.4.2   Placing Components

Click **File→New schematic**, Fig. 1.3 and the background image (which defaults to the Antikythera Mechanism) will be replaced by a grey field with a matrix of small dots, and the cursor will change to a large, thin '+'. We are now going to build a very simple circuit to illustrate Ohm's Law.

There can be up to four ways of placing a component. We shall explore these with a resistor.

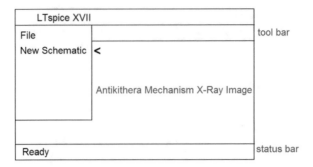

Fig. 1.3 New schematic

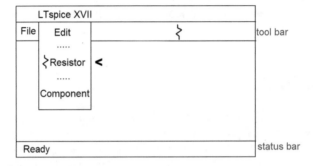

Fig. 1.4 Select resistor

1.4.2.1   Resistors

Left click **Edit** and select **Resistor**, Fig. 1.4. A large zigzag resistor will now replace the cursor. Move it to some convenient spot and left click to place it on the workspace.

A new *R2* will now replace *R1* and move with the mouse. Click '**Esc**' to dismiss it.

Example – Placing a Component

To test all the following methods, we need to remove unwanted resistors. We do that by pressing **F5**, and the cursor turns into a pair of open scissors. Move over the unwanted component and left click. The component will vanish. Right click to dismiss the scissors.

The second method is to click on the resistor symbol in the tool bar. The third is to right click on the schematic, and from the pop-up menu, select **Draft→Component** to open the **Select Component Symbol** dialogue, Fig. 1.5. The fourth, available only to some components, is to press the initial letter of their name whilst in the schematic. Thus *R* will create a resistor. We can test all these leaving just one resistor on the schematic.

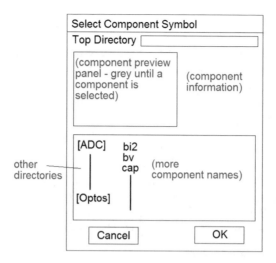

Fig. 1.5 Select comp

The 'Select Component Symbol' Dialogue
This is what we shall mostly use to select components. At the top is the full path to
the component. Below, to the left, is the symbol itself and in the space to the right
perhaps some words of explanation. Immediately below is a list box that can be
scrolled horizontally with all the accessible components and, at the start at the left,
directories for more components. We then have the choice to **Cancel** or accept **OK**.

1.4.2.2   Voltage Source

Return to **Edit,** but this time a voltage source is not available from the drop-down list
so we must select **Component** (and note the shortcut of **F2**). This also will open the
Select Component Symbol dialogue.

Scroll right, select **voltage,** and a circle with '+' and '−' signs will be seen in the
'component preview panel' in Fig. 1.5. Click **OK** and place to the left of the resistor
and level with it. Alternatively, click **[Misc]** and select **cell** or **battery**. These only
change the symbol, not the underlying functionality, so, rather bizarrely, as we noted
above, we can have an AC battery. For convenience (and simplicity), nearly all the
circuits use the **voltage** symbol, but please feel free to change.

1.4.2.3   Ground

This is accessible by all the methods for a resistor or press *G* in the workspace. This
is important: there must be a ground point somewhere in every circuit. Place it
between the voltage source and the resistor and a little below them.

1.4.2.4 Alternative Symbols

The IEC resistor, capacitor and a few others as well as cells and batteries can be accessed through **Edit→Component;** then in the **Select Component Symbol** dialogue, select **[Misc]** . They behave exactly as the default items. This is ('Ohm's Law 5.asc'). Later we can build a library of IEC symbols.

1.4.2.5 Component Names

LTspice assigns these in sequence such as *R1, R2, R3....* To give a more evocative name, move the mouse over the name, and the cursor will change to an 'I-bar', and the status line will read 'Right click to edit the Name of..' which will open the **Enter new reference designator for....** Dialogue. The names must be unique.

Component Reference Designator
This small dialogue opens when we are over the name of a component. The **Justification** options have little effect except that if we choose (not visible) we cannot recover it. The font size option applies to that instance only, as does making the text vertical.

If we just want to reposition the name or the value, press **F8** and the cursor will turn to a closed hand, left click on the element, move it, and right click to finish.

1.4.3 Connecting the Circuit

We are now ready to connect the circuit. Click on the pencil, and the cursor will turn to two dotted lines at right angles spanning the workspace. Move the intersection of the lines to one of the open square terminals of a component and left click. This will start the wire. Now move the cursor to another terminal, and note that the wire can only be drawn on a rectangular grid. Left click to finish the wire.

Draw the remaining wires and dismiss the wiring tool. You should now have the schematic ('Ohm's Law 1.asc') . Make sure the earth symbol is connected.

There is a trick to make life easier, and that is to draw straight through the components rather than connecting one end at a time. Create the schematic Fig. 1.6 of just the five resistors. (Press **Ctrl+R** to rotate the resistor before it is placed) Then select the wire tool and left click on point A. Now move straight to point *B* and left

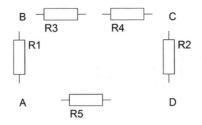

Fig 1.6 Easy wiring

click again: the wire will be trimmed to the ends of the resistor. Continue the wire to point *C* and left click again, and the wire will join up resistors *R3* and *R4*. And continuing in like fashion to point *D* then back to *A* will join up the circuit.

You can zoom in and out using the mouse wheel, and hold down the left mouse button and drag to move the circuit.

1.4.4 Adding Values

LTspice insists that every component has a value. At present, they only have labels.

1.4.4.1 Resistors

Move the mouse over the resistor and note that the status bar now has 'Right click to edit . . . .' with the name of the component under the mouse and the cursor changes to a pointing hand: now right click.

The **Resistor – R1** dialogue will appear, Fig. 1.7 with the **Resistance** edit box highlighted. Insert some number such as *100* and click **OK**. LTspice accepts all the multipliers including 'u' for 'μ', but note that 'm' or 'M' is always 'milli'. Either use scientific notation such as 1e6 for a megohm resistor, else '1Meg' is accepted.

We should also note that we can enter the tolerance and power rating of the resistor. LTspice accepts them but does not use them. We can click on the **Select Resistor** button to open the catalogue and select a resistor from there. As it happens, the **Mfg.** and **Part No.** fields are empty, but had they been filled the information would have been copied to the dialogue.

1.4.4.2 Voltage Sources

Move to the voltage source, right click again, and the dialogue **Voltage Source – V1** will appear, Fig. 1.8. Insert some value such as *10* in **DC value[V]** and click **OK**. Ignore the **Advanced** button – that is for later chapters.

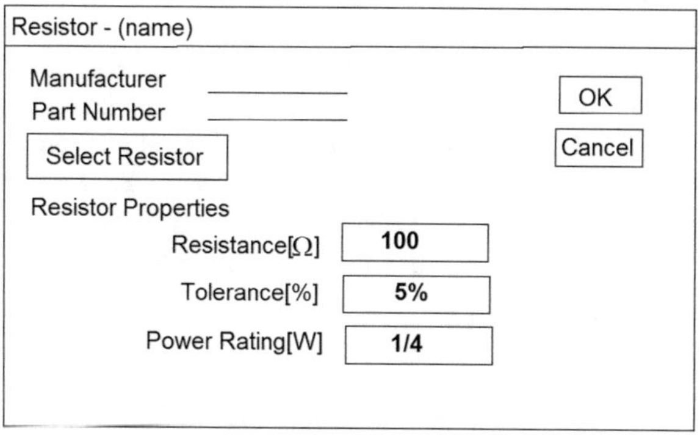

Fig. 1.7 Resistor dialogue

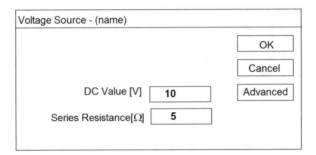

Fig. 1.8 V Source dialogue

We can also insert a series resistor to represent the internal resistance of the source. This is computationally more efficient than an external resistor but leave it empty now. The circuit is complete.

1.4.5   Editing the Circuit

With such a simple circuit, it is unlikely that much will need to be done. However, for practice, try the following, and note that, in practically every case, a left click will finish the action and **Esc** or a right click will cancel it.

To Cancel a Component
While it is still being dragged on the schematic and before it has been placed, right click or press *Esc* to cancel it.

To Remove Unwanted or Wrong Components
Press **F5** or select the scissors on the toolbar. Move the scissors over the component and left click; right click afterwards to dismiss the scissors.

To Move or Rotate a Component Without Its Connecting Wires
This can be useful for swapping components. It is also useful to be able to rotate a component through 180° since the direction of the current measured by LTspice depends on this.

Before it has been placed, press **Ctrl+R**. Once it has been placed, press **F7** or select the open hand from the toolbar, move the hand over the component, and left click. The hand will disappear and now **Ctrl+R** will rotate the component. At this time you can also drag the component without its wires to some other position. Left click to finish the move, then **Esc** to restore the cursor. If **Esc** is pressed or a right click is made, before the left click, the move is cancelled.

To Flip a Component Left-Right
While it is still being placed, press **Ctrl + E** then left click to restore the hand and **Esc** to finish. It can be dragged to the 'E' and reversed 'E' button on the toolbar, but this

can be confusing. Once it has been placed, select **F7** or **F8** then press **Ctrl + E. Esc** will undo it.

To Move a Component with Its Connecting Wires
Press **F8** or select the closed hand from the toolbar. Trying to rotate or flip a component with its wires usually creates a tangle, and it is better to select **F7** to handle the component alone. However, it can be useful for sliding components along a wire to make a more aesthetic layout.

To Move Several Components and Their Connecting Wires
Click **F8** then hold down the left mouse button to draw an enclosing rectangle. Release the key and drag. Left click to finish or right click to cancel. And **Edit→Undo** will also restore the previous setting.

To Move Just the Name of a Component, or Its Value
Click **F7** then click on the item and drag it.

To Change the Value or Name of a Component
Once the component has been given a value, move the cursor over it and it will change to an 'I-bar' and a right click will again bring up the editor. Similarly we can edit the name.

To Copy Components
For a single component, press **F6**, move over the component and left click. Now a copy can be dragged without any connecting wires. As usual, left click to place it, then right click to restore the cursor, or right click or **Esc** to cancel before it is placed.

To copy more than one component, press **F6** then hold down the left mouse button and drag an enclosing rectangle. Release the button. If the rectangle encloses wires with both ends connected, the wire will be copied as well as the enclosed components.

To Undo or Redo
Click the left or right arc to the right of the closed hand or **Edit→Undo/Redo** or right click in the schematic window then **Edit→Undo/Redo** from the pop-up menu. This works for a depth that is sufficient for any reasonable schematic. It can be used for 'what if?' explorations by changing a the value of a component (or, indeed, changing the component itself) and then restoring the original value.

To Give a Node a Label
The nodes are labelled from 001 onwards by default. It is often more convenient to give them a meaningful label. Move the mouse over a wire: right click and select **Label net**. The label with a small square box underneath will then replace the cursor. Move the box over a wire and left click to place it or right click to dismiss it.

To Remove Several Components
Select the scissors tool, move to a point outside the components to delete then drag to define a rectangle enclosing the components. Release the mouse button and they will be deleted, but **Redo** will bring them back again.

1.4.6 Annotations

These are not essential, but can be added to the schematic by two methods.

To Add Text
Press 'T' (upper or lower case) with the mouse in the schematic, or right click on any open space in the schematic and from the pop-up menu select **Draft→Comment Text**. Either will open **Edit Text on the Schematic**, Fig. 1.9.

1.4.6.1 Edit Text on the Schematic Dialogue

The options are as follows.

Comment or SPICE Directive
These two radio buttons determine how the text is handled. A comment will be shown in blue, a directive in black. Make sure **Comment** is checked. You can toggle between the two. In particular, to switch between different analysis options, right click on the command then click **Cancel** in the **Edit Simulation Command** dialogue to open the **Edit Text on the Schematic** dialogue and click **Comment**. The text is unchanged; it is just turned to blue.

Justification
This option follows the usual practice except that the text can be hidden (**not visible**) posing the problem of how to show it again.

Font
This is a global setting. And note that this and the previous are best set after the text has been entered because they close the dialogue.

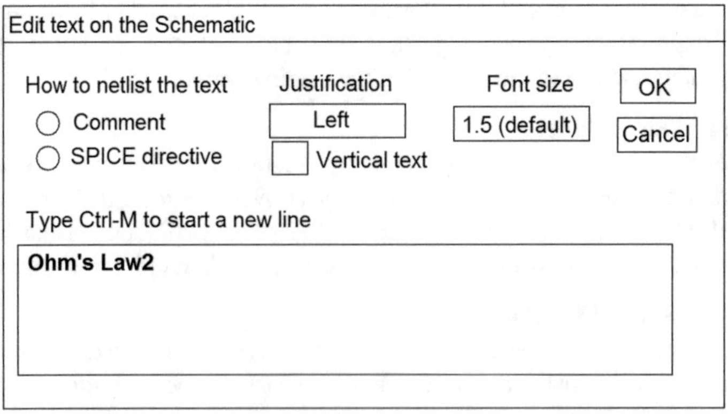

Edit text on the Schematic

How to netlist the text   Justification   Font size   OK
  ○ Comment   Left   1.5 (default)   Cancel
  ○ SPICE directive   □ Vertical text

Type Ctrl-M to start a new line

Ohm's Law2

Fig. 1.9 Annotation Dialogue

Example – Editing Text and Changing Comment to SPICE Directive
Press 'T' on the schematic to open the text dialogue and enter 'Ohm's Law 2' and click **OK.** The text will replace the cursor, move it to some convenient place, then left click to fix it. Now right click on the text and select the radio button **SPICE directive.** The text will turn blue. Of course, this is not a proper SPICE directive, so we should click again and turn it back to a comment. This is now schematic ('Ohm's Law 2.asc') with an additional comment added.

1.5   Running the Simulation and the .op Command

The *.op* command finds the quiescent operating point by setting all capacitors to open circuits and all inductors to short circuits. It is only really necessary before any small-signal simulation where an operating point must be established first. Most simulation commands involve a full analysis from DC. There are two options to do this:

From the main menu, select **Simulate→Edit Simulation Cmd,** Fig. 1.10, which will open a dialogue, Fig. 1.11. Choose the right-hand **DC op pnt** tab and notice the cursor blinking ahead of the text *.op* in the edit box. LTspice has filled in the type of simulation, which, in this case, is complete, but had we selected other simulations we would have to add more to the command edit box.
Or right click in a free space in the schematic, and from the pop-up menu click **Edit Simulation Cmd.** which opens the same dialogue.
Click **OK** and the dialogue will close. Move the text to some convenient place and left click to fix it.
Now click the running man on the toolbar, or right click on the workspace and click **Run.** After a short pause, a small window will appear with the results.

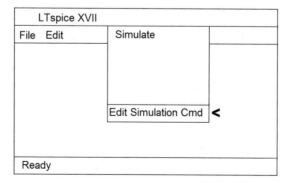

Fig. 1.10 Select Simulation

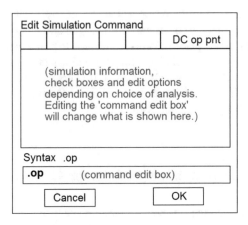

Fig. 1.11 Simulation Options Dialogue

1.5.1   Simulation Results

Some are available on the schematic, the rest are in files.

1.5.1.1   Voltage, Current and Power Probes

LTspice enables much useful information to be displayed in the status bar by moving the mouse over a component. What is seen depends on the type of analysis. If we find the operating point by the **.op** command and move the mouse over a component, the cursor will turn into a pointing hand and the status bar will show the current and power. Assuming a 10 V source and 100 Ω resistor we will see:

$$\textbf{DC operating point I(V1)} = -\textbf{100mA Dissipation} = -\textbf{1W}$$
$$\textbf{DC operating point I(R1)} = \textbf{100mA Dissipation} = \textbf{1W}$$

The negative signs are because *V1* is supplying power. If we detach *R1* using **F7**, we can rotate it twice and replace it. When a component was first placed, the text was on the right; it is now on the left. Now we run the circuit again; we find that the status bar shows the current in the resistor is **−100 mA**.

If we move the mouse over a wire, the status bar gives us useful information about it. In this case, we see:

$$\textbf{This is node N001.DC operating point V(n001)} = \textbf{10 V}$$

We can also hold down **Ctrl** and click on the wire to show its voltage.

If, however, we had used another analysis without a unique result, for example, if we had stepped the excitation voltage (schematic 'Ohm's Law 3.asc') we find the cursor is a clamp ammeter with a red arrow showing the direction of the current and the status bar shows:

'Left click to plot I(R1), right click to edit'.

LTspice also creates three files.

1.5.1.2   The Text Document (.log)

This is technical information on the simulation conditions and how the analysis was performed and need not detain us here. It can be opened by **View->SPICE Error Log** Fig. 1.12. Its contents, with numbered comments, are:

```
Circuit: * C:\Users\Me\Desktop\LTSpice Book\01 DC Circuits\Circuits
(asc)\Ohm's Law2.asc {1}

Direct Newton iteration for .op point succeeded. {2}

Date: Wed Mar 14 23:14:32 2018 {3}
Total elapsed time: 0.030 seconds. {4}

tnom = 27 {5}
temp = 27 {6}
method = trap {7}
totiter = 3
traniter = 0
tranpoints = 0
accept = 0
rejected = 0
matrix size = 2
fillins = 0
solver = Normal
Matrix Compiler1: 36 bytes object code size
Matrix Compiler2: 96 bytes object code size
```

The points to note are these:

{1}. The fully qualified name of the file, which extends to a second line.
{2}. The method used to find the solution. The Newton-Raphson method works for most circuits, but there are others which LTspice will automatically try if this fails.

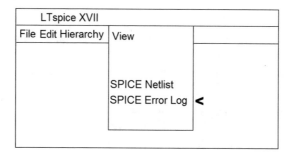

Fig. 1.12 View Netlist

{3}. The date and time.

{4}. The time to perform the analysis. In this trivial circuit, it is of no importance, and even complex circuits often solve in just a few seconds.

{5}. The nominal temperature at which the component parameters were measured. In most cases, it is 300 K (27° C), but for some, such as resistance thermometers, it is 273 K.

{6}. The simulation temperature.

{7}. This is the integration method which defaults to modified trapezoidal which we shall not change.

The remaining lines report the performance of the solver, but we should notice how small the matrix is, and even quite complicated circuits often are less than 1000 bytes.

1.5.1.3 The .NET File

This contains the circuit listing and the SPICE commands and is identical to the file in the window opened by **View->Spice Netlist** Fig. 1.12 above, with the important difference that this can be edited.

```
* C:\Users\Me\Desktop\LTSpice Book\01 DC Circuits\Circuits
(asc)\Ohm's Law2.asc {1}
R1 N001 0 100 {2}
V1 N001 0 10
* Ohm's Law {3}
.op {4}
.backanno {5}
.end {6}
```

{1}. Any line starting with an asterisk is a comment, and the second line is simply a continuation of the first.

{2}. The components are described as:

$$< \textbf{Name} >< \textbf{Node1} >< \textbf{Node2} > \ldots < \textbf{Value} >< \textbf{optional extra parameters} >$$

In this case we have a resistor *R1* of 100 Ω connected between nodes *N001* and earth and a voltage source *V1* of 10 V connected between the same two nodes, but other devices may have more nodes.

{3}. Any comments placed on the schematic. These are ignored by the compiler as they start with an asterisk.

{4}. SPICE commands all start with a full stop and so are known as 'dot commands'.

.op – finds the quiescent operating point

{5}. backanno – is associated with the raw data file and allows text to be added afterwards when the file is viewed. This is saved with the file.

{6}. end – finishes the netlist

1.5.1.4   The RAW Files

These contains the results of the simulation, and, in this specific case, it is just a text
file with the results we saw in the window that opened after the simulation finished,
often with the extension **.op**. The file can be read by LTspice itself. To open it, click
File→Open, then depending on how the computer has been set up, in the drop-down
list box to the right of the 'File name', edit select 'Waveforms (∗.raw;∗.fra), else if it
is a full-screen window, the options will be at the bottom right corner of the window.

1.6   Sweeping Voltage and Current Sources

It could be argued that the previous result was due to a fortuitous choice of voltage,
and other values would show a different relationship between voltage and current.
Therefore we shall try several voltages by sweeping *V1*.

1.6.1   DC Sweep Command

Starting with schematic ('Ohm's Law 2.asc'), again try **Simulate→Edit Sweep
Cmd,** only this time select the **DC sweep** tab, Fig. 1.13. Initially we have only **.dc**
in the edit box.

Enter *V1* for the **Name of 1st source to sweep** and select *Linear* for **Type of
sweep**.

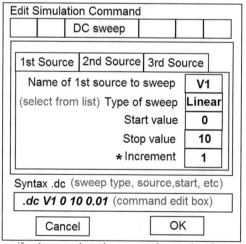

Fig. 1.13 DC Sweep Command

Enter *0* for **Start value** and *10* for **Stop value**. There is no need to add the unit *V*. And note that these values are copied to the edit box. Set the **Increment** to *1* and click **OK**.

The dialogue will close, and the edit box text will replace the cursor. Fix it at some convenient place, and note that *.op* is now preceded by a semi-colon which cancels the command, although the colour has not changed. This is schematic 'Ohm's Law 3'. This was actually set up slightly differently so the *op* command has been turned into a comment and is blue.

Click on the running man.

1.6.2   The Trace Window

A new window will open above the schematic consisting of a black background and **0 V** to **10 V** in grey on the x-axis.

1.6.2.1   Adding Traces

Right click anywhere in the trace window and select **Add Traces** from the pop-up menu. This will open the **Add Traces to Plot** dialogue where the available traces will be listed, Fig. 1.14. We need not concern ourselves with most of the options for now, just click on the traces to add, and they will appear in the **Expression(s) to add** edit. Click **OK** to finish and the traces will be drawn. Generally voltages will be drawn on the left-hand y-axis and current on the right hand.

An alternative is not to open the dialogue but to move the mouse over a component or wire in the schematic. The status bar will then show words to the

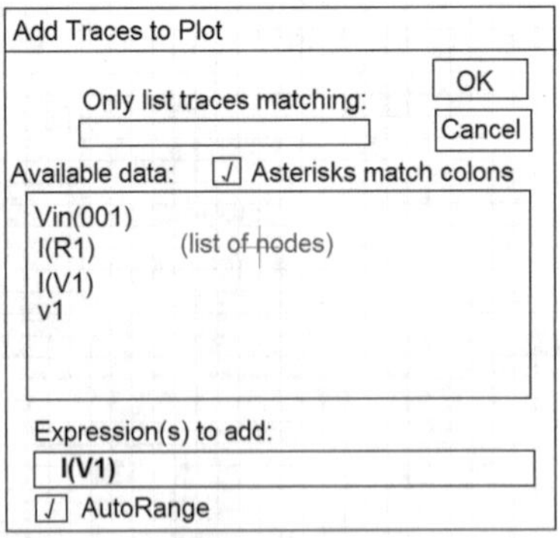

Fig. 1.14 Add trace dialogue

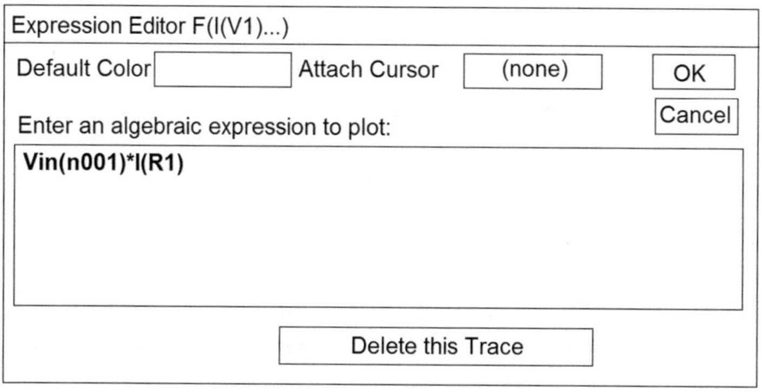

```
┌─────────────────────────────────────────────────────────────────┐
│ Expression Editor F(I(V1)...)                                     │
│ ┌─────────────────────────────────────────────────────────────┐ │
│ │ Default Color[   ]   Attach Cursor   (none)        OK       │ │
│ │                                                   Cancel     │ │
│ │ Enter an algebraic expression to plot:                      │ │
│ │ ┌─────────────────────────────────────────────────────────┐ │ │
│ │ │ Vin(n001)*I(R1)                                         │ │ │
│ │ │                                                         │ │ │
│ │ │                                                         │ │ │
│ │ │                                                         │ │ │
│ │ └─────────────────────────────────────────────────────────┘ │ │
│ │            ┌─────────────────────────────┐                  │ │
│ │            │      Delete this Trace      │                  │ │
│ │            └─────────────────────────────┘                  │ │
│ └─────────────────────────────────────────────────────────────┘ │
└─────────────────────────────────────────────────────────────────┘
```

Fig. 1.15 Expression editor

effect '**Left-click to plot I(R1) Right click to edit**' depending on what is underneath the mouse. But note – this will also remove any traces previously set up using the **Add Traces to Plot** dialogue.

1.6.2.2 Editing Traces

Move the mouse to the title bar of the traces window over the name of a trace, and the cursor will change to a pointing hand. Right click and the **Expression Editor** dialogue will open, Fig. 1.15. Here we can change the colour of the trace and also add an expression to plot such as $I(R1)*V(n001)$ to plot the power. The allowed functions follow the usual rule except that exponentiation is two asterisks $**$ not a circumflex $\wedge$. There are many more possibilities including trigonometric functions. We can also add an expression using the **Add Traces to Plot** window.

Example – Plotting Traces
So, if we have the 'Ohm's Law 2' schematic open, add traces for the current in the resistor and the input voltage. We can also enter the expression in the edit box $V(n001)*I(R1)$ to plot the power. We can either type the whole expression or a single – not a double – click on the two items and add the asterisk by hand. Note that we can only use the voltages and currents, we cannot, for example, plot $V(n001)**R1$ – the I^2R relationship using the resistor $R1$. We can use $I(R1)$ but not $R1$ itself. We can, however, use VI.

1.6.2.3 Changing the Colours

If we want to change the colour of not just one trace, we click **Tools→Color Preferences** to change the colours of everything. In particular, a black background would use lots of ink were the traces to be printed, white is better.

We can change almost every aspect of how LTspice presents results using the **Tool Box,** but let us leave that for later.

1.6.2.4   Showing the Results

- Move the cursor in the trace window, and the current values for the position of the cursor are shown in the status bar at the bottom left of the screen. A typical reading is: $x = 5.00\ V\ y = 0.547\ W,\ 4.97\ V,\ 54.70\ mA$. This is totally false. The value is only correct if the cursor is sitting on a trace, and even then it is only valid for that one trace.
- Open the **Expression Editor** window and attach a cursor to the power trace. We now see cross wires extending the full width and height of the trace screen with either *1* or *2* when the cursor is on the cross-wire depending on which we have chosen. This will open a small window at the bottom right of the screen. The **Horiz** and **Vert** dialogues will show *250 mW* at *5 V*.
- We can incrementally move the vertical cross-wire with the left and right arrow keys.
- If we right click when we are actually on a cross-wire, that is, when we see a *1* or *2*, we may find a pop-up **Cursor Step Information** window.
- Move the cursor over a wire or component in the schematic, and **Ctrl** with left click will toggle the trace of the voltage or current.

1.6.2.5   Saving the Results

The results are automatically saved in a '.*raw*' file with the same name as the schematic.

Viewing the RAW File
From **Menu→Open,** select ∗.**raw** ∗.**fra** file types. Open this file to show a blank trace window with just the voltage scale along the x-axis. Now add the traces. You can also perform any manipulation you like; adding, subtracting, multiplying and so on; you are not restricted to the original ones.

Sending Results to a File
In addition, if you want to add the simulation results to another file, you can chose:

- **Tools→Write image to .emf file** or **Copy bitmap to Clipboard**. These options are also available by right clicking on the trace window and selecting **View**.
- Right click in the trace window, and choose **File→Export data as text** which will open the dialogue **Select Traces to Export**, Fig. 1.16. Hold down *Ctrl* to select more than one, then **OK**. A new .*TXT* file with the same name as the schematic will be created which, by default, will be placed in the same folder as the schematic; else click **Browse** to change it. The opening entries below show that the measurements were made at 10 mV intervals and the selected traces

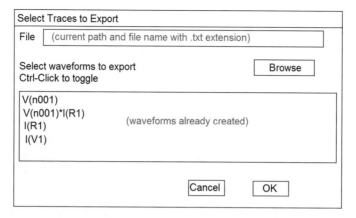

Fig. 1.16 Export traces

includes *v1*. We should note that this is a pure text file and therefore can be passed to a spreadsheet or other programme for further processing.

v1	V(n001)	V(n001)*I(R1)	I(R1)
0.000000000000000e+000	0.000000e+000	0.000000e+000	0.000000e+000
1.000000000000000e−002	1.000000e−002	1.000000e−006	1.000000e−004
2.000000000000000e−002	2.000000e−002	4.000000e−006	2.000000e−004
3.000000000000000e−002	3.000000e−002	9.000000e−006	3.000000e−004

1.6.2.6 Saving Plot Settings ('Plt' File)

This will save not the plot itself, but the axes, grid style and other information. Right click in the trace window and select **File→Save Plot Settings** and it will be save as plain text. That for schematic ('Ohm's Law 4.asc') is:

[Operating Point]

```
{
  Npanes: 1
  {
   traces: 1 {524290,0,"1/I(R1)"}
   X: ('',0,1,10,100)
   Y[0]: ('',0,0,1,10)
   Y[1]: ('',0,1e+308,1,-1e+308)
   Units: "A-1" ('',0,0,0,0,1,10)
   Log: 0 0 0
   GridStyle: 1
  }
}
```

and next time the schematic is run the same traces will be shown with the same axes. Any cursors are not saved.

1.6.2.7 Printing

The schematic and the traces can be printed.

- Right click in the appropriate window and select **File→Print preview** from the pop-up menu. This is common to both schematic and traces. Right click again on the window and another pop-up window appears with the check box **Print Monochrome** to print in colour or black and white.

However, these will expand the circuit or traces to fill the page. It is often more convenient to export them by select **Tools→Copy bitmap to Clipboard** or **Write image to .emf file**.

It can be advantageous to increase the trace thickness, change the font size and so on to make the traces clearer. All this can be found at **Tools→Control Panel** which we shall now explore.

Under Windows, the snipping tool is available and was used to create several figures in this book.

1.6.3 The Control Panel

At this juncture it is more nice to know rather than essential information. We shall discuss only three of the ten tabbed windows: these are sufficient for the next few chapters. Notice that – irritatingly – not all the settings are remembered. The effects of most of those that change the visible schematic are immediately visible.

1.6.3.1 Basic Options

The controls of interest now are as follows.

Default Windows Tile Pattern
By default this is horizontal, which is the most useful as schematics are mostly long and thin, as are traces. But it can be changed to vertical.

Save all Open Files on Start of Simulation
Depends how nervous you are and how vital the files are. If you think LTspice could crash, check it.

Automatically Delete .Raw Files
These contain the results of the simulation. They may simply be the DC operating point or traces of waveforms. Unless there is little disc space left, it is a good idea to keep them because they can be reviewed, annotated and processed, for example, by creating a new trace that is the product of two others.

Background Image
This is the Antikythera Mechanism but there are two other choices.

Toolbar Icon Size
This defaults to Large. But you can have Yuge or Normal.

Directory for Temporary Files
This is not often of great importance, but it is nice to know where they are.

1.6.3.2   Drafting Options

These check boxes are of more immediate interest.

Allow Direct Component Pin Shorts
This means components can join directly without an intervening wire.

Automatically Scroll the View
This is set by default but can be irritating because it will scroll whenever a component is near the edge of the screen. If the schematic does go off screen, using the mouse wheel to zoom out will often make it visible; else **View→Zoom to fit** will bring it back.

Mark Text Justification Anchor Points
If enabled, a small circle will be shown. Leave it on if you want to be really precise about how you align text.

Mark Unconnected Pins
This puts a small square box on them which vanishes when the pin is connected. A useful indication of wire errors.

Orthogonal Snap Wires
If checked they must all be at right angles. There are some occasions when this is not the best option. Unchecking it frees up wires to snap to any grid point.

Ortho Drag Mode
If checked, this only allows things to be dragged horizontally or vertically. Best left checked.

Pen Thickness
The default is **1**, but for printing, slightly thicker is better. This only affects the schematic, not the trace panels.

Font
These are standard font settings. The defaults are generally best.

Color Scheme
This button opens the **Color Palette Edit** dialogue to set colours for the netlist, the schematic and the traces. Adjust them to whatever pleases you. The same dialogue is accessible through **Tools→Color Preferences**.

Hot Keys
These are best left as they are – and in any case, it is difficult to remember them and they do not seem to align with the shortcuts shown on menus.

1.6.3.3   Waveform Options

Most of these options are cosmetic.

Compression Data Trace Width
The default is **1** but **2** is clearer, and perhaps **5** for printing at reduced scale. The width does not affect the accuracy of measurements made with the cursor.

Cursor Width
Increasing this to **2** from the default **1** makes it easier to find on a white background.

Use Radian Measure in Waveform Expressions
A matter of choice if you prefer degrees. But trigonometrical functions use radians.

Use XOR Type Cross Hair Cursor
By default the cursor sits on top of the trace and is easy to see; but you can change it.

Font
This is a limited choice but adequate for labelling the traces

Font Point Size
A larger size is beneficial for printing at reduced scale.

Bold Font
This is checked by default.

Directory for Raw and Log Data Files
Check **Store raw .plt ...directory** and browse to change from the default.

1.6.3.4   Compression

We shall doubtlessly mention this again: turning compression on dramatically reduces the size of *.raw* file. That is not so important here, but can save tens of kB of disc space with complex circuits. The penalty is a slight loss of accuracy, so for very precise measurements, it should not be turned on.

1.7   Changing the Value of a Component During Analysis

There are two useful methods. The first is to change it step-by-step, repeating the simulation after each change. With complex circuits, this can take a little time. The second is randomly to assign values to components within tolerance limits.

1.7.1 Using Parameters '.param'

A *parameter* allows the same value to be passed to several components without having to change each individually: it also enables an expression to be passed to a *Value* field which only accepts a single number. To create the command we right click on the workspace and select **Draft→Spice Directive** or press 'S' in the workspace and the **Edit Text on the Schematic** dialogue will appear. Ensure that the **SPICE directive** radio button is checked.

1.7.1.1 Syntax

The parameter is in two parts. The first is the definition of the parameter. This consists of a period (dot) followed by the reserved word *param* and the name of the parameter. Then follows an equals sign and the value.

$$\textbf{.param} < \textbf{parameter name} > \ = \ < \textbf{value} >$$

The name can be that of an existing component but it is less confusing if it is not. The *value* must resolve to a number before the simulation starts, so something like:

$$\textbf{Vx} = 24 * \textbf{time}$$

will throw a message that LTspice cannot resolve the parameter. This is because time is a variable of the simulation run. Equally, something like:

$$\textbf{V1} = 5 * \textbf{I}(\textbf{R1})$$

will fail because the current in *R1* is not known before the simulation starts. However

$$\textbf{Vi} = 5 * \textbf{pi}/2 + 3$$

is acceptable since *pi* is a constant.

1.7.1.2 Usage

Having created a valid *.param* statement, we then must change the *Value* field of the appropriate component to the parameter name enclosed in curly brackets as {Vx}. This is a placeholder for the actual value. The way it works is that LTspice first looks for parameter definitions and evaluates them. Then it checks to see where these parameters are used and substitutes the values for the placeholders. It is not a syntax error to have unused *.param* statements on the schematic. But if there is any parametrized component value, there must be a corresponding *.param* statement.

Creating a parameter is particularly useful with voltage and current sources which only allow a single value in most of the fields. If we open one of the 'Ohm's Law' schematics and try a DC value of **10 + 5**, we will find that it fails. However, if we create a parameter: **V1 = 10 + 5** and replace the DC value by **{V1}**, there is no problem. A point to note is that {V1}+5 will fail - the whole expression must be enclosed in curly brackets as {V1 + 5}.

And although it is not appropriate here, imagine we had several components all of the same value. If we parametrize them, we can change them all at once just by altering the parameter value.

1.7.2   Step Command '.step'

It is possible, but very tedious, to change a *Value* between runs to build a picture of how the result of a simulation changes with that *Value*. The *.step* command saves us the bother. It works in conjunction with a parameter to change the value between runs. It is accessed in that same was as a *param* and opens the same **Edit Text on the Schematic** dialogue. We can enter the complete command and close the dialogue, but if we are not sure we can just enter **.step,** close the dialogue and place the text on the schematic, and then right click on it and this time the **.step Statement Editor will open for us,** Fig. 1.17, which will show the current directive in the lower edit box and the pre-existing values, if any. The directive can be edited either value-by-value or as a whole in the edit box.

1.7.2.1   Syntax

This also is a dot command:

.step param < type of step >< parameter name >< start value >
< end value >< step size >

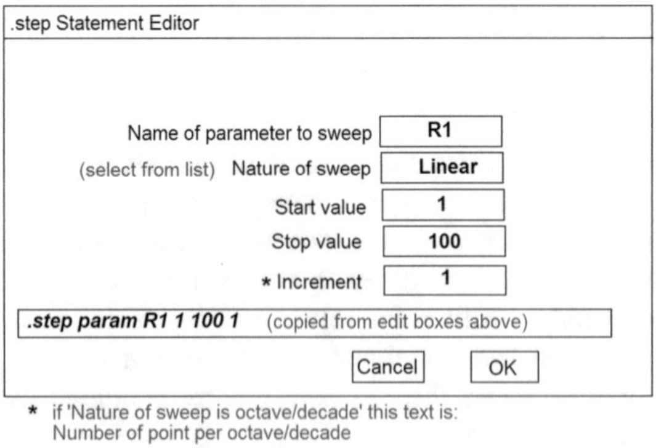

Fig. 1.17 Step editor

The parameter can be defined here as a single value or separately in a *param* statement. By default, SPICE uses linear steps. So, for example, *.step R 1 100 1* would step a resistor *{R}* in 1 ohm steps from 1 to 100. Remembering that a resistor must not be zero, we could have *.step R 1f 100 1* which is near enough zero. If the start value is larger than the end, there is no need to specify negative steps. But there should be enough steps to adequately resolve the changes: in general, *100* is about the minimum.

1.7.2.2 Usage

This is very flexible and it could be the value itself, or a temperature coefficient, or the tolerance. And it does not matter if this is the name of an existing component. And again if any value is parametrized, there must be a corresponding *.step* directive.

Example – 'Step' Parameters
So, in this case, what we do is replace the value with a name in curly brackets **{R}** then create the *.step* command. This is ('Ohm's Law 4.asc'). We can step more than one parameter at a time, schematic ('Ohms Law6.asc') where the resistance is stepped over its full range, then the voltage is stepped to the next value. We may sweep two or three sources, but that is all.

1.7.2.3 Showing the Result

A trace window will open and we can select whatever parameters we like. However, if we plot the resistor current, we will see it very quickly falls to a low value and is difficult to read unless we use the cursor or change the y-axis to logarithmic.

Explorations 1
1. Run the various Ohm's Law schematics, view the traces, check that numbers agree with theory. Explore different colours, etc.
2. Note the effect on the sign of the values shown in the status bar when a component is rotated by 180°
3. Just explore at least the essential features until you are satisfied that you understand what is happening.

1.7.3 Production Yields

One may wonder why this topic is introduced here. One reason (excuse?) is that the essentials can be illustrated even with Ohm's Law. The second is to avoid overloading the following chapter by doing the theory here.

Practical components have a tolerance; they are not all precisely their nominal value. So with a bit of luck (or careful design), the simulated circuit may meet its design objectives, but that is with components that are exact: a *100 Ω* resistor is *100 Ω*, not *101 Ω* nor *99 Ω*. The question then is that, given the components have tolerances, how many products from a production run will still meet specifications. This is a matter of considerable importance and one on which whole books have been written. Here, we shall just touch on the basics and explore a little more in the next chapter using a potential divider.

One possibility for assessing the yield provided in some version of SPICE is a *worst-case analysis*. This usually sets every component in turn to the limits of its tolerance, both positive and negative, makes an analysis and stores the result. Finally it returns the extremes of the circuit's performance. This is open to two objections. The first is if it can be proved that the extremes actually give the worst case for every circuit or if some intermediate values are more important. The second, perhaps, it can lead to overdesign. To achieve a *100%* passing rate may require expensive, close-tolerance, components. And remember, the specification may also define a range of temperatures over which the product must work, so we now add low temperature coefficients.

In a circuit of any complexity the chances of hitting the worst case are very small, often far less than 0.1%. So it could be argued that it is more economical to use cheaper components with a wider tolerance and poorer thermal performance and accept some failures. What to do with the failures is another matter.

1.7.3.1   Statistical Distribution

It is an interesting fact that if we start off with an equal probability distribution of the value of a components, when we assemble two or more in a circuit, the distribution of the their combined values is no longer equal probability.

If we take a single unbiased dice we have equal probability of throwing *1,2,3,4,5,6*. But if we take two dice, the possibilities are:

Total	2	3	4	5	6	7	8	9	10	11	12
Ways	(1,1)	(1,2)	(1,3)	(1,4)	(1,5)	(1,6)	(2,6)	(3,6)	(4,6)	(5,6)	(6,6)
		(2,1)	(3,1)	(4,1)	(5,1)	(6,1)	(6,2)	(6,3)	(6,4)	(6,5)	
			(2,2)	(3,2)	(4,2)	(2,5)	(3,5)	(4,5)	(5,5)		
				(2,3)	(2,4)	(5,2)	(5,3)	(5,4)			
					(3,3)	(3,4)	(4,4)				
						(4,3)					
Total	1	2	3	4	5	6	5	4	3	2	1

From this table, we see there are 36 outcomes of throwing the dice and only 2 are the worst case – less than 6%. If we add another dice, the probability of the worst case becomes even smaller, and, incidentally, the distribution more closely resembles the Gaussian bell-shape.

Standard Deviation

We can imagine that if we have close-tolerance components, the results of a simulated production run will all be far closer together than if we had wide-tolerance components. We need a way of measuring this. We can calculate the *mean* by adding up all the values and dividing by the total. But this will be the same for both close-tolerance and wide-tolerance components. Instead, we need to measure how far the results differ from the mean. To avoid the problems of plus and minus values cancelling out, we square the differences: $(x_n - \mu)^2$ where x_n is the value and μ is the mean. Then we add these up, divide by the total and take the square root (because we squared the difference before). This is the *standard deviation* σ.

$$\sigma = \sqrt{\frac{1}{N} \sum_{i=1}^{n} (x_n - \mu)} \tag{1.1}$$

Its importance is in estimating the prospective product yield. The three common measures, and lastly an extreme one, are:

Standard deviations	Yield
+/−1	68%
+/−2	95%
+/−3	99.7%
+/− 6	99.99966%

The first three show that, for example, if we design for a spread of $+/-3\sigma$, we can expect to reject *0.3%* of the production. The last line is especially interesting in that if we design for $+/-6\sigma$ the rejection rate is *3.4/million*. This, and other possibilities, are shown in the presentation https://www.youtube.com/watch?v=SjpUuSE8iS4

1.7.3.2   Monte Carlo Analysis

This is another way of changing component values. The name derives from the popular gambling resort in Monaco; only the house does not always win; see https://towardsdatascience.com/the-house-always-wins-monte-carlo-simulation-eb82787da2a3. The value of a component is replaced by *mc*, then its value and tolerance enclosed in curly brackets as:

$$\{ \text{ mc } (< \text{ value } >< \text{ tolerance } >)\}$$

So for schematic ('Monte Carlo.asc'), we replace the 100 Ω by *{mc(100,.05)}* giving it a 5% tolerance. And remember, incidentally, that the tolerance in the resistor defining dialogue is ignored. To change the value of the resistor, we use the dummy parameter x. This is not the value of the resistor, but a counter for the number of passes.

Example Monte Carlo Analysis

An *.op* run will produce a very jagged trace which – frankly – is of little use. However, anticipating later chapters, we can use a *.meas* command to measure the maximum and minimum values.

Simply by setting the resistor to *95 Ω* and *100 Ω* with *.meas* statements, we find the limits are *0.0105263 A* and *0.00952381 A*, respectively, which agrees with calculation. The simulation returns *0.0105254 A* and *0.00952409 A* showing that the 1000 runs have come very close to the absolute limits. But we must note that this is a very simple example, and if there are many tolerance components, we may need a longer run of 10,000 or more. This need not be a problem since the computer can be left to crunch the numbers whilst we go for coffee.

1.8 SPICE

This is a brief summary of the abilities and performance of the programme.

1.8.1 Schematic Capture

In the early days of SPICE, this did not exist. Instead, it was necessary to draw the circuit on paper, then number every node, that is, every point where two or more components joined, and transfer that information to a netlist just like the net file described in Sect. 1.5.1.3. This was tedious and prone to error, so the ability to draw the circuit on the computer screen and convert it directly into a netlist was a considerable improvement. Today this is a very mature front-end to SPICE provided by every version. We shall now discuss the desirable features.

1.8.1.1 Learning Curve

Using the schematic capture should be intuitive so that the user can productively create and analyse circuits without first wading through a handbook. Certainly more advanced features may take some time to master, but the very basics of picking, placing and manipulating components, of connecting the circuit and of stimulating it with a voltage or current source must be readily accessible.

Likewise setting up the stimulation must be straightforward although in the case of the arbitrary voltage and current sources this is by no means so, nor is it obvious how to use the other controlled sources. However, the simple voltage and current sources are easy to use and will serve for many types of investigation.

It is not so important that setting up cosmetic features such as changing colours, marking unconnected pins and the like are not so intuitive. But we have mainly covered them in Sect. 1.6.3.

1.8.1.2 The Component Palette

Obviously this must include icons, symbols, glyphs, call them what you will, to represent all the components commonly in use today. But it is also important to have the ability without too much fuss, to create new ones; for example, a relay, or alternate ones, like the IEC rectangle for a resistor as opposed to the zigzag line, or to modify an existing symbol to turn a resistor into a potentiometer.

An extremely useful feature of LTspice is the amazing flexibility of the passive components and arbitrary sources. In particular, in classic SPICE, a dummy voltage source having an output of zero volts is necessary to measure the current in a circuit; see schematic 'VISHAY NTC_stimulated_LTspice' using classic SPICE where *VSENSE* is needed to measure the current. Compare that with VISHAY NTC_stimulated_LTspice V2' and LTspice enables us directly to sense the current through resistor *R1*.

1.8.1.3 Accessing the Results

The ability to draw a graph is essential. And it will be found in later chapters that being able to change from logarithmic to linear and other scales is very important and also to zoom in on areas of interest is invaluable.

A very useful feature, which will be used extensively in later chapters, is that of being able to measure accurately, for example, the time between points on a waveform. It is also essential to have wide range of mathematical procedures that can be applied to the data after the analysis.

1.8.1.4 Saving the Results

It must be possible to save the raw data, and, to reduce the file size, the ability to compress the data can be useful. It is also important that the trace window can be saved as an enhanced Windows metafile, or printed, or pasted in a text document.

And as we have noted in Sect. 1.6.2.5 and elsewhere the netlist is saved and also the analysis settings which can be seen by **View→SPICE Error Log.** It is also desirable that the data from a Monte Carlo analysis can be saved in a text format readable by a spreadsheet.

In short, the greatest flexibility in processing and saving the results is very important.

1.8.2 SPICE Analysis

To process the netlist, it is first turned into a matrix. This is a popular way of solving circuit equations and is very widely used by computers since it lends itself to a repetitive process.

The matrix, however, has many more empty or zero elements than those with data and is called a *sparse matrix*. It is clearly inefficient to implement matrix multiplication where most of the time we shall be multiplying *(something)x(nothing)*, so the first task is to condense the matrix. This is done by every version of SPICE to a greater or lesser extent. One possible pitfall is that of using time-hungry elaborate techniques to reduce the matrix when it would have been faster to use a simple method and just solve the larger matrix.

1.8.2.1   Numerical Integration Methods

These are tweaked by programme authors so the following description is very basic. The point is that we need to solve the differential equations. The difficulty is that, more often than not, there is no nice analytical solution. One popular method was invented by Euler. In recent years, more have been added. Many are described briefly at http://qucs.sourceforge.net/tech/node24.html. There is an extensive analysis by Charles A Thompson docs.lib.purdue.edu/cgi/viewcontent.cgi?article. The following discussion also assume that these are ordinary differential equations so that the first derivative of the Taylor Series is linear. In the first case, we start with the known equation and the solution at one point. Many work by linearizing the function using a Taylor's Series where the value after a small increment h in x is:

$$y(x + h) = y(x) + hy'(x) + \frac{h^2 y''(x)}{2!} + \frac{h^3 y'''(x)}{3!} + \cdots \qquad (1.2)$$

The Euler Method
We retain only the first two terms in Eq. 1.2 to give:

$$y(x + h) \approx y(x) + hy'(x) \qquad (1.3)$$

where h is a small increment. It is worth considering this for a moment. The first term, $y\{x\}$, is the current position. The Taylor Series in Eq. 1.2 sets out how to move to the new value $y(x + h)$. But Eq. 1.3 also describes the same move by discarding every term after the second. We must look at how serious is the error. The increasing factorial progressively will reduce the important of the higher terms. In addition, the higher derivatives in some cases will diminish, for example, the derivatives in order of $x^3 \rightarrow 3x^2 \rightarrow 6x$. But more important is that because h is raised to ever higher powers, if it is much less than *1* that will be more decisive. For example, even with $h = 0.1$, the third term will have *0.01 h*, the fourth *0.001 h* and so on. So provided we keep to small values of h, the truncation is not so cavalier.

It is also important to note that by definition, y' is the tangent to the curve and this has replaced the original function. This is illustrated in Fig. 1.18 where we start at (x_0, y_0). A small step h moves us to x_1. We now use the linear function to move to

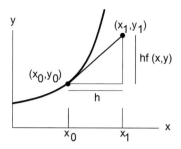

Fig. 1.18 Euler Method

point (x_1, y_1). This is an *explicit* method making explicit use of the current position to move to the next.

This is well explained with an example at http://calculuslab.deltacollege.edu/ODE/7-C-1/7-C-1-h-c.html, and several nice examples can be found at http://tutorial.math.lamar.edu/Classes/DE/EulersMethod.aspx where we should note that the solution does not always get closer to the true answer. Reducing the step size usually helps but that takes more computing time, so generally the step size gets shorter as the graph gets steeper.

The advantage of the method is that the equation for tangent only has to be found once from Eq. 1.2. It is then a matter of substituting the values for each step into a linear equation, and this can be solved quickly. So it is fast. The disadvantage is accuracy and if it will converge at all to a stable solution.

Example – Euler Method
This is a trivial example that can be solved easily:

$y' = 6x$
We can compare the estimate and the solution which is $y = 3x^2 + 5$ shown in square brackets after each calculation.

We take $x = 0$ $y = 5$ as our starting point. If we take a step $h = 1$, we have $x = x_0 = h = 1$. The new value of y will be 5 plus the original x_0 multiplied by 6, that is, $6x0 = 0$ so y is unchanged and $(x_1, y_1) = (1,5)$ [6] and this is our new starting point.

We now move to $x_2 = x_1 + h = 2$ and then $y = 5 + h(6 \times x_1) = 5 + 1(6 \times 1) = 11$ and $(x_2, y_2) = (2, 11)$ [12]. For the third move, we have $y = 11 + 6(2) = 23$ [27] and the error is getting larger. But then $h = 1$ is excessive as we can see from Eq. 1.2.

The Backward Euler Method
This is a single-step method that works in a similar fashion but uses the next point to calculate the previous from it, and it again truncates the Taylor Series at the second (linear) term. Then similar to the Forward Euler method we have:

$$y(x_n) = y(x_{n+1}) - hy'(x_{n+1}) \tag{1.4}$$

which we rearrange as:

$$y(x_{n+1}) = y(x_n) + hf(x_{n+1}, y_{n+1}) \tag{1.5}$$

and now we need to find the roots of Eq. 1.5 by using, for example, the Newton-Raphson method. This is an iterative process briefly outlined below, so finding the solution is more time-consuming than the previous. However, it is extremely stable and allows larger time steps, so the time penalty may not be that great. But it is no more accurate than the previous.

The Trapezoidal Method
This is the method we all learned at school for finding the approximate area under a graph, Fig. 1.19. This means that if we have a differential equation first we must integrate it then find the value at (x_n, y_n) and (x_{n+1}, y_{n+1}) then:

$$y_{n+1} = y_n + \frac{1}{2}\left[f(x_n, y_n) + f(x_{n+1}, y_{n+1})\right] \tag{1.6}$$

This, either modified or unmodified, is usually the best integration method. It is very stable; the time penalty is that of calculating the underlying function each time. This also is a single-step method in effect combing the two Euler methods to give greater accuracy.

The Gear Method
This truncates the Taylor Series after higher terms. In particular, if it truncates after the third instead of the second, we have a parabolic and not a linear interpolation. Then the formula is:

$$y(x_{n+1}) = \frac{4}{3}y(x_n) - \frac{1}{3}y(x_{n-1}) + \frac{2}{3}hf(y(x_{n+1})) \tag{1.7}$$

where we should note that because it needs the value at n-1 as well as the value at n it is not self-starting since there is no previous value and an alternative method is needed to initiate it. There are formulae for using higher orders. This method is lossy so, for example, the oscillations in an LC tank circuit starting with a charged capacitor will decay rather than continue for ever. Also it attenuates high frequencies. Therefore it should be used with caution if all else fails.

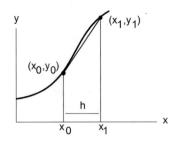

Fig. 1.19 Trapezoidal Method

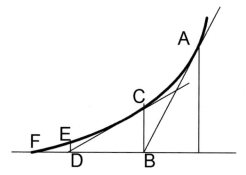

Fig. 1.20 Newton Raphson

The Newton-Raphson Method

At school we all learnt how to solve quadratic equations: $ax^2 + bx + c = 0$ but the real world is not so well-behaved, and we can find ourselves faced with equations more like $ax^{2.31} + ax^{-0.4} + c = 0$ or differential equations. This is not a new problem: it had been around for several hundred years and was solved by Newton.

We need to find the point where the function is zero, that is, some point on the x-axis where y is zero. In essence, the method consist of first guessing at a point A on the graph of the expression, Fig. 1.20. We then draw a tangent from A to meet the x-axis at B. Next we calculate the value of the function when $x = B$ marked as point C. Now we draw a new tangent from C to meet the x-axis at D, then again we calculate the corresponding value of the function E and draw a new tangent to F. And so we continue with the intercept getting ever closer to the true answer until it is close enough. This raises the question what is 'close enough'. LTspice has defaults that can be accessed by **Tools→Control Panel** and clicking the **SPICE** tab. These are briefly described in the 'Help' file under *.OPTIONS*

1.8.3 Performance

The crucial considerations are accuracy and speed.

1.8.3.1 Accuracy

This depends both upon the accuracy with which the components are described and the performance of the SPICE engine. The palette of passive components and sources we have explored above is common to all versions of SPICE. However, things change when once we move to more complex components: for example, a thermistor is not included as a SPICE component, even if we find a symbol for it, that will call a sub-circuit which is constructed from parts, likewise a varistor. We shall see in subsequent chapters that, although in some cases there is general agreement on how to construct the model, this is not always true and we shall encounter examples where the performance of the model is a far cry from the data sheet parameters. This, of course, is not the fault of SPICE, but of the author of the model.

The accuracy of the simulator depends upon how well it is written. As users, we have the crass test of whether the simulation iterates to a stable solution. In this respect, if there is a problem, we should always first try shorter time steps in a transient run and maybe switch to Gear. The more subtle errors can only be seen by carefully comparing the simulation results with reality.

1.8.3.2 Speed

There are many factors that contribute to the speed with which SPICE returns an answer. The improvements are not open to examination by us but, naturally, are private to the programme's author.

The Circuit Matrix
The Numerical Integration Methods calculate the next state from the present. Therefore this difference can be used to find the difference in current through an inductor or capacitor, and as Charles A Thomson explains in the article quoted above, we can replace them by a current source in parallel with a resistor so they will fit nicely in the circuit matrix.

Access Times
The speed of access to data in a computer depends on where it is stored. At the slowest it is the hard disc. It takes about 7 ms to find the track and then data is slowly read at the rate of about 1 Mb/s. Therefore disc access should be reduced to the minimum.

RAM usually has a clock speed of the order of 100 MHz, but as it takes more than one clock cycle to read the data, see https://en.wikipedia.org/wiki/Dynamic_ran dom-access_memory explaining the process: a pessimistic access rate is 10 Mb/s.

The fastest access is the CPU cache which works at the CPU clock speed of a few GHz, so access is more than 100 Mb/s.

These are global figures, but they suggest that, as modern computers have at least 1 GB of RAM, it might be possible to load the whole SPICE programme into memory. In this we are fortunate that LTspice is not bloated with fatuous graphics and the complete folder is only 556 MB of which 429 MB is the library.

Pipe Line
Given the relatively slow speed of RAM access compared to the CPU cache, all operating systems use a pipeline. Generally, the instructions in a programme are accessed linearly, that is, given the present instruction, the next to be executed is likely to be the next in sequence. Of course there are jumps and conditional statements (if...then), but in the main, it will be the next instruction. So whilst the CPU is busy processing the present instruction, the next is *pre-fetched* from memory and is immediately available without the overhead of reading from RAM.

This has been extended in LTspice to pre-fetch data. This is a novel introduction: although the pre-fetched instruction saves time in getting what to do next, there is still the overhead of getting the data. And this can waste several clock cycles. So also caching the data is a significant improvement in speed.

Comprehensive Passives
Capacitors and inductors all incorporate parasitic components such as series and shunt resistance and inductance or capacitance in their description. It is faster to use these rather than external discrete parts except that they do not accept the full range of LTspice qualifying parameters.

1.8.3.3 Omissions

In passive components, we find that items such as tolerance, power rating and RMS current can be entered but are not acted upon. So LTspice will not demur if you want to pass 1000 A through a 1 Ω resistor rated at ¼ W. The more serious omissions, common to most version of SPICE, arise in semi-conductors where breakdown voltage is not modelled. This is shown in the schematic 'No Breakdown.asc'.

1.8.3.4 Enhancements

LTspice has by far the most flexible description of a component. As well as the usual temperature coefficients, we shall see in later chapters that without too much fuss we can make voltage-dependent resistors, and inductors whose inductance depends upon the current.

A minor improvement, if one may so call it, is that to create a source controlled by the current in another net, it is no longer necessary to introduce a dummy voltage as it is in other versions of SPICE. This we saw in the Vishay schematics mentioned in 'The Component Palette' and will be used in later chapters.

1.9 Summary

In this chapter, we have covered the basics of LTspice by exploring Ohm's Law. The important points are:

- Components can be picked from the tool bar or from the drop-down menu **Edit→Component** then scroll and click to preview; click **OK** to place.
- Components can be rotated and flipped by **Ctrl-R** to rotate **Ctrl-E** to flip before they are placed, or by **F8**, then click on the component, and **Ctrl-R** or **Ctrl-E** as before.
- A DC analysis is performed from the main menu by **Simulate→Edit Simulation Cmd;** then select **DC op pnt;** then click on the running person.

- By enclosing a component value in curly brackets {**value**} and adding the SPICE directive **.step < value > <step type > <start value > <end value > . < increment>**, we can make many runs to test the effects of changes.
- Useful information can be found on the status bar when the mouse is over a component or wire.
- Equal probability tolerance distributions tend to a Normal or Gaussian distribution if several components are involved.
- *99.7%* of the output of a long production run are within +/− standard deviations and only *3.4* in a million are outside *6* standard deviations.
- A Monte Carlo analysis of at least *1000* runs will return reliable statistics on simple circuits.
- The *modified trapezoidal* method is the preferred one with the *Gear* option if all else fails.

Chapter 2
DC Circuits

2.1 Introduction

At first sight it would seem that there is little need to go to the bother of using SPICE on DC circuits since, with the exception of passive filters and those circuits thought up by malevolent examiners to tax the minds of students, in real life they generally consist of but a handful of components and are perfectly amenable to well-established manual techniques without too much effort.

Whilst this is true, nevertheless it will be found that once the driving instructions for SPICE have been mastered from the previous chapter, 'what if' design questions can be answered very rapidly, and, in particular, the effects of temperature changes and component tolerances can be explored. Furthermore, the rather more awkward analysis of circuits with non-linear components can be handled easily. We shall also introduce the LTSpice *.meas* directive which saves us from having to drag the cursor around the Trace Panel to find parameter values.

So in his chapter, we shall start with Kirchhoff's Laws – the basic analytical tools for circuit analysis – and use them to analyse combinations of resistors and some interesting circuits, including converting a moving coil meter into an ammeter, voltmeter and ohmmeter, the Wheatstone bridge and specialist derivations from it. Certainly, some of them are more of historic interest now we have digital meters, but they are still object lessons in ingenious problem-solving. From there we shall study other analytical tools and apply them to an important class of circuits – attenuators. Finally we shall discuss the conversion of a 'Delta' or 'Pi' connection of three resistors into a 'star' or 'Tee' equivalent and the reverse.

2.2 Kirchhoff's Laws

These are fundamental for solving circuit problems. These build on Ohm's Law formulated a couple of decades earlier, still in the infancy of electrical engineering.

© Springer Nature Switzerland AG 2020
C. May, *Passive Circuit Analysis with LTspice*®,
https://doi.org/10.1007/978-3-030-38304-6_2

2.2.1 Resistors in Parallel and Kirchhoff's Current Law

This states that the sum of the currents at a node must be zero, taking into account their directions into or out of the node. Thus in Fig. 2.1a:

$$I_0 + I_1 + I_2 + I_3 = 0 \qquad (2.1)$$

Now remembering that the wires have no resistance, all the resistors in Fig. 2.1b are connected electrically as in Fig. 2.1a. We draw loops to define the direction and so we can write:

$$V_B = I_1 R_2 \qquad (2.2)$$

$$V_B = I_2 R_2 \qquad (2.3)$$

$$V_B = I_3 R_3 \qquad (2.4)$$

$$I_1 R_1 = I_2 R_2 \qquad (2.5)$$

$$I_2 R_2 = I_3 R_3 \qquad (2.6)$$

The easiest solution is to substitute for the currents in Eq. 2.1 using Eqs. 2.2, 2.3 and 2.4:

$$I_0 = \frac{V_B}{R_1} + \frac{V_B}{R_2} + \frac{V_B}{R_3} \qquad (2.7)$$

or:

$$I_0 = V_B \left(\frac{1}{R_1} + \frac{1}{R_2} + \frac{1}{R_3} \right) \qquad (2.8)$$

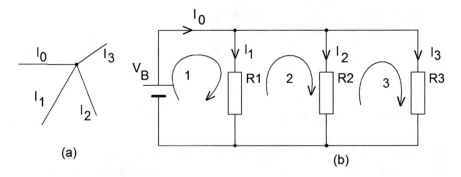

Fig. 2.1 Resistors in parallel

and by comparison with Ohm's Law:

$$I_0 = \frac{V_B}{R_{eq}} \tag{2.9}$$

where:

$$\frac{1}{R_{eq}} = \frac{1}{R_1} + \frac{1}{R_2} + \frac{1}{R_3} \tag{2.10}$$

A special case is when there are just two resistors then:

$$R_{eq} = \frac{R_1 R_2}{R_1 + R_2} \tag{2.11}$$

and a quick check on one's mathematics is that R_{eq} is larger than the smaller resistor and smaller than the larger.

The attentive reader might have noticed that we have not used Eqs. 2.5 and 2.6. This often happens: we have more equations than we need, and we chose those that make life easier.

2.2.1.1 A Moving Coil Ammeter

Although digital meters are widely used, the humble moving coil meter still has a place, especially for viewing waveforms that are fluctuating such as sound level (although often these moving coil meters are designed to respond quickly to an increase in level but decay rather slowly).

We can use a moving coil meter to measure current. If it has a full-scale deflection (fsd) of 1 mA, then we can measure currents up to that value. But to measure larger currents, we must divert some of it around the meter so only 1 mA goes through it. Suppose we want to measure 10 mA, then 9 mA must be diverted and 1 mA allowed through the meter. This is done by a resistor in parallel with the meter, Fig. 2.2, where the numbers are for the Example below. We should note that these are non-standard resistors and of very low value and should be of high accuracy, better than 1%. Also they must have a low temperature coefficient of resistance.

Example – Converting Moving Coil Meter to Ammeter
A meter has an fsd of 1 mA and a resistance of 100 Ω. The meter and its internal resistance are shown in the dotted rectangle. Convert it into an ammeter of 50 mA fsd.

With the maximum current of 1 mA though the meter, the voltage across it will be 100 mV. The parallel resistor must pass 49 mA at that voltage so it is *100 mV/ 49 mA = 2.041 Ω*. And if we want an fsd of *1 A*, the resistor is *0.100 Ω*. This is schematic ('mA Shunt.asc').

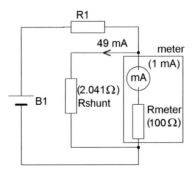

Fig. 2.2 Converting 1 mA fsd meter to 50 mA fsd

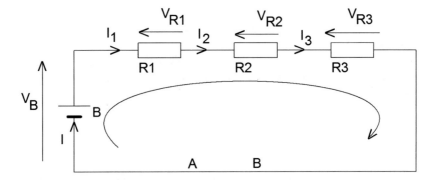

Fig. 2.3 Resistors in series

2.2.2   *Resistors in Series and Kirchhoff's Voltage Law*

A simple circuit consisting of a battery and three resistors in series is shown in Fig. 2.3. We now need to find the voltages across the resistors.

First, we assume that the current is the same through all the circuit, that is,

$$I_1 = I_2 = I_3 \tag{2.12}$$

This is a reasonable assumption else charge carries are falling out of the circuit or mysteriously finding a way in at the junctions of the resistors. It is very easy to get the sign of volt drops wrong. A reliable procedure – which can be simplified with practice — is the following:

- Assign a direction for the loop current. In the figure, it is clockwise, since, in this case, it is clear that the battery drives the current in the direction shown. But in more complicated circuits, it may not be clear. And in any case, this is quite arbitrary and counterclockwise is just as good. All that will happen is that the current will be in the reverse direction.

- For every component in the loop, follow the direction of the current and draw an arrow with the head towards the more positive end of each component. Taking the circuit as shown, the current enters the left-hand end of *R1* so that is more positive.
- Move round the loop starting from some arbitrary point such as *A*, and add up the volt drops until we arrive at *B*. If we first encounter the head of the arrow, it is positive, else the volt drop is negative. This, too, is arbitrary, and we could choose that the head of the arrow is negative.

This leaves us with the question whether the potential at *B* is the same as *A*. We can answer this by noting that the conductor is assumed to have zero resistance, So if there is any potential difference between these two points, there must be infinite current. This is clearly impossible so the potential at *B* must be exactly that of *A*. Hence, following the procedure described above, we write:

$$-V_B + V_{R1} + V_{R2} + V_{R3} = 0 \tag{2.13}$$

which is Kirchhoff's Voltage Law (usually abbreviated to KVL) that states that the sum of the volt drops around a closed circuit is zero. Now, by replacing the volt drops with the current and resistances and rearranging:

$$V_B = IR_1 + IR_2 + IR_3 \tag{2.14}$$

or:

$$V_B = I(R_1 + R_2 + R_3) \tag{2.15}$$

and we can replace the three resistors by a single resistor *Req*:

$$R_{eq} = R_1 + R_2 + R_3 \tag{2.16}$$

provided, of course, we do not want to know the volt drops across the individual resistors.

Example – Resistors in Series   Use the schematic ('Resistors in Series.asc') to try various values. Note that we can draw a line from *A* to *B* straight through all three resistors and LTspice will correctly make the connections. This is a tip from: https://www.analog.com/en/technical-articles/ltspice-connecting-the-dots.html by Gabino Alonso which we mentioned in the previous chapter.

Example – Resistors in Series and Parallel   These laws are sufficient to analyse resistive circuits with only one voltage source. How we do the analysis depends on what we want to know. Let us take Fig. 2.4 and ask what is the current in resistor *R1*.

We note that the total current flows through this resistor, so we must find the equivalent resistance of all the resistors. First we merge *R2* and *R3* in parallel as 20.30/50 = 12 Ω and put that in series with R1 so the total is 22 Ω and hence the current is 0.45 A.

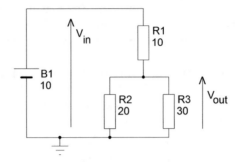

Fig. 2.4 Resistors in Series and Parallel

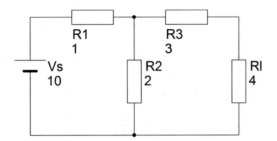

Fig. 2.5 Kirchhoff practice circuit

Example – Kirchhoff Practice A second example is Fig. 2.5 where again we want the current in R1. We merge R3 and R4 in series as $R34 = 7\ \Omega$. Next we put it in parallel with $R2$, so we have $R234 = R2*R34/(R2 + R34) = 14/9 = 1.56\ \Omega$ Then finally we add $R1$ in series to give $R1234 = 2.56\ \Omega$, and the current therefore is 4.9A. Use LTSpice to build the circuit and confirm the result.

2.2.2.1   A Moving Coil Voltmeter

The meter we met before can be used to measure voltage. As it stands, it can measure only 100 mV between its terminals because, with that voltage and an internal resistance of $100\ \Omega$, it will pass its maximum allowed current of $1\ mA$. To increase the voltage range, we must add resistance in series, Fig. 2.6. And these also are non-standard values and must be of high accuracy.

Example – Convert a Moving Coil Meter to a Voltmeter To convert the meter of 1 mA fsd and 100 Ω internal resistance to a voltmeter of 25 V fsd. We now need a total series resistance of $25\ V/\ 1\ mA = 25\ k\Omega$, so we must add an external resistance of $24.9\ k\Omega$. To convert it to an fsd of 100 V, the total resistance must be $100\ k\Omega$ and we add $99.9\ k\Omega$.

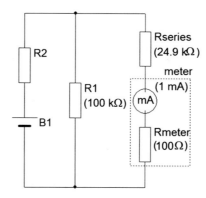

Fig. 2.6 Converting a 1 mA fsd meter to a Voltmeter with 25 V fsd

2.2.2.2 Meter Loading

We can note that the necessary total resistance of our meter is 1 kΩ for every volt of fsd. This is a useful figure of merit. Were we to use a meter with 200 μA fsd, the figure becomes 5 kΩ/V.

If we use the voltmeter with 25 V fsd to measure the voltage across $R1$ in Fig. 2.6, we now have the resistor $R1$ in parallel with the meter resistance and the total is $(25 \times 100)/125 = 20\ k\Omega$, and this will seriously alter the circuit conditions. If we turn the meter to the 100 V fsd range, the loading now is $(100 \times 100)/200 = 50\ k\Omega$. But if we turn our 200 μA meter into a voltmeter of 100 V fsd, the meter resistance will be $5\ k\Omega/V \times 100 = 500\ k\Omega$, and the loading now is only $(500 \times 100)/600 = 83\ k\Omega$, so it is better to use a sensitive basic meter, perhaps with 25 μA fsd, but unfortunately these tend to be rather delicate.

On the other hand, if we use the meter to measure current, we insert extra resistance in series with the circuit as we saw in Fig. 2.2, and it can be better to use a less sensitive range with a small shunt resistance to make a measurement.

2.2.2.3 A Moving Coil Ohmmeter

We can connect a moving coil meter in series with a cell or battery and a variable resistor to make an ohmmeter, Fig. 2.7. Initially R_{zero} is adjusted so that the meter reads fsd (1 mA) with $Rtest = 0\ \Omega$. Then we insert the test resistor and measure its value. This is a reciprocal scale because $I = V/R$.

Example – Converting a Moving Coil Meter to an Ohmmeter If we have a 1.5 V cell and a meter with $fsd = 1\ mA$ with a resistance of 100 Ω, then we set $Rtest = 0\ \Omega$ and adjust $Rzero$ so that the meter reads fsd, which will be when $Rzero = 1.4\ k\Omega$ and the total resistance is 1.5 kΩ. Now suppose $Rtest = 100\ \Omega$ and the meter will read $1.5\ V/1.6\ k\Omega = 0.938\ mA$. And if $Rtest = 1k\Omega$, the reading will be $1.5\ V/2.5\ k\Omega = 0.6\ mA$.

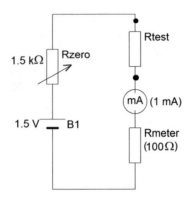

Fig. 2.7 Converting a 1 mA fsd meter to an Ohmmeter

The question then is what happens if the battery voltage falls to, perhaps, 1.3 V. Then we must set $Rzero = 1.2\ k\Omega$ for fsd with $Rtest = 0\ \Omega$, and now with $Rtest = 100\ \Omega$ the meter will read $1.3\ V/1.4\ k\Omega = 0.929\ mA$, and if $Rtest = 1\ k\Omega$ the meter will read $1.3\ V/2.3\ k\Omega = 0.565\ mA$.

If we arbitrarily say that a current of 0.1 mA represents the highest resistance we can read with acceptable accuracy then at that current the total circuit resistance is $1.5\ V/0.1\ mA = 15\ k\Omega$ and then with $Rzero = 1.5\ k\Omega$ to set fsd we find $Rtest = 13.5\ k\Omega$. If, however, the cell voltage is only 1.3 V, the total resistance is $1.3\ V/0.1\ mA = 13\ k\Omega$ and to set fsd we now have $Rzero = 1.3\ k\Omega$ and $Rtest = 11.7\ k\Omega$ and we find an error of around 10%.

Explorations 1
1. Build the previous circuits and check the simulation results against calculations.
2. Set new resistor and source values in the circuits and calculate the currents and voltages for all the resistors and check by simulation.
3. Exchange the voltage source and each of the resistors in turn and repeat the previous calculations and simulations.
4. Given a meter of 1 mA fsd and resistance 100 Ω, convert it into meter to have an fsd in turn of 10 mA, 100 mA 1 A, 10 V 100 V and 1 kV, and note its loading on circuits.
5. Think of a way of creating an ohmmeter where the falling battery voltage does not reduce the accuracy and test it.
6. Build a few more circuits with one voltage source and several resistors.

2.3   Some Useful Circuits

These have but one voltage source and just a few resistors and hence can be handled easily by Kirchhoff's Laws.

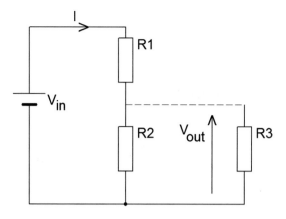

Fig. 2.8 Potential divider

2.3.1   The Potential Divider

This is a useful circuit when we have a DC voltage supply but need some lower voltage at a small current, typically less than a few milliamps. Rather than going to the expense of building a new voltage supply, we can use two resistors as shown in Fig. 2.8, ignoring *R3*.

Now we have a simple series circuit so:

$$V_{out} = IR_2 \tag{2.17}$$

and for the current:

$$I = \frac{V_{in}}{R_1 + R_2} \tag{2.18}$$

so:

$$V_{out} = V_{in} \frac{R_1}{R_1 + R_2} \tag{2.19}$$

And we can obtain a reduced output voltage, for example, 5 V in Fig. 2.9. Of course, Eq. 2.19 has two unknowns, so we cannot exactly fix the resistor values, only their ratios. However, a potential divider usually has to supply a load current which may vary, so a useful rule of thumb is to make the current I in Eq. 2.18 ten times the load current. To test the loading effect, we need to connect an appropriate resistor *R3* to draw the load current.

Current and Voltage Probes
Once the circuit has been run, moving the mouse over a component causes the cursor to turn into a pointing fist or a clamp ammeter as we saw in the first chapter. The

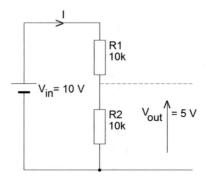

Fig. 2.9 5 V Potential divider

status bar will show the DC operating point current and the dissipation. Moving the cursor over a wire and the status bar will show the DC voltage.

Ctrl and left click on a wire and the voltage will be written above it.

Example – Potential Divider Design

Given a 10 V supply, we want an output of 3.6 V with no external load and not less than 3.3 V when there is a load current of 10 mA.

Using our rule of thumb, we draw a current of 100 mA, and from the previous equations we find $R1 = 64\ \Omega\ R2 = 36\ \Omega$, and this is a slight error.

The full analysis is surprisingly awkward – tedious rather than profound – and one pitfall is to assume either that the current in $R1$ increases by 10 mA or that the current in $R2$ falls by 10 mA when the load is applied and we need to start with Eq. 2.19 for both the unloaded and loaded cases:

$$3.6 = \frac{10R_2}{R_1 + R_2}\ (a) \quad 3.3 = \frac{10R_2'}{R_1 + R_2'}\ (b) \quad \text{hence from (a)} \quad R_1 = 1.78R_2$$

where $R2'$ is the parallel sum of $R2$ and the 330 Ω load resistor $R3$ to draw 10 mA. We rearrange Equations (a) and (b) to leave 10 on the right-hand sides, substitute for $R1$ and equate them: $10 = \frac{1.85R_2 + 3.3R_2'}{R_2'}$ then $R_2' = 0.876R_2$, $R_2 = 46.8\Omega$, $R_1 = 83.2\Omega$.

This is schematic ('Pot Design.asc'). These are not standard resistors, and their values are given to three figures. The change in circuit current is only some 3 mA. If we try the preferred values of $47\ \Omega$ and $82\ \Omega$, there is a slight error.

Explorations 2

1. Given a DC supply of 12 V, create a potential divider to have an output of 5 V with no added load.
2. What load can be added to the previous circuit so that the output is at least 4.5 V?
3. Assuming the resistors are 2% tolerance and there is no external load, what are the limits of the output voltage?

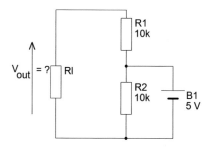

Fig. 2.10 Reverse potential divider

4. Build and test the circuit of Fig. 2.10, and see if you get an output of 10 V if *Rl* is
   very large.

This seemingly trivial circuit takes only a minute and one line of working
(Eq. 2.19) to arrive at the output voltage. However, the situation becomes less
clear if we give all the elements some tolerance and then ask what are the correct
resistor values and tolerances to use so that the output voltage remains within
prescribed limits despite changes in the load current.

Example – Toleranced Potential Divider Design
As a concrete example, let us ignore resistor tolerances and suppose we have a 12 V
supply with a 2% tolerance and the circuit is to have an output of 4.0 V with a 6%
tolerance despite load changes from zero to 2 mA, and to consume as little current as
possible.

This design may seem a little artificial, but the point about wasting as little power
as possible is very real and precludes the obvious approach of making *R1* and *R2* in
Fig. 2.8 very small to minimize the loading effects.

As a first step, for a nominal output of 4 V, we require $R1 = 2*R2$, and the highest
output voltage will occur with no load resistor and the supply 2% high. That is:

$$12.24 = 3R_2I_1 \tag{2.20}$$

for the current through both resistors, $I1$, and the output voltage is:

$$4.24 = I_1R_2 \tag{2.21}$$

when the supply is low and a load current is taken, we really need to go through the
procedure of 'Example 3 - Potential Divider Design':

$$11.76 = (I_2 + 2mA)2R_2 + I_2R_2 \tag{2.22}$$

and:

$$3.76 = I_2R_2 \tag{2.23}$$

From 2.20 and 2.23, we have a relationship between $I2$ and $I1$. Thence we can
find *R2* in terms of $I1$ and finally arrive at 188 Ω for *R2*.

The schematic is ('Pot Divider Problem.asc') showing the load resistor *R3* disconnected. If we run the circuit and note the limits of voltage, then add the load resistor and repeat, we find the output voltage is within tolerance. Remembering that numbers were rounded, the agreement is good and implies that non-standard resistors will be needed. Notice, incidentally, that the current of about 22 mA agrees with the rule of thumb of ten times the load current. An interesting point on the schematic is to rotate *R2* through 180°, and it is found that the current is in the opposite direction.

Example – Monte Carlo Potential Divider
We have also assumed that the resistors are exact, but suppose they are 2% tolerance, and the problem is more complicated. However, there is a way that we can probably get the right answer, and that is to make a Monte Carlo analysis. The key word is 'probably' because even with a run of 10,000, we cannot be certain we have included the extremes.

The strategy is to get the correct lower limit with the load resistor in place and then make it into some huge number like *1e22* and check the upper limit. This is schematic ('Pot Divider Problem MC.asc') where the nominal resistor values have been changed slightly. The output voltage is shown as 1000 lines, one for each stepped value of the dummy parameter *x*. We can judge by eye what the maximum and minimum values are, but it looks as though the circuit will work. However, for an accurate assessment we can use the **measure** directive.

2.3.2   The '.measure' (.meas) *Directive*

The 'measure' directive is a powerful and flexible method of measuring values precisely where the information is available after a run. We shall here explore the general outline of the directive leaving some important aspects for later chapters. Scripts are placed on the schematic and executed after the analysis has been made. The results – rather oddly – are seen by **View→SPICE Error log**. Here we shall only use two commands to find the limits of the output voltage of the potential divider. Two more will be added a little later. We should also note that it is not necessary for the parameter be shown on a trace: LTspice preserves all the measurements after a run.

If we press *S* on a schematic, it opens the same dialogue we used for *.param* and *.step* commands. It is possible to continue with the whole command, but, as it is easy to make an error, we can just type *.meas* and click **OK** and place it on the schematic as we did with the *.step* directive. If we now right click on it, the dialogue **'meas Statement Editor** will open.

Caution – Do Not Leave Spaces
This is not relevant just now, but later we shall see that it is possible to include formulae in measure directives. Something like:

$$Vmin * 2$$

will work, but:

$$Vmin \ * 2$$

with a space will ignore the *2.

2.3.2.1 Measure MAX, MIN

At its simplest we just want to know the maximum value of a parameter. For this the syntax is:

.meas < Applicable Analysis >< Result Name > MAX < Measured Quantity >

Only the applicable edit boxes are shown in Fig. 2.11. There will be more figures illustrating the use of the rest. They are:

Applicable Analysis
The drop-down list consists of the tabs of the **Edit Simulation Command** dialogue and allows us to choose which analysis will activate the measurement. We can leave it at **(any)** so every analysis will activate it.

Result Name
This can be any meaningful name to identify the result. This is important because we can have several .*meas* statements on a schematic and the results of one can be used by others.

Genre
This defines the type of measurement. We can see that this permits differentiation and integration and other interesting possibilities.

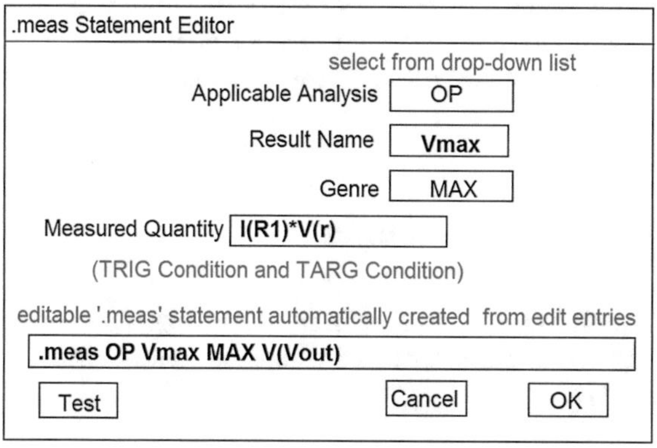

Fig. 2.11 .*meas* Statement Editor

Measured Quantity

This can be just a simple variable or any function of variables. Below this are other edit boxes, not used here, and which we leave empty so that the measurement spans the full range of the analysis.

The script is shown in the edit box at the bottom of the dialogue. The appropriate syntax - including parameters we have not used - is shown just above it. Text can be entered here or in the edit boxes, each updates the other. Attempting to test before a run results in the massage **No waveform data available** or failure.

Trig Condition, Targ Condition

Example – Toleranced Potential Divider Using Measure MAX, MIN

In schematic ('Pot Divider MC(2).asc'), we have chosen a **DC op pnt** analysis, so here in Fig. 2.11, we want **OP** for the **Applicable Analysis,** or we could leave it at the default of **(any)**. An obvious **Result Name** is *Vmax,* and for the **Measured Quantity,** we want **V(out)**. And the **Genre** is **MAX**. After a run, pressing the **Test** button reveals the full script *Vmax: MAX(Vout) = 4.10732 FROM 1 TO 999* which saves us the trouble of opening the error file because it is the same. Else we get the depressing **Measurement < name > FAIL'ed.**

We add a second script for the minimum value, schematic 'Pot Divider MC(2). asc', and now the limits are 3.87 V and 4.107 V. But can we be sure? Increase the run to 10,000, and the results are 3.86 V and 4.11 V. At 100,000 runs – which take a minute or two – the outcome is essentially the same, 3.86 V and 4.118 V differing by 0.2% and incidentally showing that a different seed is used each time for the random numbers. This large run takes a few minutes. The progress can be followed on the status bar.

And whilst this is a fairly trivial example, we can appreciate that for a circuit with many toleranced components, this is a powerful method for finding the limits or 'worst case'.

Explorations 3
1. Check that the voltages, currents and powers agree with manual calculation.
2. Is it possible to make the circuit work with 2% tolerance resistors (you will need to recalculate the values); if not, what resistor tolerance will work?
3. Given a supply of 10.0 V, build and test a potential divider to give an output of 5.0 V when a current of 100 μA is drawn. What is the no-load output voltage?

2.3.3   Maximum Power Transfer

The circuit schematic is simply that of Ohm's Law, Fig. 2.12. Note that the source resistance of *V1* (3 Ω) has been subsumed within the component as **Series resistance (Ω)** and is denoted by *r*.

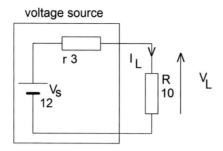

Fig. 2.12 Maximum Power Transfer

2.3.3.1   Theory

Every textbook explains how, given a voltage source with a certain internal resistance, the maximum power transfer occurs when the load resistor is equal to the internal resistance of the source. The proof is not difficult and relies on writing down the equation for the load voltage as:

$$V_L = V_s \frac{R}{r+R} \tag{2.24}$$

and the load current as:

$$I_L = \frac{V_s}{r+R} \tag{2.25}$$

whence it follows that the load power is:

$$P_L = R \frac{V_s^2}{(r+R)^2} \tag{2.26}$$

and as the source voltage V_s is constant, we can omit it. So then writing Eq. 2.26 as $P_L = R(r+R)^{-2}$, we differentiate by parts with respect to R, and as the differential of $(r+R) = 1$ we have:

$$\frac{dP_L}{dR} = (r+R)^{-2} - 2R(r+R)^{-3} \tag{2.27}$$

$$\frac{dP_L}{dR} = \frac{(r+R) - 2R}{(r+R)^3} \tag{2.28}$$

and finally:

$$\frac{dP_L}{dR} = \frac{(r-R)}{(r+R)^3} \tag{2.29}$$

which has a turning point when $r = R$.

All this is probably familiar stuff and not too difficult to do manually; why then invoke SPICE? The answer in this case is to see how sensitive the maximum power transfer is to changes in the load resistor.

2.3.3.2   The '.*step*' Command

We have used this before. In the next chapter we shall expand upon it. The basic syntax is:

.step param < parameter name >< start value >< end value >< step size >

so

.step param R 0.001 6 .01

will step *R* from *0.001 Ω* to *6 Ω* in linear steps of *0.01 Ω*.

It can be entered by pressing *S* on the schematic to open the **Edit Text on the Schematic** dialogue, and make sure the **SPICE Directive** button is pressed and there is a space between the values: the purists might prefer 0.01 to .01 but LTspice does not mind.

Example – Maximum Power Transfer Using the .*step* Command
Given a 10 V supply with a 3 Ω internal resistance, find the optimum load resistor for maximum power transfer. You can find this circuit as 'Max Power LoadR.asc' where the 3 Ω source resistance is included in the voltage source.

A simple **.op** analysis with $R1 = 1\,\Omega$ correctly gives 2.5 V across the resistor and a current of 1 A through it. But to find the optimum, we need to step the resistor. As it cannot be zero, we start at 0.001 Ω and go to 6 Ω in 0.01 Ω steps.

First add traces for *V(in001)* and *I(R1)* by right clicking on the waveform panel and selecting **Add Traces** or **View→Visible traces** to open **Select Visible Waveforms.** Holding down *Ctrl* whilst clicking will enable both to be selected. Note that the maximum of *Vin* is 6.667 V – which is correct – but it is not possible to probe the internal voltage of the source.

To plot the power dissipated in the resistor, there are two options: (a) right click in the trace window, then select **View→Visible Traces** to open the **Select Visible Waveforms** dialogue, then **Alt + double click** to open the dialogue **Expression Editor,** and type *I(R1)*V(n001),* and this will be added to the list of available traces. Select it and the current and voltage as before.

(b) Right click in the trace window, and select **Add Traces** to open the **Add traces to Plot** dialogue as before. Type *I(R1)*V(r)* in the edit box and click **OK**.

The optimum resistor is 3 Ω and the power of 8.33 W. This is very easy to find by hand. What is not so easy to find is the shape of the power curve where we see that for resistors greater than the optimum the reduction of load power is less than with resistors smaller than the optimum. In fact, for a power of 90%, the resistor limits are 1.56 Ω and 5.78 Ω.

We might note that after the schematic has been run, when the cursor is over a component, it changes into a current clamp meter surrounding a small red arrow denoting the direction of the current.

Example – Maximum Power Transfer Using Measurement MAX
The circuit is the same as above, schematic ('Max Power Transfer.asc'). If we have used the dialogue to enter the values, the edit box at the bottom will show

$$\text{.meas maxP MAX I (R1)} * V(r)$$

We step the resistance by: **.step param R 0.001 6 0.01,** and then after a run **View→SPICE Error Log**, we find:

$$maxp : MAX(i(r1) * v(r)) = 8.33333 \; FROM \; 0.001 \; TO \; 6$$

where **maxp** is the measurement name of **i(r1)∗v(r))** which is the formula for the power and we see that the maximum power measured in that interval is *8.33333 W*. We can also see this by opening the editor and clicking the **Test** button.

2.3.3.3 The .*meas* Directive (TRIG,TARG)

If we want to find that range of load resistors for which the load power is at least 90% of the maximum, rather than dragging the cursor around over the traces, we can measure the limiting points. The new controls are shown in Fig. 2.13 where the values relate to the **TRIG,TARG** directive below. The new fields of interest are:

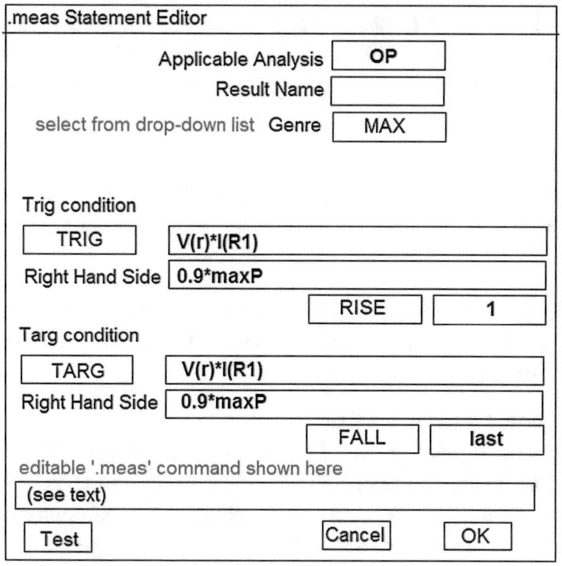

Fig. 2.13 Measure editor TRIG and TARG

TRIG

This triggers the measurement when the Right Hand Side condition is met and works with the TARG option. There is a second option **FROM** which we ignore for now.

Right Hand Side

This is the other half of the equation started by **TRIG**. It can contain the result of a previous *.meas* statement such as **maxP** if, for example, we have:

$$\textbf{TRIG} \qquad\qquad \textbf{V(r)} * \textbf{I(R1)}$$

$$\textbf{Right Hand Side} \qquad \textbf{0.9} * \textbf{maxP}$$

and assuming *Result Name* = *p90%* we find the point at which the power first reaches 90% of its maximum and in the bottom edit box we would see the incomplete directive:

$$.\textbf{meas p90\% TRIG I(R1)} * \textbf{V(r)} = \textbf{0.9} * \textbf{maxP}$$

RISE, FALL, CROSS

There two open identical drop-down lists, one for TRIG and the other for TARG which give three conditions we can select to trigger the measurement. The first is **RISE** when the rising TRIG value equals the Right Hand Side. The second is **FALL** which is the same only it applies to a falling value, and the third is **CROSS** which applies when the value crosses the Right Hand Side in either direction.

WHEN

This is the edit box to the right of both the TRIG and TARG edits wherein we enter when the trigger should occur. We can enter **first** or **1** to cause it to trigger the first time the Right Hand Side is met, or any number such as **3** that will trigger the measurement the third time; or **last** to trigger it the last time the Right Hand Side condition is met. Note that we must supply a condition; we should not leave it empty.

TARG

This is the target condition. It can be the same as the trigger, or different, and again we have a Right Hand Side and a choice of RISE, FALL, CROSS and When.

TD

We shall not use it, but just for completeness, we can enter a delay time, for example, to prevent the trigger or target from being activated by some start-up disturbances until the delay time has elapsed.

Example – Measure TRIG, TARG

Let us take schematic ('Max Power Transfer V1.asc') to find the 90% limits of the load power. If we name the result *P90%*, we can set the trigger to be when the resistor power *V(r)*I(R1)* is 90% of the maximum. We can enter this directly *7.5* (watts) or as a formula *0.9*maxP* using the measured maximum script on the schematic. We also need the first RISE and the last FALL. The edit box will now show the complete directive:

$$.\textbf{meas p90\% TRIG I(R1)} * \textbf{V(r)} = \textbf{0.9} * \textbf{maxP RISE}$$
$$= \textbf{first TARG I(R1)} * \textbf{V(r)}$$
$$= \textbf{0.9} * \textbf{maxP FALL} = \textbf{last}$$

We might note that the **Test** button says **Measurement "p90%" FAIL'ed** 'but after a run we find:

$$\textbf{maxp} : \textbf{MAX(i(r1)} * \textbf{v(r))} = \textbf{8.33333 FROM 0.001 TO 6}$$

$$\textbf{p90\%} = \textbf{4.21637 FROM 1.55849 TO 5.77485}$$

The first confirms that the maximum power is 8.33333 W as before. The second line is not the 90% power value but the range of resistance for which it is at least 90% and is from 1.55849 Ω to 5.77485 Ω – a range of 4.21637 Ω. And this agrees with running a cursor over the trace - so the initial error message is false.

Explorations 4
1. Move the cursor to the *I(R1)*V(r)* text in the waveforms panel title bar of schematic ('Max Power Transfer.asc') so that the cursor changes to a pointing hand. Click on it and in the **Expression Editor** dialogue, go to **Attach cursor,** and select *1st.* Use this cursor to find the value of resistor for maximum power transfer, and confirm if it agrees with theory and the measured values.
2. Estimate the limits of resistance so that the load power is within 70% of the maximum. Check the results by measuring.
3. Try other combinations of components and limits of power, for example, 95%.
4. Set the load resistor R to 3 Ω, remove the internal resistance of the source, and insert a separate series resistor so that it can be swept from 0.01 Ω to 6 Ω to represent the source resistance. Hence explore the optimum source resistance for maximum power transfer to the 3 Ω load. This is schematic ('Max Power SourceR.asc').

2.3.4 The Wheatstone Bridge

This was invented by Samuel Christie and developed by Charles Wheatstone (who also invented the concertina) for the precise measurement of resistance. The circuit essentially consists of two potential dividers as shown in Fig. 2.14. An excellent exposition, including the effect of detector sensitivity, can be found at http://pioneer. netserv.chula.ac.th/~tarporn/311/HandOut/BridgePPT.pdf

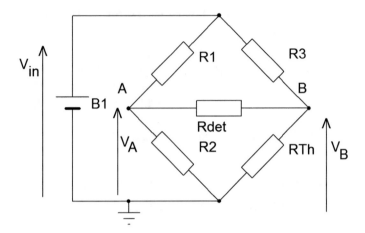

Fig. 2.14 Wheatstone bridge

The bridge is balanced when there is no voltage across the detector *Rdet*, which is true if $V_A = V_B$ from which it follows that:

$$V_{in} \frac{R_2}{R_1 + R_2} = V_{in} \frac{R_3}{R_3 + R_4} \tag{2.30}$$

and the balance is independent of the bridge supply voltage. If we expand Eq. 2.30:

$$R_2(R_3 + R_4) = R_3(R_1 + R_2) \tag{2.31}$$

And after cancelling the term $R_2 R_3$ on both sides, we recast Eq. 2.31 to give:

$$R_4 = \frac{R_3 R_2}{R_1} \tag{2.32}$$

In its original application for measuring resistance, R_3 consisted of precision wire-wound resistors switched in three decades, typically 0–9 Ω, 0–90 Ω and 0–900 Ω covering the range 0–999 Ω in 1 Ω steps. These values were multiplied by the ratio arms R_2/R_1 each of which consisted of more switched precision wire-wound resistors of 1, 10, 100 and 1000 Ω so that the total span of the bridge was from *1/1000 × 1 = 1 mΩ* to *1000/1 × 999 = 999 kΩ*. The clever point about the design is that it does not use excessively high resistor values (difficult to make using wire) nor very low values, equally difficult to make accurately. However, care must be taken in the construction of high-precision versions of the bridge to avoid dissimilar metals as spurious thermal emfs would give false results.

The bridge is often used today with resistive sensors. The bridge is balanced in the inactive state so that changes in the sensor unbalance the bridge. A good example is a strain gauge. Typically, a gauge has an unstrained resistance of 120 Ω and a change of resistance of 0.12 Ω at maximum strain. Hence we can use a Wheatstone bridge which is balanced when there is no strain, and then a small voltage appears across *Rdet* when the gauge is strained. A good discussion can be found at http://elektron. pol.lublin.pl/elekp/ap_notes/NI_AN078_Strain_Gauge_Meas.pdf.

However, we shall see in the section 'Wheatstone Bridge Sensitivity' that such small differences are difficult to measure using a simple moving coil meter for the detector. Up until 1960 or so, the detector was usually a spot galvanometer consisting of a very sensitive (and very delicate) moving coil meter with a small mirror instead of a pointer, and the mirror reflected a spot of light onto a scale some 15 cm long at the front of the galvanometer.

2.3.4.1 The Kelvin Double Bridge

If we try to measure resistances less than 1 Ω with a Wheatstone bridge, we find that the resistance of connecting leads becomes significant. The Kelvin double bridge obviates this problem, Fig. 2.15. The resistors *R1* and *R2* are 'normal' bridge resistors as before, carrying little current, so there is no problem with lead resistance. On the other hand, the unknown resistor *Rx*, the lead resistance *Ry* and the standard resistor *Rz* may all carry the same current - which can be quite substantial. This is well described in https://instrumentationforum.com/t/kelvin-double-bridge-circuit-principle-and-derivation/5600/4from which the following analysis is derived. A slightly different version of the bridge using *four terminal resistors* can be seen at https://en.wikipedia.org/wiki/Kelvin-bridge.

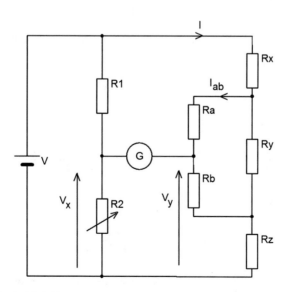

Fig. 2.15 Kelvin double bridge

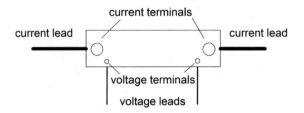

current terminals

current lead

current lead

voltage terminals

voltage leads

Fig. 2.16 Four terminal resistor

Four Terminal Resistors

These are precision low resistance standards, typically 1 Ω or less having two terminals for connecting the current leads and two more for the voltage sensing leads which are positioned so that the resistance between them is exactly the standard, often to an accuracy of 0.01% or better, Fig. 2.16. The point is that the resistance between the current terminals is not defined and could change by a few milliohms just by tightening the screws holding the wires. On the other hand, no current is taken by the voltage leads, so there is no added volt drop due either to the screw connections or the length of the leads. More can be found at https://www.allaboutcircuits.com/direct-current/chpt-8/kelvin-resistance-measurement

Balance Conditions

The analysis is tedious rather than profound, and it is very easy to get into a muddle with convoluted equations. At balance $V_x = V_y$ and whilst:

$$V_x = V \frac{R_2}{R_1 + R_2} \tag{2.33}$$

it takes a little more work to find V_y. As the galvanometer draws no current at balance, the current through Ra is the same as Rb and we start with:

$$V_y = IR_z + I_{ab}R_b \tag{2.34}$$

then:

$$I_{ab}(R_a + R_b) = (I - I_{ab})R_y \quad \text{so} \quad I_{ab} = I\frac{R_y}{R_a + R_b + R_y} \tag{2.35}$$

substituting in Eq. 2.34:

$$V_y = I\left[R_z + \frac{R_b R_y}{R_a + R_b + R_y}\right] \tag{2.36}$$

We now need to equate $V_x = V_y$. For this we need V in terms of I so that we can substitute in Eq. 2.33. From the right hand arm we have:

$$V = I\left(R_x + R_z + \frac{(R_a + R_b)R_y}{R_a + R_b + R_y}\right) \qquad (2.37)$$

we substitute in Eq. 2.33 for V and equate it to Eq. 2.36 and cancel I:

$$\frac{R_2}{R_1 + R_2}\left(R_x + R_z + \frac{(R_a + R_b)R_y}{R_a + R_b + R_y}\right) = R_z + \frac{R_b R_y}{R_a + R_b + R_y} \qquad (2.38)$$

We now isolate the term in R_x and write $\left(\frac{R_1}{R_2} + 1\right)$ for $\frac{R_1 + R_2}{R_2}$:

$$R_x = \left(\frac{R_1}{R_2} + 1\right)\left[R_z + \frac{R_b R_y}{R_a + R_b + R_y}\right] - R_z - \frac{(R_a + R_b)R_y}{R_a + R_b + R_y} \qquad (2.39)$$

expanding and cancelling the two terms in R_z and the two terms in $R_b R_y$:

$$R_x = \frac{R_1 R_z}{R_2} + \frac{R_1 R_b R_y}{R_2(R_a + R_b + R_y)} + \frac{-R_a R_y}{R_a + R_b + R_y} \qquad (2.40)$$

or:

$$R_x = \frac{R_1 R_z}{R_2} + \frac{R_b R_y}{R_a + R_b + R_y}\left(\frac{R_1}{R_2} - \frac{R_a}{R_b}\right) \qquad (2.41)$$

If we make:

$$\frac{R_1}{R_2} = \frac{R_a}{R_b} \quad \text{we are left with } R_x = \frac{R_1 R_z}{R_2} \qquad (2.42)$$

which is the same balance condition as the Wheatstone Bridge.

Explorations 5

1. Build a Wheatstone bridge with a strain gauge of resistance 120 Ω replacing R3 so that it is balanced with no strain. The resistance changes linearly with strain to a maximum increase of 0.12 Ω. Is the detector voltage linear with strain? Since the bridge is no longer balanced, how sensitive are the measurements to changes in the supply voltage? Use the .step command to change the gauge resistance. We shall take up this theme a little later when we have developed more analytical tools.
2. Insert a resistance in series with the battery B1 to represent its internal resistance and show that it does not affect the balance condition.
3. Explore the improvements of using two or more gauges, in particular, if the test piece is a bar, putting one gauge on one side (tension) and another on the opposite side (compression), so the resistance of one increases and that of the other decreases. (See the website mentioned above).
4. The Carey Foster bridge is well described in Wikipedia. Try simulating it.

2.3.4.2   The Murray and Varley Loop Tests

To put it mildly, it is extremely annoying to find there is a fault in several kilometers or, even worse, several tens of kilometers of underground cable where one or more of the conductors has short-circuited to earth, although to generalize, we have included Re as the resistance to earth. The prospect of digging up the whole lot to locate the problem is one to engender sleepless nights. If there is a fault in one or more conductors in a cable having at least one conductor without a fault, or if there are two cables laid together where one only is faulty, then enter Mr. Murray with the answer.

The situation is shown in Fig. 2.17a where the free end of the good conductor is strapped to the faulty by a piece of stout copper wire. The fault is located by the modified Wheatstone bridge shown at Fig. 2.17b where the DC supply is connected to two variable resistors and to earth.

The resistance to earth Re does not affect the balance, just like the internal resistance of the battery, and the bridge is balanced when:

$$R_x = \frac{(R_g + R_x)R_2}{R_1} \tag{2.43}$$

And notice that always $R1 > R2$ because always $(Rg+Ry) > Rx$. We also have:

$$R_x + R_y + R_g = 2R_c \tag{2.44}$$

where the total resistance of each conductor is Rc. Substituting in Eq. 2.43:

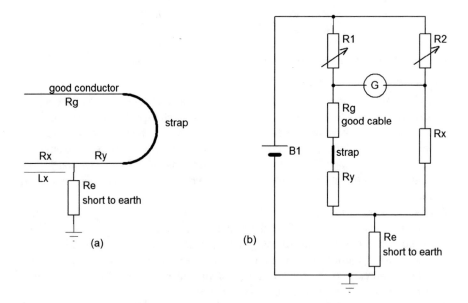

Fig. 2.17 Murray loop test

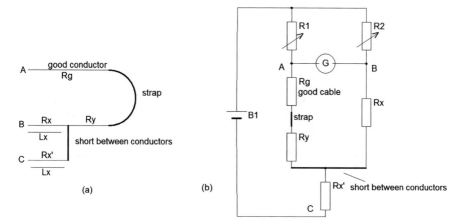

Fig. 2.18 Murray Short-Circuit Test

$$R_x = \frac{(2R_c - R_x)R_2}{R_1} \tag{2.45}$$

and on collecting terms:

$$R_x\left(1 + \frac{R_2}{R_1}\right) = 2\frac{R_c}{R_1} \tag{2.46}$$

or:

$$R_x = 2R_c\frac{R_2}{(R_1 + R_2)} \tag{2.47}$$

and as resistance is proportional to length:

$$L_x = 2L\frac{R_2}{(R_1 + R_2)} \tag{2.48}$$

where L is the length of the cable. Thus, if the fault is at the mid-point, $Rx=Ry$ and $R1=3R2$ hence $Lx = 1/2$.

If there is a short-circuit between conductors, and we have at least one without a fault, we can proceed as Fig. 2.18. We connect the bridge supply not to earth, but to the end of one of the shorted pair, point C in Fig. 2.18b. It does not matter which since point B is the same distance from the short and therefore has the same resistance. The analysis is the same.

Example – Murray Loop Test

A cable is 300 ft. long and a test has a bridge balance with $R1 = 520\ \Omega$ $R2 = 351\ \Omega$. Locate the fault. From Eq. 2.48:

$$L_x = 600\frac{351}{871} = 241.8ft \text{ from the start.}$$

and we do not need to know the resistance of the wire.

Varley Loop Test

The sensitivity of the Murray test can degrade if the fault resistances are high. The Varley loop test overcomes this problem. The circuit is Fig. 2.19. This has two fixed resistors and a new variable resistor $R3$ and a switch $Sw1$.

In the first instance with the switch in position B, the bridge is balanced when:

$$R_{3A} + R_x = \frac{R_1(R_g + R_y)}{R_2} \tag{2.49}$$

where R_{3A} is the balance setting for the variable resistor.

Now the switch is changed to position A and at balance we have R_{3B}:

$$\frac{R_{3B}}{R_1} = \frac{R_g + R_x + R_y}{R_2} \tag{2.50}$$

Extracting $R_g + R_y$ from Eq. 2.50:

$$R_g + R_y = R_2\frac{R_{3B}}{R_1} - R_x \tag{2.51}$$

and substituting in Eq. 2.49:

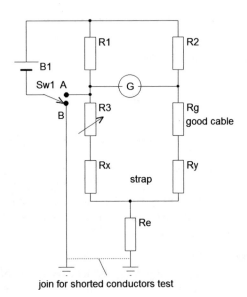

join for shorted conductors test

Fig. 2.19 Varley loop test

$$R_{3A} + R_x = R_{3B} + \frac{R_1}{R_2} R_x \qquad (2.52)$$

Finally:

$$R_x = R_1 \frac{(R_{3B} - R_{3A})}{R_1 + R_2} \qquad (2.53)$$

And similarly for a short-circuit, remove the earth connection and short-circuit the wire from switch pole B to the bottom of Re.

2.4 More Analysis Methods

As we have seen, Kirchhoff's Laws are adequate for many simple circuits, and apart from analogue filters, most consist of about half-a-dozen components. There are three more popular methods of attack, which will be explained in turn. But first let us note that matrix methods are a powerful tool for solving circuit networks (after all, that is what SPICE uses), the main difficulty being in solving them.

2.4.1 Superposition

This works by replacing every voltage source but one in turn by a short circuit. If there are current sources, these are replaced by open-circuits. Then calculate the current in every branch due to that one remaining source. The final result is the algebraic sum of all the currents.

Example – Circuit Analysis Using Superposition
We take Fig. 2.20 as an example where the sense of the voltage across each component is shown by an arrow.

First we short-circuit $B2$. The total resistance is: $R_2 + \frac{R_1 R_3}{R_1 + R_3} = 20 + 7.5 = 27.5\Omega$.

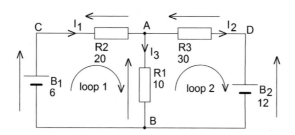

Fig. 2.20 Superposition

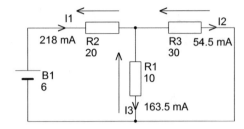

Fig. 2.21 Superposition example A

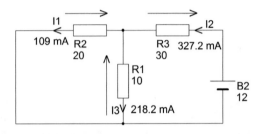

Fig. 2.22 Superposition example B

from which the current is $6\ V/27.5\ \Omega = 218\ mA$ and divides as shown in Fig. 2.21. Similarly, the current due to *B2* alone is *327.2 mA* and divides as shown in Fig. 2.22. Taking account of the polarities, we arrive at:

Current in R1 = 381 mA; Current in R2 = 109 mA; Current in R3 = 273 mA

We can also, of course, solve this circuit using Kirchhoff's Laws.

$$6 = 20I_1 + 10I_3 \qquad \text{for the first loop} \tag{2.54}$$

$$10I_3 = 30I_2 + 12 \qquad \text{for the second loop} \tag{2.55}$$

$$I_1 = I_2 + I_3 \qquad \text{for the currents} \tag{2.56}$$

From (2.54) and (2.56) substituting for *I1* we have:

$$6 = 20(I_2 + I_3) + 10I_3 + 3$$

or:

$$3 = 20I_2 + 30I_3 \tag{2.57}$$

re-arrange (2.55):

$$12 = -30I_2 + 10I_3 \tag{2.58}$$

multiply (2.57) by 2: so that it equates to (2.58)

$$40I_2 + 60I_3 = -30I_2 + 10I_3 \tag{2.59}$$

so

$$I_2 = \frac{-5}{7} I_3 \tag{2.60}$$

use (2.60) to substitute in (2.55):

$$10I_3 = \frac{-5}{7} 30I_3 + 12$$

and finally $I3 = 0.382A$, $I2 = 0.273A$ and $I1 = 0.109A$ (remembering the directions of the currents). It is your choice which method to use.

2.4.2   The Thevenin Model

This works best if we are only interested in the currents and voltages of one or two components rather than the whole circuit. If we really do want to know what is happening everywhere, we must wade through Kirchhoff's Laws or superposition or use matrix analysis.

Thevenin's Theorem states that a circuit consisting of resistors and voltage sources can be reduced to a single resistor in series with a single voltage source. We can get an intuitive feel for this by noting that any circuit consists of meshes each of which is described by a linear equation, and where there is more than one voltage source in a mesh, their total effect is found by adding or subtracting, not by multiplying or dividing; and where more than one current flows through a component, the voltage across it depends on the sum or difference of the currents, not their product. Hence we would expect the network as a whole to behave in some way determined by the sum of the contributions of the meshes (as we saw using Kirchhoff's Laws), and this will be another linear equation of the form $V = IR$.

But – and this is a very important point – we must first define two nodes where we 'look into' the circuit. And the result is generally not the same if we pick another pair of nodes. We create the model by:

- Removing the component of interest.
- Calculating the open-circuit voltage between the terminals it was connected to using Kirchhoff's Laws or superposition.
- Replacing all voltage sources by short-circuits.
- Calculating the resistance of the resulting network.

Then, having created this Thevenin equivalent, we can see the effects of re-connecting the component to it, and it is a simple matter to work out the voltage across it and the current through it.

Example – Thevenin Model
We will do this for $R1$ in Fig. 2.23 following the above steps:

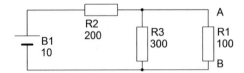

Fig. 2.23 Thevenin example - original circuit

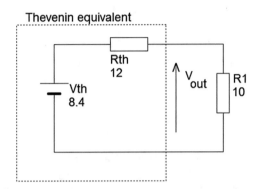

Fig. 2.24 Thevenin example - equivalent circuit

- We first remove *R1*, and we are left with the potential divider formed by *R2* and *R3* so the voltage across terminals *AB* is *10 * 300/500 = 6.0 V*.
- Now we replace *B1* by a short circuit, so we have the two resistors in parallel so: *Rth = 200*300/(200 + 500) = 120 Ω*.

If we now reconnect *R1* in Fig. 2.24 the voltage across *R1* is *6 * 100/220 = 2.73 V*

But look carefully at the Thevenin equivalent circuit enclosed in the dotted rectangle. It is a single voltage source and resistor, and so it can be described by a straight line. If the load is disconnected, the output voltage is just V_{Th}, whilst if we short-circuit the output (not always something to be done in practice!), we find a current of V_{Th}/R_{Th}. Thus we can easily fix the two ends of the line, Fig. 2.25, and the slope is $-1/R_{Th}$, where R_{Th} is the *output resistance* of the circuit – a term we will frequently use.

Conversely, this means that if we have to deal with some electronic circuit or an instrument such as a power supply, or a signal generator, or a meter, or an oscilloscope, with two output terminals (one of which is often ground), we do not need to know the internal circuitry, but can represent it by a Thevenin model of a voltage source in series with a resistance (or, more generally, an impedance). We can measure the open-circuit voltage, and whilst it is usually not safe to short-circuit the output terminals, we can apply various loads and hence plot a graph to find the output resistance.

Explorations 6

1. Replace *R1* in Fig. 2.23 by 10, 30, 300 and 1000 Ω in turn, record the currents and voltages, and hence plot the load line graph.
2. Find the Thevenin models with *R2* and then *R3* removed in turn.

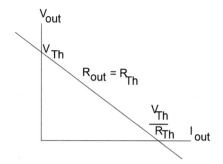

Fig. 2.25 Thevenin loadline

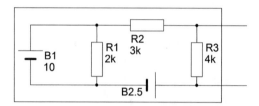

Fig. 2.26 Thevenin Example 2

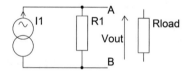

Fig. 2.27 Norton model

3. Construct the circuit of Fig. 2.26, and create Thevenin models viewed from every resistor in turn.
4. Using Thevenin's Theorem, repeat the analysis of the potential dividers in Examples 3 and 4. Is this easier than using Kirchoff's Laws?
5. Try a few more circuits – but remember that for most of analogue electronics, excluding filters, 'Thevenin Example 2' is about as bad as it gets.

2.4.3   *The Norton Model*

Let us start with the simple circuit of Fig. 2.27 of a current source in parallel with a resistor. If there is no load connected between terminals *A* and *B* the output voltage is just $V_{out} = I1R1$. If we short-circuit the output terminals, the current is *I1*.

If we connect a load resistor *Rload* to draw current, we again find that the readings fall on the straight line joining the open-circuit voltage and the short-circuit current, Fig. 2.28, just as the Thevenin model.

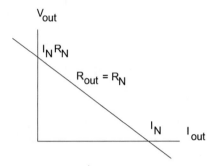

Fig. 2.28 Norton loadline

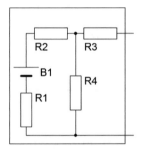

Fig. 2.29 Norton example

So the Norton model is an alternative to the Thevenin model, and we can use whichever is most convenient. However LTspice solves the Norton form faster. We only require the open-circuit voltages and short-circuit currents to be the same or:

$$V_{o/c} = I_N R_N = V_{Th} \tag{2.61}$$

and:

$$I_{s/c} = I_N + \frac{V_{Th}}{R_{Th}} \tag{2.62}$$

from which we find that:

$$R_N = R_{Th} \quad V_{th} = I_N R_N \quad I_N = \frac{V_{TH}}{R_{Th}} \tag{2.63}$$

and we can replace a Thevenin model by a current source V_{TH}/R_{th} in parallel with a resistor $R_{th,}$ and we can replace a Norton model by a voltage source $I_N R_N$ in series with R_N.

Explorations 7
4. Construct the Norton equivalent for the circuit of Fig. 2.29.

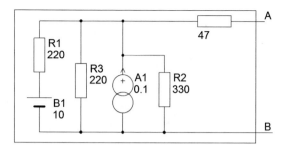

Fig. 2.30   Example 1 for Thevenin and Norton

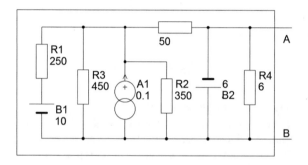

Fig. 2.31   Example 2 for Thevenin and Norton

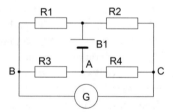

Fig. 2.32   Wheatstone Bridge unbalanced

5. Replace the batteries in Figs. 2.24 and 2.26 by current sources (choose your own values) and create the Norton equivalent.
6. Create Thevenin and Norton models for Fig. 2.30 and (if you really want to!) Fig. 2.31.

2.4.3.1   Wheatstone Bridge Sensitivity

We can now analyse the effect of the galvanometer sensitivity if the bridge is not balanced by removing the galvanometer and finding the Thevenin equivalent of the bridge and power supply, Fig. 2.32. To find the Thevenin equivalent resistance, remember that *B1* is a short-circuit, so we have:

$$R_{Th} = \frac{R_1 R_3}{R_1 + R_3} + \frac{R_2 R_4}{R_2 + R_4} \qquad (2.64)$$

and if we take point A as the voltage reference, we have a potential divider to point B so:

$$V_{BA} = \frac{R_3}{R_1 + R_3} \qquad (2.65)$$

and similarly:

$$V_{CA} = \frac{R_2}{R_2 + R_4} \qquad (2.66)$$

and the Thevenin voltage is the difference:

$$V_{Th} = V_{BA} - V_{CA} \qquad (2.67)$$

Example – Wheatstone Bridge Sensitivity

A Wheatstone bridge has resistors $R_1 = R_2 = R_3 = 500\ \Omega\ R_4 = 501\ \Omega$. Because the bridge could be out of balance in either direction, we need a centre-zero meter rather than a conventional one with zero at the left. Suppose our 1 mA, 100 Ω meter is a centre-zero type, calculate the meter reading if the bridge is excited by a 2 V supply.

The Thevenin resistance is $R_{th} = 250 + 250.25 = 500.25\ \Omega$. The voltage is $V_{th} = 1.000 – 0.999 = 1\ mV$. Inserting the meter we have $I = 1\ mV/600.25 = 1.67\ \mu A$ or a deflection of 0.167% of fsd – which is undetectable (schematic 'wheatstone Bridge unbalanced.asc'). If we replace the meter by one with an fsd of 100 μA, we will find a deflection of 1.67% which could just be seen.

What this tells us is that we can only detect a difference of resistance of around 1% – which is not really adequate. We could improve the sensitivity with a higher power supply but at the risk of self-heating of the resistors causing their values to change. Otherwise we need a genuine spot galvanometer (rather delicate) or an electronic voltmeter: most hand-held types can read to 0.1 mV, so then we can hope to see a difference of resistance of 0.1%. Or we use a low-frequency AC excitation and amplify the out of balance voltage.

Exploration 8

1. Build a Wheatstone bridge and explore its sensitivity, noting the effect of the galvanometer resistance as well as its fsd.
2. Build a Murray loop test bridge inserting different lengths of various size wires (use wire tables freely available on-line) with both earth and short-circuit faults, and estimate the precision with which the fault can be located.
3. Build a Varley loop test bridge, and compare its performance with the Murray if the fault resistance is high rather than just a few ohms.

2.5 Attenuators

An attenuator is a circuit designed to reduce the level of a signal. In the main, those constructed only from resistors are used at low frequencies, but we can equally well explore them with DC. The potential divider is an example and is often referred to as the 'L' attenuator. These are explained in great detail in https://www.electronics-tutorials.ws/attenuators/l-pad-attenuator.html and the following pages.

We may also note that the same configurations, with some or all of the resistors replaced by capacitors or inductors, are used for AC filters.

The Transfer Function '.tf'
This is a useful option found by **Simulate→Edit Simulation Cmd, and** then pick the fifth Table **DC transfer function.** This computes the small-signal DC transfer function between a node voltage or the current in a component and a source voltage or current. However, we must be consistent and not to try to measure the current compared to a voltage. This function will automatically find the operating point.

The syntax is:

$$\textit{.tf } \textbf{V}(< node >) < \textit{Voltage source} > \text{ or}$$
$$\textit{.tf } \textbf{I}(< component name >) < \textit{Current source} >$$

But the two edit boxes will create this for us and in the case of voltage they are filled in as:

$$\textbf{Output V} < \textbf{name of node} >$$
$$\textbf{Source} < \textbf{name of voltage source} >$$

We should note that the voltage source must not have any internal resistance.

2.5.1 The L-Attenuator

This is shown in Fig. 2.33. It is simply a potential divider with $R1$ turned sideways so giving it the appearance of a letter 'L' flipped left-right and inverted. It is an example

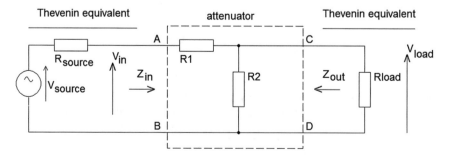

Fig. 2.33 L-attenuator

of a *two-port network* having an input port between terminals A and B and an output port between terminals C and D. We will meet two-port networks time and time again in later chapters. They need not be constructed only from resistors but can include capacitors, inductors and voltage and current sources. Therefore we shall use the general identifier Z for an *impedance* although here we shall concern ourselves only with resistors.

We can 'look into' the network from the left and the impedance is Z_{in}, whilst from the right, we see Z_{out}. We model the input, which could be the output of an amplifier or a signal generator, as a Thevenin equivalent circuit consisting of V_{source} *and* R_{source}. In the same way, we can model whatever load is connected to the attenuator as another Thevenin equivalent circuit of resistor R_{load} where we assume that the load does not generate a voltage. Thus we have:

$$Z_{in} = R_1 + \frac{R_2 R_{load}}{R_2 + R_{load}} \tag{2.68}$$

and:

$$Z_{out} = \frac{R_2 (R_1 + R_{source})}{R_1 + R_2 + R_{source}} \tag{2.69}$$

The point to note here is that the impedances are greatly different.

Example – L-Attenuator

Suppose we have a 600 Ω audio output *Rsource* and a load *Rload* also of 600 Ω and we want to reduce the voltage measured at the load by 50%. This is NOT the open-circuit voltage of the source V_{in}. If we make $R1 = 300\ \Omega$ and $R2 = 600\ \Omega$, Z_{in} will be 600 Ω and the voltage between C and D will be half the input across A–B. However, looking back into the attenuator, Z_{out} is R2 in parallel with R1 and R_{source} in series or only 480 Ω.

If we want to reduce the output to 20%, the mismatch becomes worse. Now we have $R1 = 480\ \Omega$, $R2 = 150\ \Omega$ and $Z_{in} = 600$ and $Z_{out} = 131\ \Omega$.

Example – L-Attenuator with Large Attenuation

The situation becomes dire if we want a large attenuation, for example, an output of only 1% of the input. Writing Z for the parallel combination of R_2 and R_{load}, we have:

$$\frac{Z}{R_1 + Z} = 0.01 \quad \text{and} \quad R_1 + Z = 600\Omega$$

which solves as $R1 = 594\ \Omega$ and $R2 = 6.06\ \Omega$ and now $Z_{out} = 6\ \Omega$. Moreover, the circuit is very 'lossy' in that most of the power is dissipated in $R2$ and wasted. Or we can work the other way so that Z_{out} is correct and then Z_{in} will be wrong.

This shows that if we need a high input impedance, it is better to use the circuit as shown with the input to $R1$. The schematic is ('L-Attenuator 1.asc').

The website mentioned above has helpful formulae for calculating the resistors. We define $K = \frac{v_{out}}{v_{in}}$ then for Z_{in} to be correct:

$$R_1 = R_{source}\left(K - \frac{1}{K}\right) \quad R_2 = R_{source}\left(\frac{1}{K-1}\right) \tag{2.70}$$

whilst for Z_{out} to be correct:

$$R_1 = R_{load}(K-1) \quad R_2 = R_{load}\left(\frac{K}{K-1}\right) \tag{2.71}$$

Example – L-Attenuator Transfer Function
Using the transfer function *.tf V(vload)* **Vs for** schematic ('L-Attenuator 1.asc') we find:

Transfer function $= 0.1$.
Input impedance $= 1200\ \Omega$.
Output impedance at V(vload) $= 108\ \Omega$.

The figures agree with the circuit analysis.

2.5.2 The 'T'-Attenuator

There are a number of applications where it is desirable to have R_{source} and R_{load} the same, and there are three common values: 50 Ω and 75 Ω for high-frequency work, often using coaxial cables, and 600 Ω for audio and PA systems. Then the requirement is that both Z_{in} and Z_{out} should equal the source and load impedances. This is done by adding another resistor R2, Fig. 2.34. It is symmetrical if we make $R1 = R2$ so that if the input impedance is correct, the output impedance will also be correct.
If we define Z_x as the parallel sum of R_3 with $(R_2 + R_{load})$ in series :

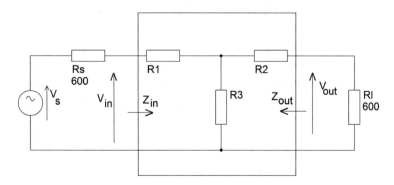

Fig. 2.34 T-attenuator

$$Z_x = \frac{R_3(R_2 + R_{load})}{R_2 + R_3 + R_{load}} \tag{2.72}$$

and remembering $R_{source} = R_{load}$ and $R1 = R2$ we have:

$$Z_{in} = Z_{out} = R_1 + Z_x \tag{2.73}$$

and:

$$V_{out} = V_{in}\frac{Z_x}{R_1 + Z_x} \tag{2.74}$$

Once again, the above website has useful formulae. If the attenuator is symmetrical, and where Z is its impedance:

$$R_1 = R_2 = R_3 + \frac{Z(K-1)}{K+1} \quad \text{and} \quad R_3 = \frac{2ZK}{K^2 - 1} \tag{2.75}$$

Example – T-Attenuator

Let us take as an example an attenuator that will reduce a signal by the factor 0.3164 (this is an attenuation of 10 dB), and we need to be clear that this is the ratio V_{out}/V_{in} and NOT V_{out}/V_s; see Fig. 2.35. And we shall add to the specification that the attenuator must have input and output resistances of 600 Ω – a common standard for audio work. Now as the correct attenuation is obtained when R_{source} and R_{load} are both 600 Ω, and as $R1 = R2$, we can solve the equations to find $R1 = R2 = 311$ Ω and $R3 = 422$ Ω. This is schematic ('T-Attenuator 1.asc'). An .op analysis shows the correct attenuation.

And we can note from Eq. 2.75 that as the attenuation increases, the series resistors $R1$ and $R2$ increase whilst $R3$ decreases.

Example – T-Attenuator Transfer Function

If we measure .tf V(out) Vs using the previous schematic, the results are that the output impedance is 299.851 Ω, the input impedance is 1199.4 Ω, and the transfer

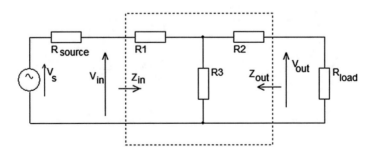

Fig. 2.35 T-attenuator example

function is *0.158368*. The problem lies with the source resistance because we are measuring from V_s and not V_{in}. If we subtract the source resistance from the input impedance the answer is correct at 600 Ω, and likewise the transfer function is exactly half of the correct value. However, the output impedance is correct.

We cannot subsume R_{source} in the voltage source resistance because for a transfer function analysis, this must be zero. But as we need to measure at the input terminals V_{in}, we can just remove R_{source} (or reduce it to an insignificant value). We do this in ('T Attenuator 2.asc') where the transfer function and input impedance are correct but the output impedance is still wrong. To get the correct result, we would need to apply a voltage at the output, but as the circuit is symmetrical, there is no need.

We can explore the effects of a load mismatch by stepping the load resistor.

2.5.2.1 Impedance Matching

This is possible to some extent with the 'L'-attenuator, but both attenuation and impedance matching cannot be achieved together. Here, using a 'T' we have input and output impedances to match, and the attenuation, and three resistors at our disposal, so it is possible to satisfy all the conditions. The equations are:

$$R_1 = Z_{in}\left[\frac{K^2+1}{K^2-1}\right] - R_3 \quad R_2 = Z_{out}\left[\frac{K^2+1}{K^2-1}\right] - R_3 \quad R_3 = 2\sqrt{Z_{in}Z_{out}}\left[\frac{K}{K^2-1}\right] \quad (2.76)$$

where again K is the attenuation.

Example – T-Attenuator from 75 to 50 Ω
We have a source impedance of *75 Ω* and want an attenuation *a = 0.5 or 6 dB* with a load impedance of *50 Ω*. We find $K = 10^{a/20} = 10^{6/20} = 1.9953$ $K^2 = 3.98$ from Eq. 2.76 we have: $R_3 = 2 \times \sqrt{3750} \times 1.9953/2.98 = 82.0$ Ω then $R_1 = 75 \times 4.98/2.98 - 82.0 = 43.3$ Ω $R_2 = 50 \times (4.98/2.98) - 82.0 = 1.56$ Ω this is schematic ('T Attenuator Match.asc'). The input impedance is correct, and it looks as though the impedance 'looking in' from the load is about 75 Ω. However, the transfer function is not correct, nor is it correct in schematic ('18 dB T-pad V2.asc') using values taken from the above website.

2.5.2.2 The Bridged-T Attenuator

This adds another resistor bridging input and output, Fig. 2.36. The point of this is that we can have a variable attenuation without altering the input and output impedances provided input and output impedances are identical. We require $R1 = R_{source} = R_{load}$ leaving only $R2$ and $R3$ at our disposal.

The analysis is easiest if we anticipate the results of the 'Start-Delta' conversions to turn the 'T' into a 'Π' using Eqs. 2.104 and 2.105, Fig. 2.37a. We first remove $R3$

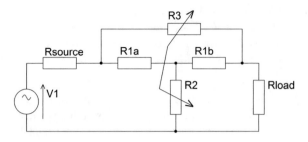

Fig. 2.36 Bridged-T attenuator

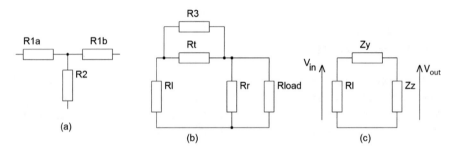

Fig. 2.37 Bridged-T attenuator analysis

and ignore the source and load resistors leaving just the 'T'. To aid identification, we add a suffix a,b to the two identical resistors. From Eq. 2.104, we define a left resistor R_L for the Π as:

$$R_l = R_{1a} + R_2 + \frac{R_{1a}R_2}{R_{1b}} = R_1 + 2R_2 \tag{2.77}$$

similarly we define top and right resistors:

$$R_t = R_{1a} + R_{1b} + \frac{R_{1a}R_{1b}}{R_2} = 2R_1 + \frac{R_1^2}{R_2} \qquad R_r = R_{1b} + R_2 + \frac{R_{1b}R_2}{R_1 b}$$

$$= R_1 + 2R_2 \tag{2.78}$$

When we come to find the attenuation and the input resistance, we omit the source resistance because we shall be 'looking in' to the circuit after the source. Thus we now restore only the load resistor and $R3$, Fig. 2.37b, and create a final Π circuit '(c)'.

$$Z_z = \frac{R_R R_{load}}{R_R + R_{load}} = \frac{(R_1 + 2R_2)R_{load}}{R_1 + 2R_2 + R_{load}} \tag{2.79}$$

$$Z_y = \frac{R_T R_3}{R_T + R_3} = \frac{\left(2R_1 + \frac{R_1^2}{R_2}\right)R_3}{2R_1 + \frac{R_1^2}{R_2} + R_3} = \frac{R_1 R_3(2R_2 + R_1)}{R_1(2R_2 + R_1) + R_2 R_3} \tag{2.80}$$

Then:

$$V_{out} = V_{in} \frac{Z_x}{Z_x + Z_y} \tag{2.81}$$

The input and output impedances are identical. The input impedance is:

$$Z_{in} = \frac{R_L(Z_y + Z_z)}{R_L + Z_y + Z_z} = \frac{(R_1 + 2R_2)(Z_y + Z_z)}{R_1 + 2R_2 + Z_y + Z_z} \tag{2.82}$$

This is not a nice equation to solve, but again, from the above website we have the design equations which are:

$$R_2 = \frac{R_1}{K - 1} \quad R_3 = R_1(K - 1) \tag{2.83}$$

and we can note that these are in opposition that as R2 increases, R3 decreases at the same rate. So to vary the attenuation, we can use a double-gang potentiometer with R3 in inverse connection, Fig. 2.36.

Example – The Bridged-T Attenuator

Given a 50 Ω source and load impedance, we can calculate the resistor values for an attenuation from zero up to 90%. For zero attenuation R2 is irrelevant and we require R3 = 0;

For an attenuation of 1/10, we need $R2 = 50/(10–1) = 5.55 \ \Omega \ R3 = 50 (10–1) = 450 \ \Omega$. This is schematic 'Bridged-T Attenuator.asc'. A transfer function analysis gives the correct output voltage but not the correct impedances since they include the load and source resistors.

The website has tabulations of the resistor values for different attenuations from which we find that for 50 Ω we require $R2 = 2500/R3$. This is schematic ('Bridged-T Attenuator 2.asc') where the resistor values go in pairs. It can be seen that the input impedance is constant and the output voltage is proportional to the reciprocal of R3 by plotting 1/V(out) We can use a two-gang 1 kΩ potentiometer for R2 and R3.

2.5.3   The Pi-Attenuator

This can give much higher levels of attenuation than the T-attenuator. The circuit is shown in Fig. 2.38.

This, too, is symmetrical if we make R1 = R3, and then merging R3 and R_{load}, we have:

$$Z_x = R_2 + \frac{R_3 R_{load}}{R_3 + R_{load}} \tag{2.84}$$

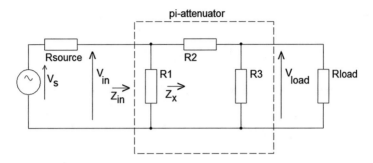

Fig. 2.38 Pi attenuator

$$Z_{in} = \frac{R_1 Z_x}{R_1 + Z_x} \tag{2.85}$$

and:

$$\frac{V_{load}}{V_{in}} = \frac{Z_x - R_2}{R_2 + Z_x - R_2} \tag{2.86}$$

and the formulae are:

$$R_1 = R_3 = Z\left(\frac{K+1}{K-1}\right) \quad R_2 = Z\left(\frac{K^2-1}{2K}\right) \tag{2.87}$$

Example – The Pi-Attenuator

Given source and load impedances of 75 Ω, design an attenuator to reduce the output to 20% of the input.

From Eq. 2.86 we require $R1 = R3 = 75(6/4) = 112.5\ \Omega$ and $R2 = 75(25–1)/10 = 180\ \Omega$. This is schematic ('Pi Attenuator.asc'). The attenuation is correct and the transfer function shows an input impedance of *150 Ω* but that includes R_{source} of 75 Ω and is correct

Explorations 9

1. Explore the practical limits of T- and Pi-attenuators by trying 10, 20, 30 dB, etc. of attenuation using 600 Ω source and load impedances.
2. Repeat for 50 Ω and 75 Ω.
3. Run schematic ('Bridged-T Attenuator 2.asc') first by stepping the resistor values and then by uncommenting the tabulated values for the resistors from the above website.
4. Design an attenuator to give 20 dB attenuation with a source of 50 Ω and a load of 75 Ω.

2.6 Delta-Star Conversion

It is also known as the 'Delta-Y' conversion. This is a useful technique for converting between two forms of a circuit consisting of three components. It is not one that we need very often, but it is good practice in circuit analysis.

The circuit is shown as Fig. 2.39 where the 'star' has been drawn inside the 'delta' in red to clarify the relationships between the elements. A more familiar form is the topologically identical circuit of Fig. 2.40 which is yet another name for the conversion – 'Pi Tee'. We may either chose to convert the delta to a star or the other way round. And we should note that the 'Y' adds an extra node M.

2.6.1 Delta-Star Conversion

We start with terminals 1–2 in the delta. The resistance between them is R_a in parallel with $R_b + R_c$ in series, and this equals $R_x + R_z$ in series between terminals 1–2:

$$R_{12} = \frac{R_a(R_b + R_c)}{R_a + R_b + R_c} = R_x + R_z \qquad (2.88)$$

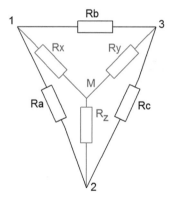

Fig. 2.39 Delta-star Conversion

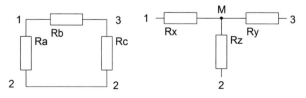

Fig. 2.40 Delta to Star Conversion

Similarly between terminals *2–3*:

$$R_{23} = \frac{R_c(R_a + R_b)}{R_a + R_b + R_c} = R_z + R_y \tag{2.89}$$

and between *1* and *3*:

$$R_{13} = \frac{R_b(R_a + R_c)}{R_a + R_b + R_c} = R_x + R_y \tag{2.90}$$

From this we can write pairs of equations where $R_T = R_a + R_b + R_c$:

$$R_x = \frac{R_a(R_b + R_c)}{R_T} - R_z \quad R_x = \frac{R_b(R_a + R_c)}{R_T} - R_y \tag{2.91a, b}$$

$$R_y = \frac{R_c(R_a + R_b)}{R_T} - R_z \quad R_y = \frac{R_b(R_a + R_c)}{R_T} - R_x \tag{2.92a, b}$$

$$R_Z = \frac{R_c(R_a + R_b)}{R_T} - R_y \quad R_Z = \frac{R_a(R_b + R_c)}{R_T} - R_x \tag{2.93a, b}$$

Substituting for R_z in Eq. 2.91a from Eq. 2.93a, we have:

$$R_x = \frac{R_a(R_b + R_c) - R_c(R_a + R_b)}{R_T} + R_y \tag{2.94}$$

Then substituting for R_y from Eq. 2.92b:

$$R_x = \frac{R_a(R_b + R_c) - R_c(R_a + R_b) + R_b(R_a + R_c)}{R_T} - R_x \tag{2.95}$$

Expanding we finally have:

$$R_x = \frac{R_a R_b}{R_a + R_b + R_c} \tag{2.96}$$

We can similarly show that:

$$R_y = \frac{R_b R_c}{R_a + R_b + R_c} \quad R_z = \frac{R_a R_c}{R_a + R_b + R_c} \tag{2.97}$$

Example – Delta-Star Conversion
Starting with the *Π* side of schematic ('PiTee.asc'), we substitute in the above equations to find:

$$R4 = 200/60 = 4.33\Omega \quad R5 = 600/60 = 10\Omega \quad R6 = 300/60 = 5\Omega$$

Assuming a *1 V DC* input, we calculate that $I(R2) = -0.02$ A, $I(R3) = 0.02$ A, $I(R1) = 0.1$ A $I(V1) = 0.12$ A, $V(2) = 0.6$ V. With $V1 = 1$ V we can check these results and find that they agree. We might notice that the currents in R2 and R3 are opposite. This is due to the orientation of the resistors. If we use **F7** to cut out R3 then turn it through 180°, we find the currents are both negative. So we must be careful how we use the direction of the currents. But at least, ignoring directions, they agree. Also the resistance between nodes *1* and *ground* is *1 V/0.12A* = *8.33* Ω and agrees with the calculation. As there is no excitation for the right hand circuit, the currents are zero. However, if we turn to schematic 'PiTee1.asc', we can test both forms by commenting and uncommenting the transfer function where we find the transfer function to node *1* is 0.33333, the input impedance is *15* Ω, and the output impedance is *6.6667* Ω. Taking node 2, there is a very slight difference in the output impedance.

2.6.2 Star-Delta Conversion

This is just the reverse. It is a little more complicated and easy to lose a factor or two. Several websites just quote the result or omit the derivation. The clearest – which is followed here – is at http://www.sakshieducation.com/Engg/EnggAcademia/ElectricalEngg/ElectricalCircuits/STAR%E2%80%93DELTATRANSFORMATION.pdf. Starting from the star, we write down in turn the resistance between the terminals and compare them to the corresponding resistances in the delta:

$$R_x + R_z = \frac{R_a(R_b + R_c)}{R_T} \tag{2.98}$$

$$R_y + R_z = \frac{R_c(R_a + R_b)}{R_T} \tag{2.99}$$

$$R_x + R_y = \frac{R_b(R_a + R_c)}{R_T} \tag{2.100}$$

We now form the sum of products of the combinations of the terms using Eqs. 2.96 and 2.97.

$$R_x R_y + R_x R_z + R_y R_z = \frac{(R_a R_b.R_b R_c) + (R_a R_b.R_a R_c) + (R_b R_c.R_a R_c)}{R_T{}^2} \tag{2.101}$$

Collecting terms:

$$R_x R_y + R_x R_z + R_y R_z = \frac{R_a R_b R_c (R_a + R_b + R_c)}{R_T{}^2} = \frac{R_a R_b R_c}{R_T} \qquad (2.102)$$

Substituting for R_T in Eq. 2.102 using Eq. 2.96

$$R_x R_y + R_x R_z + R_y R_z = R_a R_b R_c \frac{R_x}{R_a R_b} = R_c R_x \qquad (2.103)$$

Then dividing by R_x and swapping left side for right side:

$$R_c = R_y + R_z + \frac{R_y R_z}{R_x} \qquad (2.104)$$

and similarly:

$$R_a = R_x + R_z + \frac{R_x R_z}{R_y} \qquad R_b = R_x + R_y + \frac{R_x R_y}{R_z} \qquad (2.105)$$

A number of rather improbable practice circuits can be found at the above website and also at https://brilliant.org/wiki/transformation-of-resistances-star-to-delta-and/ One is shown as schematic 'PiTee Test.asc'.

Example – Star-Delta Conversion
This is schematic 'StarDelta.asc'. We find the transfer functions agree. There are two points to note: the first is that the resistor values in the delta are all larger than the star; and the second is that the conversion is not symmetrical: if we move the inputs to *s* and *d* and take the outputs from *alt s* and *alt d*, we do not get the correct results. The differences become clearer if we use a larger range of resistors in the star.

Explorations 10
1. Select some sets of three resistors, and convert them from 'delta' to 'star' and vice versa.
2. Note the effect of source and load resistances applied to the circuits.
3. The 'star' has a centre connection. Suggest how this could be used.
4. Provide an input to schematic 'PiTee Test.asc' and test it.

2.7   The Thermocouple

These are popular, small, low-cost temperature sensors which generate very small DC voltages, so we include them in this chapter.

The physics behind their operation need not detain us beyond noting that they use the Seebeck effect, which in turn consists of the Peltier voltage created by the junction of dissimilar metals and the larger Thompson voltage created by the temperature gradient along the wires. A good, non-mathematical explanation can

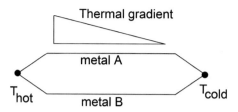

Fig. 2.41 Thermocouple

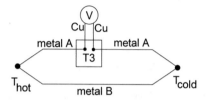

Fig. 2.42 Thermocouple meter

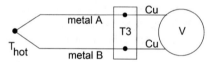

Fig. 2.43 Thermocouple meter 2

be found at http://www.capgo.com/Resources/Temperature/Thermocouple/Thermo couple.html, whilst for the mathematically minded, the Wikipedia article is very good.

The basic requirement, then, is for two wires of different composition to be joined at the ends to make two junctions, Fig. 2.41. A voltage then will be generated depending on the difference in temperature of the two junctions. Traditionally, the cold junction was held at $0\,°C$ and is the reference temperature. As ice can be colder than $0\,°C$, an equilibrium mixture of ice and water was used. The purists may insist that this is the triple point of water and is actually $0.01\,°C$, but as thermocouples are only accurate to about $1\,°C$, we can ignore this minor error.

The problem begins when we want to measure the voltage. If we insert a meter, Fig. 2.42, we introduce another pair of junctions between metal A and the copper wire to the meter. If both of these new junctions are at the same temperature T3, as shown in the figure, one will cancel the other so their contributions to the voltage is zero. This leads us to the improvement that we can dispense with the cold junction, provided we know T3. Therefore we join the copper wires of the meter to the thermocouple wires in a metal block and measure temperature T3 with a platinum resistance thermometer (not another thermocouple), Fig. 2.43.

All this is the practical stuff on how to use a thermocouple: now we need to know how to simulate the voltage versus temperature graph. First, there are many types of thermocouples using different pairs of metals, often alloys, sometimes pure metal.

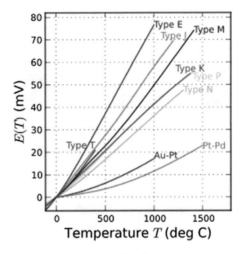

Fig. 2.44 Themocouple Voltages (original work by Nanite)

There are a number of standard pairs of metals each denoted by a letter. But as metals may contain impurities, or alloy compositions may not be exact, there are international tables for all the thermocouple types giving the voltages at 1 K steps from the minimum to the maximum that each type must produce. The five-figure tabulation of their voltages can be found at https://www.thermocoupleinfo.com/thermocouple-reference-tables.htm

   A graph of some, courtesy of Nanite - Own work, CC0, https://commons. wikimedia.org/w/index.php?curid=30047220 in Wikipedia is shown as Fig. 2.44 showing that there is a range of upper temperature limits (often set near the melting point of the metals) and corresponding limits to the voltage output. However, all can be simulated in the same way since, generally, the graph can be represented by a power series of the form:

$$y = a_0 + a_1x + a_1x^2 + a_2x^3 + \ldots\ldots a_nx^n \qquad (2.106)$$

within prescribed limits: beyond those limits, the value can vary wildly. In the case of thermocouples, the coefficients depend on the materials used. For example, the tabulation for a J-type thermocouple (and many others) can be found at: https://www.thermocoupleinfo.com/type-k-thermocouple.htm so any thermocouple purporting to be J-type must generate all of these voltages. To take a few points from the table, at $-210\,°C$ (its lowest working point), it is -8.095 mV; at $0\,°C$ it is 0 mV (obviously, as this is the reference temperature); at $100\,°C$, it is 5.269 mV; and at the maximum temperature of $1200\,°C$, it is 69.553 mV.

   This table is what must be modelled in SPICE. To do so, manufacturers use up to eight coefficients in Eq. 2.106. But rather than create a new model for every type, they use a base structure to which the appropriate parameters are inserted depending on the thermocouple type. Remembering that the voltage is proportional to the

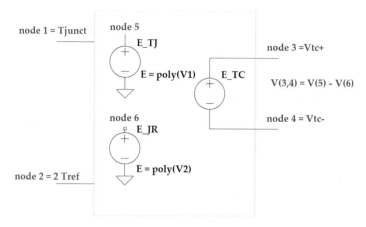

Fig. 2.45 SPICE thermocouple

difference in temperature between the two junctions, this model uses a polynomial for the voltage controlled voltage source E_TJ, Fig. 2.45, representing the junction temperature whose value is determined by the voltage applied to terminal 1. The same polynomial is used for the second voltage-controlled source E_JR controlled by the voltage at terminal 2 representing the reference temperature. The difference between these two voltage sources is applied to E_TC to produce a voltage proportional to the temperature. This is described in http://www.ecircuitcenter.com/Circuits/Thermocouple/Thermocouple1.htm. In fact, the voltage sources give volts, so the output is divided by 1000 to get millivolts. This can also be found as schematic 'SPICE Thermocouple.asc'.

This is the generic SPICE model. And it has two problems. The first is that it is a four-terminal device, not two. The second is that the temperature must be represented by a voltage; we cannot directly apply the parameter *temp*. All this is necessary because some poor mutts have laid out good money to buy inferior versions of SPICE when they could have kept their bank balance intact and used LTspice for free and created better, simpler models! We shall now see how.

Example – Thermocouple

If we take a restricted range of just 0 °C to 100 °C for a J-type thermocouple, we can try a cubic equation using the published SPICE parameters. For this we need an 'arbitrary behavioural voltage source' – that's all. This is quite simply a voltage source whose voltage can be set to depend on almost anything (as we have seen before). In this case we want a voltage that is a function of temperature. So we can take the SPICE parameters and replace the voltages by the temperature. But to satisfy the SPICE requirements, we must add a ground and a dummy parallel resistor to complete the circuit, ('Simple thermocouple.asc'). As it stands, the output is in volts rather than millivolts. It will be found that a cubic only gives 5.257 mV at 100 °C instead of 5.269 mV, but adding the fourth term and it is 5.270 mV. The schematic has the full equation as a comment which can be copied and pasted into the voltage source definition.

The attentive reader will have noticed that only one temperature is given, and there is no reference temperature. In this specific schematic, it is not needed because, as we saw before, the global 'temp' is used. However, if we had other components in the circuit and wished to change their temperatures, there would be a problem. So, because of that, and to save the bother of having to enter the coefficients every time, it is better to create a sub-circuit.

Explorations 11

1. These are the full coefficients for a J-type thermocouple:

.E_TJ 5 0 VALUE = {
+   0.503811878150E-01 $*$ V(1) + 0.304758369300E-04 $*$ V(1) $**$2+
+ − 0.856810657200E-07 $*$ V(1) $**$3 + 0.132281952950E-09 $*$ V(1) $**$4+
+ − 0.170529583370E-12 $*$ V(1) $**$5 + 0.209480906970E-15 $*$ V(1) $**$6+
−0.125383953360E-18 $*$ V(1) $**$7 + 0.156317256970E-22 $*$ V(1) $**$8   }

Continue to add them to ('Simple themocouple.asc'), and check the output against the table. Decrease the exponentials by 3 to get millivolts rather than volts; hence the first is 0.5038....E-04.

2. Experiment with other thermocouple types.
3. Use a Wheatstone bridge to measure the temperature.
4. A search of the online literature for thermocouples shows that often two or three or more are used in a bridge to avoid errors due to long leads. Explore.

2.8   Metrology

This is rather a specialist topic and concerts the realization of accurate standards. A brief summary can be found at https://www.nist.gov/sites/default/files/documents/iaao/NIST_SIM-web-vsn.pdf which usefully shows the relationships between the standards and fundamental physical constants. But voltage and resistance will suffice for now.

2.8.1   Voltage Standard

The problems with a voltage standard are:

• It must be reproducible so that the same standard can be realized worldwide.
• It should have a very low EMF change with temperature.
• It should be stable and not subject to long-term degradation.
• Its EMF should not change if a small current is drawn.

Until 1990 this was the Weston cell which superseded the Daniel Cell and answers most of the difficulties except the change with temperature (which was greatly improved by a variation on it) and not being able to supply current (see http://conradhoffman.com/stdcell.htm) who states that even a few microamps change the value. In addition the internal resistance of the cell would cause voltage drops, and as the voltage is specified as 1.018638 V, even micro ohms are important. Therefore it is used in a null circuit where its EMF is exactly balanced by another. A lengthy historical discussion can be found at https://www.nist.gov/sites/default/files/documents/calibrations/mn84.pdf

The current standard is now the Josephson Junction (and the Wikipedia article about him is fascinating) where the voltage V_n is given by:

$$V_n = n \times h/2e \times \mathbf{f}$$

where h is Plank's constant, e the elementary charge, f the frequency and n the number of steps which are about 150 μV apart at a frequency of 70 GHz. These steps are exactly equal but very small, so therefore the standard requires several thousand junctions in series in a special integrated circuit held at 4 K and irradiated with microwaves and supplied with a bias current.

2.8.1.1   Equipment Voltage Standard

As the absolute standard is difficult and costly to build, secondary standards use special Zener diodes having very low drift against temperature and time. Voltage standards to be built into equipment similarly must have very low drift. The Linear Devices LT1027 is an example being an integrated circuit that has a temperature coefficient, depending on which version is chosen, down to 2 ppm and a drift of 20 ppm/month.

Example – Voltage Standard
There are schematic using the LT1021-7 and the LT1021-10 but these do not seem to implement temperature or line changes but do show a change of 1 μV for a load change from 0 to 10 mA.

2.8.1.2   Voltage Divider

This is a precise voltage divider giving 6-figure accuracy and incorporated in some high-accuracy measuring instruments. Multi-turn potentiometers are not good enough having a linearity of only 0.1% at best. In principle we could use a simple potential divider with a fixed upper resistor and a precision decade resistance box for the lower and the output voltage taken across the lower resistor but that has the disadvantage that the output voltage is not a linear function of the resistance so the resistance box could not be a standard one with decades of equal resistors.

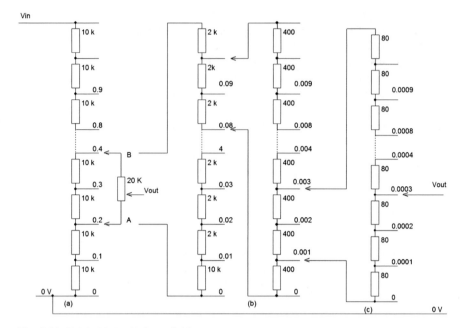

Fig. 2.46 Kelvin-Varley Voltage divider

The Kelvin-Varley Voltage Divider

This is the classic solution and consists of strings of 11 identical resistors for each decade except the last one, Fig. 2.46. Originally the resistors were precision wire-wound types and so limited to DC and low-frequency use. Today 0.01% resistors are available using resistive foil and having virtually no inductance nor capacitance.

The principle is shown in Fig. 2.46(a) where a two-pole switch connects across two adjacent resistors in the chain of 11 resistors of 10 kΩ each. However, the total resistance is not 110 kΩ because the 20 kΩ potentiometer bridging the two resistors means their sum is only 10 kΩ and the total is 100 kΩ. The numbers against the wires are the minimum fractions of the input voltage V_{in}. At the lower end of the potentiometer, point 'A', there are two 10 kΩ resistors between it and ground the voltage is *0.2 Vin*, whilst at the top there is effectively 30 kΩ so point 'B' is *0.3 Vin*. And by turning the switch, we can run from zero to *Vin*.

In practice, we need greater accuracy, and the potentiometer is replaced by an identical chain of 11 resistors of 2 kΩ each, Fig. 2.46(b) with a switch bridging two. In similar fashion the resistance between the poles of the switch formed by the next 11 resistors is 4 kΩ and the total resistance is 20 kΩ. These stages can be cascaded until finally we just have a simple stick of resistors at '(c)'.

The Hamon Voltage Divider

This is a technique for building up from a low resistance to a higher one, typically by a factor of ten, and was first described by B.V. Hamon in 1954, and is still in use today in precision measuring instruments and as stand-alone items. The original article used

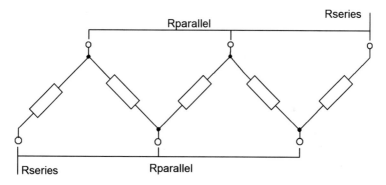

Fig. 2.47 Hamon Voltage divider

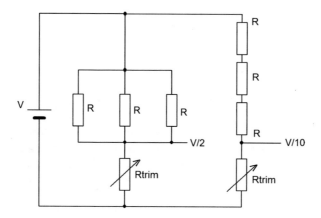

Fig. 2.48 Hamon concept

11 resistors and is not freely available online. The Fig. 2.47 shows five resistors which can be connected in series or parallel although there is no absolute requirement on how many to use. And these should be four-terminal resistors. An application of the concept is provided by Conrad Hoffman at http://conradhoffman.com/HamonResistor.html using three resistors, which we shall now follow.

At first it appears more like a magician's trick rather than electronics, but here is how it works. Suppose we have three identical resistors of value R, and we do not need to know their value, only that they are identical, and we place them in parallel as the upper arm of a potential divider, Fig. 2.48, with a lower resistor that can be trimmed so that the output voltage is exactly $V/2$. The trimmer resistor will then be exactly $R/3$. Now suppose we place the same three resistors in series without altering R_{trim} This is the right-hand circuit. The upper arm of the potential divider is now $3R$ and the output is: $V \frac{R_{trim}}{R_{trim}+3R} = V \frac{\frac{R}{3}}{\frac{R}{3}+3R} = \frac{V}{10}$ and we have a precise potential divider. And as we only need to know that the resistors are identical, we can use a Wheatstone bridge to measure them, likewise for adjusting R_{trim}. A possible circuit

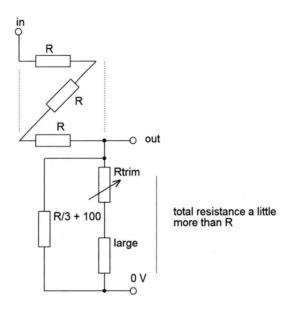

Fig. 2.49 Hoffman Hamon Voltage Divider

is Fig. 2.49 where the upper three resistors can be connected in parallel as shown by the dotted shorting wires and then connected in series by removing the two shorts.

However, the question arises as to what happens if the resistors are not exact. Suppose we have three resistors of *320 Ω, 330 Ω* and *340 Ω*. If we place them in parallel, the sum is:

$$\frac{1}{R_p} = \frac{1}{320} + \frac{1}{330} + \frac{1}{340} \quad \text{then} \quad R_p = 109.938\Omega.$$

And we note that the parallel sum differs slightly from the average of 110 Ω. We again adjust the trimmer resistor to have exactly half the voltage, but this time it is not $R/3$ but R_P.

If we now place the resistors in series, we have $R_S = 990$ Ω and the ratio $R_P / R_S = 109.9231/990 = 0.1110$ instead of the required *0.1111*, which is an error of *0.1%*. This is due to the difference between the true average resistance of *330 Ω* and the apparent average resistance found from the resistors in parallel which is *3 x 109.9231 = 329.769 Ω* and is an error of *0.07%* But remember that *Rtrim* has been adjusted against the three resistors in parallel, not the average, so if we place the three resistors in series, and given a *1 V* input, the output is:

$$V = \frac{R_P}{R_p + R_T} = \frac{109.9331}{109.9331 + 990} = 99.945mV$$

The simulation below gives a more accurate result of *99.944882 mV*, and the error is *0.05%*. This further reduction in the error is because the lower arm of the potential divider is not the average value of the resistors, but the apparent average from the three in parallel.

We can illustrate this problem very dramatically by supposing that we could suspend the laws of physics and take resistors, for example, of *20 Ω, 30 Ω* and *40 Ω*, and place them in parallel where their parallel sum is the true average instead of 'resistors in parallel' and is exactly *10 Ω*. If we now place them in series, we have *90 Ω*, and the ratio of series to parallel is exactly *1/9*, and we could use any old resistors and have an exact and precise division of 1/10.

If some kind soul can find a way to do this, they would earn the undying gratitude of everyone involved in metrology, a Nobel Prize, fellowship of the Royal Society and lots of money.

Example – The Hamon Voltage Divider
But to return to the real world, using the values of *320 Ω, 330 Ω* and *340 Ω*, we can explore the divider with schematic 'Hamon R Divider.asc' where we find the above results. As it is, this circuit reduces the 10% error of the resistors to 0.05% for the divider. In practice, of course, we start off with closely matched resistors to achieve an error of one part in a million.

Explorations 12
1. Build a Kelvin-Varley divider and note the effect of resistor errors, for example, if the second set of resistors were not exactly 2000 Ω but 1990 Ω.
2. Create a Hamon divider with a ratio of 0:1 and an accuracy of 1 in 10^6.
3. Create a Hamon divider with a ratio of 1:100 and hence suggest how non decade divisions could be made, for example, 1:60.

2.8.2 Resistance

The everyday standard is still a physical piece of metal made as a four-terminal resistor and usually immersed in a sealed oil bath to prevent corrosion and contamination. The fundamental standard now is the Quantum Hall Resistance (QHR). This is based on the effect discovered in 1879 by Edwin Hall that if a strong magnetic field is applied to current-carrying conductor, the charge carriers will be deflected sideways creating a small voltage. This effect has been widely used to measure magnetic fields.

The quantized effect was discovered in 1980 by Klaus von Klitzing using a very powerful magnet of 14 T and a very low temperature, less than 1 K. The resistance of the probe – a heterostructure of GaAs/AlGaA – showed very precise levels in steps depending on the magnetic field strength and reproducible to one part in a billion.

It has been suggested that these probes, difficult and expensive to make, could be replaced by graphene, https://www.nist.gov/news-events/news/2017/05/new-stan dard-resistance-standards.

2.8.2.1   Interrelationship of Standards

The units of voltage, current and resistance are now defined in terms of fundamental units; current is defined by force using the current balance and resistance and voltage can both be related to the elementary charge and Plank's constant. However, these units are related through Ohm's Law, so we may not define each independently.

2.9   Practical Considerations

Resistors are available in a wide range of values and power ratings. A useful guide is http://www.learnabout-electronics.org/Resistors/resistors_08.php.

2.9.1   Fixed Resistors

The figures are typical because they depend upon the resistor value and temperature. A useful point to consider if good stability and reliability are essential is to derate the resistor so that its temperature is far less than the operating maximum. Thus, if the actual power is 1/10 W, instead of a 1/8 W resistor, use 1/2 W.

2.9.1.1   Mounting Methods

There are .five

Surface Mount Devices (SMD)
For mass production and low power, we use very small chip resistors which can only be handled by pick-and-place machines. Some have very low power ratings, but in many cases a resistor only dissipates a few mW; others using thick film technology can dissipate several watts.

Through-Hole Axial Lead
These are the traditional cylindrical resistors with axial leads available in a wide variety of power ratings from 1/8 W up to 5 W. They can be supplied on reels with the wires preformed to suit the conventional 0.1 in. grid of PCBs.

TO-220
This is a popular outline for power transistors and is also used for power resistors where they can be bolted to a heat sink, just as a power transistor. This is becoming increasingly popular.

DIL

These are SMD but with leads similar to a six-lead DIL package and are used for power resistors. The legs hold the body clear of the PCB and help heat flow.

Wire-Wound Ceramic Encased

Power wire-wound resistors often have a square section ceramic covering. They are inductive, but a non-inductive option is sometimes offered.

2.9.1.2 Resistor Types

The most popular are these.

Carbon Film

These can be used at high temperatures, up to 100 °C, and have widely ranging temperature coefficients from −500 ppm/K to 350 ppm/K and a low current noise figure of 1 µV/V. The resistance change is typical ±1.5% after 1000 hrs. They have resistance values from about 1 Ω to 10 MΩ and power ratings from 1/8 W to 5 W.

Metal Film

These, too, have good long-term stability and can be used at high temperatures. They have temperature coefficients as low as 25 ppm/K. The resistance and power range is the same a carbon film.

Thick Film Power Resistors

These can dissipate up to 25 W and have temperature coefficients typically of +/- 100 ppm/K.

Special Types

A glance through the RS catalogue will show that there are exotic types including adjustable wire-wound resistors of some considerable size (and cost).

2.9.1.3 Resistor Series

Resistors for general use are not made in every possible value because of their tolerance. Although 5% and even 2% are now the norm, for ease of calculation, let us take 10% tolerance and start with a resistor of 100 Ω. This can be as high as 110 Ω, so there is no point in trying to manufacture resistors of 105 Ω because they are already randomly included in the 100 Ω one. This leads to the consideration of what is the next significant value. When it is 10% low, it should be 110 Ω. So $0.9 \times R = 110$, we arrive at $R = 122$ Ω, but as the last figure is not significant, it is 120 Ω. By the same reasoning, the next value is 150 Ω because 10% low is 135 Ω and a 120 Ω resistor 10% high is 132 Ω – close enough agreement.

Thus we build up a series of values. And one very desirable feature is that the values should repeat in decades, that is, we have started with 100,120,150... and we want the next higher to be 1000, 1200, 1500... not 1100, 1300... so there is some slight adjustment made to the preferred values. Finally we have the:

E12 series − ... **8.2, 10, 12, 15, 18, 22, 27, 33, 39, 47, 56, 68, 82, 100** ....

so-called because there are 12 values in the decade. For 5% tolerance, we have the **E24 series** which fills in intermediate values, for 2% the **E48 series** and for 1% the **E96 series** and even an **E192 series**. The values can be found at https://www.vishay.com/docs/28372/e-series.pdf.

2.9.1.4   The Resistor Colour Code

This is used on wire-ended resistors and consists of four or more bands of coloured paint; in theory near one end of the resistor, but with small low power components this is difficult. However, the sequence of the bands usually precludes reading it the wrong way round because we shall find an improbable or impossible value if we read it from the wrong end.

For the E12 and E24 series, the resistor only has two significant digits, and these are the first two bands. For higher series there are three bands.

The third band is the multiplier for the digits and the last band is the tolerance.

The colours used and their meanings are:

Black	Brown	Red	Orange	Yellow	Green	Blue	Violet	Grey	White	Gold*	Silver*
0	1	2	3	4	5	6	7	8	9	0.1	0.01

So it can be remembered as *black, brown, the rainbow, grey* and *white*. The last two marked with an asterisk are not used as digits in the resistor value.

Thus *brown, black = 10, grey, red = 82, yellow, violet = 47, red, green = 25*; however this last is not an E12 or E24 value, so perhaps we have mistaken the colour.

The previous omits the multiplier. Here the colours have the same meaning, but they are powers of 10 or, simply, how many zeros to add (apart from the last two colours). Thus:

$$brown, black, black = 10 \times 10^0 = 10, \quad grey, red, yellow = 82 \times 10^4 = 820k,$$
$$and \ \ yellow, violet, red = 47 \times 10^2 = 4700$$

This leaves the question of how to specify an 8.2 Ω resistor or even a 0.47 Ω one. Here we invoke the last two colours to divide the value so *grey, red, gold = 8.2* and *red, green, silver = 0.25*

And finally we add the tolerance using some the following colours.

Grey	Violet	Blue	Brown	Red	Gold	Silver
±0.01%	±0.25%	±0.5%	±1%	±2%	±5%	±10%

The last four are most commonly used. An extensive colour description can be found at https://www.digikey.com/en/resources/conversion-calculators/conversion-calculator-resistor-color-code-4-band and also for five- and six-band.

Depending on the colour of the resistor body, the distinction between gold and yellow and even brown can become blurred. Even so, trying to read the resistor the wrong way round will usually result in a non-standard or even impossible value.

2.9.1.5 The Resistor Letter and Number Code

This consists of a series of letters and numbers printed on the ceramic body of power resistors and SMD types. The resistor value is given in digits with the multiplier R, K and M inserted at the decimal point position. Thus $5R6 = 5.6\ \Omega$, $47K = 47\ k\Omega$ and $0R22 = 0.22\ \Omega$.

There is a trailing letter denoting the tolerance:

F	G	J	K	M
±1%	±2%	±5%	±10%	±20%

So in full we might have $5R6G = 5.6\ \Omega\ 2\%$, $47KK = 47\ k\Omega\ 10\%$ and $0R22M = 0.22\ \Omega\ 20\%$.

2.9.2 Variable Resistors

These are either *trimmer* resistors where the value is set up during test or *variable* resistors where the user can adjust the setting, usually by a front panel control such as the volume control on a radio (although solid-state alternatives are replacing them).

Trimmer controls are available as open skeleton types with a carbon track on a phenolic base. The wiper is adjust by screwdriver, but as it is not insulated, care is needed. They are cheap and acceptable for domestic equipment. Professional trimmer resistors are often encased in TO-5 format and use cermet.

Front panel controls can be single or two ganged together on a single shaft. They can have a linear variation of resistance with rotation angle or logarithmic, although this is not strictly logarithmic but consists of two linear sections of different resistivity. There are also wire-wound type able to dissipate greater power and multi-turn wire-wound potentiometers. And also specialist sine/cosine potentiometers.

2.10 Summary

In this chapter we have covered the basics of circuit analysis.

- KVL states that the volt drops around a closed path sum to zero, so define a direction around the loop, and then add volt drops remembering polarity.
- KCL states that the sum of currents at a node is zero using mesh current directions.
- To use superposition, replace every voltage source but one in turn with a short-circuit and use KVL and KCL to analyse the modified circuits, and then add the individual results. Replace current sources by open-circuits.
- To find the voltage and current in just one component, remove it from the circuit and use the previous methods to reduce the remaining circuit to one resistor in series with one voltage source (Thevenin model).
- Or reduce the circuit to the same resistor in parallel with a current source (Norton model).
- For currents up to a few milliamps, a potential divider can be used to provide a lower voltage than the supply. It is an example of an 'L-attenuator'.
- The Wheatstone bridge measures resistance to high accuracy when it is balanced. But for faults in cable, we need the Murray or Varley loop test bridges and the Kelvin Double Bridge for measuring very small resistances.
- A 'T-attenuator' can give moderate attenuation and can be symmetrical and can match imput and output impedances, and a 'Bridged-T attenuator' allows the attenuation to be varied with a dual-gang potentiometer without altering its impedance.
- A 'Π attenuator' allows higher attenuation.
- Three resistors connected as a 'T' can be converted to three resistors connected as a 'Π' and vice versa.
- Thermocouples are a cheap, convenient and accurate sensor for measuring temperature where the resistance/temperature relationship is modelled by a polynomial up to the eighth power.
- Resistors have temperature coefficient and generate noise. They also have power ratings and tolerances.

Chapter 3
Non-linear Resistors

3.1 Introduction

In the previous chapter, we simply gave a resistor a value. But LTspice has many more parameters that we shall use here. Much is rather specialist and concerns temperature-sensitive resistors like platinum resistance thermometers and thermistors and varistors which are voltage-dependent resistors. And although these many not be of direct interest, they are an introduction to the flexibility of the LTspice resistor.

There are two pages of the LTspice 'Help' file that we shall make use of. The first is *OPTIONS* which is mainly concerned with settings for the SPICE engine. Here, two are of interest, *tnom* and *temp*, and will be explained at the appropriate point. The other page is – or rather 'are' because there are three pages which all contain the same table and are **Waveform Arithmetic**, **.PARAM** and **Arbitrary Behavioural Voltage or Current Sources**, we find myriad functions that can be incorporated in the definition of a value. We shall later use just a few, but the gamut will be explored in Chap. 5.

3.2 The LTspice Resistor

We have used these before, but now we want to extend the description to non-linear resistors whose resistance is a function of temperature, current or, indeed, any parameter we can think of. It follows the general SPICE implementation with one special and important addition – an LTspice resistor can be given almost any arbitrary relationship.

At its basic level, as we saw, just right click on the component when the cursor is a pointing hand to open the **Resistor- <Resistor Name>** dialogue, and we can fill in the following:

© Springer Nature Switzerland AG 2020
C. May, *Passive Circuit Analysis with LTspice*®,
https://doi.org/10.1007/978-3-030-38304-6_3

Resistance (Ω)
This must exist, the same as *Value* in the **Component Attribute Editor** dialogue.
This can include temperature coefficients.

Tolerance (%)
This is accepted but, as we have already seen, not acted on.

Power Rating (W)
Accepted but not acted on. Dissipating100 W in 1/2 W resistor is no problem. Pity!

So tolerance and power rating really are there as reminders when specifying resistors for production.

3.2.1 The Component Attribute Editor

This is opened by holding down **Ctrl** and right clicking on the component, Fig. 3.1. It is not necessary to use this just now because we can enter the value of a component in the simple dialogue we have used so far. But it can make life easier when we come to complex value statements
The data that can be edited is contained in the edit boxes. The **Vis** property is toggled by right clicking and defines whether this attribute is shown on the schematic or not.

Open Symbol Button
The adjacent edit box has the full path and file name terminating in .asy of the location of the file with the instructions for drawing the symbol to represent the component on the workspace. By all means open it and have a look, but it is better not to try to edit it now. And it is not possible to change it here for, say, the European

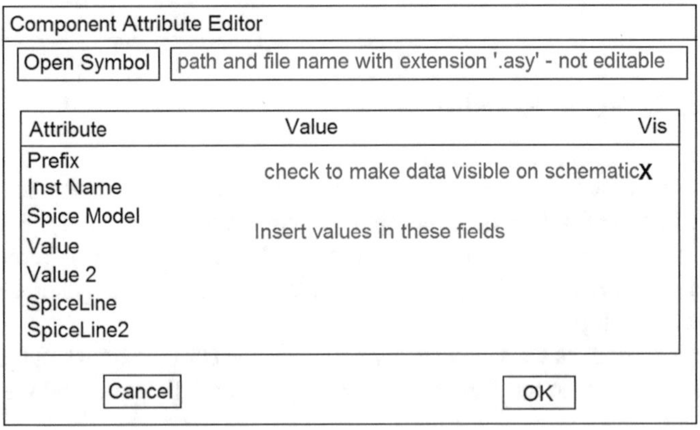

Fig. 3.1 Component Attribute Editor

rectangle resistor. In fact we can use any '.asy' file - new ones we have created or existing ones, to represent a resistor.

Prefix
This defines how it is simulated. Do not change from R; else you could end up simulating a resistor as a capacitor.

InstName
This is the name of this resistor. Duplicates are not allowed; otherwise there are few restrictions on what can be used. By default this is visible (an 'X' in the **Vis** column).

Spice Model
This is only for a predefined class of components such as diodes, all types of transistors and controlled voltage and current sources. So leave blank.

Value
This must always exist and rather confusingly has a default of 'R'. It is also visible by default. It can contain temperature coefficients as well; see Sect. 3.4. It is also possible to put the name of a SPICE model here if there is no other data in this line.

Value 2, SpiceLine, SpiceLine 2
We need not use them here. We can, if we wish, add temperature coefficients or other parameters described in later sections, but they will all end up as extensions to the resistor definition line, not as separate entities.

3.3 Variable Resistors

We used this in the previous chapter to view the effects of mismatch in the Maximum Power Transfer section by stepping the value of the resistor. To recapitulate, the syntax is:

> *.step param < type of step >< name >< start value >< end value >*
> *< step size >*

By default, SPICE uses linear steps, and the name must be without curly brackets. So, for example, **.step param R 1 100 1** would step a resistor R in 1 ohm steps from 1 to 100 Ω (remember, a resistor must not be zero). This is ('Variable Resistor.asc'). And in the waveform window, right click and then **View→Mark Data Points** will show the individual values.

Decade Sweep
To change to the decade sweep we have two possibilities

- Right click on the linear sweep text on the schematic to open the **.step Statement Editor** dialogue on the Screen.

- Hold down **Ctrl** and right click on the decade sweep text on the schematic to open the dialogue **Edit Text on the Schematic** and click 'SPICE directive' after making the changes. In this case we must remember the sequence of the terms as there are no edit boxes to guide us.

Generally, clicking on a comment or command will bring up a dialogue to edit that command. Then if we click **Cancel** to close that dialogue the **Edit Text on the Schematic** dialogue will open.

3.3.1 Potentiometers

These are three-terminal resistors having a fixed resistance between the two ends and a third moving contact which can travel from one end to the other and so split the total resistance into two parts whilst retaining the same total resistance. Lots of interesting information can be found at http://www.geofex.com/article_folders/potsecrets/potscret.htm.

Example – A Linear Potentiometer

We want the resistance between the slider and either end to be directly proportional to the position of the slider along the track. For this we need two resistors in series whose values change together and their junction is the slider.

One way of handling this is to use parameters. First we have the total resistance Rt which we set at 100 kΩ. Next we need to divide this between the resistors R1 and R2. If we try something like $R1 = Rt - R2$ and $R2 = Rt - R1$, we have circular definitions to which LTspice – quite rightly – objects. Instead we define another parameter SliderPos as the fraction of the total resistance ascribed to each resistor. We can leave this with no value, or insert a dummy one as in the schematic. But as we cannot directly change resistor values, we create two more parameters {Rx} and {Ry} in schematic ('Potentiometer.asc').

The last point to note is that as resistors cannot go to zero, we add a small padding resistor, so small that it will not affect the result, and now we can step from one end of the potentiometer to the other. The relationship between slider position and resistance is defined by the key statement of the value of Rx.

$$.param\ Rx = Rt * SliderPos + 1n$$

As this is linear, we see the output voltage at node 3 is linear.

Explorations 1
1. Run ('Potentiometer.asc') with a high value for R3, and note that the voltage at node 3 is linear. Note also that the current is constant.

2. Reduce R3 to some trivial value, perhaps 1 nΩ to convert the potentiometer into a variable resistor, and now the voltage at node 3 (of course) is effectively zero, and we find the initial current is about 6 GA. Move the cursor on the trace panel to below the x-axis so that it turns into a small yellow ruler, right click, and in the Horizontal Axis dialogue, enter a value of 50 m for the **Left**, then right click on the trace panel, and click **Autorange Y-axis**. To see a reducing current, plot the reciprocal of the current, and it is a straight line as Ohm's Law predicts.
3. Try different values of total resistance.

Example – A Sine Potentiometer

This is a specialist component often used in servo systems. We can use the previous arrangement but change the definition of the upper resistor Rx to

$$.param\ Rx = Rt * (sin\ (SliderPos * 2 * pi) + 1)/2 + 1n$$

The (sin(SliderPos*2*pi) ensures that there is one complete cycle from the beginning to the end of the potentiometer. We must divide by 2 because we have two sinusoids that add. And without the +1, the voltage across R1 goes negative after *SliderPos* = 0.5, so the voltage at node *3* will run to 15 V.

Explorations 2
1. Run 'Sine Pot.asc' and plot V(3), V(1)–V(3) and the traces are in anti-phase. Notice there is a small glitch in the current when one of the traces cross zero.
2. Note the effect of changing +1 for other values and removing the division by 2.
3. We cannot make a Fourier analysis – that only applies to a .trans analysis. But we can right click on the trace panel and select **View→FFT** then select **V(3)** to make a Fast Fourier transform analysis. And although we may not fully understand it, at least we can see that there is a large component of 11 dB at 1 Hz and the next components are at 1 kHz and higher and are much, much smaller.
4. Reduce R3 to 1 nΩ, and the reciprocal of the current in *R3* is sinusoidal, but the current runs up to a huge value when *R1* = 0. So we need to set the **Right** of the Horizontal Axis to *700 m*.

Example – A Logarithmic Potentimeter

Audiophiles, of course, will want a logarithmic and not a linear potentiometer to serve as a volume control. This is difficult to do. In fact, we really need an exponential relationship. But to make a truly logarithmic track is difficult, and manufacturers often compromise with two linear sections with different resistances. And we also can approximate to a logarithmic pot by using a linear one with a parallel resistor ('Taper Pot.asc'). This is a fairly good approximation up to 95% rotation (i.e. *SliderPos = 0.95*).

The difficulty with a true logarithmic or exponential potentiometer is these never go to zero. We can try various options, but the easiest is just **.step param SliderPos 1 10 001**. This avoids the problem of negative logarithms or an error since the result of **.param Rx = Rt * log10(SliderPos) + 1n** runs from *log10(1) = 0* to *log10(10) = 1* spanning the full range.

Explorations 3
1. Run 'Log Pot.asc' with a logarithmic x-axis and the current is logarithmic.
2. Run 'Taper Pot.asc' with different values for R3. Also try a logarithmic x-axis.

3.4   Resistor Temperature Effects

For this and indeed for resistors whose value changes with temperature, or voltage, or whatever, it is important to keep in mind KISS – Keep It Simple, Stupid – and not to change everything at once, but test each factor singly.

SPICE allows both linear and quadratic temperature coefficients to be appended to the resistor value and also an exponential – but only these. Despite the LTspice Help, cubic and higher terms are not allowed. And this is in line with SPICE in general. In LTspice these are denoted by *tc1*, *tc2* and *tce* and, like other versions of SPICE, use a default reference temperature *tref* of 27 °C (300 K). The relationship is that the resistance at any temperature temp is:

$$R_t = R_0 \left(1 + tc1(temp - ref) + tc2(temp - ref)^2 + tce\left(\frac{temp - ref}{100}\right)\right) \quad (3.1)$$

where the last 'exponential' term *tce* is the change of resistance in %/K rather than a true exponential.

3.4.1   Adding Temperature Coefficients

It is important to remember that SPICE processes the netlist – the schematic capture is just there to make it easier to create it by visually placing components and connections. LTspice has some flexibility in how we add temperature coefficients offering us four ways of doing it:

Adding Temperature Coefficients Using the 'Resistor <resistor name>' Dialogue

Move the cursor over the component, and when it turns into a pointing hand or a clamp ammeter after a run, right click, and this dialogue will open. The numbers can be directly inserted into the box **Resistance(Ω)** as shown in the simulation ('R tempco test.asc'). The coefficients must be exactly *tc1, tc2, tce* after the resistance value with no commas so, for example, a resistor of 1150 Ω could be:

$$\textbf{1150 } \textbf{tc1} = \textbf{0.001 tc2} = -\textbf{0.0004}$$

but the equal sign ($=$) is optional.

Adding Temperature Coefficients Using the Component Attribute Editor

Hold down **Ctrl**, and right click on the resistor to open the editor, and then in the **Value** edit box, enter the temperature coefficients as before.

Adding Temperature Coefficients Using the 'Enter New Value for <component name' Dialogue

Right click on the text for the value of the resistor **100 tc1**... not on its name **R1,** to open the **Enter new Value for R1** dialogue.

Adding Temperature Coefficients by Editing the NetList

There is a netlist that can be viewed by **View->SPICE Netlist** but cannot be edited. However, if we right click on it, we can choose **Edit as an Independent Netlist** from the pop-up menu. If we choose that, we can save the new netlist in the current schematics folder or navigate to another. We shall then see it in colour.

We can arrive at the same netlist after a run by **File→Open** and change the file search to *Netlists (*cir,*net,*sp)* where the netlist again will be loaded in colour.

It is now possible to change everything, for example, we can open 'R tempco Test.net' and change the resistor and temperature coefficients. The file is shown here in black and white with *tc2 = −0.004* added. We should note that it is possible to save the netlist more than once, but the name does not change.

```
* C:\Windows\system32\config\systemprofile\Documents\LTspiceXVII\R
tempco test.asc
R1 N001 0 1150 tc1=0.001 tc2=-0.004
V1 N001 0 10
.op
.step temp 0 800 10
.backanno
.end
```

However, these changes are not reflected in the schematic and will be overwritten if the schematic is run again, so save it with a different name. This is not, perhaps, the best way of changing things, and it is safer to change the original circuit.

But then, old SPICE hands who are used to dealing with netlists will have no trouble because right clicking on the netlist window will open a pop-up menu

offering the option to run the schematic which will take us back to the simulation
window except that this time we have a netlist and not a schematic.

3.4.2   Temperature Analysis

There are two methods. In either case Press *S* with the mouse over the schematic and
the dialogue **Edit Text on the Schematic** will open.

Temperature Analysis Using a List
Enter the command

$$.\textbf{temp} < \textbf{temp1} >< \textbf{temp2} > \ldots ..$$

with a list of temperatures. This gets tedious if there are more than just a few special
points of interest. It is also useful if the list cannot be converted into a *.step* command
because the increments are neither linear, octave, nor decade.

Temperature Analysis Using the '.step' Command

We have seen that this is a very flexible command having the general form of:

$$.\textbf{step} < \textbf{Nature of Sweep} > \textbf{param} < \textbf{Name of Parameter to Sweep} >$$
$$< \textbf{start value} >< \textbf{end value} >< \textbf{step size} >$$

If we cannot remember the way to enter the values, we can just type **.step** and then
right click on it to open the **.step Statement Editor** dialogue. The fields are:

Name of parameter to sweep

This is essential, but notice that rather confusingly it is not the first entry in the
command. However, the dialogue takes care of that.

Type of Step

There are four options:

- **Linear** which is the default and applies if nothing else is entered. This works with
  the **Increment** which is the linear increment to apply.
- **Octave, Decade** which advances the step in octaves or decades and now the
  Increment is the number of measurements per octave or decade.
- **List** which we have just covered.

Start value, Stop value

These are the limits of the steps.

Increment

As we have seen, this is the linear increment for *Linear* steps; else it is the number or points per octave or decade which is clearly shown when the *Nature of Sweep* is selected. If we enter the command directly as:

.step temp 0 100 1

there may be a warning dialogue **.step Syntax error** with the message *'Expected "param" Accept anyway?'*. If so, click Yes .

Example – Resistor Tempco Test
As a simple example, open ('R tempco test.asc'). This will test the circuit at temperatures of 0, 25, 50 and 100 °C with a linear temperature coefficient of 0.01, and the result will be line segments.

Explorations 4
1. Add 'tc1 = 0.01'and run the simulation to ('R tempco test.asc.'). Select 'temperature' on the waveform window, and check that it runs from 0 to 100 °C. The current will plot as a few line segments.
2. Right click on the *step temp 0 25 75 100* line, and insert a semi-colon at the start to disable it. Click the *.step 0 100 0 '*, but this time remove the semi-colon and run again. Now, of course, the graphs are smoother. In the **Add Traces to Plot** dialogue, add 'V(n001)/I(R1)' in the **Expression(s) to add** dialogue to plot the resistance. We can do this easily by clicking first 'V(n001)', and it will appear in the Expression(s) to add dialogue, and then type a forward slash for division and click on 'I(R1)'. Note that the resistance is 100 Ω at 27 °C – the global reference temperature for SPICE.
3. Confirm that the resistance increases at the rate 1 Ω/°C, and hence explain why there is an error if tc1 = 0.1 (the resistor must never be zero).
4. Add quadratic and exponential temperature coefficients, change the resistance value, and test. (See for example ('Rtempco Test 2.asc')

Changing the Resistor's Reference Temperature

It is not possible to add a reference temperature as *tnom = 52* in the resistance edit box, but we shall see that there are ways of getting round this.

Modelling Self-Heating

If we know the rate of temperature rise of the resistor with power dissipation *trp* and
its linear temperature coefficient *tc*, we can describe the resistor by:

$$R = R * (1 + Ir * Vr * trp * tc)$$

where *Ir* is the current through the resistor and *Vr* is the voltage across it. It could be
extended to quadratic and exponential temperature coefficients. It does not, however,
change the temperature of the resistor.

Example – Resistor Self-Heating
The schematic ('Resistor Self-heating.asc') has a *1 kΩ* resistor with *trp = 100 K/W*
tc = 0.01. At the end of the run the dissipation is *1.9 W* and the resistance is *2.9 kΩ*.

 We shall now move on to simulate some important temperature-dependent
resistors.

3.5 The Platinum Resistance Thermometer

This is used to define a large portion of the International Temp Scale. Clearly it is
important that basic units are the same in Boston, Bristol, Bordeaux, Brisbane,
Bangladesh and everywhere else. For mass, length and time, there are no insuperable
problems in comparing national standards to the international, although there are
proposals to change the unit of mass from the prototype kilogram to the force exerted
by an electric current. Temperature is different – we cannot carry temperature from
one place to another as we could a standard of length or mass. Certainly we can
define calibration points such as the triple point of water or the freezing point of
molten tin – the Wikipedia article lists them all. So instead we define a thermometer
whose construction can be replicated everywhere – the platinum resistance
thermometer – whose resistance defines the International Temperature Scale
(ITS-90) from 13.8023 to 1234.93 K. For those interested in such things, there is
an excellent article https://www.technology.matthey.com/article/3/3/78-87on its history. And note two things: first it is 'kelvin' not 'degrees kelvin', and the second (and
more important) is the high precision that can be obtained. Of course, for everyday
scientific and engineering use, measurements to an accuracy of 0.1 K may be
sufficient.
 One serious difficulty is that the resistance of platinum is not linear with temperature. From 0 to 850 °C the defining relationship is a quadratic:

$$R = R_0 (1 + aT + bT^2) \tag{3.2}$$

where T is the temperature in degrees Celsius, $a = 3.9083e\text{-}3$ and $b = 5.775e\text{-}7$ and R_0 is the resistance at 0 °C, usually 100 Ω.

Example – Platinum Resistance Thermometer

Using ('R tempco test.asc'), we change the Resistance(Ω) to *100 tc1 = 3.908e-3 tc2 = −5.770e-7* and rename it as ('Platinum Resistance Thermometer V1.asc').

Step the temperature from 0 to 100 °C in 1 °C intervals and on the waveform panel, right click and select Add Traces from the pop-up menu. This will open the **Add Traces to Plot** dialogue. In the **Expression(s) to add** edit box, add 'V(n001)/I (R1)' as we did in Explorations 1 task 2. This will again plot the resistance. You will find that the resistance is 100 Ω at 27 °C, not 0 °C. This is where it gets a little complicated.

As the circuit stands with no reference temperature specified for the thermometer, we have seen that SPICE adds the global reference temperature *tnom* to the temperature using the relationship 3.1 above. We can test this by changing the global reference temperature by adding *OPTIONS tnom = 50*, and now it will be found that the resistance is 100 Ω at 50 °C. So one possibility for getting the correct result is to change the global nominal temperature to 0 °C. This is ('Platinum Resistance Thermometer V2.asc'). But note that this change will affect every component in the schematic. In this case, it does not matter, but generally, this is not the best way of handling it.

3.5.1 Arbitrary Temperature Coefficient

It was said at the start of this chapter that a resistor can be given almost any arbitrary relationship. We use this now to overcome the problem of the reference temperature by not using the built-in temperature coefficients but explicitly defining the resistance as:

$$R = 100 * (1 + 3775.9083e\text{-}3 * temp - 2.5e\text{-}7 * temp ** 2)$$

where *temp* is the simulation temperature, not the nominal temperature. A single asterisk '∗' stands for multiplication and two asterisks for a power, so '∗∗2' is squared, '∗∗3' is cubic and so on, overcoming the problems of only linear and quadratic coefficients. And, although we have not used it, *exp(expression)* is a true exponential.

We can enter the above expression in the **Resistance – R1** dialogue, but it is rather difficult to keep track of where we are with this small edit box. Better is to hold down **Ctrl** and right click on the resistor to open the **Component Attribute Editor** dialogue.

LTspice will now use this equation and the temperature as defined by a *.step* or *.temp* command, and the resistance will be correct. Also, the nominal resistance has no effect. This is ('Platinum Resistance Thermometer V3.asc')

3.5.2 The Cubic Equation

From $-200\,^\circ$ to $0\,^\circ$C, the International Temperature Scale replaces Eq. 3.2 by a cubic:

$$R = R_0(1 + aT + bT^2 + c(T - 100)T^3) \tag{3.3}$$

where $c = -4.1830e\text{-}12$. This is the 'Callendar-Van Dusen Equation'. The coefficient will differ depending on the resistance and composition. Certainly, this is a small correction, but for precise scientific work, it must be included. And here we have a problem because LTspice only allows linear and quadratic temperature coefficients to be added in the value field, not cubic. But (and this seems to be unique to LTspice) we can extend the explicit definition of the resistor to include not only quadratics, as above, but cubic, exponential and, indeed, any other function. So first let us test the cubic term on its own. This is 'Platinum Resistance Thermometer cubic V1.asc' where it will be seen that the resistance at $-150\,^\circ$C is 6.5 mΩ falling to $0\,\Omega$ at 100 $^\circ$C and then becoming negative, -418 nΩ at 110 $^\circ$C. What is more worrying is the sharp spike in the current at 100 $^\circ$C. This is because the resistance then was zero. And there also remains a problem – this cubic term applies only below $0\,^\circ$C. A neat solution, posted by Harald Kapp in https://www.electronicspoint.com/resources/managing-temperature-in-ltspice.18/, is to use the unit step function $u(x)$ which is 1 if x > 0; else it is 0. So here we need to append *(1-u(temp)* and a small resistor *0.1e-12* to prevent the resistance falling to zero.

$$\boldsymbol{R = 100 * -4.1830e\text{-}12 * (temp\text{-}100) * temp * *3 * (1 - u(temp)) + 0.1e\text{-}12}$$

This resolves the problem as 'Platinum Resistance Thermometer cubic V2.asc'. Ignore the current of 100 TA – that is because the cubic term is no longer effective, so the current is limited by the 0.1e-12 Ω. So, finally, we can put it all together as:

$$\boldsymbol{R = 100 * (1 + 3.908e\text{-}3 * temp - 5.770e\text{-}7 * temp * *2 - 4.1830e\text{-}12 *}$$

$$\boldsymbol{(temp\text{-}100) * temp * *3 * (1 - u(temp)))}$$

and we no longer need the small fixed resistor since the resistance will never fall to zero. This is ('Platinum Resistance Thermometer V4.asc'). The resistance appears to change linearly, but then the coefficients are very small, so it is difficult to judge by eye. However, we can use the average temperature coefficient α from the Johnson Matthey website where:

$$\alpha = \frac{R_{100} - R_0}{100R_0} = 0.003923 \tag{3.4}$$

and is the slope of the resistance versus temperature graph from *0* to *100* $^\circ$C. We create a dummy resistor $R_2 = \boldsymbol{100*(1 + 0.003923*temperature) + 38}$, and then the curvature of the resistance *V(n001)/I(R1)* can be seen by plotting the difference two resistor values.

Reading the Temperature

Back in 1959 when the Johnson Matthey article was written, it was not a trivial matter to convert resistance to temperature, and look-up tables were often used and still can be found today: Fluke publish one on the web. Of course, now we can simply compute the result directly from the equations.

Explorations 5
1. Check by measurement from the waveforms at a few well-separated points that the resistance is correct. Even better is to use a few *.meas* directives.
2. Extend the measurements to 850 °C.
3. Convert the circuit into a Wheatstone bridge so that it is balanced at 0 °C. To avoid self-heating of the thermometer, select resistor values and the voltage supply to minimize the power it dissipates. First, assume that R_{det} is infinite, and see if the voltage A–B across the detector arm is linear with temperature. Then investigate what happens if the detector resistance is not infinite.
4. A thermometer may be connected to the bridge by a long cable, perhaps over 100 m. This will add resistance to the thermometer which does not change with temperature in the same way. Assuming the wire resistance is 0.1 Ω and its temperature is constant, find the error in the thermometer reading from 0 °C to 100 °C.
5. To minimize the effect of the connecting lead resistance, three- or four-wire configurations are used, https://www.peaksensors.co.uk/resources/resistance-ther mometer-information/. Build and test these circuits inserting small dummy resistors to mimic the lead resistance as shown in the circuits.

3.6 Thermistors

Thermistors either have a positive temperature coefficient, where the resistance increases with temperature (PTC), or a negative coefficient. There is an excellent description of both types at https://www.electronics-notes.com/articles/electronic_ components/resistors/thermistor-technology-types.php.

3.6.1 Temperature Measurement Using NTC Thermistors

One important use of small bead thermistors is to measure temperatures. Practical advice on selecting an NTC thermistor can be found at https://www.vishay.com/ docs/33001/seltherm.pdf. They are not intrusive and offer a greater change of resistance with temperature than platinum resistance thermometers, but their strong non-linearity can be a nuisance because their resistance depends exponentially upon temperature and has the general form:

$$R_{th} = Ae^{-BT} \tag{3.5}$$

where A and B are constants. The term B is the *characteristic temperature* and is a function of the material; it is usually a few thousand K. Manufacturers specify the resistance Rr at some reference temperature Tr – often 300 K.

The Beta Relationship

For most non-critical purposes, we can use the Beta relationship. The equation is:

$$R_T = R_0 e^{B\left(\frac{1}{T_0}-\frac{1}{T}\right)} \tag{3.6}$$

where R_0 is the resistance at T_0, usually 0 °C, B is as described above and T is the measured temperature. Note that the temperatures are in kelvin, not centigrade.

For an unknown thermistor, we can find B by measuring the resistance at two temperatures and rearranging Eq. 3.6 as:

$$B = \frac{\ln\left(\frac{R}{R_0}\right)}{\left(\frac{1}{T_0}-\frac{1}{T}\right)} \tag{3.7}$$

We shall take a specific example where the resistance is *100 kΩ* at *0 °C* and *B = 3380*. Although LTspice does have an exponential temperature coefficient *tce* (and we saw that this is not a true exponential), it is of no use here, and we again use the generalized resistance expression for Eq. 3.6:

R = 100k ∗ exp (− 3380 ∗ (1/(kelvin + 25) − 1/(temp + kelvin)))

where *kelvin* substitutes for −273.15 °C. The circuit is ('NTC Thermistor.asc').

This ability to define an arbitrary relationship is a powerful feature of LTspice. Otherwise, SPICE uses a convoluted approach, whereby temperature is modelled by a voltage source, http://www.ecircuitcenter.com/Circuits/therm_model1/therm_model1.htm shown as Fig. 3.2. If you go to the web page, you will see that the thermistor is defined by a sub-circuit consisting of a voltage source *ETHERM,* whose value is controlled by the current through the dummy voltage source *VSENSE,* which is set to 0 V and so does not contribute to the output. This is a trick of standard SPICE: current can be measured by the current passing through a voltage source. The *ETHERM* voltage opposes the externally applied voltage and therefore behaves as a resistor. These two voltage sources are in series and connected to nodes *1,3,2* in that order. But there are two other nodes, *4.5* whose voltage simulate the temperature. The sub-circuit, using different thermistor parameters than ours, is:

'Standard' SPICE Thermistor Model

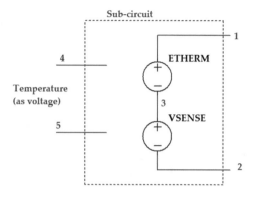

Fig. 3.2 SPICE thermistor

```
.SUBCKT NTC_10K_1 1 2 4 5
ETHERM 1 3 VALUE = { I(VSENSE)*10K*EXP( 3548/(V(4,5)+273) - 3548/(25
+273) ) }
VSENSE 3 2 DC 0
.ENDS
```

To be sure, it is only a couple of lines of code, but LTspice is simpler. And – worse – this NTC is a four-terminal resistor.

However we model the thermistor, to obtain the temperature, we divide Eq. 3.7 by R_0 and take natural logarithms to arrive at:

$$\ln\left(\frac{R}{R_0}\right) = B\left(\frac{1}{T_0} - \frac{1}{T}\right) \tag{3.8}$$

thence:

$$\frac{1}{T} = \frac{1}{T_0} + \frac{1}{B}\ln\left(\frac{R}{R_0}\right) \tag{3.9}$$

Example – NTC Resistance

We can use the data from a MuRata thermistor to compare the measured resistance with that predicted by Eq. 3.9. We take the NCxxxXH103 as an example with R0 = 10 kΩ and B = 3380.

Temperature (^{0} C)	−40	−20	0	20	25	40	60	80	100	120
Data sheet resistance (kΩ)	195.652	68.237	27.219	12.081	10.000	5.832	3.014	1.669	0.947	0.596
Simulation resistance (kΩ)	169.0	80.92	34.78	13.07	10.000	4.16	1.068	0.207	0.0247	0.00214

From this it is clear that the Beta relationship is only roughly accurate over a limited range. We only have Beta at our disposal, so we can tune that to the best fit, and no matter what value, the resistance at 25 °C will be correct. MuRata offers other values for different temperature ranges.

Steinhart-Hart Equation

This is an excellent example of the difficulty of fitting a non-linear device to an analytic equation so that it can be modelled in SPICE. For accurate work, the Beta relationship is not good enough, and the more complicated Steinhart-Hart equation is usually quoted. In essence, this is a curve-fitting approach where:

$$\frac{1}{T} = A + B.\ln(R) + C.(\ln(R))^3 \tag{3.10}$$

where A,B,C are the Steinhart-Hart coefficients. They dropped a quadratic term because it was claimed that it was less important that the cubic. However, this has been challenged, particularly in the Steinhart-Hart Wikipedia article 'Talk' page where it suggests that a quadratic expression:

$$\frac{1}{T} = A + B.\ln(R) + C.(\ln(R))^2 \tag{3.11}$$

is as good or better fit. Alain Stas of Vishay has developed a 'generic' cubic model using four terms:

$$R = R25 * exp\,(A + B/(273.15 + V(Ttot)) + C/(273.15 + V(Ttot)) * *2 + D/(273.15 + V(Ttot)) * *3)$$

shown in ('VISHAY NTC_stimulated_LTspice.asc') and here as:

$$R_T = R_{25}e^{\left(A+\frac{B}{T}+\frac{C}{T^2}+\frac{D}{T^3}\right)} \tag{3.12}$$

which essentially is a rewriting of the Vishay expression above. This is a simplified teaching model but still accurate; the full commercial model is more complicated; see below.

Self-Heating

The Steinhart-Hart equation takes no account of self-heating. The power dissipation of the thermistor may be very small, just a few microwatts, but if the thermistor itself is very small, just a tiny bead, then it is important. This can be handled by the small detached circuit shown in the above Vishay schematic. This is neatly done and repays careful study.

The current is supplied by *Imeas* and flows through the thermistor resistor *R1* and the voltage source *VSENSE*. This is a dummy source having no voltage and therefore not affecting the current in the thermistor. Its purpose is to transfer the current to the time constant circuit on the left.

The circuit can be run in two modes. The first is to measure the resistance against temperature for three different values of current. As these are not linear, nor decade nor octave relationships, they are given as a list – which can be extended. In this mode we see the steady-state resistance after the thermistor has settled to the correct final value for the given current.

If we run the circuit in this first mode, we will see three decay curves for the voltage across the resistor all tending to zero volts at high temperature showing that the resistance is decreasing exponentially. We can measure the resistance by probing V(1)/I(R1), and we see that it falls from 33 kΩ to 674 Ω and is the same for all currents – which is what we should expect.

If we change the y-axis to logarithmic, the graph is almost a straight line. And if we expand it, we see there is a very slight change of resistance with current due to self-heating.

The self-heating circuit consists of capacitor *C1* and resistor *R2* in series which model the thermal time constant. The output of voltage source *B2* represents the rate of rise of temperature because it is the power dissipated by the thermistor *Imeas*$*$*V (1,2)* divided by the thermal dissipation factor of the thermistor *Dth* which is the power in watts required to raise the temperature of the thermistor by 1 K. We should note in passing that this is not the same as the dissipation constant which has the power in milliwatts. Thus a thermistor with a large thermal capacity will have a smaller rate of temperature rise than one with a lower thermal capacity.

The capacitor *C1* represents the thermal capacity of the thermistor which is the number of joules required to raise its temperature by 1 K and is charged through *R1*, and so its voltage *Ttot* represents the resistance of the thermistor at any point in time. However, this is referred to 0 °C, so the ambient temperature is added by the voltage *Vamb* whose DC value is the parameter *AMBIENT*. Thus the total temperature *Ttot* is the sum of the temperature rise due to self-heating and the ambient and is applied to the thermistor equation. But the equation is referred to 0 K, the absolute temperature, so that must be added to the equation for *R*.

To see this we need a transient run, so we comment out the previous lines on the schematic and uncomment those starting with *.tran 10*, and what we see is that it takes nearly *8 s* to reach equilibrium as seen by plotting the resistance. It is of interest to plot *V(ttot)* which shows the temperature increases from the ambient of 25 °C by 0.009995 °C due to self-heating with a current of 100 µA.

We should note that the thermistor itself is a two-terminal device between ports *1* and *2*. Port *3* allows an external temperature to be injected, so there also is a three-terminal symbol. The circuit that we see is the expansion of a sub-circuit, and if we place an actual thermistor on a schematic, we shall only be presented with these two terminals, not 3, nor the capacitor and other items.

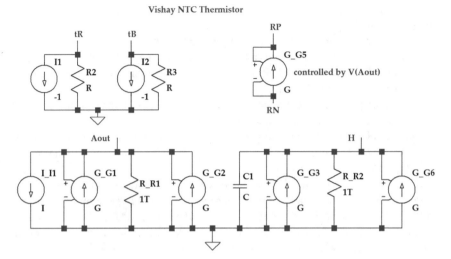

Fig. 3.3 Vishay NTC thermistor.asc

The full listing of many Vishay thermistor models using the far more complex model, applicable to positive and negative temperature coefficients, and again the work of Alain Stas, can be found in ltspice_ntc_tfpt_pts_model.zip. These are two-terminal devices using a base sub-circuit that includes thermal modelling into which the specific values for individual thermistors are inserted Fig. 3.3. This is not a functioning schematic but is merely to show the complexity of the model and that it is a phenomenological model that makes no attempt to mimic the actual structure of the device (a passive component can scarcely have eight current sources), only to reproduce its electrical characteristics. The functioning of the sources will be discussed in a later chapter. The external pins are *Rn* and *Rp*. The small circuits *tR* and *tB* handle the tolerances of the resistance at 25 °C and β at 85 °C through the resistors *R2* and *R3*, respectively. The model adds more parameters *X,Y* and *Z* plus thermal parameters *GTH1* and *GTH2*, whilst *C* and *D* set the thermal time constant as above. The good news is that, as users, we need not concern ourselves too much about the internal structures of the models because all these details are hidden when we import them into our circuit.

A point to keep in mind is that the Vishay models include tolerances with a uniform distribution whose values are assigned using the Monte Carlo method so that there is a different answer every time the circuit is run and for every instance of the device. For good statistics we need a large number of runs – at least 1000 – to have a reasonable chance of finding the extreme limits, but even with just 100 runs, we should be fairly close.

Measure FIND

The schematic uses this function to accurately record the resistance to six figures. This is necessary because the changes are small, and it is not easy to measure using the cursor. We first type *S*, and then if we just type .meas and right click on it, we will open the **.meas Statement Editor** dialogue; else we can type in directly on the schematic. The syntax is:

$$.meas < \textbf{Applicable Analysis} >< \textbf{Result Name} >< \textbf{Genre} > \textbf{FIND}$$
$$< \textbf{Measured Quantity} > \textbf{AT} < \textbf{condition} >$$

Note that FIND cannot be used over an interval.

Measure AT, WHEN

These apply to points along the x-axis which could be time, temperature or whatever. The *AT* is simplest and only requires a condition. We find on the schematic:

$$.meas\ Rt\ find - V(1)/I(11)\ at = 85$$

which will find the value of $-V(1)/I(11)$ at $85\ °C$ and save the result as *Rt*. The temperature is not needed.

Then **WHEN** option opens an edit box to its right. Enter the test parameter or formula here. Enter the desired condition in the **Right Hand Side** edit without an equal sign – that will be inserted automatically. For example:

$$.meas\ R60\ FIND\ V(1)\ WHEN\ AMBIENT = 60$$

will report the voltage at node *V(1)* in the result *R60*. This can be seen in VISHAY NTC_stimulated_LTspice V2.asc').

It is important that compression is turned off else **WHEN** may fail because the exact condition has been approximated. If you do not want to turn off compression, just add a small offset as:

$$.meas\ R60\ FIND\ V(1)\ WHEN\ AMBIENT = 60\text{-}0.0001$$

Explorations 6

1. Open 'VISHAY NTC_stimulated_LTspice.asc', and measure the resistance against temperature for the different currents.
2. The VISHAY model is generic for all versions of SPICE. We can simplify somewhat by using the flexible LTspice resistor whose value can be set directly from the current in *R1* so we can delete VSENSE. And LTspice also defines the constant kelvin for 273.15 so we do not need to remember the kelvin zero. This is ('VISHAY NTC_stimulated_LTspice V2.asc').
3. Compare the previous with ('VISHAY NTC Test.asc').
4. It is well worth downloading the Vishay files. LTspice has no symbol for a thermistor, and we shall find two here used with many different devices. Navigate to simulation_LT_therm_Vishay.asc and run it. This shows the spread of performance over 100 Monte Carlo runs.

3.6.2   Temperature Measurement Using PTC Thermistors

These come in two types. The ones for temperature measurement in fact are doped semiconductors whose resistance increases somewhat more than linearly with temperature but nowhere near as dramatically as the exponential relationship. The data for some AMWEI types is at http://www.amwei.com/views.asp?hw_id=65. The data for type AM-LPTC600 is in the following table.

Ambient temperature Centigrade (C)	AM-LPTC600
−40 C	359 ohm
−30 C	391 ohm
−20 C	424 ohm
−10 C	460 ohm
0 C	498 ohm
10 C	538 ohm
20 C	581 ohm
25 C	603 ohm
30 C	626 ohm
40 C	672 ohm
50 C	722 ohm
60 C	773 ohm
70 C	825 ohm
80 C	882 ohm
90 C	940 ohm
100 C	1000 ohm
110 C	1062 ohm
120 C	1127 ohm
130 C	1194 ohm
140 C	1262 ohm
150 C	1334 ohm
160 C	1407 ohm
170 C	1482 ohm
180 C	1560 ohm

The other type of PTC thermistor is discussed later in over-current protection.

Example – PTC Thermistor Equation
The above table shows the resistance is not a linear function of temperature. To use the standard SPICE resistor, it would be nice if we could fit it to a quadratic rather than some other function. The easiest way is to use https://mycurvefit.com/ where we find:

$$R_t = 498.0647 + 3.917955(temp) + 0.011009815(temp)^2 \qquad (3.13)$$

and it is an exact fit. That is fortunate. However, this does not include self-heating.

Linearizing the Probe

Now that we have thermistor models of greater or lesser accuracy, the problem remains of how to turn the highly non-linear resistance of an actual, real thermistor into a temperature measurement. This can be solved in software or hardware or a combination of both. In software, for example, https://www.lpi.usra.edu/lunar/ ALSEP/pdf/LinearThermistorsUseAsTempSensors%20-%20ATM-1108.pdf   of unknown date (but looks old) describes a computer programme. We can also wade through https://pdfs.semanticscholar.org/1d1c/43645cee053010695704e94ddfba6c 6945d3.pdf which is quite accessible.

In hardware, if we add a resistor in parallel, Fig. 3.4, and one in series, we can make the response more linear over a limited range of temperatures, generally about 50 °C or 100 °C. For that we take the average temperature T_{av} and the resistance at the middle of the range R_{av} (which we can find from ('NTC Thermistor.asc')).

Then we use:

$$R_p = R_{av}\left(\frac{B - 2T_{av}}{B + 2T_{av}}\right) \tag{3.14}$$

to find the parallel resistor. This is from https://en.tdk.eu/download/531110/5608e 4b12153bb12af2808fbedc5a55b/pdf-applicationnotes.pdf. The series resistor Rs is not critical and can be 1/10 of R_p. These figures are approximate and can be trimmed on the fly. The result is that the graph now has a shallow S-shape and we try to fit the point of inflexion to the middle of the temperature range. The rate of change of resistance with temperature, from the same website, is:

$$\frac{dR}{dT} = \frac{-R_{av}}{\left(1 + \frac{R_{av}}{R_P}\right)^2}\frac{B}{T^2} \tag{3.15}$$

Explorations 7

1. The schematic based on the thermistor above is ('Linearised NTC Thermistor. asc'). Plot V(vth)/I(R3) and note the sensitivity to the value of R2. Can a better fit be found so that the temperature is correct at the mid-point, 50 C?

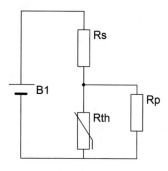

Fig. 3.4 Linearised thermistor

2. Is there a better fit including the series resistor R3 and plotting V(n001)/I(R3)? Try different values using the above equations as a guide.
3. The website http://www.ecircuitcenter.com/Circuits/therm_ckt1/therm_ckt1.htm suggests using just a parallel resistor. How well does it work?
4. https://www.edn.com/design/sensors/4429105/Linearize-thermistors-with-new-formula gives an alternative technique for linearization. Is it any better?
5. Search the Internet for other linearization techniques and compare them. Do they give better linearity, and over what range of temperatures?
6. Construct the model for the AM-LPTC600 and test it. Try linearizing it.
7. Is a Wheatstone bridge a good way of measuring temperature with a thermistor?

3.6.3   Circuit Protection

The idea is threefold. The first is to limit the in-rush current when the circuit is first switched on. The second is to limit harmful mains overvoltages from damaging the circuit, and the third is to prevent a fault in the circuit creating consequent faults elsewhere. This last application has traditionally been effected by a fuse or circuit breaker. We should note that any protection device can only protect 'downstream', so, in the extreme, the household mains supply where it enters the premises must have either a large fuse or a circuit breaker to protect against faults between there and the switchboard.

Residual Current Devices, Circuit Breakers and Fuses

The first two are usually found on the mains supply switchboard, the last inside equipment. We shall discuss them briefly here to contrast them with non-destructive electronic protection.

Residual Current Devices (RCD)

These are connected to both line and neutral of the supply. If there is a difference in the two currents, it is assumed that some current has found a path to earth either through a fault or because something (usually a human) has created an earth path. If the difference exceeds some low value, typically 50 mA, the RCD opens in a few milliseconds. It is therefore a valuable safety device to prevent electrocution since it is generally considered that 50 mA may be unpleasant but not lethal. More sensitive RCDs can be found, as low as 30 mA, but they can be prone to *nuisance tripping*.

Circuit Breakers

These are sized to the current-carry capacity of the wiring they protect, and therefore the most commonly used are rated 6 A 10 A 16 A 25 A or thereabouts, but they can be found down to 0.5 A. They often have both fast magnetic tripping for fault currents three times the rated current or more with trip times ranging from 100 ms to several seconds depending on the overload and being faster with the higher overload. In addition they incorporate thermal tripping using a bimetallic strip which is much slower but will trip on a small overload of 30% above its rated current.

Fuses

These are made in a number of types and sizes, but for electronic applications where the fuse is located inside the equipment, they are commonly found in three speed ratings **F** (fast acting), **M** (medium speed) and **T** (Time delay) and as cylinders with metal contact end caps in two sizes 5 mm × 20 mm and 6.3 mm × 32 mm (previously 1/4 in × 1.1/4 in) and with either a transparent glass body or opaque ceramic and in a myriad of ratings from 31 mA to 13 A.

The critical decisions (and these apply to every type of circuit protection) are the prospective fault current and the necessary clearing time so that the fuse 'blows' before electronic components are damaged. In other words, the fuse protects the components and not the other way round. This is a topic that demands careful study, and there are several websites giving detailed information which we shall summarize briefly here.

The first line of defence is at the input of the equipment. Here the danger is of a dramatic short-circuit, perhaps the primary of a mains transformer or a cable fault. The prospective fault current is limited only by the impedance of the mains supply and can be tens of amperes. It always takes some time for a fuse to clear, and during that time a very large amount of energy can be dumped in the fuse. For this reason ceramic fuses are used since glass would shatter because of the heat generated. And even so, smaller ceramic fuses can have their end caps blown off. The larger fuses seem able to contain the energy.

Once inside the equipment and 'down wind' of the mains transformer, we need to look carefully at what could go wrong and what the consequences might be. In particular, we must look for 'knock-on' effects where one fault, for example, a diode that has gone short-circuit, could create more faults by overloading another diode.

So we now come to the all-important question of how long does it take for a fuse to 'blow'. And there is no simple answer. For a fuse to blow, first the fuse wire has to get so hot that it melts. And as it melts, it breaks, and an arc forms across the break greatly heating whatever is in the fuse (and for this reason high-current fuses are filled with dry sand not air), and the current only stops once the arc has quenched. The parameter of interest is I^2t which is the total energy. A typical graph is Fig. 3.5 where the minimum clearing time is a few ms.

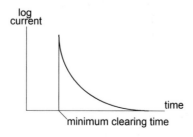

Fig. 3.5 Fuse blow time

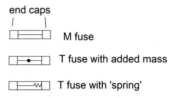

Fig. 3.6 Fuse types

The distinction between the different fuse types is stamped on the end cap but visible with some glass fuses. The slow fuses either have a small mass added to the wire to increase the thermal capacity, or one end is attached to a spring which increases the thermal capacity but also allows the wire to expand without breaking, Fig. 3.6.

In-Rush Limiting

Many circuits offer a low impedance when first switched on. Capacitors are unchanged, motors are not turning, heaters are cold, and so on. So it is desirable to limit this initial surge of current. NTC thermistors, usually in the form of ceramic discs, can be used. They have a relatively high resistance at the reference temperature of 25 °C, typically less than 10 Ω. The thermistors warm up rapidly, and the resistance falls sharply to some low value. The EPCOS Application Note 2013 to be found at https://en.tdk.eu/download/528070/f5be4fca9d1f66204de9cf37891e5265/pdf-inrush currentlimiting-an2.pdf is very informative. The time for the resistance to fall to some low value depends on the magnitude of the in-rush current and thermal capacitor of the thermistor, so, knowing the likely inrush current, the thermistor size can be found to give enough time for the protected circuit to settle. The thermistor will then continue to consume a small power due to its ON resistance which is typically just a few percent of its cold resistance. This power keeps the thermistor hot enough to maintain its low resistance and therein lies a possible problem that an excessive load current will overheat the thermistor and destroy it, so its maximum current rating must not be exceeded.

Example – In-Rush Current Limiting

A model based on the Beta relationship, ('Inrush Limiter Beta.asc'), consists of a conventional low-resistance NTC thermistor on the right and the thermal modelling circuit on the left where the voltage *V(Ttot)* corresponds to the thermistor's temperature as above. The heat transfer is handled by the voltage *B2* which consists of the power in the thermistor *I(R1)∗V(1)* divided by *Dth* as previously for self-heating, only this time we are concerned about the time for the circuit to act and the energy absorbed by the thermistor. In this we have contradictory requirement of fast response and a large thermal capacity – which implies a longer time to heat up. With the schematic we find an initial resistance of 100 Ω found by $-V(1)/I$ *(V1)* measured after the *0.01 ms* delay falling linearly to *4.2 Ω* at *55 ms* and thereafter rapidly falling to a few milliohms.

Now although the voltage across the thermistor is almost constant for the first 50 μs, being in series with only 1 Ω, plotting $V(1)*I(R1)$ gives the power dissipation, and also, by **Ctrl** and left click on the trace label, the energy is *114.8 mJ* even though the average power is *1.148 kW*.

Over-current and Over-voltage Protection

These are PTC thermistors which are not used to measure temperature but rather that their resistance depends upon their power dissipation – which (of course) increases their temperature. They consist of a polycrystalline ceramic containing ferromagnetic salts. The principle is that they consume little power in the normal state, but their resistance changes dramatically during an overload so that they restrict the current. This is because when the temperature rises to the Curie point, the ferromagnetism vanishes and the resistance rises sharply. They, too, can be found as discs and also as thin film devices, and they are all limited by the energy they can absorb, and so the duration of the abnormal condition also is strictly limited.

Their characteristics are subject to wide tolerances, and these PTC devices do not obey a nice mathematical relationship, Fig. 3.7, and hence the Vishay models of Alain Stas include a long tabulation of data points to define the resistance.

The graph shows typical data definition points, in particular that the resistance falls to R_{min}. The effect of this will be seen in subsequent simulations where the current or voltage initially actually increases with an overload. The trip temperature T_s is taken at twice the minimum resistance.

Adding Models to the Library

The folder ltspice_ptcel_cl_tl by Vishay contains models for their positive temperature coefficient thermistors and four very interesting schematics. They can be found at http://www.vishay.com/docs/29184/ltspice_ptcel_cl_tl.zip and downloaded. It is possible to run them directly from the download position if the folder is kept intact with models and simulations all together. Otherwise, copy the whole download file (at the time of writing this is 29184LTSPICE_PTCEL_CL_TL) as a new subfolder to the .*lib* folder. It does not matter that it contains simulations and text; these merely bloat the folder and can be removed if you wish. Now the path to the new subfolder

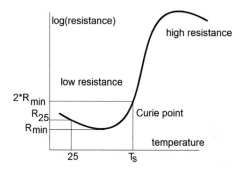

Fig. 3.7 PTC characteristic

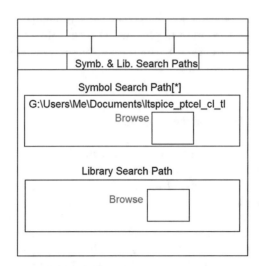

Fig. 3.8 Control panel search paths

must be added to LTspice's search paths. Do this from the main menu
Tools→Control Panel to open the window Fig. 3.8 then select the **Sym. & Lib
Search Paths** tab and right click in the panel then select **Browse** from the pop-up
menu and navigate to the download file.

Device Characteristics
The behaviour of the following circuits is best understood by starting with the
simulation ('test_PTC_res.asc'). This is a later addition and not part of the Vishay
download. It shows an initial drop from the starting resistance of 1.3 kΩ to 908 Ω just
before the dramatic rise at 0.3 A to 8.35 kΩ. And although the x-axis is time rather
than temperature, there is general agreement with the shape of Fig. 3.7 except for the
very sharp knee at maximum resistance.

If we change to a voltage input pulse *PULSE(0 1000 0.1 1f 1f 1)* and examine the
time from *99* to *110 ms*, we see an increase in current from the initial *9.25 A* at *100 ms* to
9.53 A at *100.5 ms* and the very steep decrease starting at *101 ms* and is only *62 mA* at
120 ms, 27 mA at *150* ms and *9 mA* at *300* ms. Thus the current can change dramatically
in just 1 ms. The resistance starts at *108 Ω* falls to *105 Ω* at *100.5 ms*, rises to *112 Ω* at
101 ms and then rises steeply to *1.2 kΩ* at *102 ms* and *109 kΩ* at *300 ms*. This ability of
the resistance and current to change in less than 1 ms explains the heavy distortion seen
in the decay simulation, and the continuing rises in resistance after the dramatic
increases explain the subsequent slow reduction in current amplitude.

test_PTC_AC_current_decay
The battery is not needed and can be removed: it is only required with some other
versions of SPICE. In our case, we can measure the thermistor current directly. The
simulation runs quickly up to the point where current limiting begins and then more
slowly. The slight rise in current up to some 100 ms is explicable from the previous

simulation and Fig. 3.7 showing the slight fall in resistance. The trip time reduces with temperature because the required self-heating is less.

'test_PTC_resettable_fuse.asc'

This has a mains voltage that changes in steps and is therefore given as a list: right click on *V1* to see them. This drives a current through a 2200 Ω resistor in series with a thermistor type PTCEL13R600LBE and a zero volt battery – the typical standard SPICE trick for reading a voltage but not needed with LTspice. This simulation takes some time, so for a quick assessment, we can cancel the ten runs and increase the step from 0.001 to 0.01.

During the normal operation up to time 1000 s, the input voltage is 220 V (lower panel), the PTC resistance is about 36 Ω (middle panel), and the thermistor current is 100 ma (top panel). The voltage across the PTC is effectively zero (bottom panel).

At 1000 s the lower trace panel shows the input voltage jumps from 220 to 500 V, and the upper trace shows that the load current in *V3* jumps to 223 mA.

At approximately 1150 s, the PTC switches, the current drops to just over 5 mA, the resistance of the PTC jumps from to over 90 kΩ and almost all the supply voltage appears across the PTC. This is a long delay and we would normally expect a much faster response.

At time *1300 s* (see the list for *V1*), the voltage *V1* returns to the normal 220 V, and the voltage across the PTC also drops; the difference between the two is the volt drop across the thermistor whose resistance has fallen to a little less than 14.5 kΩ and thus is still switched. The load current therefore increases to some 13 mA which accounts for the difference between the supply voltage and the voltage across the thermistor shown in the lower panel.

At 1301 s the supply voltage drop to 10 V, and now the thermistor switches back to its low-resistance state at about 1850 s as measured from the middle trace showing it takes nearly a minute for it to dissipate its heat.

At 200 s the supply is restored to its original value and we are back to normal.

What all this shows is that the PTC thermistor can be used as a fuse, but it takes far longer than a conventional fuse to operate. However, this simulation is deliberately set to separate the times, and the next simulation shows that times of 0.2 or less are possible.

'test_PTC_trip.asc'

This is a statistical analysis applying a 60 V DC step to a PTCTL7MR100SBE in series with another battery of zero volts (which is redundant with LTspice). The transient run is repeated 30 times using the internal tolerance of the thermistor itself – there is no external Monte Carlo parameter. From the plot, it appears that the initial current ranges from 5.1 to 7.6 A showing a wide tolerance in resistance. The current falls to between 0.2 and 0.4 A after 0.5 s.

From the main menu, we choose **View->SPICE Error Log.** In passing we can note the iterations of *Gmin* and that this can be stopped by *noopiter*. We scroll down to find the maximum current over the whole measurement range of 0–0.5 s for each run; this ranges from *5.1* to *7.6 A*. Scrolling further and we find the trip time to the point where the current is half the maximum is less than 0.2 s and – the last set of data – the energy is less than 50 joules.

It does not appear from this simulation that the trip time depends upon current; we could deduce it from the fact that the thermistor trips when it gets to a certain temperature, and this depends on the power it dissipates, and so we expect a longer time as the over-current is less, and hence the power is exactly the same as a conventional fuse – only the thermistor is not destroyed.

We should note that this simulation also makes good use of the *.meas* directive first to measure the maximum current as *Imax* which is then used in:

$$.meas\ Ttrip\ when\ I(V3) = Imax/2$$

to find the time when the current is half the maximum, and secondly we have;

$$.meas\ energy\ integ(V(PTC) * I(V3))\ from\ 0\ to\ Ttrip$$

to find the energy.

test_PTC_AC_current_decay
The time for the resistance to trip depends upon its temperature so that starting with a low temperature, and for the same power input, it will take longer than starting at a high temperature. This is illustrated in this simulation. If we plot the power, it is highest for the low temperature. If we comment out the temperate stepping and plot the resistance of the thermistor, we see it switching from around 1 kΩ for the first 100 ms, rising to peaks of 8 MΩ at 300 ms and more than 18 MΩ if we extend the time to 0.5 s.

test_PTC_vi
The left-hand circuit makes 100 simulations as a fair estimate of the limits of the parameters, and we see the spread with the data points marked as small circles. The thermistor behaves as a resistor up to an input of between 7 and 10 V when it switches and the resistance falls. The single trace in a different colour is from the right-hand circuit where *meas* contains the extensive table of measured data for comparison with the simulation. The simulation takes some time to run; the *noopiter* option does not make it faster.

These simulations are made with different devices so it is not possible to compare switching times.

Explorations 8
1. Open ('Inrush Limiter Beta.asc') and note the currents and voltages. Try different values for the thermal properties of the thermistor.
2. It is worth making a more detailed examination of simulation ('test_PTC_res. asc') with different thermistors and comparing the results with the data sheets.
3. Change the input voltage of 'test_PTC_AC_current_decay' to 1 kV and 600 V. Extend the simulation time to 0.5 s, and note how the final current reduces with temperature. Make the voltage a parameter *.step param Vx 600 1000 100* and set

the temperature to *25 °C*, and the product of (power)∗(start of decay) is constant at about *33 J* where the start of decay is estimated from the time at which the voltage begins to fall. Note also that the time to steady state is more than 1 s and agrees with the data sheet.

4. The simulation ('test_PTC_trip time.asc') is derived from 'test_PTC_trip.asc and replaces the voltage source with a current source. The trip times are read from the x-axis for currents of 10, 8, 6, 4, 2 and 0 amps. These are in approximate agreement with Fig. 3 of the application note 'Simulation Notes for SPICE Modelling' included in the download as 'LTSPICE_PTCEL_CL_TL as an HTML document. Simulation ('TEST_PC_res.asc') has shown that the resistance is almost constant before the trip point, so if we multiply the trip time by the current squared, we find an approximately constant figure showing that for a given ambient temperature, a fixed amount of energy is needed to heat the device to its trip point.

5. Run the simulations listed above and explore different input settings. Compare the simulations with the data sheet values.

3.7 Voltage Variable Resistors (Varistors)

In contrast to thermistors, these change their resistance due to voltage, not heating, Fig. 3.9. They have a high resistance at low voltages which falls rapidly when the clamp voltage is reached, either positive or negative, and are used to protect circuits from brief over-voltage surges. They are much faster than a fuse, with a typical response time of less than 100 ns, and it is non-destructive, so that after the pulse has ended, the circuit returns to normal operation. However, they do have a limited life. Most are made as sintered metal oxide discs, hence the alternative name of Metal

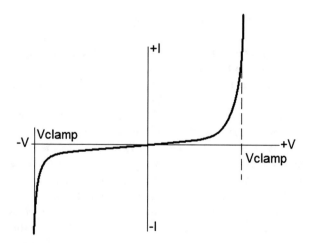

Fig. 3.9 Varistor characteristic

Oxide Varistor (MOV). Two diodes back to back in series will serve the same purpose where the clamp voltage is the diode breakdown voltage.

A good introduction to the devices can be found at https://www.electronics-notes. com/articles/electronic_components/resistors/metal-oxide-varistor-mov.php, whilst the more comprehensive http://www.littelfuse.com/~/media/electronics_technical/ application_notes/varistors/littelfuse_varistors_design_notes.pdf has data and applications.

Models range from very basic to highly complex and depend upon whether the voltage should be clamped or reduced when the critical voltage is reached.

Example – Varistor Model
We can build a very simple varistor model using LTspice's extremely flexible resistor. The definition is

$$R = if(abs(V(Vin)/150), 10, 1e6)$$

using the 'if' relationship which can be found in the Help file as *PARAM – User-Defined Parameters*. If the absolute value of *V(Vin)/150* exceeds 0.5, the resistance is 10 Ω; else it is 1 MΩ, so the change occurs at 75 V and is a function of the input voltage. The schematic is 'MOV mk1.asc' where we should note that if no maximum step time of 1 us is given and LTspice is left to choose its own, the breakpoint will be at about 73 V because the default time step is too long to resolve the sharp resistance change correctly.

In this simple model, the breakpoint is defined in terms of the input voltage, whereas it would be nice if it were a parameter. A simple change to:

$$R = if(abs(V(Vin) > Vclamp), 10, 1e6)$$

ensures a constant breakdown voltage of *Vclamp* defined as:

$$.param \; Vclamp = 75$$

in ('MOV mk2.asc'). The '.*param*' directive is extremely useful and allows a new parameter and its value to be defined or to set the value of an existing one. Here it allows us easily to change the breakdown voltage. And this is computationally more efficient than the equivalent *V(Vin)/(2∗Vclamp)*.

However, if *V(mov)* is used instead of *V(Vin)*, LTspice fails where the resistance first changes and all the usual workarounds to help SPICE converge do not work – adding a small series resistance between *R1* and *R2* to soften the very sharp change, or a parallel capacitor or an inductor, or changing the settings in the **Spice** tab of the **Control Panel**. It works only if the voltage is measured directly across the input. We can understand this because if we use the voltage across the resistor itself and then as its resistance begins to fall the voltage across it also starts to fall because it forms a potential divider with *R2*. And if the voltage falls below the trip point, *R1* reverts to the high-resistance value which means that the voltage across it rises again, and so it will trip once more because the voltage is above the trip point. So poor old LTspice is thoroughly confused.

3.7.1 Basic Models

These are for non-critical applications and do not handle turn-on and turn-off times;

The LTspice Model

LTspice has a symbol in the 'Misc' folder but no model. However, in the educational folder, there is an example, but it is not the same, as it has four terminals. It defines a breakdown voltage *IN* in the schematic ('varistor.asc') in the 'educational' folder and a clamp resistance *Rclamp*. Having a separate *IN* terminal means that the breakdown voltage need not be constant (as it is for a real varistor) but can vary with time as a triangular wave as it does in this example.

LittelFuse 'Pulseguard'

This model consists of a 1 nH inductor in series with a 0.05 pF capacitor with two Zener diodes and switches in series across the capacitor, Fig. 3.10. The values for the

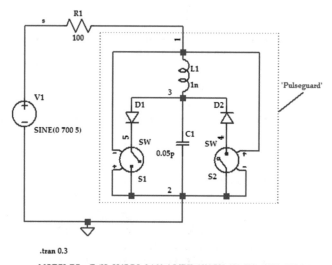

.tran 0.3

.MODEL DZ   D (IS=321F RS=0.1 N=1.5 IBV=10N BV=150 CJO=0.001p TT=1p)

.MODEL SW SW(RON=1 ROFF=2G VT=265 VH=235)

Fig. 3.10 Littelfuse 'Pulseguard'

diode and switches are from the PGB1010603. The switch hysteresis values mean that it will turn on at $VT + VH = 265 + 235 = 500$ V. We can follow the operation of the circuit from 'LittelFuse Pulseguard.asc'.

The input is applied to the control terminals of both switches. At a time of 25 ms, this is 500 V, and as $S1$ can break down at $VT + VH = 265$ V $+ 235$ V $= 500$ V, it does, and at that instance, there is 360 V across the switch terminals of $S1$ since diode $D2$ has dropped 150 V.

From then until the time of 93 ms, $D2$ conducts as seen in the bottom trace. The voltage $V1$ is the 150 Zener voltage of diode $D2$ plus the volt drop across its internal resistance of about 1 Ω, so it peaks at about 155 V. As diode $D1$ is forward biased, the voltage at node 3 follows node 3 minus the volt drop across $D1$. At time 93 ms, the supply voltage Vs has fallen to 150 V, and the voltage at node 3 also has fallen to 150 V, diode $D2$ stops conducting and the voltage $V1$ in the upper panel follows the supply until 125 ms. During that same interval, there is no current in either diode, but the voltage at node 5 follows the supply to 0 V at 100 ms and then remains at 0 V until the supply reaches -150 V at 107 ms, then it breaks down into Zener mode and the voltage at node 5 follows the supply minus 150 V until time 125 ms when switch $S1$ breaks down and diode $D1$ conducts.

Thus the supply voltage is clamped to 150 V plus a small increment due to the volt drop across the diodes.

This model still does not model the actual characteristics of the varistor; in particular the switching is instantaneous, although these devices do switch in less than 20 ns, except for a very short damped ringing which can be seen if we set the input to:

$$PULSE(0\ 700\ 1n\ 1e\text{-}15\ 1e\text{-}15\ 1n\ 2n)$$

to give a delay of 1 ns before a 700 V pulse with a rise time of $1e\text{-}15$ s and a duration of 1 ns and a period of 2 ns. We also set the analysis to:

$$.tran\ 0\ 5n\ 0\ 10f$$

which gives time steps of 10 fs – short enough to resolve the ringing lasting 0.1 ns which is due to $L1$.

The LittelFuse SPICE Varactor

This is based on a complicated voltage source E_non_lin running to six coefficients whose value is multiplied by the toleranced value that defaults to 1 – meaning no tolerance effects – and a second polynomial voltage source E_x_zero, schematic ('LittleFuse SPICE Varactor.asc'). This is a general template into which the values are plugged for the individual device types. The template is the extensive listing with default values:

```
.SUBCKT DAMOV 1 2 PARAMS: T=1 C=1pF L=1nH a1=1 a2=0 a3=0 a4=0 a5=0 a6=100u
a7=100u
......
ENDS.
```

The specific device V131DA40 is as follows:

```
.SUBCKT V131DA40 1 2 PARAMS: TOL=tx
X1 1 2 DAMOV PARAMS: T={1+TOL/100} L=20nH C=10nF a1=250.3 a2=9.855 a3=-
3.115 a4=0.2016 a5=3.161 a6=2.405E-3
.ENDS
```

where it can be seen that although the parameters are not in the same order
(L before C) because they are identified by name, there is no confusion.

The AVX SMD Varistor MAV Series

These are three small surface mount varistors designed to bidirectionally clamp a
voltage to *200 V* at a current of *1 A*. The three members differ mainly in their
capacitance and slightly in their ability to adsorb pulse energy. They have a low loss
when non-conducting.

AVX have managed to pack a lot of functionality in just seven components,
Fig. 3.11. The two parallel circuits of a resistor and inductor in series also in series
with the capacitor *CP* model the frequency response. We anticipate a later chapter to
say that at frequency f_r where:

$$f_r = \frac{1}{2\pi\sqrt{L.CP}} \qquad (3.16)$$

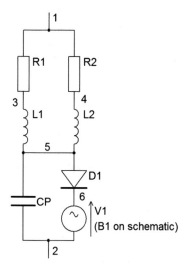

Fig. 3.11 AVX SMD varistor

the inductor and capacitor cancel each other, and we are just left with the resistance, and there is a 'V'-shaped notch in the frequency response. The width of the notch increases with the resistance. So, removing each inductor in turn in the schematic, we find a wider notch when $L1$ is removed because it is in series with the larger resistor.

The clamp voltage is cleverly modelled by the diode and the voltage-controlled voltage source $B1$. Suppose the voltage $V(B1) = 0$. We are then left with almost a short-circuit from node 1 to node 2, consisting only of the two resistors in parallel in series with the diode, and we should expect only a volt or two between these nodes with a current of 1 A. However, $B1$ creates a reverse voltage proportional to the voltage across the diode described by:

$$\mathbf{V} = \mathbf{V(5,6)} * \mathbf{B} - \mathbf{V(1,5)}$$

where, for the moment, we can discount $V(1,5)$ because, at low frequencies, we have only the voltage dropped across the two resistors in parallel, and this is just a volt or so, whereas the factor B, which is nearly 200, and the diode volt drop of typically $1/2$ V mean that V is around 100 V and this is the voltage we see between the terminals.

We can quantify this if we look more carefully at the diode characteristics. This is the SPICE diode which makes a smooth transition from non-conducting to forward-conducting at $+0.6$ V. But under reverse bias, the diode normally would not conduct until a substantial voltage caused breakdown. However, the sub-circuit gives the diode a reverse breakdown voltage of only 0.24 V so that it conducts symmetrically with a final slope voltage of RS. This is sketched in Fig. 3.12 and is shown in schematic ('AVX Diode Test.asc').

To return to the circuit, we can write the voltage between nodes 1 and 2 as:

$$V_{12} = I.R_p + V_{D1} + V_{D1}B$$

where R_P is the parallel sum of the two resistors and we ignore $V(1,5)$. So until $V_{D1} = \pm 0.6$ V, the diode will not conduct, and the voltage across it will rise linearly with the input, and the voltage between nodes 1 and 2 also will follow the input. But at voltages outside these limits, the diode will conduct, and the current will rise sharply with only a slight increase in voltage due to the volt drop across RS, and the voltage will be clamped.

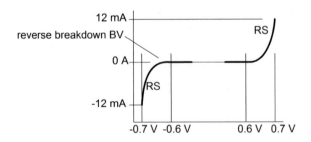

Fig. 3.12 AVX varistor diode characteristic

The term $-V(1.5)$ is to model the reduced clamping effect at high frequencies.

Example – AVX MAV0010

This has $B = 197.169$ and $RS = 39.27 \ m\Omega$. If we put the two inductors in parallel, they amount to $489 \ pH$. Putting that into Eq. 3.16, we have $f_r = \frac{1}{2\pi\sqrt{489\times10^{-12}\times13\times10^{-12}}} = \frac{10^{12}}{501} = 2GHz$, and this is what we see in schematic ('AVX SMD Varistor AC.asc').

However, the clamp voltage is too low, schematic 'AVX SMD Varistor Clamp. asc' being only $173 \ V$ and not $200 \ V$. The measured breakdown voltage at $1 \ mA$ of $129 \ V$ is within specification, and the V–I graph, schematic ('AVX SMD VI.asc'), also agrees with the published graph.

In short, as the data sheet states, the sub-circuit does not model temperature effects nor adsorbed energy, and the clamp voltage is too low.

Explorations 9

1. Use ('MOV mk1.asc') and ('MOV mk2.asc') to explore the current and voltage relationships and the effect of changing the input voltage.
2. Open ('varistor.asc') in the educational folder and explore different IN and V1 waveforms.
3. Run ('Littelfuse Pulseguard.asc') and note the effect on the ringing of increasing $L1$ and the rounding of the waveform is $C1$ is increased. Note that changing the Zener voltage changes the clamp voltage when the switches break down.
4. Test ('Littelfuse Varistor.asc'). Here we have the sub-circuit on the schematic so that it is accessible and we can change parameters. Note the large voltage spike if the rise and fall times of the pulse input are 1e-20 s. also that the varistor power is 1.1 kW but for such a short time that the energy is 55 mJ. Check that the turn-on and turn-off times agree with the data sheet.
5. Uncomment *.step param x 1 10 1* where x is a dummy variable to make a Monte Carlo run of ten simulations. Try a longer run with the actual tolerance from the data sheet.
6. Change the excitation of 4 to AC. Explore the effects of changing parameters.
7. Substitute other varistors.

So although these models can serve for approximate analysis, circuit designers will want to know the turn-on time and how large and how long an over-voltage spike can be contained, which is in effect to ask how much energy (I^2t) can the varistor absorb. For this we need to know the electrical capacitance of the device, its lead resistance and its thermal behaviour in detail. The website https://www.vishay. com/docs/29079/varintro.pdf has a comprehensive guide to their theory, parameters and applications. And it would take much time and effort to create useful models. Fortunately, most manufacturers provide libraries of better SPICE models. The one by EPCOS is more complex and is shown in Fig. 3.13. This is almost identical to the new Littelfuse sub-circuit.

The resistor R_SERIES represents the lead resistance and is not critical, often set to a small value such a 0.1 mΩ. The inductance L_SERIES represent the lead

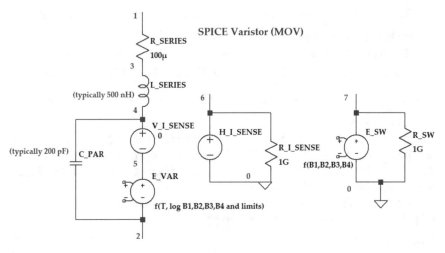

Fig. 3.13 EPCOS MOV schematic

inductance of the wires and is not critical, often being set to 20 µH. The current-voltage characteristic of the device are described by

$$\log V = b1 + b2 \times \log(I) + b3 \times e - \log(I) + b4 \times \log(I) \text{ when } I > 0$$

Automotive EMC Testing ISO 16750-2 and ISO7637-2

Modern road vehicles make extensive use of electronics, not just for entertainment or navigation but in the computerized engine control and anti-lock braking systems. It is therefore essential that these should be immune to spurious voltage spikes. The International Standards Organization has defined certain test pulses for EMC testing. The two popular ones are implemented in LTspice and discussed in https://www.analog.com/en/technical-articles/ltspice-models-of-iso-7637-2-iso-16750-2-transients.html. Both have waveforms for 12 V and 24 V systems selected from a drop-down menu by a right click on the **Spice Model** field.

ISO 16750-2
These are mainly slight over-voltage conditions where the battery voltage runs up to 14 V or so except 'Load Dump Without Suppression' which creates a 100 V spike seen by a 2 s transient run. Other slow fall/rise options may need times of 2000 s or so.

ISO 7637-2
This creates mainly single negative pulses of 100 V or more except 'Pulse 3b' which creates a train of 107 V positive pulses delayed by 1 ms.

Example – Pulse Limiting

The schematic ('ISO7637-2.asc') uses a modified MAV0010 with $B = 14.5$ to clamp the pulse. The time is delayed to $0.9\ ms$ just before the pulse to see if there is any rise time. As this is a piecewise model, to find the energy in the varistor, we must add the contributions from the resistors and the diode. It is more realistic to include a series DC of 12 V to represent the battery.

Explorations 10

1. Download models to explore pulse limiting with thermistors and varistors. In particular, note the rise time and power dissipation and clamping voltage.

3.8 Photoconductive Cells

These are widely used to detect visible light using cadmium sulphide (CDS) or cadmium selenide (CdSe), but these are now deprecated because of the use of cadmium and alternatives such as thallium sulphide (TlS) are used, but the spectral response is different.

They are not, in fact, resistors, but a film of semiconductor material, Fig. 3.14 on a ceramic base with two leads taken from the back. Because the semiconductor has a high sheet resistance, it is evaporated as a series of parallel tracks joined at alternate ends with the interdigital terminals to give a very high length-to-width ratio for the contacts and thus a reduced resistance. The front is then covered with a protective layer of transparent material.

Recent research has shown that graphene can achieve far higher sensitivity: https://spectrum.ieee.org/nanoclast/consumer-electronics/gadgets/graphene-image-sensor-achieves-new-level-of-light-sensitivity.

3.8.1 Illumination Characteristics

The article at http://www.ladyada.net/media/sensors/APP_PhotocellIntroduction.pdf is very readable and comprehensive. When light strikes the face of the photocell, it

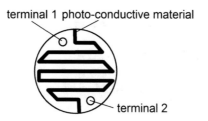

Fig. 3.14 Photocell construction

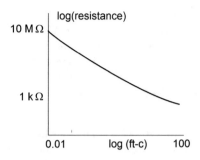

Fig. 3.15 Photocell resistance illumination

excites electrons into the conduction band which causes the resistance to fall. A typical response is shown in Fig. 3.15 showing a slight concavity rather than straight line.

Illumination Standard

We are immediately faced with the problem of an illumination standard. We can carry a standard of length and mass from one place to another, but this is a similar problem to temperature – we must define a means of realizing the standard. Originally this was illumination produced by a standard candle at a distance of one foot. This then raised the question of what is a standard candle. Clearly the wax composition, size of wick, etc. were all factors that affected the light output, so a standard pentane lamp was proposed as a reproducible standard. A fascinating historical article on the difficulties of establishing a standard and the precautions to be taken is in https://nvlpubs.nist.gov/nistpubs/bulletin/10/nbsbulletinv10n3p391_A2b.pdf being the work of E.C. Crittenden and A.H. Taylor published in 1913.

However, that alone is not enough. The light output of the lamp is not uniform over the visible spectrum. So the modern definition uses the *candela* instead which is

The candela (cd) is the luminous intensity, in a given direction, of a source that emits monochromatic radiation of frequency 540×10^{12} hertz and that has a radiant intensity in that direction of 1/683 watt per steradian

where the frequency corresponds to visible green light of a wavelength of 555 nm which is near the peak sensitivity of the human eye.

The SI unit is the *lumen (lm)* rather than the foot-candle where:

$$1 \text{ ft} - c = 1 \text{ lm/sq ft}$$

which in turn can be related to the candela as:

$$1 \text{ lm} = 1 \text{ cd/sr}.$$

As a complete sphere contains a solid angle of 4π *steradians*, a light source uniformly emitting 1 candela in all directions generates a luminous flux of *1 cd* $\times$ *4π sr = 4π cd·sr* $\approx$ *12.57* lumens.

3.8.2 Photocell Response

We saw from Fig. 3.15 that the slope is not an exact straight line and differs between different cell types. The response is also subject to a tolerance which increases at low light levels.

Spectral Response

The photocell does not respond uniformly to all wavelengths of light. Figure 3.16 is typical, peaking around visible green and is close to the human eye response. This is important if we want to assess the lighting for an office, or a room or a road, but we shall concern ourselves only with the generalities.

Dark Resistance

It takes a few seconds for the photocell to recover after being exposed to light. The resistance then is typically from 500 kΩ to more than 10 MΩ

Response Time

The response time depends strongly on the illumination; both rise and fall times are similar, falling steeply at low light levels, Fig. 3.17, then flattening out above 1 ft-c where the response become almost instantaneous.

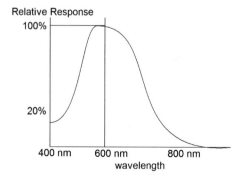

Fig. 3.16 Photocell spectral response

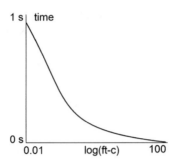

Fig. 3.17 Photocell response time

Temperature Effects

These depend upon the type of material and become significant at low levels of illumination, up to 1 ft-c where the Perkins-Elmer type 3 material at a temperature of 15 °C has a response of 80% relative to its value at 25 °C and at a temperature of 65 °C it is 125%. At 10 ft-c and above, the response is almost uniform.

3.8.3 SPICE Models

There do not seem to be any. And how we model them depends on what they are used for. If it is simply as a switch, for example, to detect if a flame is present or not in a boiler, we can use a simple switch, ('Photocell Switch.asc'), where the voltage *V2* represents the flame and the switch *S1* the photocell.

If we want to use it to make measurements, we can check the data sheets of 'LUNA' controls (who use *lux* which is 1 lm/sq m); we find that the response is linear with light intensity assuming the spectrum is constant, so a simple resistor will model it very well, schematic ('Photocell Linear.asc'), which is set to have a resistance of 1 MΩ at 0.2 ft-c using the Perkins-Elmer data. The resistance tolerance of these devices is around ±40% at 1 ft-c but improving slightly at higher levels.

3.9 Other Variable Resistors

The website http://www.engandmath.com/LTspice_Variable_Resistance.php has some useful ideas. Because of the extreme versatility of the LTspice resistor, we can create some truly improbable devices.

3.9.1   Time Variable Resistors

In the time domain we can simply use the global variable time and any arbitrary functions such as:

$$R = 10 * (5 + sin\,(2 * pi * 10k * time))$$

to see ('Time Varying R.asc') using a transient analysis.

3.9.2   Frequency Variable Resistors

In the frequency domain, we need Laplace $= 1/abs(s)$ for the time, and this can be qualified by almost any function. In standard SPICE we need a voltage-controlled current source G, ('Frequency var. R V1.asc'), but with LTspice we can use a resistor. The format is:

$$R = \;<value>Laplace = \;<function>/abs(s)$$

but note that if we use pi, it must be enclosed in curly brackets:

$$R = 1\;Laplace = 2 * \{pi\}/abs(s)$$

3.10   Summary

In this chapter we have extended the simple resistor. In particular:

- The LTspice resistor accepts all standard SPICE parameters such as linear, quadratic and exponential temperature coefficients and also arbitrary coefficients, which are a very powerful addition.
- The platinum resistance thermometer defines the International Temperature Scale and has a cubic relationship for the resistance.
- Thermistors can be used to measure temperature, but they are highly non-linear, and the Steinhart-Hart equation or extensions to it are used for NTC type.
- NTC types can be used for in-rush protection because their resistance decreases with temperature.
- PTC thermistors can be used for over-current protection.
- Voltage-variable resistors (varistors) are used for over-voltage protection and breakdown symmetrically.
- ISO pulses 7637-2 and 16750-2 are used to test EMC compatibility of automotives.
- Photocells which rely on a change of resistance with illumination can be modelled easily as a switch or a simple resistor.

Chapter 4
Models and Sub-circuits

4.1 Introduction

Much of the information in this chapter can be found in the LTspice 'Help' file. The aim here is to set this out in the order in which first a symbol and then a sub-circuit can be created. There are other exotic possibilities which are not covered being mainly for the expert user. The difficulty with this chapter is that it is somewhat like explaining how to play chess: it is possible to explain the moves and analyse certain popular openings, but in the end, it all boils down to trying it out. That is true here: how to create models and sub-circuits is gone into in some details, but, in the end, there is nothing to replace time at the computer simply exploring.

SPICE has an impressive array of components. We can divide them into the following categories.

Simple Components
These are resistors, capacitors and inductors where the minimum necessary description is just one value although often this can be qualified by temperature coefficients and sometimes parasitic elements such as the inductance of the leads, or shunt capacitance or a series or parallel resistor.

Current and Voltage Sources
We shall have a lot to say about them in the next chapter. There are three types: simple voltage and current sources that are used to excite the circuit and which we have used already to supply DC; arbitrary sources that can create virtually any waveform and relationship we care to imagine; and controlled sources where their current or voltage is controlled by another current or voltage source.

Models
Models historically were developed for components that could not be described by a single value and which therefore had complex, non-linear, relationships between currents and voltages. In the first instance, they were created for diodes and bipolar transistors, those being the elements needed for the first integrated circuits. Soon

© Springer Nature Switzerland AG 2020   143
C. May, *Passive Circuit Analysis with LTspice*®,
https://doi.org/10.1007/978-3-030-38304-6_4

after models for field-effect transistors were added, but not thyristors, DIACs and TRIACs, because those were not included in integrated circuits. And we have seen that there are no models for thermistors and varistors.

The list of models has steadily expanded to encompass some new devices, and the original models have been improved over the years by adding extra, optional, parameters, 'optional' so that legacy models without them still work. And, in any case, not many manufactures have included the new parameters in their models. But even so, if there is no model for a device, we must create a circuit from parts which can be saved as a sub-circuit.

Internally a model consists of a set of perhaps 30 parameters and equations that are used to calculate the terminal voltages and currents. The standard SPICE equations have been published but there is no compulsion for any version of SPICE to use them so long as the external connections are not changed. We, as users, do not have direct access to those equations; we only access them indirectly when we supply values for the model parameters. We may not add other parameters; else there would be utter confusion were we to import models from another source or to export our own. So we may create an instance of model only for those devices defined in SPICE and only by giving new values to the existing parameters, although we may be able to omit some and then SPICE will provide defaults.

In this text, and generally, the term 'model' is used rather loosely to mean any collection of parameters, which we shall continue to do whilst bearing in mind the specific SPICE meaning of the word.

The SPICE Alphabet

The letters of the alphabet have defined meanings to designate the component or model to be assigned to a symbol on the schematic. We have already used R for a resistor and V for a voltage source. These are SPICE standards. They are all listed in the table below from which we see that only N and Y are not used.

Letter	Type of line
*	Comment
A	Special function device
B	Arbitrary behavioural source
C	Capacitor
D	Diode
E	Voltage-dependent voltage source
F	Current-dependent current source
G	Voltage-dependent current source
H	Current-dependent voltage source
I	Independent current source
J	JFET transistor
K	Mutual inductance
L	Inductor
M	MOSFET transistor
O	Lossy transmission line

(continued)

Letter	Type of line
Q	Bipolar transistor
R	Resistor
S	Voltage-controlled switch
T	Lossless transmission line
U	Uniform RC line
V	Independent voltage source
W	Current-controlled switch
X	Subcircuit invocation
Z	MESFET or IGBT transistor
.	A simulation directive, for example, *.OPTIONS reltol = 1e-4*
+	A continuation of the previous line. The '+' is removed, and the remainder of the line is considered part of the previous line

Sub-circuits

These are an important feature of SPICE; sub-circuits can be used when more than one SPICE component is needed to represent a physical component or, indeed, to save typing a long list of coefficients as we did for the thermistor and varistor. Sub-circuits consist of a collection of standard components, including models, used to represent a component for which there is no SPICE primitive like a capacitor or a resistor and no model like a transistor or a FET. We have seen that we can use sub-circuits for temperature and voltage-dependent resistors, and it is certain that we will need others in later chapters.

We have two tasks in hand: the first is to create a symbol to represent the sub-circuit on the schematic; the second is to create the functional description of the sub-circuit. But we should note that there is no necessary correlation between a symbol and a sub-circuit: a symbol is one thing and a sub-circuit another, and how they are related is something else.

4.2 Symbols

These represent a component, a model or a sub-circuit on the schematic. LTspice has those for the components for which it has primitives such as resistors and capacitors, and for standard SPICE models like diodes and transistors, and for its extensive libraries of ICs.

The symbol consists of two aspects: the first is the collection of lines and shapes that make up the drawing; LTspice calls this an *assembly* and stores them in *.asy* files. The second is the optional association of the symbol with a component, model or sub-circuit file through its *Attributes,* but this is not essential, and we can leave the symbol as just the drawing that we can place on the schematic but which has no part in the simulation. This may seem somewhat pointless if the symbol does nothing;

one example could be a symbol for a voltmeter to show graphically that the voltage between two nodes is measured.

This section follows the order of 'Creating New Symbols' in the LTspice 'Help' file.

4.2.1 Alternative Symbols

LTspice already has a few alternative symbols: for example, there are two zigzag resistors and also the rectangular 'European resistor' in the *Misc* folder.

Example – Alternative Symbols for a Current Source

This example is a little bit esoteric, but from the main menu **Edit→Component** move to **current** and note the description at the top right beside the icon. Now move to **load** and **load2**. The description is the same. The current source can not only be used to supply current but also as an active load to sink current (we shall look at this in a later chapter). The point here is that there are alternative symbols for the current source which we can choose to make it clear how the source is being used, but this is not enforced, and we can still use the **current** symbol for everything.

In summary, it is important to remember that the symbol is only that – a picture, an image. It has no functionality and no essential binding to a component, model or sub-circuit.

Example – Opening a Symbol

Use any of the methods we have described to open the **Select Component Symbol** dialogue and place an **npn** transistor on the schematic. We might note that there are two more symbols **npn2** and **npn3** which can be used instead – cosmetically different, just like the current source, that is all. We can show this graphically as Fig. 4.1 that the one component may be represented by more than one symbol. We should note, however, that **npn4** has an extra terminal and is not an alternative.

Then with the mouse on the symbol, hold down **Ctrl** and right click to open the **Component Attribute Editor**. For the moment, ignore everything else and just click on the button **Open Symbol** at the top, to the left of the path to the symbol *.asy* file, and you should see the npn symbol. If it is not visible, look at the status bar at bottom left and see if the coordinates are negative: if so scroll the symbol into view.

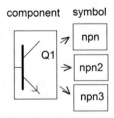

Fig. 4.1 Component to Symbol Relationship

Fig. 4.2   NPN transistor symbol

We can see that the symbol is composed of straight lines. The end of each is shown by a small red circle, Fig. 4.2. Note that there are two more small circles locating the position of the text **Qnnnn** and **NPN**. The terminals have a small box around them.

4.2.2   Creating the Drawing

Now we know what a symbol looks like, we can discuss how to create one. There are four possibilities: the first is to use an existing symbol unchanged, the second is to modify an existing one, the third is to create one from scratch, and the fourth is to download one from a website.

4.2.2.1   Use an Existing Symbol

The example above showed how we could access an existing symbol. This works for the basic components and for some of the **Misc** but not all. So we could press a resistor into service for a platinum resistance thermometer. But we should exercise a little care as to how far we go with this; else we may cause confusion if the behaviour is greatly different from that expected from the symbol.

In other folders we will find that a symbol is irrevocably linked to a specific component. For example, if, from **Edit→Component**, we select the **[Opamps]** folder and place any of the **ADxxx** or **LTxxx** or **RHxx** or any other named component on the schematic, then **Ctrl** and right click do not offer the option of **Open Symbol** because each is tied to a specific opamp and it is not possible to edit it. We shall later see how to stop a symbol being edited; for the present, we just note that not all symbols can be reassigned. This is the opposite of Figs. 4.1 and 4.3. So if we want to add other opamps, we need another symbol. LTspice offers us three: **opamp, opamp2** and **UniversalOpamp2**. These are not committed to specific opamps, and now we have access to the assembly which we can assign to whatever we like. These are also the symbols used by the various LTspice opamps.

Example – Editable and Non-editable Symbols

Opamps are not of direct interest to us in the context of this book. Nevertheless, it is interesting to note the different versions of the LTspice symbol. If we navigate to the **Opamps** library and place any of them on the schematic, **Ctrl+** right click opens a different dialogue where we can go to the test jig or get the data sheet, but we cannot

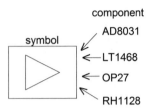

Fig. 4.3 Symbol to Component

open the assembly. However, if we scroll right to the end of the opamps folder, we find an *opamp*. This has no power supply pins and is provided for us to use with very simple opamp sub-circuits.

On the other hand, *opamp2* is the symbol used by most, if not all, of the opamps in this folder to link directly to the opamp sub-circuit. And *UniversalOpamp2* is derived from it where the intention is not to link to a sub-circuit, but to use it with a phenomenological description of the opamp without going to the trouble of creating a full sub-circuit.

The point to take from all this is the flexibility of symbols that they may, or may not, be linked to a sub-circuit.

4.2.2.2   Modify an Existing Symbol

We must first create a new schematic. There are two options: the first is easiest which is to place a symbol on the schematic as above. As we have just seen, this does not work for named opamps and also *opto* and some other components where the symbol is tied to a sub-circuit. The second option is to go to **File->Open and** then navigate to **LTspiceXVII->lib->sym** and then in the bottom right choose **Symbols (asy)**. If the one you need is not listed, try another folder. This works for every symbol.

So first open the appropriate symbol, place an instance on the schematic and then **Ctrl+** right click will open the **Component Attribute Editor** and click the **Open Symbol** button at the top left to open the symbol design window.

Adding Shapes to a Symbol
See also 'Drawing the body' in the LTspice 'Help' for a colour picture. Right click to open the pop-up menu, Fig. 4.4, click **Draw** and select what you want to draw. The '+' cursor will change to a small bold grey circle with a small, thin cross. Click anywhere on the workspace to start because shapes do not have to start or end on a grid point. There is no need to hold down the mouse key. A thin grey line or the outline of the shape will follow the cursor. The starting positions may be offset, but that is not a problem because we can move it later. Note also that **Circle** encompasses ellipse, so it is up to us to make sure that it is a true circle. The corners of the bounding rectangle are shown by the two bold small grey circles containing a '+'. An **Arc** is slightly different: we first draw a circle, then click on the circumference at the

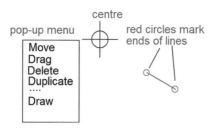

Fig. 4.4 Draw symbol

desired starting point and drag the mouse to define the end point; the mouse does not have to move on the circumference.

Cancelling a Shape in Progress
Right click or press **Esc** and the shape will be cancelled and the cursor revert to a large '+'.

Finishing a Shape on a Symbol
Left click and the small bold grey starting circle will turn to red. A rectangle will be filled in in yellow; if we don't like that, we can draw a rectangle from lines, and it will not be filled. The control points of the shape will turn to red. The shape selection is still active, so, in the case of a line, it is possible to chain one line after another. The shape will be drawn in the currently selected line style.

Deleting a Shape on a Symbol
The scissors **F5** work as usual and will cut out anything, but we must cut on the shape itself, not on the anchor points. We can also drag a rectangle with the scissors, and anything inside it will be deleted. **Undo** will restore it.

Moving a Shape on a Symbol
To move the shape as a whole, click **F7** or **F8** and then click anywhere on the body of the shape rather than a join; else it is uncertain which shape will move. But once it is free, we can click on the ends as well.

Moving One End of a Shape on a Symbol
Click **F8**, move close to the end red circle and then hold down the left mouse button to draw a small rectangle encompassing it. Release the mouse button, and the end will follow the mouse. Left click to finish.

Changing the Line Style of a Symbol
This can be set for future shapes by right clicking in the pane and then **Draw->Line Style**. For an existing shape, right click on it. The line thickness only applies to solid lines including wires on the schematic and is set globally by **Tools->Control Panel->Drafting Options** and then edit **Pen Thickness**.

Changing the Fill-in Colour of a Shape
From the main menu **Tools->Color Preferences** then select the **Schematic** tab, and from **Selected Item**, choose **Component Fill-in**, but this is global and will apply to everything, including the schematic.

Duplicating a Shape

Click on **Duplicate** or *F6*, and the cursor will change to two overlapping pages. Click on the shape to duplicate, and its control circles will turn to grey. Move the mouse, and the outline will follow; left click to place it. The duplicate method is still active, so click again on the shape to make another copy.

Example – IEC Trimmer Resistor Symbol

The European resistor already exists, but it would be useful to have a two-terminal trimmer resistor where we can alter its total value but without a tapping point like a potentiometer. Of course, the standard resistor can be varied; this is just an alternative to make it clear that a user should expect to alter the value. It will have the same functionality as the standard resistor. In other words, the value would have to be changed in the same way as a standard resistor. In real life such resistors are usually adjusted with a screwdriver when the circuit is set up and is not accessible from the outside. The conventional symbol is a resistor with a 'hammer' through it, top left in Fig. 4.5. And we can also have a variable resistor with an arrow through it to indicate that this is a front-panel control available to the user, top right in the same figure.

We start either by putting a resistor on a schematic and then **Ctrl** and right click to open the assembly or from the main menu select **File→Open→lib→sym→Misc** and then **European Resistor**.

There is an immediate problem that the hammer will conflict with **Rnnn,** so click *F7* and drag the text to the left. Draw an angled line through the rectangle and a short line at right angles. We can leave the existing pins and do not change the **Rnnn** because this is just another resistor. And we have finished.

Anticipating the Sect. 4.2.5, it is better to create a new folder for it such as **IEC Symbols**. Then save it as **IECtrimmerR** and put a new instance on a schematic – the text **R1** is too far to the left. Go back to the assembly and right click on **Rnnn** and select **Justify Right**. to fix it.

Example – Variable Capacitor

Place a fixed capacitor on the schematic, then **Ctrl** and right click to open the assembly, and then add a diagonal line and the arrowhead. Save it in an appropriate folder.

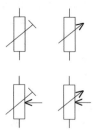

Fig. 4.5 Variable resistors

Explorations 1

1. As with many programmes, much can be learned just by practising. So select a few symbols in turn and try adding extras lines and shapes, but when you close the window, be sure not to save it or save it in a new folder.
2. Use one or more of the resistor assemblies to create symbols for trimmer and variable two-terminal resistors – not potentiometers. These will simply be aliases for the standard resistor

4.2.2.3 Downloading from an Internet Site

There are a few sites that offer ready-made symbols that can be downloaded. A tutorial on creating them and some new ones can be found at http://www.zen22142. zen.co.uk/ltspice/newsymbols.htm. It is a good idea to create a new folder in **LTspiceXV11→Mysym** as we have done for the IEC resistors to avoid contention with existing symbol names.

4.2.2.4 Automatically Creating a Symbol

If we have a sub-circuit, Gabino Alonso has pointed out, https://www.analog.com/en/technical-articles/ltspice-simple-steps-to-import-third-party-models.html, that it is easy to get LTspice to create a symbol. However, this is just a plain rectangle and is a *block* not a *cell*.

Example – Auto-Creation of a Symbol

Click **File→open** and then in the bottom right select *All Files*. We should find a **.TXT** file in the *Thermostat* group of files. Open it and we will see a small sub-circuit (which we shall create again later) and right click on *.subckt Tstat c0 c1 Vtemp = 50* to open a pop-up menu:

```
* Thermostat SubCct List
.subckt Tstat c0 c1 Vtemp=50 ← right click on this line
S1 c0 c1 v+ v Ststat
B1 v+ v- V=temp
.model Ststat SW(Ron=1n Roff=1T Vt=Vtemp Vh =1)
.ends Tstat
```

Towards the bottom of the menu, we find the symbol of a two-input AND gate and the option **Create Symbol**. We will be asked if we wish to create a symbol. If one already exists, we will be invited to overwrite it. The symbol is just a simple rectangle placed in the **sym-AutoGenerated** file. This may - or may not - be the best. In this case, the switch assembly is a more accurate representation.

4.2.2.5   Drawing a New Symbol

From the main menu, select **File→New Symbol**. This will open the symbol window that we have just been using. There are no size markers on the workspace, but the grid is 16 × 16 pels. It can be helpful to open an existing symbol as we described above to get a feel for the size.

There is no restriction on what lines or shapes we use, nor on where we place them. Figure 4.4 shows the essentials we need just now. The centre is marked by a cross containing a small circle. This is not important, it is just the (0,0) point, and it is not the centre of rotation if we flip or rotate a symbol on the schematic – that is set by the position of the closed hand on the symbol when we select it.

A point to note is that if some part of the shape is to be a terminal pin, it must be on the grid; otherwise there is no restriction. We draw lines and shapes to create the symbol as described above, and all the editing methods are applied as before.

4.2.3   Adding Pins

Now we have a shape; we must add pins for external connections. This is very important since it defines how the symbol is connected in the netlist. If the symbol is completely novel and not a variation on an existing one, there are no restrictions on the number and order of the pins. If the symbol is a variation on an existing primitive such as a resistor or capacitor, it must have exactly the same number of pins: resistors and inductors are bidirectional, so there is no difference between the two ends. The same is true of capacitors since, although there is a symbol for a polar capacitor, its capacitance as defined in SPICE is the same either way round.

If the new symbol is a variation on a predefined model such as a diode or a transistor, it must follow the same pin sequence, but it is possible to add extra pin(s). Likewise if the symbol is a variation on a sub-circuit, it must follow the same pin sequence, but we can add more.

We can open the LTspice 'Help' file entry 'Adding the Pins' but ignore the 'low-level schematic' bit. The pins must be on the grid, but there is no requirement that a line or shape should connect to them to the rest of the assembly.

Example – Viewing the Pinout

If we have any of the npn transistor assemblies except **npn4** open, right click anywhere, and on the pop-up menu, select **View→Pin Table**, and we find in order *C,B,E*. If we create a new symbol, perhaps this one but with an enclosing circle, we absolutely must follow this pin order; else chaos follows.

If we swap the symbol for **npn4**, we find the same order for the first three pins and an extra pin number *4* which is labelled *S*. The meaning of this need not detain us: the point is that we must follow the prescribed order for the symbol which we are modifying and add any new pins at the end.

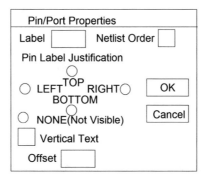

Fig. 4.6 Pin edit dialogue

4.2.3.1 Placing a Pin

Right click in the window, and in the pop-up menu, select **Add Pin** which will open
the **Pin/Port Properties** dialogue, Fig. 4.6. The **Netlist Order** field is an ascending
series of numbers and will have been filled in by LTspice with the next number. The
Label will be blank and can be left so. Click **OK**, and a small bold circle will appear
enclosed in a square. This will hop from one grid point to another as the mouse is
moved. To place the pin, left click; to cancel it, **Esc** or right click before it is placed;
else once it has been placed, use the scissors.

The label and pin order can always be edited after the pin is placed either by right
clicking on the pin to reopen the dialogue or by right clicking anywhere on the
window to open the pop-up menu, and then **View->Pin Table** opens a small
dialogue listing all the pin names and their order.

4.2.3.2 Pin Labels

These are not essential and are there to aid correct connection because they can be
used in a netlist or sub-circuit instead of the pin numbers, in which case, if the
symbol is an extension of an existing one, it must follow the same pin names just like
the **npn4**. But if the new symbol is going to be associated with a model or sub-circuit
that is functionally different, for example, a potentiometer instead of a two-terminal
resistor, we change the names and pin order, if we so desire. It is the pin order that is
important; the names are mainly there to remind us of the function of the pin. They
can be left empty or filled with a helpful letter or name as in Fig. 4.7 where *C,B,E*
identify collector base and emitter. If we have already placed the pin, then move the
mouse over the pin small square and right click. This will open the above dialogue,
and the name of the pin will be shown in the **Label** edit box.

The text can also be written vertically or hidden. For most symbols the names are
hidden since it is fairly obvious what they are – collector base, emitter of a transistor.
But a few special symbols like the capmeter do have them visible.

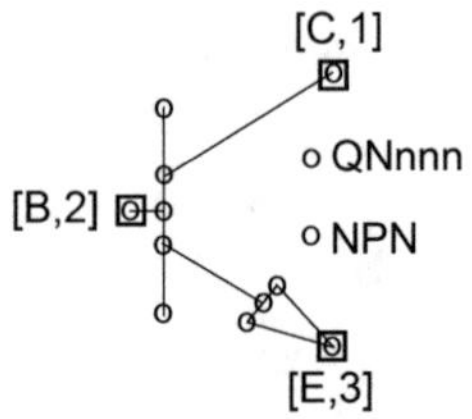

Fig. 4.7 NPN Pinout

Below that is a square of radio buttons defining where to place the name relative to the pin and the offset from the pin in pels.

Example – Placing Pin Names

Navigate in LTspice to **lib→sym→SpecialFunctions** and open **capmeter**. We can see the names of the pins inside the rectangle, and they can be moved by the four radio buttons or hidden. These, however, seem to work the other way round than we should expect that justification **Bottom** places the text above the pin. Pin names are unrestricted although there are conventions such as *e,b,c* for a bipolar transistor, but these are not enforced.

Showing Pin Labels

It is a good idea to have an instance of the symbol placed on a schematic because the offset on the assembly is quite different and we can trim it interactively. We may also find that the default text size is too large. This cannot be changed, so the best thing is to hide the label and right click on the symbol window and select **text** from the pop-up menu. We can then select font size.

4.2.3.3 Pin Order

This defines the order in which the pins are connected to the external circuit. LTspice will assign pin numbers in the order in which the pins are created, but this can be edited. When we create a sub-circuit, after the name comes the pins in the order they are connected. This is very important for components like diodes, transistors and opamps. Or take an optoisolator as an example.

Example – Pin Order

Chose from the main menu **File->Open** then navigate to **LTSpice->lib->sym->opto** and open 4N25 which is an optoisolator, although that need not concern us now. We can see the pins one by one as before, Fig. 4.8, or, in the window away from the symbol, right click to open the pop-up menu and then select **View->Pin Table**, and we will see an editable list of the pins and their names in order. These are:

A   1
K   2
C   3
E   4
B   5

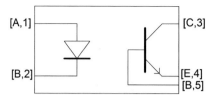

Fig. 4.8 4N25 Pinout

where the helpful letters are used: **a**node, **k**athode, **c**ollector, **e**mitter and **b**ase. This table is a summary of what we found by examining each pin in turn.

To understand the importance of the pin order, we look at the sub-circuit definition found from the main menu by **File->Open->lib->sub** where we have cut out all the interesting parts to leave only the start and close of the definition we have:

```
* Copyright © Linear Technology Corp. 1998, 1999, 2000. All rights
reserved.*
.subckt 4N25 1 2 3 4 5
........
.ends 4N25
```

The first line is a comment, and the second line defines it as a sub-circuit **subckt** with its name and the pins in order 1,2,3,4,5. We could equally well use the pin names A,K,C,E,B. But note that the sub-circuit expects the pins in their order of declaration. If we swap the order around to something like *B,E,A,C,K*, we can be certain of chaos because the base pin will be connected to the node that should be the anode, the emitter will be connected to the node that should be the cathode, and so on.

Equally, when we place a 4N25 on a schematic, the user will expect the anode to be the top left pin and the cathode (or 'kathode') the bottom left pin. If we had changed this to the above order either in the **Pin List** of the symbol or in the sub-circuit, the top left pin will now be the base and the bottom left the emitter: once again, confusion.

On the other hand, if we have not made any changes when we place an optoisolator on the schematic, there is no problem because the pins and their order are already assigned, so when we make connections to it, they will be correct. It is only when we want to use it with a sub-circuit that we need to be careful; if we create a netlist invoking this symbol where, by mistake, we write **.subckt 4N25 A K E B C**, we will have problems.

4.2.4   Symbol Attributes

These define how the symbol is associated with a component, model or sub-circuit. This is not essential, and the symbol can be saved without filling in any leaving that

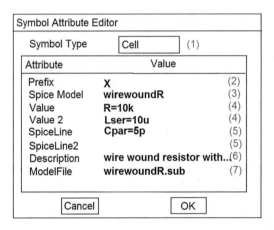

Fig. 4.9 Symbol Attribute Editor

to be filled in later if necessary. Much of this only makes complete sense after we have explored sub-circuits: see the LTspice 'Help' file 'Adding Attributes' and 'Attribute Visibility'.

Some of the combinations are not immediately obvious. Right click and select **Attributes->Edit Attributes** from the pop-up menu. This will open the **Symbol Attribute Editor** dialogue, Fig. 4.9 which is almost identical to the **Component Attribute Editor**. We can also reach it from the schematic by **Ctrl** and right click on the component and then the **Open Symbol** button at the top of the **Component Attribute Edit**. The bracketed numbers in the following headings refer to this figure. We shall illustrate these attributes by examining a few examples.

4.2.4.1   Symbol Type (1)

By default this is **Cell**. Leave it at that. The other option is **Block**.

4.2.4.2   Prefix (2)

The first letter defines the symbol type. If it is to be used with one of the components listed in the introduction, that should be used, **C** for a capacitor and **D** for a diode. The trailing **nnnn** is ignored. The instances will be numbered in turn as they are placed on the schematic, for example, **R1, R2, R3**. . . . Each must be unique.

If the symbol will be used with a sub-circuit, it must be **X**, and the prefix will be shown as **Unnn**. This is slightly confusing since the table at the head of this chapter says that **U** is for a uniform RC line.

4.2.4.3   Spice Model (3)

This works in conjunction with the **Model File** which can contain more than one model. For example, we could have a model file *MyPRT.lib* for platinum resistance thermometers. This file could have models for different types:

```
.model PRT100 1 2
definition>
.ends

.model PRT200 1 2
definition>
.ends
......
```

From this it follows that we are here picking a specific model within the *MyPRT.lib* file, and because the models do not have extensions, only the containing file has, there is none here. For example, we could enter a default model as:

Spice Model PRT200

and from this the library file will pick the sub-circuit *PRT200* provided (and this is important) we have entered **ModelFile = MyPRT.lib**; else LTspice has no idea where to find the file containing the sub-circuit. This Spice Model will also be passed to the **SpiceModel** field of an instance placed on a schematic where it can be changed for another model within *MtPRT.lib*. – provided we remember its name.

There is an important exception, and that is we can give the full name including the extension to make a component that cannot be edited.

What is important is the structure of the Model File: it must be pure text with no enhancements. The **Spice Model** allows the symbol to be used by different sub-circuits provided the pin-out is the same as we saw for the plethora of opamps.

Taking the neon bulb assembly found in the [Misc] folder, if we have **Spice Model neon bulb** without an extension but no **Model File** and no **Value 2**, we find in the netlist of the schematic that **neon bulb** has been added to the description, but there is no call to the library containing it:

```
XU1 NC_01 NC_02 neonbulb Vhold=50
.op
{Missing library call → .lib neonbulb.sub}
.backanno
.end
```

We can still use the symbol, but we will need to give the full library path on the schematic as:

.lib.neonbulb.sub

which is effectively supplying the information missing from the neon bulb assembly. This adds flexibility in that we can associate the neon bulb with a different library file provided it contains a sub-circuit named *neon bulb*. Generally, to avoid utter confusion, it is a good idea to associate an assembly with a library file. (But remember LTspice does not enforce the .lib extension.)

4.2.4.4 Value and Value2 (4)

These two values and the following **SpiceLine** and **SpiceLine2** work together in many cases. If we are not adding a model name, we can distribute the parameters over the two lines as we please. When we place an instance on a schematic, the **Value** field is visible by default. Simple components and models by default use the letter prefix of the table at the start of this chapter, so we have **R** for a resistor which forces us to insert a value every time we invoke one. However, we can change this to a more useful default, such as **R = 10 k**. Sub-circuits may show their name, so the user can see what type it is.

   Value is also used to name a model which must be declared on the schematic. The voltage-controlled switch is an example. For example, **Value MySwitch** would expect **.model MySwitch Sw(Ron = ...)** on the schematic.

Example – Value and Value2
If we open the neon bulb assembly, we may move the values **Vstrike = 100 Vhold = 50** between the two value fields. If we check the netlist, we find in both cases that the information has been appended on the same line:

$$XU2\ N002\ 0\ neonbulb\ Vstrike = 100\ Vhold = 50\ Zon = 2K\ Ihold$$
$$= 200u\ Tau = 100u$$

Hiding 'Value' of a Symbol
If the **Value** field is visible, it must contain something, and attempting to delete the text will fail. However, it can be removed with the scissors. If it is not visible, its contents can be deleted. Visibility can be restored by right click in the assembly window and then **Attributes→Attribute Window** and click on **Value**, but the text will be smaller than before. Right click on the text to edit it.

'Value' Order of Precedence
A value entered on the symbol on the schematic takes precedence over one in the symbol file, and this takes precedence over any default value on the sub-circuit itself. If the value on the symbol is blank, the default value on the sub-circuit will be used. However, it is a good idea to have one on the symbol to remind the user of its name as in $R = 10\ k$.

'Value2' of a Symbol

Optionally it gives additional information but does not appear on the schematic. Both values will appear in the netlist, but note the following exception.

Exception 'Component Cannot Be Edited'

Taking the neon bulb as an example, if we add a **Spice Model** this time with the extension of **lib, sub** or whatever and also add the same model name to **Value 2** but without the extension, the component cannot be edited, and we can no longer change any value if we place it on a schematic:

```
Spice Model neonbulb.sub {model name with extension}
Value Vstrike=100 Vhold = 50 {or whatever else we like}
Value2 neonbulb {model name without extension}
Model File {leave empty}
```

If in the schematic we try **Ctrl** and right click on the symbol, we find a small window telling us that this is a *parametrized neon bulb* and *this component cannot be edited*. The same holds if we leave these two fields empty as in the 'fixedind'. Of course, we can always open the assembly file and make changes there.

4.2.4.5   SpiceLine and SpiceLine2 (5)

These allow more default parameters to be specified. It does not matter greatly how we distribute them over the previous two *value*s and these two lines. It makes it slightly easier to change the values of an instance on the schematic if we keep to one value per field, but they will run into just one line in the actual netlist.

Example – Non-editable Component

Navigate to **File→Open→lib→sym→opamps** and open the **AD549,** and we see **ADI1.lib**. Open it and then right click and select **Atrributes→Edit Attributes,** and we find:

```
Prefix X
Spice Model ADI1.lib
Value AD549
Value2 AD549
SpiceLine
SpiceLine2
Description Precision Low-Cost HS BiFET Dual Op Amp
Model File
```

This tells us that the sub-circuit for an AD549 is to be found in the ADI1 library. Note the extension is included because we also have **Value2,** so it cannot be edited, and its name in **Value** means this will be visible on the schematic.

Example – Using Value and SpiceLine Fields

Navigate to **File→Open→lib→sym→Misc** and open the 'neonbulb.sub':

```
Prefix X
Spice Model neonbulb
Value Vstrike=100
Value2 Vhold=50
SpiceLine Zon=2K Ihold=200u
SpiceLine2 Tau=100u
Description Parameterized Neon Bulb
Model File neonbulb.sub
```

This uses a sub-circuit that has five parameters. On the schematic they will all be shown as one line extending from the sub-circuit as:

$$XU1\ NC + 01\ 0\ neonbulb\ \ Vstrike = 100\ Vhold = 50\ Zon = 2K\ \ Tau$$
$$= 100u\ Ihold = 200u$$

and on the symbol, we can decide what information to show on the schematic by using **Attributes→Attribute Window** when we edit the symbol.

Example – Using Value and SpiceLine Fields (2)
Even more dramatically, if we open the 'UniversalOpamp2.asy' we find the following:

```
Prefix X
Spice Model level 2
Value
Value2 Avol=1Meg GBW=10Meg Slew=10Meg
SpiceLine ilimit=25m rail=0 Vos=0 phimargin=45
SpiceLine2 en=0 enk=0 in=0 ink=0 Rin=500Meg `
Description Universal Opamp model that allows 4 different levels of
simulation accuracy. See ./examples/Educational/UniversalOpamp2.asc
for details. En and in are equivalent voltage and current noises. Enk and
ink are the respective corner frequencies. Phimargin is used to set the
2nd pole or delay to the approximate phase margin for level.3a and
level.3b. This version uses the new, experimental level 2 switch as the
output devices.
Model File UniversalOpamps2.sub
```

The **Value** is empty, so nothing is shown on the schematic because, unlike the AD549, this is not a unique opamp. We can give it a value (in effect, a name) when we place it on the schematic. The sub-circuit name is found in the **Model File** field where we can deduce that there is more than one version because the **Spice Model** field gives it as **level 2**. We need not worry about the meaning of the parameters, just to note that they are spread out over **Value2, SpiceLine** and **SpiceLine2** fields, and if we place an instance on a schematic, we will see them formatted in exactly this way.

If we navigate to **File→Open→lib→sub→Opamps**, the first lines are:

```
.subckt level.2 1 2 3 4 5
S1 5 3 N002 5 Q
S2 4 5 5 N002 Q
A1 2 1 0 0 0 0 N002 0 OTA G={Avol/Rout} ref={Vos} Iout={slew*Cout} Cout=
{Cout} en={en} enk={enk} in={in} ink={ink} incm={incm} incmk={incmk}
Vhigh=1e308 Vlow=-1e308
```

And we see that the values are simply converted to a long string, not assigned to separate lines.

All of these fields can be made visible on the schematic when we have the symbol assembly window open by **Attributes->Attribute Window** and click whatever value we want and then it will be highlighted. Now click **OK,** and it will be dragged onto the symbol. Once it is placed, right click to open the **Symbol Attribute** dialogue where we can change the font size and justification and have the text running vertically if we so desire. The scissors will cut out anything we do not want.

4.2.4.6   Description (6)

This is brief (or not so brief) information that will appear on the right at the top beside the icon after we select a symbol from the main menu by **Edit→Component**. For opamps this can be a useful guide to their salient properties and applications. It does not appear in the **Component Attribute Editor** when we hold down **Ctrl** and right click the component in the schematic.

4.2.4.7   ModelFile (7)

This field is not visible on the schematic. What it does is to add a line to the netlist giving the file that contains the model, but does not specify a Spice Model to use within that file – only where to look for it, so it must include the extension. This also ties the symbol to this particular file. But we can use the same assembly with a different ModelFile to create a variety of sub-circuits all using the same assembly. This is what is done with the many opamps.

If this field is left empty, it is up to the user to find the correct file when an instance is placed on the schematic, which raises the rather improbable possibility that there is more than one file containing this SpiceModel. We can preserve the greatest flexibility by leaving the **SpiceModel** and **ModelFile** fields empty.

Example – ModelFile and SpiceModel

Place a neon bulb on a schematic and connect one end to earth, right click on it, and open the assembly file. Delete the SpiceModel and save the model. Now run the schematic, and we get the message **Unknown sub-circuit called in**. This is what we see by **View→SPICE** Netlist:

```
XU1 NC_01 NC_02 Vhold=50 {no Spice Model added}
.op
.lib neonbulb.sub {Added by ModelFile including extension}
.backanno
.end
```

Where SPICE has added the file to search but there is no specific model given within the file.

Model File Extensions

The extension is usually **.lib** or **.sub** or **.inc,** but there is no restriction, and sometimes **.txt** will be used. The difference is that a **.sub** file may be a binary file that cannot be edited and is not plain text. These files may contain the sub-circuits of several components. What matters is that the file is in the correct place and has the correct format.

The **Model File** is a file that will always be used when an instance of this symbol is placed on the schematic. It is not the same as the Spice Model which can be changed for different instances. In effect it locks this symbol to that file, so only models contained in that file are accessible. As we saw above with the neon bulb, it is possible to leave this field empty and provide the library name on the schematic.

4.2.5   Saving the Symbol

If it is derived from an existing assembly, we must save it with a different name. We can even create a new folder such as **Mysymb**. If this is placed as a folder within the **sym folder,** it will be visible from **Edit→Component**.

It is also a good idea to test the symbol which we can do by saving it and then invoking an instance on a schematic. However, if we make other than changes to the appearance, we must delete the symbol on the schematic and place a new one. Much can be learned this way by viewing the netlist and noting how the symbol parameters alter its contents.

Example – IEC Variable Resistor

Returning to the IEC resistors, we can make another with an arrowhead in place of the hammer and save it as **IECvariableR**. Of course, to vary the value, we have either to turn the value into a parameter or edit it on the schematic as we have done before. We can also add a default value instead of plain **R** something like **1Meg** would be suitable, but do not forget that '1 M' is 'one milli'.

Explorations 2

1. Create trimmer and variable capacitors, a fuse, an IEC current source (two linked circles) and a loud-speaker – and anything else you like.

4.3   Sub-circuits

Now we are ready to handle the other half of creating a new component. All text can be in upper- or lowercase or a mixture. They are useful in reducing the visual complexity of a circuit by turning commonly used circuits into sub-circuits. This is essentially the same philosophy that we find in programming languages where we create methods, functions or procedures. Here, we assemble the circuit, test it and decide which values are fixed and which should be parameters that can be changed. By building the sub-circuit on the schematic, the netlist for the body of the sub-circuit is automatically created for us.

4.3.1   Sub-circuit Structure

Apart from the first line, the structure is quite flexible although there are accepted conventions.

4.3.1.1   The First Line

The first line must either be blank or a comment; it will not be read as part of the sub-circuit itself. It can be continued with other comment lines identified by an initial asterisk '*'.

4.3.1.2   Sub-circuit Identification

The first non-comment line must identify the sub-circuit. It does so by the reserved word **.subckt**. It must be exactly this with a leading 'dot' and then upper- or lowercase followed by **name** which can be anything within reason but avoid predeclared names. The syntax is:

> .subct < name >   < pin 1 >   < pin 2 >   < pin 3 > ..... PARAMS
>   :   < optional component values >

Pin 1, pin 2... are the external connections to the sub-circuit and are usually 1,2,3 and 4, but any numbers can be used. It does not matter if these nodes are already used in the main schematic; SPICE will simply reassign them when the sub-circuit is expanded. Thus **.subckt MySub 15 99** has internal nodes *15* and *99*, but these could become *3* and *5* when added to a schematic.

PARAMS: LTspice seems to be fairly tolerant of the existence or not of this prefix to a list of values.

Optional component values these apply to components of the sub-circuit, so if a sub-circuit contains a resistor **Rx**, then **.subckt MySub 1 2 3 Rx = 100** would set it to 100 Ω.

4.3.1.3   Sub-circuit Body

The sub-circuit identification is followed by the body of the sub-circuit. This can consist of simple components in any order where each line is of the same form as the netlist:

<component name> <node1> <node 2> <node 3> .... <value>

where the elements are:

component name the reference of primitive component such as **R2**
nodes the nodes within the sub-circuit to which it is connected
value which may contain temperature coefficient and other qualifiers or the name of
   a parameter in curly brackets whose value can be set externally

```
. C2 5 0 10u
Lseries 0 3 {Lx}
D1 2 3 MyDiode
Rout 4 2 100 tc1=0.0002
```

This will set the value of **C2** at **10u**, and that cannot be changed. **Lseries** is declared as a parameter, and a value must be supplied. **Rout** also is fixed at *100 Ω* and so is its linear temperature coefficient of *0.0002*.

4.3.1.4   Models and Other Sub-circuits

The sub-circuit may also contain models and calls to other sub-circuits using the same convention as the netlist. By convention these are placed at the end of the listing.

.model <model name> <model type> <model parameters>

.model that of an existing SPICE model, usually a semiconductor device
model name as found in the model database and which has been invoked previously
   in the list such as 'MyDiode' above although it is not an error to add a model that
   is not actually used – it just bloats the sub-circuit.
model type used to specify what sort of model SPICE will use
model parameters enclosed in brackets to change them from the default model
   values
e.g. .model MyDiode D (Is = 1.2e-12)

4.3.1.5   Ends

The sub-circuit is concluded with **.ends <name>** although LTspice does not insist on the name.

4.3.1.6   Exclusions

A sub-circuit may not contain SPICE directives (dot commands), so it cannot specify a type of analysis such as transient, nor measure statements.

Example – Opto-Isolator Sub-circuit
The sub-circuit of the 4N25 optoisolator is:

```
* Copyright © Linear Technology Corp. 1998, 1999, 2000. All rights
reserved.
*
.subckt 4N25 1 2 3 4 5
R1 N003 2 2
D1 1 N003 LD
G1 3 5 N003 2 .876m
C1 1 2 18p
Q1 3 5 4 [4] NP
.model LD D(Is=1e-20 Cjo=18p)
.model NP NPN(Bf=610 Vaf=140 Ikf=15m Rc=1 Cjc=19p Cje=7p Cjs=7p C2=1e-
15)
.ends 4N25
```

This shows that it has a 2 Ω resistor between nodes N003 and 2, a diode **D1** of type **LD** which is defined by the model call where **LD** is declared to be a standard diode **D** with two parameters changed from the default. It also includes a voltage-dependent current source **G1** whose value is 0.876 m, a small capacitor of 18 pF and a transistor **Q1** which has a model **NP** defined as an **NPN** transistor followed by a list of parameters. We should also note that all the components have at least one node connected to an external pin, although that is not often the case.

4.3.2   Downloading Sub-circuits

Most device manufacturers have libraries of components, and some have symbol assemblies as well. If they are zip files, they need to be unzipped. Check the contents of the folders because sometimes they contain more than one type of file.

4.3.2.1 Symbol Files

It makes sense to have these easily visible so that we can add them to a schematic. So collect all the symbol files. These will have an AND gate icon and the type **LTspice symbol** although the extension may not be visible. Create a new folder in the **sym** folder and move the files to it. This new folder will now be visible from **Edit→Component**.

4.3.2.2 Library Files

If we place our new sub-circuits in the LTspice **sub** folder, we can access them with no more ado. But that folder already has hundreds of opamps and special function devices, so it is cleaner under **LTspiceXVII→lib** to create a new folder such as **MySub** and put them there. The same applies to downloaded libraries. Then we must supply a path to the folder from the main menu. We chose **Tools→Control Panel** and then the tab **Symb.&Lib. Search Paths** and right click in **Library Search Path,** select **Browse,** and navigate to the **MySub** folder.

4.4 Example Sub-circuits

It we are building systems is very useful to create sub-circuits of function blocks such as filters, loud speakers and amplifiers to reduce complexity.

There are two routes to building a sub-circuit. The first, the only practical one for extended circuits, is to create a schematic. We can also include signal sources, load resistors or anything else to enable us to test and edit the circuit. Then **View→SPICE Netlist** to show it, but as this cannot be edited, we right click and select **Edit as Independent Netlist** to save it. Now we can easily remove everything that is not part of the sub-circuit, add an initial comment line followed by the opening **.subckt** defining line and a closing **.ends** and save it as a text file or with the extension *.sub* which helps to avoid confusion with other text files. We need to be careful about where it is saved: it is a good idea to create a sub-folder **Mysub** or whatever in the **sub** folder and save it there to simplify access. And we will need to add the path in the Tool Box.

There is an alternative, best suited for small sub-circuits, of building it directly on the schematic by pressing S to open the **Edit Text on the Schematic** dialogue and then entering the sub-circuit line by line. It is then invoked by setting the **Value** field of an instance to the name of the sub-circuit, which we shall do with the wire-wound resistor.

4.4.1   A Wire-Wound Resistor

Most of these are inductive because they are effectively a coil of resistance wire. They also have capacitance. The standard SPICE resistor symbol, or the IEC one, will serve to show it on the schematic, but we must change it to a sub-circuit because we have added other components. We could just have used an inductor and added its parasitic capacitance and resistance, but we cannot give them temperature coefficients. We could, if we like, add something to the symbol, such as the text 'WW', to show that it is not just any old resistor.

The only fields filled in on the symbol are **Prefix 'X'** to identify a sub-circuit, **Value R** which we will change to the name of the sub-circuit and **Description 'A WW Resistor(IEC graphic)'**.

4.4.1.1   NetList on the Schematic

Having placed an instance of the symbol on the schematic, ('WW resistor fixed. asc'), we now enter the text for the sub-circuit on the schematic or copy and paste the following:

```
* wirewound resistor with series inductance and parallel capacitance
.subckt wirewoundR 1 2
R 1 3 10k
L 3 2 10u
C 1 2 5p
.ends
```

and change the **SpiceModel** field of the instance to **wirewoundR**. The previous rules of precedence apply, so any **SpiceModel** or **ModelFile** on the assembly will be overridden by a value on the sub-circuit drawn on the schematic, and this in turn will be superseded by a value on the instance itself. We can now easily change any parameter and immediately see the effect. But once the sub-circuit is saved, these are fixed values which will be used by every instance.

We first save the complete circuit as 'WWR fixed.cir' where the first line is filled in by LTspice with the location of the schematic:

```
* G:\Users\Me\Desktop\Spice Book\04 Subcircuits\Circuits(asc)\WW
resistor fixed.asc
V1 1 0 SINE(0 10 10k) AC 1
R1 N001 0 1
XU2 1 N001 wirewoundR
.ac dec 10 10k 1000Meg
* wirewound resistor with series inductance and parallel capacitance
.subckt wirewoundR 1 2
R 1 3 10k
L 3 2 10u
C 1 2 5p
```

```
.ends
.backanno
.end
```

We now edit the circuit to leave only the sub-circuit and then save the sub-circuit as described in the section after the following.

4.4.1.2   Sub-circuit from Parts on the Schematic

The alternative is to create the sub-circuit from parts which will automatically create the netlist. This is schematic ('WWR parts.asc'). We save it as 'WWR parts.cir'. Where the extension *.cir* is supplied by default. It seems reasonable to save it in the same folder as the *.asc* files since it is a usable schematic. Then after saving, it reappears ready to be edited:

```
* G:\Users\Me\Desktop\LTSpice Book \Subs\Schematics (asc) \Draft20.asc
R1 N002 0 10k
L1 N001 N002 10µ
C1 N001 0 5p
V1 N001 0 AC 1
.ac dec 1e4 100 1e8
.backanno
.end
```

The first line is the location of the original schematic which we can leave if we like or change it as we do here. We must now edit the rest by adding the line *subckt WWR2* giving it a different name so that we can compare the two versions and the two nodes *n002 0* to:

```
* wirewound R
.subckt WWR2 n002 0
R N002 0 10k
Lser N001 N002 10µ
Cpar N001 0 5p
.ends
```

and this is essentially the same as the sub-circuit built on the schematic apart from the node names.

4.4.1.3   Saving and Testing the Sub-circuit

In both cases we save the circuit in the **MySub** folder with the extension *.txt*. We must make sure the path to the sub-circuit is shown in the **Control Panel**.

To test the first model, we must add **wirewoundR** as the **SpiceModel** in the assembly, and this will be passed to every instance on the schematic and cannot be edited. We must also add either the full path to the sub-circuit in the **ModelFile** field or just the file name if the path has been added in the **Control Panel**. If we do that, we arrive as schematic ('WWR test.asc'), and it will be seen that the **ModelFile** has been added to the netlist.

4.4.2   Potentiometer

This follows on from the previous chapter where we created one from two resistors. Now we merge all the *.param* statements into a sub-circuit.

4.4.2.1   Symbol (Assembly)

We can chose between the *Potentiometer* symbol of a diagonal line through the zigzag resistor symbol in the default **Sym** directory or the *IECPotentiometer* symbol in the **IEC Symbols** directory {assuming we have created them).

4.4.2.2   Netlist on the Schematic

The model consists of two resistors whose total is constant but whose relative values change:

```
* Potentiometer
.subckt Potm 10 20 30 Rt=5k {SliderPos}=0
R1 10 30 R = Rt*{SliderPos}
R2 30 20 R = Rt*(1-{SliderPos})
.ends Potm
```

And we should note that an attempt at **R2 30 20 R = Rt − R1** fails. If we have a value for **SliderPos** on the symbol and not a parameter, that will be fixed and not variable. The file is *potentiometer.sub*, but the actual model is *Potm*, not potentiometer, so slightly confusing. This is fine if we are going to add other potentiometer sub-circuits, perhaps with fixed values, and then it would be clearer to call it *potentiometer.lib*.

As it stands, the relevant symbol fields are:

Spice Model   Potm
Model File   potentiometer.sub

Accessing Sub-circuit Values
The sub-circuit parameters are not accessible as are those of models. So to change them, we must declare the values as parameters.

Example – Step Model Parameter

Even though the resistance and temperature are parameters, these are local to the sub-circuit. What we must do on the instance in the schematic is write **Value2** **wpos = {wp}** or something similar. Then **.step param wp 0 1 0.01** will work.

4.4.2.3 Platinum Resistance Thermometer (PRT)

We saw how to define a PRT in the previous chapter. Here we shall create a sub-circuit ('Platinum Resistance Thermometer Subcircuit.asc'). To create it, place a resistor on the schematic, right click on it and enter 'X' for the **Prefix**. This is essential to denote that it is a sub-circuit. Next, add an **InstName** of whatever takes your fancy, and finally **Value** must be the sub-circuit name which below is **PRT100**.

Press 'S' to open the **Edit text on the Schematic** dialogue and make sure the **SPICE directive** radio button is checked. The easiest is to copy the sub-circuit definition below and paste it in the dialogue. Otherwise enter the sub-circuit statement, the name *PRT100* and nodes *1* and *2*. Any numbers can be used for the nodes *20* and *21* would be equally acceptable provided the resistor is attached to the same nodes and not *1* and *2* as shown below.

Move to the next line by 'Ctrl+M', give a name to the resistor, connect it to the two nodes, and then add its value. To prevent it stretching right across the screen, we can break the line by **Ctrl+M** – and this is essential – start the next line with a '+' sign to show that it is a continuation and not a new directive. The final line is just . **ends** and the name *PRT100*. The final sub_circuit is:

```
.subckt PRT100 1 2
Rt 1 2 R=100*(1+3.908e-3*temp
+ -5.770e-7*temp**2
+ -4.1830e-12*(temp-100)*temp**3*(1-u(temp)) )
.ends PRT100
```

And now we have the sub-circuit completely at our disposal to add, remove and edit, in short to do anything we like. We complete the circuit with a *1 V* DC source and an earth connection. An *.op* command will show a current of just over *9 mA*. We can now add, if all is well, a temperature step analysis:

$$.step\ temp\ \ -100\ 100\ 1$$

will show it is 100 Ω at 0 °C. If, by mistake, the sub-circuit has been created as a comment in blue text, it will not work, and it will be necessary to right click on the text and check the **SPICE directive** radio button. If we plot $-V(in)/I(V1)$, we find a straight line.

But what if we use a thermometer with a different resistance, perhaps 200 Ω or 1000 Ω? We could, of course, make a new model for each. The easiest change is to give the resistor a name and add a default resistance value of *100 Ω*, so the first two lines on the schematic become:

```
.subckt PRT 1 2 Rref = 100
Rt 1 2 R=Rref*(1+3.908e-3*temp ......
.ends
```

It is not necessary, but the name has now been changed to PRT without the 100, and this must be changed on the resistor. A temperature sweep gives the same results as before.

To change the resistance, right click on the resistor to open the **Component Attribute Editor** and add a new resistance to follow the sub-circuit name in the **Value** field; for example, we could have:

PRT Rref = 200

as in ('Platinum Resistance Thermometer Subcircuit RvarV1.asc'). If in the wave-form window we now **Add Trace to Plot V(n001)/I(Ix(prt:A)**, we probe the supply voltage and current in the thermometer and find the resistance is 200 Ω at 0 °C.

The other possibility is to make a slight change to the sub-circuit so that the resistance is a parameter. Remove *100* and replace it by *{Rref}*. This takes us to ('Platinum Resistance Thermometer Subcircuit RvarV2.asc') where we add to the schematic: **.param Rref = 200** to change the resistance. The penalty for this method is that we must always add this directive to set the resistance: there is no default, and if we insert, for example, '.param Rref = 100' in the sub_circuit, the 200 Ω will be ignored.

Note also the new resistor *R1* in the schematic. The point of this is that it has a temperature coefficient, and its resistance also changes. What if we only want to change the temperature of the thermometer?

We can do that by another parameter; let us call it *{tprt}* and use it to replace every instance of *temp* in the sub-circuit. Now we have ('Platinum Resistance Thermometer Subcircuit Rvar tempvar.asc'), and it will be seen that only the thermometer temperature changes.

We can, if we so wish, create a sub-circuit for positive temperature just using the two temperature coefficients *tc1* and *tc2*, but then the temperatures are referred to 27 °C, and we will need the directive *.OPTIONS tnom = 0* to correct it and that will apply to everything.

Creating the PRT Sub-Circuit

We can move directly from the schematic by **View→SPICE Netlist**. This will show the complete circuit, but we cannot edit it. To do so we right click on the netlist and select **Edit as Independent Netlist** from the pop-up menu. We will then be presented with a screen where we can navigate to a folder and save the netlist with the default extension *.cir*. We then see the circuit listing in colour, and we remove everything except the final sub-circuit listed above, and for a change, we save it as a text file in the **sub** directory.

Then in the symbol if we enter:

Spice Model   PRT
Model File   PRT.txt

and save that. And as we noted previously, the file extension is not important, only its structure.

Example – PRT Test

The schematic ('PRT Test.asc') illustrates how the temperature of the PRT can be changed independently of everything else by setting its temperature to a constant $27\,^{\circ}C$ by:

$$param\ tx = 27$$

where the temperature of the PRT $tprt$ is declared as a parameter (tx):

$$tprt = \{tx\}$$

An **.op** analysis of:

$$.step\ param\ tx\ 0\ 100\ 1$$

shows the change of resistance of $R1$, whilst that of the PRT remains constant. If we uncomment **.step param tx 0100 1** and convert the previous **.step** command to a comment, we see only the resistance of the PRT changes.

4.4.2.4   Thermostat

This is a small sub-circuit that opens a switch when a certain temperature is reached. The symbol is simply a circle with two connections and two lines inside representing the switch. Or we can use an 'auto-generated' rectangle.

Schematic

The thermostat switch is represented by a voltage-controlled switch whose control terminals are connected in parallel with an arbitrary voltage source $B1$ whose voltage is set to be equal to the temperature. The thermostat built from parts is shown in schematic ('Thermostat.asc') which we save and edit.

Thermostat Sub-circuit

The sub-circuit is a *.txt* file. The hysteresis of the thermostat is set to **Vh = 1** meaning a two degree (actually $2\ V$) differential. This could be brought out as another user parameter. The default temperature is 50 °C which can be overridden on the schematic:

```
* Thermostat
.subckt Tstat c0 c1 Vtemp=50
S1 c0 c1 v+ v Ststat
B1 v+ v- V=temp
.model Ststat SW(Ron=1n Roff=1T Vt=Vtemp Vh =1)
.ends Tstat
```

The test circuit is ('Thermostat Test.asc'). We can change the trip temperature in ('Thermostat values.asc') where we have three instances with different trip temperatures. We can see that these are *Vtemp + Vh*. As it does not seem to be possible to create an up-down temperature ramp, we cannot test the hyteresis. And trying a negative temperature step to start at 100 °C does not work because LTspice converts it into a positive ramp.

4.4.2.5 A Single-Pole Change-Over Relay

This has five terminals: the two ends of the coil (C0,C1), the common contact or pole (P) of the switch and the normally open (N0) and normally closed (NC) contacts.

Schematic
Since the voltage-controlled switch is either open or closed, we need two to effect a change-over. We also need to supply the resistance of the relay coil and a nominal operating voltage. However, in reality, this is by no means exact, so we can add some hysteresis *Vh = Vdiff* to the switches to account for that. We must also take into account that the relay coil is inductive, so we can use an inductor with a substantial series resistance, and then we shall be able to see the potentially damaging voltage spike if the current in the coil is interrupted quickly. The schematic of the relay built from parts is 'SPCO Relay.asc. The resistors *R1* and *R2* are there only to provide a current path for the switches and are not part of the relay.

The final sub-circuit, after the usual editing but with added comments in italics in curly brackets – which are not part of the sub-circuit, is:

```
*SPCO Relay
.subckt SPCORelay c0 c1 NO NC p
S1 p NC C0 C1 ncSW { switch definition below, This is NC
S2 p NO C0 C1 noSW { ditto identical but Ron/Roff swapped
L1 c0 c1 {Lcoil} Rser={Rcoil} { inductance and resistance of coil set
externally
.param Vdiff = 0.15*Vcoil { hysteresis fixed at 15% of coil voltage
.param Vlatch = Vcoil-Vdiff { in practice, latching voltage is less than
Vcoil
.model noSW Sw(Ron=1n Roff=1T Vt=Vlatch Vh=Vdiff)
.model ncSw SW(Ron=1T Roff=1n Vt=Vlatch Vh=Vdiff)
.ends
```

The inductor interprets the first parameter as its inductance, but we have to make clear that the second is its series resistance and not capacitance or anything else. The parameter *Vdiff* is to give the switches some hysteresis that is a fraction of the relay coil operating voltage.

There are no default values in the sub-circuit: the fields are left empty. Neither are there any in the symbol assembly, they are only supplied in the instance on the schematic. This is not the best way of doing things because when the symbol is placed, there is no indication as to the number or nature of the values that must be supplied.

The schematic ('SPCO Relay Test.asc') shows the relay switching. Provided we set a very short step time over an interval of about 8 μs when the relay changes over one output, voltage rises whilst the other falls.

4.4.2.6   Changing Values in a Sub-circuit

The essential point is that all the components must have a value.

Fixed Values Sub-circuit
If we supply actual values directly in the definition of the sub-circuit as we did with the wire-wound resistor, these cannot be changed:

```
* wirewound resistor with series inductance and parallel capacitance
.subckt wirewoundR 1 2
R 1 3 10k
L 3 2 10u
C 1 2 5p
.ends
```

And if we open the schematic ('WWR test.asc') we cannot even see what the parameters are.

Default Values on the Sub-circuit
These are the deepest level and will be used by every instance if they are not overridden. For the thermostat we had:

$$\text{.subckt Tstat c0 c1 Vtemp} = 50$$

which sets a $50\,^{\circ}C$ trip temperature. If we open ('Thermostat test.asc'), this value is not seen on the instance nor on the assembly, but a run will show the thermostat switching at that temperature.

Value on the Assembly
We may add in the attributes of the assembly:

Value   Vtemp = 30

and if we save the assembly file, this will override the sub-circuit value, and the thermostat will close at *30 °C*. But if we already have an instance on the schematic, we must excise it and replace it for the new setting to be used consistently. And this new value will apply to all later instances.

Value on an Instance
We may also change *Vtemp* on every instance, and this will override the previous two. Thus we see a hierarchical structure from the sub-circuit text to the individual instance.

Example – Value Hierarchy
The schematic ('Thermostat Values.asc') has three thermostats in parallel, and they all close at the default *51 °C*. If we now change the **Value** on the assembly to *30 °C*, it will still close at *50 °C*. But nothing has changed on the instances. But if we remove them all and replace them, we now find this value has been passed to all three. If we remove *Vtemp* from one instance, it now closes at the default temperature. And if we add V*temp* = *60* to one of the other two, we find it closes at *60 °C*, and we can see in these three instances (a) the default behaviour; (b) a general behaviour defined in the assembly; (c) and a specific behaviour defined in an instance.

Parametrized Sub-circuit
We do this by enclosing all (or some) of the parameters in curly brackets. Somewhere there must be the values for them. For example, we could declare the wire-wound resistor sub-circuits as:

```
* wirewound resistor with series inductance and parallel capacitance
.subckt wirewoundR 1 2
R 1 3 {R}
L 3 2 {Lser}
C 1 2 {Cpar}
.ends

.subckt SPCORelay c0 c1 NO NC p
L1 c0 c1 {Lcoil} Rser={Rcoil}
...
.ends
```

In both of these cases, we have defined the values as parameters but not given default values. The order of precedence of the values is the same as before, and we can add default values at all levels starting from the sub-circuit itself, for example, we could have for the wire-wound resistor and the relay:

```
.subckt wirewoundR R=10K Lser=10u Cpar=5p
.......
.ends
```

```
.subckt SPCO_Relay c0 c1 NO NC p Lcoil=0.1 Rcoil=100 Vcoil=6 { these will
be used....}
L1 c0 c1 {Lcoil} Rser={Rcoil} { .. for the coil if not overridden }
...
.ends
```

Alternatively, we can omit the default values in the sub-circuit first line and add them as parameters in the body of the definition as:

```
* wirewound resistor with series inductance and parallel capacitance
.subckt wirewoundR 1 2
R 1 3 {R}
L 3 2 {Lser}
C 1 2 {Cpar}
.param R=10k
.param Lser = 10u
.param Cpar = 1n
.ends
```

In all cases we can override these default values on the schematic provided we can remember their names: so entering **Value2 Lx = 1 m** will be ignored.

The important point is that these parameters are accessible globally, which is very useful if we have several instances because we can declare the parameter on the schematic as *.param Cpar = 50p,* and this will apply to every instance. And we must take a little care not to use the same parameter in another sub-circuit or component.

Example – Parametrized Values
Schematic ('SubCct Params.asc') uses a parametrized version of the wire-wound resistor, and the parameter *Cpar* is also used for the inductor's parallel capacitance. Thus the two change together. As there is no default capacitance for the resistor, and if none is given on the assembly or the instance on the schematic, we supply the value by the *.param* statement. We can profitably explore various scenarios on the use of parameters.

4.4.2.7   For Later Chapters

A sub-circuit is a convenient way of encapsulating the inner structure of filters, such as the RIAA equalizer. It is also a good way of handling loudspeakers where the same circuit is used for treble mid and bass, and only the values differ. It is also useful for encapsulating delay lines of many repeated LC stages. These and others will appear at the appropriate place in later chapters.

4.5 Summary

In this chapter we have covered most of the ground in making new symbols and sub-circuits. The chief points are:

- A symbol consists of an assembly of lines and shapes which can optionally have the names of a file to search for the corresponding sub-circuit in the **Spice File** fields and the name of the sub-circuit within that file in the **Spice Model** field.
- The symbol can have default parameters for the sub-circuit which will override those on the sub-circuit.
- If a sub-circuit file is named in **Spice Model** with its extension and repeated in **Value2** without its extension and **Model File** is left empty, that symbol cannot be edited on a schematic.
- It is a good idea to keep new symbols in subfolders of the **sym** folder where they will be visible by **Edit→Component**.
- A sub-circuit must have a comment or blank first line. The second line must be **. subckt <model name> <pins in order of declaration><optional default parameters>**. Then follow simple components, each on a new line, with either fixed values or the parameter in curly brackets. Then come any model calls and finally **.ends <model name>** although the model name is not enforced.
- The order of precedence of default values is first on the sub-circuit listing, next on the assembly, finally on an instance on a schematic.
- Sub-circuit files can contain more than one model and usually have the extension **sub, lib, inc** and **txt** although others can be used provided the file is pure ASCII without any enhancements.
- It is a good idea to group models in folders and give the search path in **Tools→Control Panel**.

Chapter 5
Voltage and Current Sources

5.1 Introduction

SPICE in all its form has a panoply of voltage and current sources with varied and interesting relationships between input and output. There are some features not of immediate interest, such as the Laplace option; but, for completeness, we shall deal with them here and use their full power in later chapters. In LTspice all are accessed through **Edit → Component**. We can divide them into three groups:

Independent Voltage and Current Sources
These are the ones we have used up until now. There are two: one for voltage and one for current. They can create continuous sine and exponential waveforms, a frequency modulated sine wave and piecewise waveforms. Their DC values can be swept.

Arbitrary Sources
These have the most flexible output and can reproduce all the waveforms of the independent sources and many, many more. The examples we shall use are trivial, just to show the waveforms.

Controlled Sources
Their output is controlled by the input. There are four: a voltage-controlled voltage source, a voltage-controlled current source, a current-controlled current source and a current-controlled voltage source. In general the output merely scales the input.

We must be a little careful in selecting them since the same symbol can be used for more than one, so we must look carefully when we pick them from the main menu and **Edit → Component**.

© Springer Nature Switzerland AG 2020   179
C. May, *Passive Circuit Analysis with LTspice*®,
https://doi.org/10.1007/978-3-030-38304-6_5

5.2   Independent Voltage and Current Source

The ones to be discussed in this section are extremely flexible and can provide DC and AC and pulse waveforms. There is a symmetry in that the functions applicable to a voltage source also apply to an AC current source except for the **Table** option; and although, in real life, current sources are difficult to find, we shall invoke them here. Both are found from the main menu by **Edit → Component** and then **voltage** or **current**. We have used them before, but only as a DC sources.

5.2.1   DC Source

A small dialogue **Voltage Source-Vn** or **Current Source-In** opens when we first place the component and right click it where n is its number. There we can enter the DC value. For a voltage source, we can also add a series resistance representing the internal resistance of the source, but note that this cannot have a temperature coefficient nor a tolerance. We may also sweep the DC value by **Simulate → Edit Simulation Cmd** and then select the **DC Sweep** tab; any other settings, sine, pulse or whatever, will be ignored.

To access the other waveforms, we click the **Advanced** button as we shall see later. This will open the **Independent Voltage Source-Vn**, Fig. 5.1. The area where waveform parameters can be entered will be filled with appropriate edit boxes when one of the **Functions** is selected.

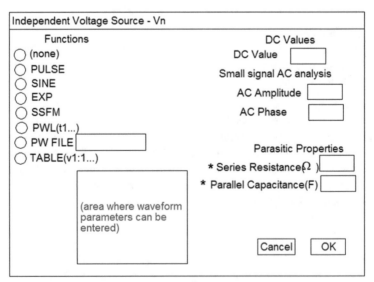

*In the case of a current source, the 'Series Resistance'
   is replaced by 'This is an Active Load' and the
   'Parallel Capacitance' is absent

Fig. 5.1 Voltage Source Dialogue

Current "Load"

This option is applicable to DC and AC analysis by appending the word **load** to the value or by checking **This is an active load** in the **Independent Current Source** dialogue Fig. 5.1 and converts the current source into a sink whose value is fixed for positive or negative voltages between its terminals provided we declare the load to be positive and negative, respectively. There are two synonyms **load** and **load2** which have different symbols but still use the current source.

Example – Current Source as a Load

The schematic ('Isource load.asc') consists of a voltage source *V1*, a current source *I1* and a resistor *R1* all connected in parallel. Its behaviour is shown in Fig. 5.2. The voltage source has a DC values of 5 V, but this is ignored when it is stepped *.step V1–2 10 2* in 2 V intervals from −2 V to 10 V . The current source is **2 load** and the resistor is *1 Ω*.

In the period (a) when the voltage from *V1* rises from −2 V to 0 V, the current in the resistor increases linearly from −2 A to 0 A, and the current in *I1* rises linearly from −16 A to 0A. The currents obey KCL.

The second period (b) from 0 to 2 V with linear increases is fallacious and is due to the large time step. This stands as a warning against being too eager to interpret a trace – always try a reduced time step first. With a step size of *0.1 V* schematic ('Isource load small step.asc') and the current follows the linear rise to *0 A* at *0 V* then blends smoothly to *2 A* at *0.5 V* and the current in *I1* remains constant thereafter. By changing the load current, we find that *I1* has an apparent resistance of $R_{I1} = \frac{0.25}{load}$. However, if we make the load negative, it still works correctly for negative input voltages with the same apparent resistance for forward voltages.

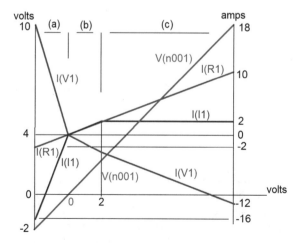

Fig. 5.2 Isource load

Step Load

The LTspice 'Help' file explains that this works in conjunction with the *step* option of the *trans* analysis. The syntax is:

Ixxx < n1 >< n2 >< default value > step(V1, V2, V3) load

The first three items **Ixxx <n1> <n2>** are set when we place the current source on the schematic and connect it. To set the *step* of the current source, it seems we must add the text **step (V1,V2,V3)** after the DC value, in this case of *1.5 A*. If the *trans* analysis has the **Step the load current source** check box checked, the command . **tran <stop time> step** will be shown on the schematic. And in theory this should work. The netlist is shown below where the description of the current source conforms exactly with LTspice.

```
* G:\Users\Me\Desktop\LTSpice Book new\09 Passive Filters\Schematics
(asc)\Draft1.asc
I1 N002 0 1.5 step(2,3,4) load
V1 N001 0 10
C1 N002 0 100n Rser=0.1
R1 N002 N001 1
.tran 5 step
.backanno
.end
```

but we get the error message **'Don't know how to detect this circuit's steady state.'** This is schematic ('Step Load fail.asc').

However, all is not lost. We have another option, and that is to create a PWL load omitting **step** from the **.tran** analysis and then it works, schematic ('Step Load PWL. asc').

5.2.2 AC Analysis

This is controlled by two edit boxes shown towards the top right of Fig. 5.1.

AC Amplitude

There must be a value for an AC analysis to be performed. Setting it to 1 V is best if we wish to measure the voltage or current gain because then we only have to measure the voltage at a node or the current through a component rather than having to divide the output by the input. We may step the DC of the source at the same time. So we could have *AC = 1 V* for source *V1*, and then the directives.

.ac dec 1000 1 1e8

.step V1 0 5 1

will perform an AC analysis from *1 Hz* to *1e8 Hz* six times giving *V1* a DC value
from *0 V* to *5 V* in six steps.

Phase

This can be left empty. If we set a value, the AC waveform is divided into a real and
an imaginary part on the left and right axes, respectively.

Example – Viewing AC Phase

With a setting of *AC 10 45* which is *10 V* at *45°* we correctly find for node *n001*, it is
10 V on the left and *45°* on the right, schematic ('AC Analysis.asc'). The trace is a
horizontal line because the resistor and input are constant.

Changing Between Analyses

Quite often we want to make several different analyses. LTspice allows us to place
them on the schematic and then convert all but one to a comment. To do that, right
click on the command you wish to disable. This will open the **Edit Simulation
Command** dialogue. Click **Cancel** and the small **Edit Text on the Schematic**
dialogue will open. Select the **Comment** radio button to disable the command.
You can also click on a disabled command to open the same small dialogue and
click **Command**. This also works with other commands such as *.meas*. However, if
we place a new command on the schematic before turning the old one to a comment,
LTspice will place a semi-colon in front of the old, which also converts it to a
comment.

   LTspice also allows us to define one of the waveforms in Fig. 5.1 as well as an AC
and DC source.

Explorations 1

1. It is worth experimenting with the active load option for a current source. It will
   be found that the behaviour of the sink depends only on the DC value.
2. Try running an active load with an AC source.
3. Open ('AC Analysis.asc') and explore different phase settings. Note the use of the
   dotted 'phase' trace.

5.2.3 Voltage Source: Parasitic Properties

On the right we have an edit box to add series resistance to represent the internal
resistance of the source and parallel capacitance. LTSpice says that including them
with the source is more computationally efficient than adding them as external

elements. However, they cannot be toleranced nor given temperature coefficients. Note the check box to show this information on the schematic.

5.2.4 Functions

On the left, we now have the radio buttons for all the excitation functions which we shall explore in turn. They are used with a transient analysis.

(None)

This cancels any of the following but retains the DC setting from the opening dialogue. The **DC Value** edit is only available for this option and will show any value we entered in the original dialogue and which we now can edit.

Pulse

By careful choice of the parameters, we can make repetitive rectangular, square, trapezoidal and triangular waveforms. This is Fig. 5.3. Click on this and a number of edit boxes appear below the text with values in voltage or current. These are as follows:

Vinitial(V) or I1(A) is the starting value or offset that applies to all the pulses. This can accept a parameter enclosed in curly brackets as **{x}** . It does not accept *time*.
Von(V) or I2(A) is the final value after the delay time. This also can accept a parameter such as **{x + 1}**, and note that the whole expression must be inside the brackets.
Tdelay(s) is the time from the start to the beginning of the rise time.
Trise(s) is the linear rise of current or voltage from its initial value to its 'on' value.

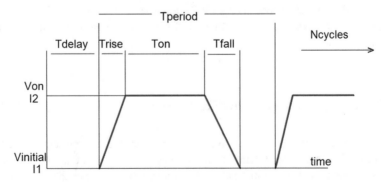

Fig. 5.3 Pulse Waveform

Tfall(s) is the linear fall of the current or voltage from its 'on' value to its initial value. Notice that these values should not be left empty; else the default rise and fall times are quite appreciable and can give rise to false results by missing fast transients because the slopes are too slow. Insert something like *1 p* for a true vertical rise and fall.

Ton(s) is the time after the voltage or current has reached its 'on' value until the start of the fall time, in other words, the duration of the flat top.

Tperiod(s) is the total time before the next repetition. If this is more than the total of the rise, on and fall times, the current or voltage will revert to the initial value, and there will be a delay until the next rise time. If, on the other hand, the period is less than the total, LTSpice will calculate the period and use that.

Ncycles is the number of time the waveform repeats, and if left blank it continues for the transient time interval.

Example – PULSE Waveform
A typical setting is:

$$PULSE(10\ 20\ 1m\ 1m\ 2m\ 5m\ 10m)$$

meaning an initial offset of 10 V and a pulse that starts after 1 ms with a rise time of 1 ms to a voltage of *20 V*, a duration of 5 ms and a fall time of 2 ms and an off time of 1 ms – which is derived from the period time of 10 ms minus the sum of the delay, rise, fall and on times.

These parameters are used in schematic ('V Pulse Source Test.asc'). Notice the check box **Make this information visible on the schematic** which is checked by default. Uncheck it so that the settings cannot be seen.

Explorations 2
1. Place a voltage source and a 1kΩ resistor in series and explore various settings. Note that if the settings are visible, they can be altered by right clicking on the text **PULSE.....** but there is no hint to the meaning of each term. Better to right click on the source.
2. Reduce the times to µs with rise and fall times set to zero, and note the slope of the rise and fall is 0,5 µs. Now set them to 1 fs; the rise and fall are vertical. This is important, zero does not work - a short time is needed.
3. Repeat with a current source.

Sine

This is the fundamental waveform from which any other can be constructed by adding sine waves of the correct frequency and amplitude. Once again, click on this, and a number of edit boxes will appear. If you have just come here from the previous explorations, you will find these have been rather unhelpfully filled by the previous values. The edit boxes are as follows:

Voff(V) or **(A)** is similar to the *Vinitial* of a pulse and is the offset. Any DC value set on the initial dialogue is ignored.

Vamp(V) or **(A)** is the peak value of the sinusoid which is added to the DC offset, so a 1A peak sine wave with a 1 A offset would run from 1 A to 2 A.

Freq(Hz) is the inverse of the period, so a frequency of 50 Hz is a period of 20 ms.

Tdelay(s) from zero up to this time the output is constant and is the offset plus the contribution from *phi* (see below).

Theta(1/s) is the *damping factor*. Its effect is to successively reduce the amplitude of the waveform. A large number, at least 100, is needed to see a change over just a few cycles.

Phi(deg) is the phase angle, that is, the point on the cycle at which the waveform starts. By default it is zero. In this case we have used current where up to the end of the delay time the voltage or current is:

$$I = I_{off} + I_{amp} \sin\left(\frac{\pi\phi}{180}\right)$$

where Φ *(phi)* is the phase angle in degrees: setting it to 90 results in a cosine wave. The total current or voltage at any time is given by:

$$V(t) = V_{off} + e^{-(t-Td)\theta} V_{amp} \sin\left(2\pi f(t - Td) + \frac{\pi\phi}{180}\right)$$

where Θ (theta) is the damping factor, f is the frequency and *Td (Tdelay)* is the delay time. What this shows is that if Θ is zero the exponential term is $e^\circ = 1$, and there is no damping. Otherwise, the amplitude undergoes an exponential decay, Fig. 5.4, where the envelope of the waveform is the exponential term. Offsets and delays apply, as the equation shows.

Ncycles is the number of cycles and need not be an integer. If none is given, the waveform will repeat for the duration of the transient time.

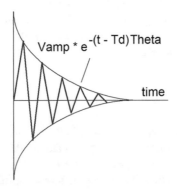

Fig. 5.4 Voltage Source Damping

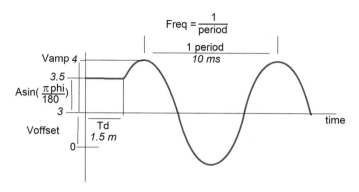

Fig. 5.5 Sine Waveform

Example – Delayed Sine Wave

If we take a sine wave of 2 V_{pk} and add an offset of 2 V, we find a waveform running from 0 V to 4 V, Fig. 5.5. If we set $\Phi = 48.6°$ the waveform starts at 3.5 V and has a period of 1/1 kHz = 10 ms, and the delay has been set at 1.5 ms in schematic ('Delay3,asc'.)

Example – Decaying Sine Wave

Run schematic ('Sine wave decay.asc'). This has a current source with theta = 200 and phi = 30. Measure the starting angle at *a* and remember the decay starts at that point so the first peak is only 1.93 A not the full 2A. The easiest point to take to check the value is the peak of the waveform, for example, at time *b*. Then after taking natural logarithms we have:

$\ln(0.581/2) = -(7.163-1).10^{-3}$ then theta $= -0.5369/-6.163\ 10-3 = 200$

Example – Current Sine Wave

The setting for the current source in schematic 'I Sine Source.asc' is:

$$SINE(3\ 1\ 2\ 0.1\ 0\ 30)$$

which is a sine wave with an offset of 3 A, a peak amplitude of 1 A and a frequency of 2 Hz with a delay of 0.1 ms, no decay but a phase angle of 30°. Therefore we expect the starting value to be 3 A *(offset)* + 0.5 A (1 A x sin(30)) = 3.5 A and the positive peak value to be 3 A + 1 A = 4 A and the negative peak to be 3 A-1 A = 2A.

Example – Fourier and Time Step

A point of some interest is that the command:

$$.four\ 2\ I(R1)$$

makes a Fourier analysis of the current in R1 with a fundamental frequency of 2 Hz. By default it returns the first nine harmonics, which is good enough here. With the default time step, this is some 0.238% – which seems unlikely given that this is a pure sine wave generator. If the time step is reduced to 10 μs, it reduces to 0.0446%, and at 1 μs, it is a mere 0.00016%. The error arises from the length of the time step and is important if we are making measurements on hi-fi systems. And we must also be careful to turn off compression.

Explorations 3
4. Open ('I Sine Source.asc') and confirm that the waveform starts at 30° after a 0.1 s delay.
5. Open schematic ('Damped Sine Wave.asc'). This is eight cycles of the waveform with a damping factor of 200. Measure the decay and check against calculation. Try some different values.
6. Experiment with different frequencies, amplitudes and delays.
7. Repeat with a voltage source.

Exponential

The exponential is of the form: $v(t) = V_0(1 - e^{-kt})$. At time $t = 0$ the exponential is 1 so $v(0) = 0$. As t increases, the exponential decreases, so v tends to an asymptotic value of $V = V0$. This option creates a single pulse with an exponential rise and fall. The parameters are as follows:

Vinitial (V) or **(A)** is the starting value without *Vpulse* which applies until the expiration of Rise Delay.

Vpulse(V) or **(A)** is the asymptotic amplitude of the pulse including the initial value. So if *Vpulse = Vinitial*, there is no pulse.

Rise Delay(s) is the delay from the start before the exponential comes into play.

Rise Tau(s) is an exponential that in theory never reaches its final value. It is usual to define a *rise time* as the time when $kt = 1$ and thus $v = Vo(1–0.3679)$ or approximately 2/3 of the asymptotic value. From time until **Rise Delay** until **Fall Delay**, the voltage or current is given by:

$$V(t) = V_1 + (V_2 - V_1)\left(1 - e^{\frac{-(t-Td1)}{Tau1}}\right)$$

showing that the output (voltage in this case) is the initial value V_1 plus the difference in voltage between the pulse voltage V_2 and the initial voltage multiplied by the exponential function where $-(t-Td1)$ is the difference in time between the current time and the Rise Delay $Td1$, Fig. 5.6. As this is a negative exponential, its value decreases with time so the voltage tends to *Vpulse*.

Fall Delay(s) is the total time from the start before the beginning of the fall. If this is 4 or 5 times *Rise Tau*, the final voltage or current will be within a few percent of the asymptotic value. After this time the voltage or current is described by:

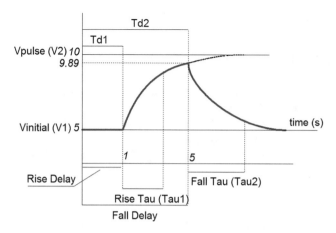

Fig. 5.6 Exponential Waveform

$$V(t) = V_1 + (V_2 - V_1)\left(1 - e^{\frac{-(t-Td1)}{Tau1}}\right) - (V_2 - V_1)\left(1 - e^{\frac{-(t-Td2)}{Tau2}}\right)$$

where *Td2* is the Fall Delay. After that time the second exponential subtracts from the first.

Fall Tau(s) is the converse of the *Rise Tau*, and the amplitude will decay towards the initial value for as long as the transient analysis lasts.

Example – Exponential Wave
This is shown in Fig. 5.6 and in schematic ('Exp Wave.asc') where the voltage setting is:

$$EXP(5\ 10\ 1\ 1\ 5\ 2)$$

which is a pulse having an initial value of *5 V*. Then after a delay of *1 s*, it increases towards *10 V* with a rise time of *1 s*. After a time of *5 s* from the start, it decreases towards the initial voltage of *5 V* with a fall time of *2 s*.

Explorations 4
8. Open ('Exp wave.asc') and measure the waveforms.
9. Experiment with different rise and fall times, and the transient run duration, and check against calculation.
10. Repeat with a current source.
11. Schematic ('B1.asc') has a rather fanciful output - try it.

SFFM (Single-Frequency Frequency-Modulated Source)

We just mention this in passing. It is a frequency modulated constant amplitude voltage or current waveform that we shall not need just yet. The parameters are given first following the radio button SFFM which are (Voff...Fsig) and then the text preceding the edit boxes:

Voff(V) or DCoffset(V) or **(A)** which are the offset voltage or current as before.
Vamp(V) or **Amplitude (V) or** (A) as before.
Fcar(Hz) or Carrier Freq(Hz) is the sinusoidal carrier frequency. For FM radio this
   is of the order of 100 MHz.
MDI is the **Modulation Index**, which can be set to any value, unlike amplitude
   modulation where 0.8 is the practical limit.
Fsig(Hz) is the signal frequency **Signal Freq(Hz)** which in real life is a very small
   fraction of Fcar.

The waveform is:

$$V(t) = Voffset + Vamp\sin(2\pi Fcar.t) + MDI\sin(2\pi Fsig.t)$$

showing we have the offset plus the carrier term in V_{amp} plus MDI sin $(2\pi Fsig.\ t)$, and then using the identity:

$$\sin(x)\sin(y) = \tfrac{1}{2}\left[\cos(x-y) - \cos(x+y)\right]$$

we see that this gives us the sum and difference between the carrier and modulation frequencies.

Example – SFFM Voltage
Out of interest, open schematic ('FM Waveform.asc') where the setting is:

$$SFFM(1\ 5\ 1k\ 20\ 25)$$

which has a DC offset of *1 V* and an amplitude of *5 V*, so the waveform ranges between *−4 V* and *+ 6 V* at a constant amplitude. The carrier frequency is *1 kHz*, and the *25 Hz* signal has a modulation index of *20*.

Explorations 5
1. Open ('FM Waveform.asc') and explore different settings

PWL, PWL File

This enables us to create a piecewise linear waveform in current or voltage with straight line interpolation between the data points using a transient analysis. The table is in pairs of (time,voltage) to which more can be added by clicking **Additional PWL Points**. If the transient analysis continues beyond the last time, that final value is used.

Example – PWL Linear Waveform
A simple current waveform is:

$$PWL(0\ 1\ 1\ 2\ 4\ 2\ 5\ 3\ 6\ 0)$$

At time *0 s* the current is *1 A* which rises linearly to *2 A* at *1 s*, then remains constant to *4 s* and rises to *3 A* at *5 s* before finally falling to *0 A* at *6 s*,

A useful feature is that we can repeat the waveform by **repeat for < no. of times >** and closing with **endrepeat** so the following will repeat it twice.

$$PWL\ repeat\ for\ 2(0\ 1\ 1\ 2\ 4\ 2\ 5\ 3\ 6\ 0)endrepeat$$

A slight variation is **repeat forever**, and then it continues up to the end of the transient run.

We can also chain the current or voltage by adding on from the previous time. The directive below produces a linear ramp to *6 A* at *6 s* by adding *1* at each second and stays at *6A* for any time after *6* s.

$$PWL(0\ 0 + 1\ 1 + 1\ 2 + 1\ 3 + 1\ 4 + 2\ 6)$$

Saving the PWL Table

If the same waveform is often used, it can be saved as a file and called back in to save entering the data each time. After a run, right click on the trace panel and select **File→Export Data as Text**. This will open the **Select Traces to Export** dialogue and select the current. Do not add *time* or any other text. This will save the data as a text file of two columns separated by tabs and will automatically include the time as the first column. The text file for the setting *PWL(0 2 1 3 4 –1 6 2.5 7 0)* is:

time	I(I1)
0.000000000000000e+000	2.000000e+000
1.005859375000000e+000	2.992188e+000
4.000000000000000e+000	-1.000000e+000
6.000000000000000e+000	2.500000e+000
7.000000000000000e+000	0.000000e+000
651367187500000e+000	0.000000e+000
9.993164062500000e+000	0.000000e+000
1.000000000000000e+001	0.000000e+000

where LTspice has added the last two entries to make up the total time for the transient run of *10 s*. If we reduce the run time to *7 s* and make a run, we find the data is:

0.000000000000000e+000	2.000000e+000
1.002050781250000e+000	2.997266e+000
4.000000000000000e+000	-1.000000e+000
6.000000000000000e+000	2.500000e+000

(continued)

6.769042968750000e+000	5.773926e-001
6.994628906250000e+000	1.342773e-002
7.000000000000000e+000	0.000000e+000

and the values are not quite the same as before because LTspice has to calculate the currents and times step by step, including some intermediate points which we can see by right clicking on the trace panel and then **View → Mark Data Points**.

We can make a more pleasing table by using a text editor such as Notepad instead of LTspice to create the PWL file with the data entered with just so many decimal places as we want and tabs used to separate the two columns.

Loading the PWL Table

We can open it with LTspice with **File → Open** but change the file type to **All Files** (∗.∗), and then look for a **.txt** file and open. Then remove the header:

$$time I(I1)$$

and save it again. This is what has been done with the second set of data above. If we now click **PWL FILE** and **Browse,** we will be offered a list of possible files, but note that this can include files with more than two columns and with the header row. If we click **OK** we will find:

$$PWL\ file = "PWL\ Waveform.txt"$$

or whatever name we gave it. Sadly, this does not fill in the PWL table, so we cannot edit it here on the schematic. However, we can edit it with LTspice but *repeat* is not accepted, only a straight list of pairs of data. The answer is to enclose the file within a *repeat* block:

$$PWL\ repeat\ for\ 3(file = "PWL\ Waveform.txt")endrepeat$$

Notice that once we have loaded a PWL table, we can no longer get back to the **Current Source – Ix** dialogue to change to another waveform, for example, an exponential. Instead by clicking on the source, we open the **Component Attribute Editor.** If we delete the PWL entry above, we can now return to the **Current Source – Ix** dialogue.

Explorations 6

1. Open ('Piecewise Waveform.asc'). We cannot have two values at the same time, so to make a step change, we give the next a very small extra increment of time.
2. Open and run ('PWL Waveform.asc'). Then replace the data by loading **PWL Waveform.txt**. Open the *txt* file in Notepad and edit the data, save the file, and then reload it.
3. Explore repeating the data points with the *repeat for* statement.
4. The file ('PWL active load.asc') has sinusoidal excitation and uses an active load. The resistor *R1* draws a peak sinusoidal current of *5 mA* as one would expect. Open the PWL file, and compare the load trace with the piecewise waveform.

WaveFile

These can be loaded and saved. Wave files are used for high-quality lossless audio, unlike lossy MP3 format, at a standard rate of 44,100 Hz for stereo with 16 bits per sample which converts to 96 dB giving a very good dynamic range. However, other data rates can be used and are specified in the file header. The file can also include multiple channels, not just the two for stereo.

Loading a WAV File

A **.wav** file can be used as input. This is not shown on the **Independent Voltage Source** dialogue nor in the one for current. LTspice is very restricted on how it searches for a file. Usually we need the full path as:

$$wavefile = ``G:\backslash Program Files(x86)\backslash OpenOffice4\backslash share\backslash gallery\backslash sounds\backslash laser.wav"$$

and we can add an optional **chan = <channel no.>**. LTspice uses the first channel by default if none is specified, or if the specified channel does not exist.

We can keep sounds altogether in a **Sounds** folder in LTspice:

$$wavefile = ``G:\backslash Program Files\backslash LTC\backslash LTspiceXVII\backslash lib\backslash Sounds\backslash cow.wav"$$

but LTspice will not search for it. Pity!

The file always has a range of -1 V to $+1$ V or -1A to $+1$A, and by adding data points in this instance and expanding the trace, we find *22* from *1.5332 s* to *1.5352 s* giving a sampling interval of *87 μs* or a sample rate of *11.0 kHz* which, allowing for the estimation errors, is close to the standard 11,025 Hz.

It is not possible to drag or duplicate the waveform so editing must be done elsewhere.

Saving a WAV File

To save data we use a SPICE directive:

$$.wave < ``filename.wav" > < bits per sample > < sampling frequency > < data source >$$

There is no restriction on the filename except that it must have the extension **.wav** and be enclosed in double quotation marks. The audio standard is 16 bits per sample at 44,100 Hz with two channels for stereo. But as the above examples show, any rate can be used, and there is no need to remember it because this is contained in the file header. A typical file is:

$$.wave ``C:\backslash test.wav" 16\ 44100\ I(R1)$$

PWL files can be saved as wav. files, but remember that the amplitude of the *.wav* files is limited to $+/-1$ V or $+/-1$A. So we can scale the PWL data to not exceed that by:

PWLvalue_scale_factor = **0.33333(0 1 1 2 4 2 6 0 8 1.5 9−1 1 0 3)**

then: *.wave "C:\test.wav" 16, 20 k I(R1)* will save it.

Example – Wav File
Do we need such a high data rate? The answer is found by loading the file, and it will be seen that there are no data points along the flat region from *1 s* to *4 s*. Clearly the data rate is not constant and depends upon the rate of change of the data. Test this by:

"C:\test.wav" 16, 20 I(R1)

and the data is correct.

Explorations 6
1. Open and explore ('Isource wav_in.asc') and ('Isource wave_out.asc').
2. Try scaling time as well as value by **time_scale_factor = .**
3. Repeat with a voltage source.
4. Create various input waveforms, and explore the necessary sampling frequency to preserve the data.
5. Try saving with more than one channel. For example, using schematic 'Isource wave_out test.asc using the file *wave "C:\test.wav" 16, 20 I(R1) V(12)* or wherever you have stored it. We scale the data as before so that the current range is correct, but there is no problem with the 1.6 kV. And both traces are recovered correctly.
6. Enjoy experimenting with other circuits and saving other values.

Current Source-Table

This defines the current by the voltage across the source, not the time. For example, schematic 'I Source Table.asc' has a voltage source in parallel with the current source, and the values which must be in ascending order of voltage of the form <voltage><current> and are:

TBL(3 0.01 5 0.05 10 0.02)

meaning that until the voltage reaches 3 *V*, the current is *10 mA* and then rises linearly to *50 mA* at *5 V* and reduces to *20 mA* at *10 V* The data set will be reused repeatedly for the duration of the transient analysis. If the first value is out of range, the output is constant at the first value, and the remaining data pairs are ignored.

Example – Current Source Table
We shall take the above table of values and a sine wave input of *12 V* amplitude at a frequency of *1 Hz* and place it in parallel with the current source, schematic ('I source Table.asc') The resulting waveform for the first 250 ms is shown in Fig. 5.7.

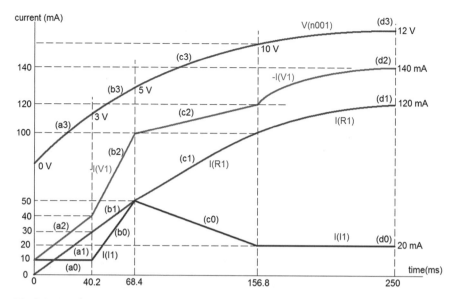

Fig. 5.7 Current source Table

For the first 40 ms, the output of the current generator $I1$ is *10 mA*, so we have the horizontal line (a0). In the same interval, the voltage $V1$ – measured at node *001* – rises from *0 V* to *3 V*, (a3) which is the time at when the next pair of data points become active. As the load resistor $R1$ is *1 kΩ* we find its current, (a1) increasing with $V1$ to *30 mA* And because KCL must be obeyed the voltage source $V1$ must sink *40 mA*, (a2).(Note that it is *-I(Vi)*)

At *68.4 ms* the input voltage (b3) has reached *5 V*, the next marker. During this interval the current in $I(I1)$ has increased to *50 mA* (b0) as required by the table. The current in resistor $R1$ follows the input voltage and has also reached *50 mA* (b1), so the voltage source (b2) must sink a current of *100 mA* at the point.

From *68.4 ms* to *156.8 ms*, the current $I1$ decreases to *20 mA* (c0), whilst the current in $R1$ (c1) follows the input voltage (c3) to *100 mA*, and the voltage source sinks a total of *120 mA* (c2).

In the period ending this 1/4 cycle, the current generator output has fallen to *20 mA* (d0), and the load resistor current (d1) increases to *120 mA* following the input voltage $V(n001)$ (d3), so the voltage source must sink *140 mA* (d2).

The following 1/4 cycle is almost a mirror image where the time from 330 ms to 530 ms is shown in Fig. 5.8. The current agrees with the voltages; just one point to note, and that is with a step time of 1 ms, the current is *49.7 mA*, but reducing the step time to *1 μs*, we get *49.94 mA*.

If we just make the change *(15 0.02)* to the last entry in the table and the voltage will next rise high enough that after 67 *ms* the current in $I(I1)$ has reached *50 mA*, after which the current source is no longer controlled so as the voltage continues to rise and the current through the load resistor rises with it, Fig. 5.9, the current source

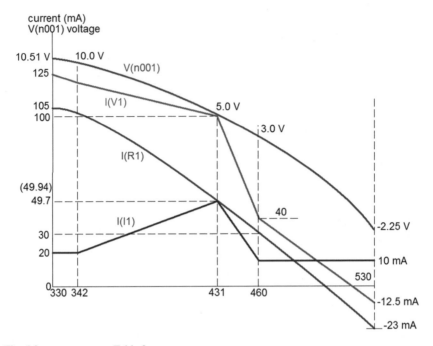

Fig. 5.8   current source Table 2

I1 adjusts its current to ensure that the total current is zero and does not follow the straight line (a) as before but the shallow curve.

Explorations 8
1. Explore the schematic ('Isource Table.asc'). In particular, check that the currents are correct.
2. Turn the current source through 180°, and now the current source applies to the negative half-cycle.
3. Rotate the resistor through 180° .
4. What happens if an AC voltage is also given to the current source?

Current Source – Step

This is applicable only to a current source and is an extension of the simple **load** option. We discussed this above.

Current Source – Active Load

This check box applies only to the current source and can be enabled for every option except Table.

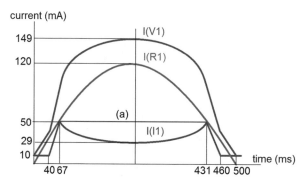

Fig. 5.9 Isource Table 3

5.3   Arbitrary Sources (B)

These are current or voltage sources whose value is defined by a function. This is very flexible and can range from a simple DC value to quite complex expressions, similar to the LTspice resistor. As a rough guide, the function can contain the arithmetic and logic statements found in programming languages as well as node voltages and currents. Therefore they are able to create waveforms that are not possible with other voltage and current sources. They are widely used in modelling non-linear devices and are briefly described in the 'Help; file. There are three:

Bi a current source
Bi2 the same but with reversed polarity
BV a voltage source

   In every case they will be identified on the schematic as **Bnnn.** The different designations are simply an aid for the user and can be overridden on the schematic so that a voltage source becomes a current source by replacing $V = ...$ with $I = ...$

5.3.1   Constant Power

The source can be declared as dissipating a constant power by $P = $ **<value>.** The only difficulty is that for correct operation, the input must not be less than ±1 V, schematic ('Constant Power.asc'), so the directive:

$$.param\ Vprxover = 0.1$$

has no effect, but the corner point *Vprxover* can be increased to higher voltages than 1.

Example – Constant Power Source
Schematic ('Constant Power.asc') shows *I(B1)*∗*V(n001)* is constant at 4 W except between −*1 V* and *1 V*. If we change to a current source, schematic ('Constant Power

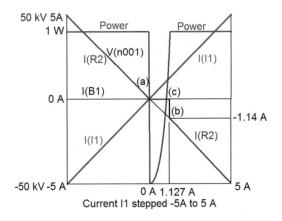

Fig. 5.10 Constant Power V2

V2.asc') and Fig. 5.10, we find that as current source *I1* is stepped from −5 A to 0 A, the voltage V(n001) starts at 50 kV in order to drive the 5 A through the load resistor R1 and progressively decreases to 0 V. The current from *I1* appears to be 0 A. Not so. As the voltage is 50 kV, it only requires 20 µA to dissipate the 1 W, and if we use the cursor, that is what we find, the current increasing as the voltage falls.

The current directions are somewhat arbitrary and depend on the orientation of the sources rather than KCL. This can be seen by moving the mouse over the components after an analysis when it changes to a current clamp meter with a red arrow showing the direction of the current. In this case the current from *I1* is negative, but if we turn it round (by *F7* and then **Ctrl R** twice), it is positive, and in either case careful measurements show that the difference between it and the resistor current exactly accounts for the current in source *B2*.

When the current from *I1* becomes positive, the current in the load resistor follows the path *(a)* to *(c)* at 0 A before dropping steeply to −1.14 A at *(b)*. This steep fall is spurious and is due to the current step: decreasing it to 0.001A removes it. Also the current at *(b)* depends upon the power: if it is reduced to 0.5 W, it is only 0.57A .

After point *(b)* the load current continues in a straight line to 5A. And during the interval *(a)-(b)*, the current in *B1* falls linearly from *(a)* to *(b)* and then rises steeply to *(c)* at 88 µA and then continues slowly falling to 20 µA. Also during the interval *(a)-(b)* the load power falls to zero and then recovers to 1 W along the curved path showing that the constant current function is inoperative.

Explorations 9
1. Open ('Constant Power.asc') and explore the effects of different power settings and *Vprxover*. Increasing the voltage does not reduce the minim limit of *Vprxover*.
2. Open ('Constant Power V2.asc') and reduce R2 to 10 Ω, and note that the inoperative region extends from −0.63 A to 1.20 A. The reason is that there is not enough current to provide the power. The voltage at that point is 3.3 V, so the load resistor consumes 0.33A leaving only 0.3A for B1 – so it has just enough power, 3.3 V x 0.3A = 1 W, but any lower voltage and it fails.

3. We can explore that by increasing the power for *B1*.
4. Try inserting a resistor between *I1 I* and *B1*. Indeed, it is worth exploring what complexity of circuit will work.

Function

These sources do not have selectable waveforms such as a sinusoid or pulse like the independent sources. When we right click on an instance, we are presented with:

$$out = F(\ldots\ldots\ldots)$$

where *out* is either current or voltage. We must then enter the function without the enclosing brackets although some functions discussed below do themselves need to be contained in brackets. The format is:

$$V\,or\,I < expression > [< IC = ? > < tripdv = ? > < tripdt = ? > < Laplace = ? >$$
$$< Rpar = ?\,(for\,current\,source) >]$$

with the optional qualifiers in square brackets.

Expression
This is essential and can be a simple value such as *V = 5.6* or a function. A trivial example is: *I = time∗5* where the current is just the time multiplied by 5. It can contain the current through devices, the voltage at a node or the voltage between a pair of nodes and a large number of trigonometrical, exponential and relational functions described in the 'Help' file and which we shall now explore.

Optional Parameters

IC This is an optional Initial Condition set before the start of the simulation. This can be used with most components.

tripdv<value>,tripdt<value>
These two optional qualifiers work together. If the voltage across the source changes by more than **tripdv** volts in **tripdv** seconds, that step is rejected. This is useful in rejecting sudden spikes in the response but we need to be careful – perhaps they were genuine.

Laplace
This option opens up the powerful possibility of complex relationships. It can be applied meaningfully to many of the functions explored below. This also can be used by other sources and will be described briefly there.

Rpar
Is an optional parallel resistor across a current source.

Rounding Functions

These modify the output by changing it when a predetermined level is reached.

ceil(x), floor(x), round(x)
These convert the signal from a continuously variable into steps. The first **ceil(x)** converts the value to the next higher integer so that *5.6* becomes *6*. The second, **floor (x),** converts the value to the lower integer so that even *5.99* becomes *5*. The last, **round(x),** rounds the value to the nearest integer so *5.51* becomes *6* and *5.49* becomes *5*.

u(x),uramp(x),inv(x), buf(x)
The first, **u(x),** converts the input to a unit step if the value is greater than zero, else it is zero. It is useful as a switch function to detect a positive value. This is in contrast to the second **uramp(x)** where the output follows the input if the input is greater than *0*; else it is zero. The third, **inv(x),** is not quite the inverse of the previous since the switch point is *0.5* and the output is *0* if the input is greater than *0.5*; else it is *1*. The last **buf(x)** is *1* if the input is greater than *0.5*; else it is *0*. These are subtle but important differences and are illustrated in schematic 'UrampInvU.asc'.

abs(x), int(x)
The first, **abs(x),** returns the absolute value so that negative values become positive. The second, **int(x)**, returns the lower integer value so that *0.99* is *0* and *1.001* is *1* and so on. Schematic 'AbsInt.asc'.

Limiting Functions

These change the output when a limit is reached. An overview of their behaviour is in schematic 'LimUplimIf.asc'.

min(x,y),max(x,y)
The two values are tested in either order so **max(x, y)** is the same as **max(y, x),** positive or negative, integer or fractional, and each can contain functions.

Example – max
Given a voltage ramp, the function $V = \textit{max}\,(5, 7 * \cos\,(10 * V(s)))$ gives an output of *5 V* whilst the input is less than *5 V* and an output of *7*cos(10*V(s))* once that value has been exceeded: in other words, just the peaks of the cosine wave superimposed on a *5 V* level, schematic 'MinMax2.asc'. The values can be negative depending on the orientation of *B1*

Example – max, min
Again given a voltage ramp, the function $V = \textit{max}\,(5, min(7, 10 * \cos\,(10 * V(s))))$ now creates a window between *5 V* and *7 V* where the output is the cosine wave; else it is *5 V* or *7 V*. As a corollary, change the function of *B1* from voltage to current. $I = \textit{max}\,(5, min(7, 10 * \cos\,(10 * V(s))))$.

if(x,y,z)

The function is *<value> = (x, y if x > 0.5, else z)*. If the value of *x* exceeds *0.5*, the output is value *y*; else it is value *z*. So if we wanted the change to happen for some variable *r = 0.1*, we could handle it by *if(r + 0.4,y,z)*. But we must also remember that LTSpice defines *true = 1, false = 0*, so more elegantly we can set up a simple binary test instead as *if(r > 0.1,y,z)* which will be *0* until *r > 0.1* and *1* thereafter.

Example – if

The expression *V = if(0.1 * V(s), V(s) + 1, -V(s))* tests *0.1*V(s)*, and if it is more than *0.5*, the output is *V(s) + 1*; else it is *-V(s)*. This is seen in schematic 'LimUplimIf.asc' where the input is a ramp from *0 V* to *10 V* over a time of 10s. The output runs from *0 V* to *-5 V* at a time of 5 s because this is less than *0.1*V(s)*, but at this point the input is *5 V*, so *0.1*V(s) = 0.5 V*, and the output changes to *V(s) + 1*.

Example – If, And

The **if** function can accept quite convoluted expressions. A modest one is:

$$V = if\ (time > 0.2\&time < 0.6, 5, -4)$$

so that if the time is between *0.2 s* and *0.6 s*, the output is *5 V*; else it is *−4 V*. This is schematic ('IfAnd.asc').

dnlim(x,y,x)

This is the same as **max(x,y)** except the the transition is smooth, not abrupt, over the range of **y** volts. Thus *V = dnlim(1.5 * V(s) − 1, 7, 2)* will have a value of *7 V* until *V (s)* reaches *4.67 V* when *1.5*V(s) = 7 V* then it blends smoothly by *9 V* to an upward sloping straight line.

uplim(x,y,z)

The format is *<value> = uplim(<limit1><limit2><transition distance>)* so it is essentially the same as **max** except that instead of a sudden break, the output changes smoothly over the transition distance **z**.

Example – uplim

The expression *V = uplim(0.9 * V(s), 5, 0.5)* will cause the voltage to follow whichever is larger of *0.9*V(s)* and *5 V* and then make a smooth transition over the next *0.5 V* from the breakpoint. This is shown in schematic ('LimUplimIf.asc').

lim(x,y,z)

This is of the form: *Value = (limit1, value, limit2)* where the value is constrained by the two limits, taken in either order. There is no restriction on the values or functions.

Example – lim

The expression *V = limit(3, 5, V(s))* will cause the output to follow *V(s)* between the limits *3 V* and *5 V*; else it is either of the voltage limits, schematic ('Limit3.asc'). And if we interchange *3, 5*, it is the same.

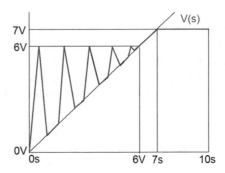

Fig. 5.11 Limit 2

Example – lim (2)
The schematic $V = limit(6 * sin\ (V(s) * 20), V(s), 7)$ is used in conjunction with the
voltage source **PULSE(0 10 0 10)** which is a pulse rising from *0 V* to *10 V* over *10 s*.
This is schematic ('Limit 2.asc'), and the output is sketched in Fig. 5.11. This shows
that for the first *6 s*, the output cannot go lower than *V(s)*, and so the sine wave can
rise to its *6 V* peak but is clipped by the current value of the ramp voltage. From *6 s* to
7 s, *V(s)* exceeds the peak of the sine wave so the graph is just a straight line. Then at
7 s the upper limit comes into play and the voltage is limited to *7 V*.

Explorations 10
1. Open and run the schematics and Examples mentioned above.
2. Open and run schematic ('Limit 3.asc'). Note that the expression $V = limit$
   $(8 * sin\ (V(s) * 20), -V(s), 7)$ constrains the output to a maximum of *7 V*, so
   the positive and negative peaks of the sine wave are rounded until *8 s* after which
   the lower peak is not clipped, only the upper.
3. Exchange the parameters in *max* and *min* statements, for example, so that
   Example 6 is $I = max\ (\ min\ (7,10 * cos\ (10 * V(s))), 5)$.
4. Build a circuit with a *10 V* ramp over *10 s* and a **B** with the function $V = limit$
   $(5 * sin\ (V(s) * 20), V(s), 3 * sqrt(V(s)))$, and for the first 5 s the lower part of the
   sine wave is clipped by the ramp and the upper part by the square root. This is
   schematic ('Limit 4.asc').

Power Functions

There are three that behave slightly differently. None of them require integer inputs.
All calculate *x* to the power *y*.

pow(x,y), pwr(x,y), pwrs(x,y)

The format is **value** $= x^y$ where **y** can be positive or negative, integer or fraction. The first, **pow,** returns the real part, so an attempt to find the square root of -1 will return zero. The second, **pwr,** returns the absolute value meaning it is always positive, so although $pwr(-3,2) = 9$, it will return 27 for $pwr(-3,3)$ not -27. The third, **pwrs,** is the same as the previous only the sign applies so negative values raised to a power return a negative result.

We should note that negative powers will go to infinity if $x = 0$. These are illustrated by the schematic ('PowPwrPwrs.asc').

exp(x)

This is essentially the same as the previous; only now it is to the base e. It is a function that will occur time and time again in the following chapters.

Example – exp

Another example is schematic ('exp(sine).asc') where we have a B voltage source whose function is $V = $ **exp (sin(100 $*$ time))**. There are two interesting points to note. The first is the rounding of the negative half of the voltage $B1$ because a negative exponential is less than unity. The second point to note is that the voltage source uses radians, but if we add a trace **sin(5.7 k$*$time)**, this is in degrees.

A quick check is to move the cursor over the first peak at *15.73 ms*. This is *1.573 r*, and converting that to degrees gives 90. Working in reverse, if that time represents 90°, we must multiply it by *5.72 k* being the number of degrees to a radian and remembering the time is in ms. We can also note that the *1 V* peak of *V(s)* becomes *2.72 V* from *B1* being the exponential of 1.

sqrt(x)

This only applies to positive values; else the result is zero.

Random Numbers

There are three similar functions. In every case the amplitude is a random number independent of x which determines the rate at which values are produced.

Bandwidth

These functions produce a new value every time x changes. In other words, the rate at which values are generated is the rate at which x changes. Thus a simple $V = $ **rand (100)** will produce one value only since x is constant. To create a train of values, we can use time as $V = $ **rand (100 $*$ time),** and because time increases linearly (pace Einstein), this will create a steady stream of *100* values every second, that is, *100 Hz*. However, because the values are not of uniform amplitude, the waveform consists of a spectrum of frequencies, and *100* is approximately the upper 3 dB point.

Spectral Density

Instead of *time*, we may use any other varying quantity such as $V = random(V(s))$ where $V(s)$ is a varying voltage. This too will create values at a variable rate depending on how the voltage changes: if it is a steady ramp, then the density of value will be constant. But we can also have $V = random(V(s) * * 2)$ which creates values whose density increases linearly with voltage.

Resetting the Seed

By default, these functions repeat the same sequence of random numbers. We can reset it from **Tools→Control Panel→Hacks!** and then check **Use the clock to reseed the MC generator.**

rand(x), random(x)

Both return a number between *0* and *1:* at rate *x*. so if Vs is a ramp voltage $V = rand (10 * V(s))$ creates 10 values per second. The difference is that **random(x),** instead of creating step changes, smooths the transition somewhat.

white(x)

Here the value ranges from -0.5 to $+0.5$ and is slightly smoother which we can see if we add *0.5 V* to the *white* trace in the schematic below so that it overlays the *random* trace. It can be used as a white noise source with a spectral width limited by *x*. We can see these three in schematic ('RanRandWhite.asc').

Example – white

Run the schematic ('RanRandWhite.asc'), and an FFT analysis shows that the spectrum for *white* is limited to 10 Hz, which is the rate at which values are created, and the amplitude falls off quickly above just a few Hz. On the other hand, the spectrum for *rand* extends to some 100 Hz because the square pulses are rich in higher harmonics, although the decline in amplitude is the same.

But if we increase the pulse amplitude to *500 V* and add a minimum time step of *10u*, we see the spectrum is flat to over 100 Hz before it falls off quickly. This shows that it requires care if we want a really broad band white noise source: a pulse of 50 kV giving a rate of 500 k sample per second is more than enough for audio work. Schematic ('RanRandomWhite.asc') uses different functions to create the noise.

Note that if we increase the transient time beyond the end of the ramp, the outputs remain constant.

Logarithmic Functions

In many ways these are the converse of power functions. The logarithm of a number is the power to which the base must be raised to get the value. There are two commonly used bases: *10* and *e*.

log10(x)

This is the standard notation for a logarithm to the base 10 and works for any value.

log(x), ln(x)
The first usage is not standard and in most contexts is the logarithm to the base ten. These are essentially the same as the previous except that it uses the base *e* instead of *10*, so now we have different values in that: ln(10) = 2.303 and not *1*.

Example – log
The schematic ('log(sine).asc') traces both logarithms with a *1 V* sine wave input. The trace runs swiftly when the input is positive, but the 'Marching Waves' crawl slowly during the negative portion as **log10(sin(V(s)))** plunges to −6e+023. But logarithms are not defined for negative numbers.

Explorations 10
1. Open and run ('PowPwrPwrs.asc'). Notice that the power functions change the shape of the wave, not just its amplitude.
2. Open and run ('PowPwrPwrs.asc') with logarithmic axes. Note that the traces are straight lines, but if we extend the run from *10 s* to *100 s*, the time greatly increases. Admitted, there are five sources, but even so, this is a warning that these functions are very costly in time. Measure the slopes of the traces and check that they agree with the input voltage.
3. Explore other input waveforms, including some that go negative.
4. Explore further schematic ('exp(sin).asc') by finding the exponent of other functions.
5. Note the waveforms in schematic ('RanRandWhite.asc') and the FFT distribution. Try different functions instead of a ramp input.
6. Run ('log(sine).asc'). We might wonder about the logarithms of negative numbers.

Trigonometrical Functions

Let us remind ourselves of the simple trigonometrical relationships which we can illustrate with Fig. 6.3 from the Chap. 6. By definition:

$$\sin(\phi) = \frac{a}{c} \quad \cos(\phi) = \frac{b}{c} \quad \tan(\phi) = \frac{a}{b} \tag{5.1}$$

from which it follows that:

$$\tan(\phi) = \frac{\sin(\phi)}{\cos(\phi)} \tag{5.2}$$

sin(x), cos(x), tan(x)
The simulation is in schematic ('SinCosTan.asc') where the angles are in radians – a little confusing because we enter delay angles and so forth in degrees in the various dialogues. We see that the cosine and sine traces run from −*1 V* to +*1 V*. However, the tangent becomes infinite at multiples of 90°, and we see the graph spiking at hundreds of volts. We can visualize this in Fig. 5.12 where we start with a right-

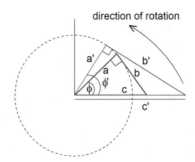

Fig. 5.12 SinCosTan

angled triangle with angle *Φ*. We now rotate the radius *a* counter-clockwise to *a′* where we see that to preserve the right angle, the two sides *b* and *c* must grow to *b′* and *c′*. The result is that the tangent *b′/a′* rapidly increases towards infinity, whilst the sine *b′/c′* tends to unity at 90°.

Zoom in on the area between the spikes to see the familiar tangent graph. The height of the tangent spikes at multiples of 90° depends upon the time step of the transient run. If we reduce it to *0.001 s*, the picture changes: now they can be resolved to tens of kV.

Example – Trigonometrical Functions

We are not limited to simple trigonometrical functions. We can have **V = sin (x) ∗ cos (x)**. Using:

$$\sin(x)\cos(y) = \frac{1}{2}(\sin(x+y) + \sin(x-y)) \text{ we have } \sin(x)\cos(x) = \frac{1}{2}(\sin(2x))$$

so we expect a result at twice the frequency and half the amplitude. This what we see in schematic ('sin(x)cos(x).asc').

Example – Trigonometrical Functions (2)

An interesting example – but probably of no use – is schematic ('SinCostan2.asc') where the arbitrary voltage sources are **V = sin (V(in))**, **V = cos (V(in))** and **V = tan (V(in))** and where *V(in)* is a 1 V sine wave of 1 Hz. We find that *V = sin(V (in))* only reaches 0.84 V at 90°, whereas we might have expected 1 V since if *Vin = 1 sin(90) = 1*. But remember that **V = sin (V(in))** and the like calculate the sine, cosine and tangent from the amplitude of the input voltage converted to radians. Thus at 90° the input amplitude is *1 V*, so the function finds *sin(1r) = 0.84* and likewise *tan(1r) = 1.557*.

This also is the explanation for the reduced amplitude cosine wave because *cos (1r) = 0.54* then *cos(0) = 1* and *cos(−1r) = 0.54*, so we have double the frequency at a reduced amplitude and a DC offset.

Explorations 11

1. Open ('SineFunct.asc'). This is a sort of voltage-to-frequency converter where the sine wave frequency depends linearly on the amplitude of the ramp. Try different amplitudes and frequencies remembering these are measured in radians per second.

2. There are lots more interesting relationships that can be explored, double angles, sums and differences of angles, $\cos^2 + \sin^2 = 1$ and so on. You could use ('SinCosTan2.asc') as a starting point, always keeping in mind the point about degrees and radians.
3. Although not mentioned, we can look at cot (= 1/tan), sec (=1/cos) and csc (=1/sin).

asin(x), acos(x), atan(x)

These return the real part of the angle in radians corresponding to the trigonometrical function. The imaginary part is ignored. These functions are sometimes written as sin^{-1} (x), cos^{-1} (x) and tan^{-1} (x).

The y-axis in schematic ('aSinaCosaTan.asc') is the angle in radians and the x-axis is the sine, cosine or tangent of the angle. The values are restricted to the range over which these are real: for sine and cosine, this is the range $1 <= x <= 1$ since, for example, a value of 1.2 is complex, not a real number. In the case of **asin(x)**, this corresponds to the range π to $+\pi$. Outside this range, the function returns $-\pi/2$ for negative values and $\pi/2$ for positive ones. For **acos(x)** the range is 0 to π because cos $(180°) = -1$ and $cos(0°) = 1$. And the function returns 0 for positive values and π for negative ones outside the range. There is no restriction on **atan(x)** which can range from $\pm\infty$.

Hyperbolic Functions

sinh(x), cosh(x),tanh(x)

We should first explain the difference between hyperbolic and circular functions. This is shown in Fig. 5.13. On the left we have the 'usual' sine and cosine functions where they are the projections of the vertical and horizontal ordinates of the point T on the unit radius R. The hyperbolic functions are similarly defined by the projection of the vertical and horizontal ordinates H on the unit vertex V.

From this we can see that at $x = 0$ $cosh(x) = 1$ and $sinh(x) = 0$, but as x increases both tend to the same value and approach infinity at high values of x. From this it follows that tanh will tend to a constant value of 1. The Wikipedia article is useful

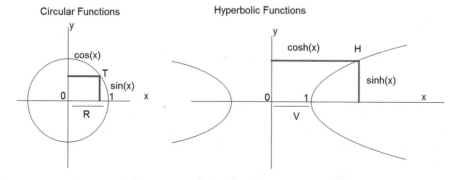

Fig. 5.13 Circular and Hyperbolic Functions

and there is an animated explanation at http://www.sosmath.com/trig/hyper/hyper01/hyper01.html

Example – Hyperbolic Functions

If we run ('SinhCoshTanh.asc') with **.step param x 0 1 0.01**, we can see the two values beginning to coalesce towards the end. If we increase the time to **.step param x 0 10 0.1,** the two traces finally are almost indistinguishable.

We can also note that **tanh** tends to a constant value since it is defined similarly to Eq. 5.2 as

$$\tanh(x) = \frac{\sinh(x)}{\cosh(x)} \qquad\qquad (5.3)$$

Example – Sinh as Exponentials

Sinh is defined as:

$$\sinh(x) = \frac{e^x - e^{-x}}{2}$$

We can see that in schematic ('SinhAsExp.asc'), the two traces of $\sinh(x)$ and the exponentials coincide.

asinh(x), acosh(x), atanh(x)

These return the angle corresponding to the argument x.

We can understand this by setting the x-limit of ('SinhCoshTanh.asc') to something like 2.0. We can now read, for example, y = 2.35 V x = 1.5 for cosh. Then on ('SinhCoshTanh.asc') when y = 1.5, we have approximately x = 2.35.

Explorations 12

1. Run ('SinhCoshTanh.asc') with negative values and note the symmetry about the y-axis
2. Create exponential functions for cosh and tanh by modifying ('SinhAsExp.asc')
3. Check Eq. 5.3 by building all three functions
4. Explore multiples such as $cosh(5*x)$

Calculus

LTspice has the derivative of x against time.

ddt(x)

This finds the time derivative of x. We can have a lot of fun here testing whether the product rule and others are followed. Note, however, that we can only differentiate using time, so **.step param x**... will not work. We can simply place a **B** source on the schematic, and then **V** = **ddt(time** * * **3)** will correctly give 3*(time)**2 . This is schematic ('Differentiate t 3.asc').

We can also differentiate the voltage or current from another source.

Example – ddt

If we place a **B** source on a schematic and set its value as $V = 5*(time)**2$ which we call **V(in)**, we may apply this as the input to a second **B** source set to $V = ddt(V(in))$ which we call **V(out)**. After 10s we find *V(in) = 500 V(out) = 100 V*. This is schematic ('Differentiate f(x).asc').

On the other hand, we may have a simple input such as a sine wave **V(s)** from a voltage or current source and want to differentiate some function of it such as $V = ddt(0.2 * V(s))$. This is schematic ('Differentiate sine2.asc'). And $V = ddt$ $(0.2 * V(s) ** 5)$ is interesting.

In schematic ('Differentiate sin 1.asc'), we have a sinusoidal input of *5 V 1 Hz* which we differentiate and find the output has a peak of *31.41 V* because the differentiation uses radians so *1 Hz = 2π rad*, and therefore when we differentiate, we have $5*2\pi*cos(1) = 31.41\ V$.

idt(x), idtmod(x)

These integration functions offer additional optional conditions. But first let us try a simple integration.

Example – idt

To test **idt(x)** we place a **B** source on a schematic ('Integrate.asc') and set the value to: $V = idt(time)$. A transient run of 1 s produces a nice quadratic curve of the integration starting from 0. To test the initial condition as **idt(x,ic)**, we change the value to $V = idt(time, 0.1)$; we find the same curve but offset by 100 mV on the y-axis. Finally to test the reset condition *a* in **idt(x,ic,a)**, we use $V = idt(time, 0,$ **time > 0.8**) , and now the integration falls sharply to zero at *time = 0.8 s*. And note that we must include some initial condition, even if it is only zero.

idtmod(x)

This is the similar but has an optional offset for the output – the *constant of integration*. The first three parameters have one difference from the previous: with $V = idtmod(time, 0.1, 0.3)$ the output will start at *100 mV* and rise to *300 mV* when it will reset to zero, but this time it will start to integrate again, not remain at zero, schematic ('Integrate idtmod.asc'). The output offset is added to all the integrations so $V = idtmod(time, 0.1, 0.3, 0.2)$ will start at *300 mV (100 mV initial conditions + 200 mV output offset)*, rise to *500 mV (300 mV reset point + 200 mV output offset)* and finally fall to *100 mV (the output offset)*.

Explorations 13

1. Open the other differentiation schematics and run them. Set up other functions and test them.
2. Test the schematic ('Integrate Sine V.asc').
3. Try differentiating and integrating more complex functions.

Miscellaneous Functions

We have omitted four that do not fit easily into the previous categories. They are illustrated in schematic ('Miscellaneous.asc').

absdelay(x,t), delay(x,t,[tmax])

This delays the start of x until time t which must be positive and greater than zero. The delay can be stepped, so *.step param d 0.001 1 0.25* will step the delay d in $V = absdelay(V(s) - 2, d)$ from *1 ms* to *1.25 s*. Up to the delay time, V is zero and then follows *V(s)-2* volts. The optional **tmax** seems to have no effect.

Table(x,a0, a1,b0, b1)

The values must be separated by commas. Whenever x reaches the first value of the pair *a0,b0*. . ., the output switches to the second value in the pair. It does not throw an error – just makes life confusing – if the values are not in ascending order of the first in each pair. The table is read repetitively, so given a sinusoidal input $V = Table(V(s), -3, 3, 1, 3, 3, 1, 4, -4, 6, 8, 8, 5)$, we see a waveform with pairs of 'ears'.

hypot(x,y)

This calculates $\sqrt{(x^2 + y^2)}$ so negative values are accepted. And it is not restricted to simple numbers: $V = hypot(-V(s), 5 * sin (V(s)))$ will produce an interesting mountain range.

However, we must remember that although we can input angles in degrees, the computer uses radians, so a frequency of *1 Hz* is 2π rad/s or 6.283 rad/s, and this means that differentiating a *1 Hz* sine wave gives a *6.283 V* cosine wave.

Explorations 14

1. Open schematic ('Differentiate t squared.asc'), and add a linear value so that $B2 = a.time + b.time^2$.
2. Open ('Differentiate t 4.asc') and experiment with different powers, including negative. $Time^{-4}$ produces a sharp change at the origin. Resolve it by **Start saving data** *0.1* in the **Edit Simulation Command**.
3. Open ('Differentiate sine 1.asc'), and note the relationship between the voltages. Try different inputs, and confirm that the voltage of *B1* is 6.283 times the input.
4. ('Differentiate sine2.asc') joins the two sources. Note the phase differences between the currents and voltages, and confirm that KCL is obeyed for the instantaneous values.

5.4 Dependent Sources

These are four terminal sources whose output depends upon the conditions at their control terminals. They are often used to transfer voltages and currents from one part of a circuit to another.

5.4.1 Voltage-Controlled Voltage Sources (E,E2)

These are four terminal voltage sources where the output voltage is controlled by the voltage between the two control terminals. *E2* is the same as *E* but with reversed polarity. It has several uses.

VCVS Voltage Scaling
It can be used simply to scale the voltage between the control terminals. Thus, **Value** **10** will give an output ten times larger than the input, schematic ('VCVS Gain.asc').

VCVS Look-up Table
This uses pairs of data points delimited by commas. This is of the format:

$$(Vin1\ Vout1, Vin2\ Vout2, Vin3\ Vout3 \ldots)$$

Until the input reaches *Vin1*, the output remains at zero. From then until the input is *Vin2*, the output rises linearly from *Vout1* to *Vout2* and so for the remaining values. These must be in ascending order of input. A sequence *(1,0 3,5, 2 1)* will cause confusion. If the first input voltage is not reached, the table fails and the output remains at zero. A simple table is:

$$table = (1\ 0, 1\ 10, 1.5\ 5)$$

Until the input reaches *1* V, the output remains at zero; at that point the output rises immediately to *10* V and then falls to *5* V at the time when the input reaches *1.5* V. This table of values will continue to be applied for the whole transient period. An example is the schematic ('VCVS Table.asc').

VCVS Laplace
This is an extremely valuable tool in association with an AC analysis, not transient, which we shall use in later chapters. This replaces $j\omega$ with s to analyse the frequency response of the circuits. The example schematic ('VCVS Laplace.asc') has the value:

$$Laplace = (s^2 - 4s + 7)/(s^2 - 7*s + 50)$$

and the response is a very sharp notch at 220 mHz and a very unusual frequency response at higher frequencies. Note the use of '^' for exponentiation, not '**'.

VCVS Value
This is an alternative to the **BV** voltage source. However, the 'LTSpice Help' says that it is better to use a G source with a large parallel resistor as there are fewer convergence problems and it will compute faster. The **Value** option is a little awkward to use; there are two things to take care of. The first is that the expression must be in curly brackets as:

$$value = \{5*time\}$$

The second is that it must be a two-terminal source, so the control terminals must be removed. We can do that by opening the netlist to edit it by **View→Spice Netlist** then right click in the netlist and select **Edit as Independent Netlist,** and save it with some appropriate name. We are then free to edit it however we please, for example:

```
* G:\Users\Me\Desktop\LTSpice Book new\03 AC\Circuits(asc)\VCVS
Value.asc.
R1 N001 0100.
E1 N001 0 value = {5*time + sin(pi*10*time)}
.tran 5
.backanno
.end.
```

We can then right click and select **Run** to run the schematic from the netlist. We can close the schematic because that it is only a visual guide to helping us create the netlist. We can add and remove components from the netlist. What we cannot do is to construct a visual schematic. That is reasonable since the netlist contains no information on the position and orientation of the components.

Explorations 15
1. Open the schematic ('VCVS Gain.asc'), and explore different values, integer, fractions, positive and negative.
2. Open the schematic ('VCVS Table.asc'), and create and save some interesting tables. Explore scaling value and time.
3. At our present knowledge, the working of the schematic ('VCVS Laplace.asc') is opaque. Nevertheless, experiment with different numerators and denominators.
4. The' VCVS Value.net' file can be edited. Amuse yourself by doing so.

5.4.2   Current-Controlled Current Source (F)

There is a classic SPICE usage where a dummy voltage source set to zero is often used to measure current, schematic ('CCCS Classic.asc'). The current from source *I1* passes through voltage source *V1* which has neither AC nor DC voltages and is only there to pass the current to *F1*.

CCCS Current Transfer
This is a useful component for transferring current from one part of a circuit to another where the current through a voltage source is passed to this component and multiplied by the gain of this source. It is not possible to use the current through a resistor, only a simple voltage source or an *E* voltage source. The syntax is **Value** *<name of source> <gain>*. For example, the *Value* field *V1 3* takes the current through voltage source *V1* and multiplies it by *3*, in the schematic ('CCCS Test.asc'). Note that there is no equals sign or other text, just the two values.

CCCS as Alternative to Voltage Source B
A second use is as an alternative to the current source where, like the *E* source, the output is defined by *value*; only this time the control only has two terminals so that:

$$value = \{\, exp\,(time)\}$$

will run with no more ado.

5.4.3   Voltage-Controlled Current Source (G,G2)

This has the same functionality as the voltage-controlled voltage sources *E,E2* and is an alternative to the **BI** source. It has the same problem with the **Value** option.

5.4.4   Current-Controlled Voltage Source(H)

This is a two-terminal current dependent voltage source and is an alternative to the voltage source **BV**. It has the same functionality as **F**.

Explorations 16
1. Open the schematic ('CCCS Classic.asc', and create a larger circuit. Use the source *F1* to transfer the current through it to another passive circuit. Compare to the schematic ('CCCS Test.asc').
2. Explore the schematics ('CCCS Value.asc') and ('VCCS Value.asc').
3. Build a test circuit for an **H** source.

5.5   Summary

This chapter has covered the functionality of the various current and voltages sources of LTspice.

- The independent voltage and current sources can provide DC and AC as a sine wave, or a pulse, or an exponential, or an FM modulated wave, or from a table of time/voltage values.
- The voltage source can have a series resistance and a parallel capacitance.
- The current source can have a parallel resistance and can be a sink rather than a source.
- The output of the arbitrary current and voltage sources **bi, bi2, bv** can be virtually any function.
- The dependent sources **E, F, G,H** either have either a constant gain applied to the input or a look-up table of pairs <control value, output value> or a Laplace function which is only applicable to an AC analysis.

Chapter 6
AC Theory

6.1 Introduction

Most of the interest in electronics lies in the response of circuits to AC signals. This is not to say that a DC analysis is unimportant, for it is essential to establish the correct working conditions for many electronic components such as transistors.

We may treat of a signal either in the time domain or the frequency domain. These are radically different in the information they report. The former enables us to view the shape of waveforms by making point-by-point measurements of the circuit conditions at certain intervals of time. This is a *transient analysis* which we shall use in this chapter. Most circuits are not truly linear; in the extreme, clipping occurs if the amplitude of the signal exceeds the power supply voltage, but more generally, harmonics will be introduced by non-linearities in the transfer function from input to output. All this can be seen in a transient analysis of the circuit. It is possible to repeat the analysis at different frequencies using the *.step* command: the practical limit is about ten traces before they become too confusing.

On the other hand, we can sweep the frequency over as wide a range as we like and plot, not the shape of the waveform itself, but the magnitude and phase of voltages and currents as a function of frequency. This is an *AC analysis*. And here is the problem: this assumes that the circuit is linear because SPICE first evaluates the DC operating point and then takes the slopes of the characteristics of the devices – in effect making the circuit linear – which are used to calculate the frequency response. We can thus find absurdities such as an output of millions of volts at frequencies well beyond visible light. This stands as a warning that we must never use SPICE blindly but always bear in mind the limitations of the physical circuit.

From this we see that the two analyses go hand in hand: the transient analysis to check the linearity (or non-linearity) of the response and the AC analysis to find the frequency response, which in turn may lead to the need for more transient analyses if we change the circuit to improve the frequency response. However, in this chapter we shall focus on the transient analysis.

© Springer Nature Switzerland AG 2020
C. May, *Passive Circuit Analysis with LTspice®*,
https://doi.org/10.1007/978-3-030-38304-6_6

Of necessity, the AC sections are entirely mathematical. But it is necessary because once we are dealing with circuits dissipating more than a milliwatt or two, and especially with power amplifiers, thyristors and power supplies, we can find ourselves encountering strange waveforms and the ever-present danger of exceeding device ratings, be they voltage, power or current.

The later sections of this chapter cover the series and exponential forms of sines and cosines and the constant *e*. The mathematics is extensive rather than profound, and it is worth at least reading through and taking note of the very important Euler's relationships that derive exponential forms for sines and cosines from which trigonometrical identities can be derived.

We shall also make extensive use of *.meas* rather than using the cursor: the results are more accurate and are reported again each time we change the parameters so we shall now discuss the new ones.

6.1.1   Some More '.meas' Methods

The following section describes the measurement statements that we will need and gives simple illustrations by applying them to a sine wave. Several measurement options were discussed in the chapter on 'DC Circuits'. We shall here make extensive use of most of the remainder. These are accessed from the drop-down **Genre** list in the *.meas* **Statement Editor**. There are two points to remember: first, turn off compression; second, if the calculated results diverge from the simulation, try a smaller minimum time step in a transient run. Also, if we cannot remember the syntax, and repeating what has been written before we can:

- Press *S* on the schematic to open the dialogue **Edit Text on the Schematic.**
- Type *.meas* including the leading dot.
- Press **OK** and drag the text to an appropriate place.
- Left click to place the text.
- Right click on *.meas* to open the *.meas* **Statement Editor.**

Transient Analysis and Compressed Data

The values can be significantly in error – several percent – unless we inhibit data compression which is a lossy process. We must go to **Tools->Control Panel→Compression,** and uncheck **Enable 1st Order Compression.** However there is a penalty, and that is that the *.raw* data file is very much larger than before. The difference can be quite amazing, ranging from a few tens of kilobytes for the compressed file to a megabyte for uncompressed. For general viewing, or uncritical measurements, the compressed file is adequate.

As an alternative, we can compress the data but use very short steps in a transient analysis: the results are identical. And in any case, if we use compressed data and the default time step, sometimes the error may be less than 1%.

. meas DERIV AT

This finds the derivative of the **Measured Quantity** and has two forms. The first is to measure the derivative at some point on the x-axis. For this we keep the default AT in the edit box (1). The syntax is:

.meas < Result Name > DERIV < Measured Quantity > AT < condition >

thus:

.meas slope DERIV I(V1) AT 0.2 s

which would find the derivative of *I(I1)* at a time of *0.25 ms* (although the *s* is not needed) and place the result in *slope*. This is shown in Fig. 6.1 where the items marked with an asterisk are not visible.

Example – .meas DERIV AT

The schematic ('meas DERIV.asc') is just a voltage source and parallel resistor. The measurement directive is shown on the schematic and is set to time *0.25 s*. If we click on this, the edit window opens but the TIME field is not filled. If we run the

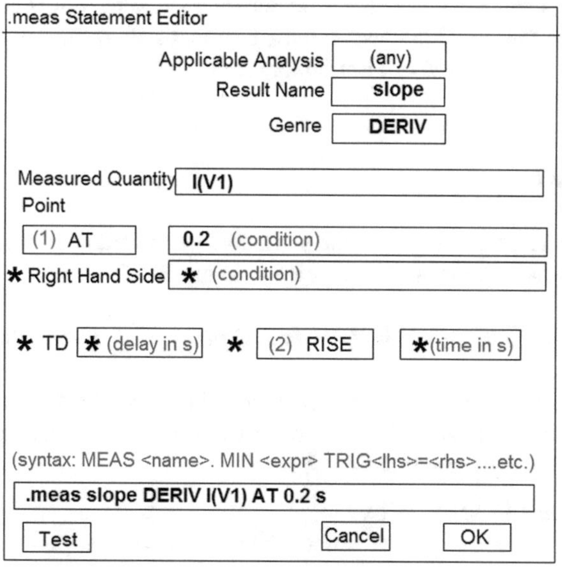

Fig. 6.1 Measure Editor DERIV

simulation, we find the slope is of the order of *e-008*. If we reduce the time step to, say, *10 μs* and turn off compression, we find the correct answer of *0*.

.meas DERIV WHEN

The second use is to measure the derivative when some condition is met:

.meas < Result Name > DERIV < Measured Quantity > WHEN < parameter >
< Right Hand Side >

The edit box to the right of *WHEN* now contains the expression used for finding the derivative. Very often it is the measured quantity but it need not be, for example *V(out)*. Two other edit boxes appear below. The edit box (2) to the right of *Right Hand Side* contains the condition such as *10*. Thus:

.meas slope DERIV I(R1) WHEN V(out) = 10

would measure the derivative of *I(R1)* when *V(out) = 10*. Note that it is an error to include the equals sign. The last edit box *TD* is used for a delay time and we leave it blank. The drop-drop list has options of RISE, FALL and CROSS as we saw in Chapter 2.

Example – .meas DERIV WHEN
This is schematic ('meas DERIV WHEN.asc') and uses the above setting. Notice that WHEN is used to calculate the time that the condition is met and turns it into a DERIV AT and the **SPICE Error Log** shows a slope of *0* at *0,249953* seconds. Again note that compression must be turned off and a short time step used; else the result will not be zero, or it will not find a result.

.meas FIND WHEN

But we cannot find the time when the derivative has a certain value, for example, a turning point of a waveform. The solution is to use FIND as:

.meas turnpt FIND I(R1) WHEN tan (V(out)) = 0.1 TD = 0.1m RISE = 1

which will find the first time that the tangent to *V(out) = 0.1* and save the result in *turnpt*. The *RISE = 1* (or whatever number we want) is essential, and *TD = 0.1* is needed to avoid a false result at *time = 0*.

Example – .meas FIND WHEN
Using the above directive '.meas turnpt FIND...' in schematic ('meas FIND WHEN. asc'), it is important to note that the tangent is found by interpreting the value of the

voltage *V(out)* as an angle; it is not the phase angle of the voltage. If we trace *tan(V (out))* and use the cursor, we find this is *0.1* at time *96.78 ms* and *V(out)* = *5.647 V*. The measure statement gives slightly different, more accurate, figures of *I (R1)* = *0.00571 A* at a time *0.09676 s*, and the same result occurs at *time* = *1.0967 s*.

If we change to a tangent of zero we should note that, depending on the step interval, a delay of *0.1 ms* may not be long enough and 1 s is better which causes the initial value at *0 s* to be ignored and the first time the rising waveform meets this condition is at *1 s*. The current in the resistor is reported as *0.89637e-013 A*. Using the cursor on the trace, we find just a few microamps as it is difficult to get exactly *1.000000 s*.

.meas PARAM

This is a convenient way of processing other measurement results where we can apply almost any mathematical manipulation to the data and save it as the named parameter where only the **Measured Quantity** edit is visible.

Example – .meas PARAM

The schematic ('meas PARAM.asc') uses two statements to measure the current and voltage and combines the two in the *result* parameter. Note that the two measurements are made at different times. We can also note that we could have:

$$.meas\ current\ FIND\ I(R1) **2\ AT\ 0.25$$

rather than squaring the current in the *PARAM*. And note the two asterisks for power. Something as fanciful as:

$$.meas\ Fun\ PARAM\ I(R1) * sin\ (time) - 50/cos\ (time) **2$$

is legal but not very helpful because it will be noticed that there is no measurement time specified. However, LTspice will resolve it using the last data point of the simulation.

.meas AVG, RMS, PP

These all will open the full editor. The essential information is:

$$.meas < Result\ Name > AVG < Measured\ Quantity >$$

and this will measure the average over the whole time run. Likewise:

$$.meas\ rmsV\ RMS\ V(out)$$

will measure the RMS value, but we can define an interval to find the RMS voltage
and return the result in *rmsV* by changing TRIG/TARG to FROM/TO as:

$$.meas\ rmsV\ RMS\ V(out)\ FROM\ 0.2\ TO\ 0.5$$

and similarly for the average and peak-to-peak values.

Example – .meas AVG, RMS, PP
The schematic ('meas AVG RMS PP.asc') illustrates these measurements where the
time limits are left open and the full run time is used. With compression turned on,
the average is only microvolts, with compression turned off it is femtovolts.

.meas INTEG

This integrates the area under the trace respecting the sign and so is similar to the
AVG except it takes functions so that we can integrate, for example, V^2. This is
also illustrated in the above Example where the current and voltage are multiplied
over $1\ s$.

Explorations 1
1. Run the simulations listed above, but try different measurement conditions and
   check the results by calculation.
2. Notice the effect of using delay times coupled with the RISE number.
3. Explore other functions applied to the waveforms.

6.2   AC Basics

In this chapter, we shall concern ourselves almost exclusively with sine waves. So
we may well ask what is so important about them. To that there are two answers. The
first is that the vibrations or oscillations of many physical systems are sinusoidal; the
oscillation of a clock pendulum is a good approximation, and the vibration of a mass
suspended by a spring is exactly so. This is known as *simple harmonic motion* which
we shall now explore.

6.2.1   Simple Harmonic Motion

This is well described in the Wikipedia article. Imagine a mass at rest freely
suspended vertically from a spring. If we now pull the mass down, the upward
force exerted by the spring will increase, thus trying to restore the mass to its rest
position. And according to Hook's Law, this force F will be directly proportional to
the displacement x or:

$$F = k.x. \tag{6.1}$$

where k is a constant. But if the mass is free to move, then Newton's Law of Motion states that we can equate the acceleration a of the mass m to the force. So we can replace F in Eq. 6.1.

$$m.a = k.x \tag{6.2}$$

Remembering that velocity is dx/dt and acceraration is the rate of change of velocity, combining the two constants m and k into c and using calculus notation:

$$\frac{d^2x}{dt^2} = c.x(t) \tag{6.3}$$

and this describes the position and motion of the mass at any time t.

This is a *differential equation*. We can solve it by noting that we have both time and the second derivative of time. And as any solution must be dimensionally consistent, we cannot have time equated to t^2 or $\sqrt{t}$ (whatever meaning those might have). There are two functions which might serve; one is an exponential: if we differentiate e^{at}, we get ae^{at} and that is still just a function of time only multiplied by a dimensionless constant a, and differentiating it a second time still leaves us with time. However, an exponential either increases to infinity or decreases to zero depending on the sign of a. Neither seems to fit out experience of a clock pendulum.

The second function is a sine or cosine. If we differentiate $sin(at)$, we get $acos(at)$, which is still a function of time multiplied by the dimensionless constant a, and differentiating again brings us back to a sine, once more a function of time, not t^2 or any other power but just time with a^2. This looks more promising. So let us start with our proposed solution of $\frac{d^2x}{dt^2} = Asin(at)$ where A is another constant, the amplitude. And if we integrate we have:

$$\int Asin(at) = -\frac{A}{a}\cos(at) + C \tag{6.4}$$

where C is a constant – not a function of time – which we can assimilate in the solution as a constant offset b so that: $x(t) = Asin(at + b)$.

At this point it is worth exploring the nature of this sine wave a little more closely. We can think of it as a vector in Fig. 6.2 that rotates anti-clockwise at a constant angular velocity ω starting from 0 degrees at time zero which is drawn as OA. After a little time, the vector has rotated by an angle α to point B where the vertical height h is the amplitude which we project onto the time axis on the right at time B'. In similar fashion we can project every point swept out by the tip of the vector as it passes through points C, D, E, F and G in turn, and the result is a sine wave with the negative peak equal in magnitude to the positive peak.

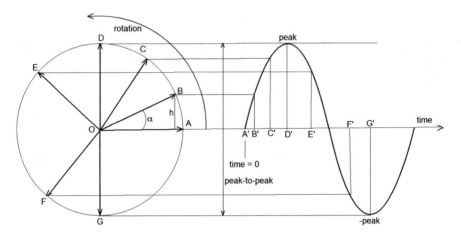

Fig. 6.2 Sinewave development

6.2.2 Waveform Synthesis

This is the second answer why sine waves are important and one of greater interest in electronics: it is that any repetitive waveform can be synthesized from a series of harmonics.

$$y(t) = a_0 + a_1 f_1 + a_2 f_2 + \ldots a_n f_n \tag{6.5}$$

where y is voltage or current, a_0 is the DC component and f_1, f_2.. are the harmonics of amplitudes a_1, a_2... Intuitively, we can see that these higher frequencies must be exact multiples of the fundamental so that each cycle of the fundamental contains the same number of cycles of the higher frequencies. At this point, we merely note the fact as a justification of the importance of sine waves. We shall say more about this later in the chapter. One realization is merely to connect voltage sources in series. Each generating a sinewave of the appropriate harmonic and amplitude. The second - more elegant- option is to use the flexible 'Bv' source.

Example – Waveform Synthesis Using the 'Bv' Source

We can create an approximation to a square wave very easily by using the arbitrary voltage source and adding terms until we get tired, schematic ('Squarewave from Harmonics.asc'. Each is of the form *1/a∗sin(a∗time)*. This is an alternative to the first exploration of creating many voltage sources.

$$V = \sin(time) + 1/3 * \sin(3 * time) + 1/5 * \sin(5 * time) + 1/7 * \sin(7 * time)$$
$$+ 1/9 * \sin(9 * time) + (1/11) * \sin(11 * time) + (1/13) * \sin(13 * time)$$
$$+ (1/15) * \sin(15 * time)$$

Explorations 1

1. The schematic ('Waveform Synthesis.asc') consists of five voltage sources connected in series. It produces a lumpy sawtooth waveform. Right click the voltage sources to change the frequencies and amplitudes. A series of odd harmonics $af_1 + \frac{a}{3}f_3 + \frac{a}{5}f_5 + \frac{a}{7}f_7 \ldots$ will create a square wave. A similar series of even harmonics will give a triangle.
2. Add a few more voltage sources. Note that to access the frequency and amplitude, we must click the **Advanced** button.
3. Experiment with different phase settings, **Phi(deg)**.
4. Repeat using current sources in parallel.

6.2.3 Sine Wave Parameters

It is worth reviewing the basic theory so that manual calculations can be compared to SPICE results. This section is quite mathematical, but the concepts are not too difficult. At any rate, we should remember the results.

Trigonometric Functions

Before we embark on a study of sine waves, it is worth reminding ourselves that there are others; Fig. 6.3 shows a right-angle triangle. For virtually almost all of the time, that is what we shall need because analytical methods resolve a value into two components at right angles. In the following chapters, we shall see that these are conventionally named the real and the imaginary parts.

From the figure:

$$\tan(\phi) = \frac{a}{b} \quad sine(\phi) = \frac{a}{c} \quad cosine(\phi) = \frac{b}{c} \tag{6.6}$$

from which we see that:

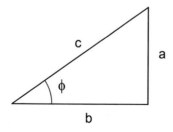

Fig. 6.3 Angles

$$\tan(\phi) = \frac{a}{c} \times \frac{c}{b} = \frac{\sin(\phi)}{\cos(\phi)} \tag{6.7}$$

Using the Cursor to Measure the RMS and Average Values

This is an alternative to the *.meas* statement. In the waveform panel, move to the desired parameter name above the plotting area, and the cursor will change to a pointing finger. Hold down the Ctrl key and left click, and a small dialogue will appear showing RMS and average values.

Example – '.meas' to find V$_{PP}$ and Frequency
The schematic ('Sine wave.asc') has:

.meas Ipp PP I(I1)

and returns *3.9953 A*. Then to find the frequency, we first find the period by:

.meas period TRIG I(I1) = Ipp/2-0.001 RISE = 1 TARG I(I1) = Ipp/2-0.001 RISE = 2

where we subtract **0.001** to be sure that the value will be reached. In point of fact, it does not matter what value we pick because we are testing against the same one for both RISEs. Finally:

.meas freq PARAM1/period

returns *1000 Hz.*

Average Value

From Fig. 6.4, we can see that a voltage or current waveform is of the form:

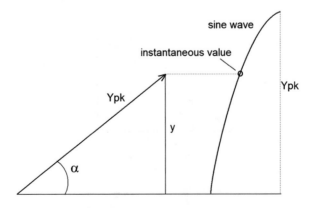

Fig. 6.4 Instantaneous values in terms of peak value

$$y = Y_{pk} \sin \alpha \tag{6.8}$$

where y is the *instantaneous value* at any point on the waveform, be it voltage or current, and Y_{pk} is its peak or maximum value which, for a pure sine wave, has the same absolute value for both positive and negative half-cycles and α is the phase angle measured in radians. However, it is more usual to think in terms of the frequency of a waveform rather than angle and then:

$$y = Y_{pk} \sin \omega t \tag{6.9}$$

where ω is the angular *velocity* and t is the time. This is the circular equivalent of $s = vt$ for linear motion so at any time t the phase angle is $\alpha = \omega t$. This is a repetitive waveform with a period T of $T = 2\pi/\omega$.

To find the average, we first integrate over one cycle and then divide by 2π in the case of Eq. 6.8 or the period T if Eq. 6.9 is used and we have either:

$$Y_{av} = \frac{1}{2\pi} \int_{\alpha=0}^{2\pi} Y_{pk} \sin\alpha.d\alpha \tag{6.10}$$

or:

$$Y_{av} = \frac{1}{T} \int_{t=0}^{T} Y_{pk} \sin\omega t.dt \tag{6.11}$$

In either case, the integral of the sine between the limits evaluates to zero, so it is immaterial which form we take. For example, using Eq. 6.10, the result is:

$$Y_{av} = \frac{1}{2\pi} Y_{pk} [-\cos\alpha]_{\alpha=0}^{2\pi} = 0 \tag{6.12}$$

We can, in fact, use Eq. 6.12 as some sort of definition of pure AC irrespective of the waveform, be it sine, square, triangle or anything else – that is, that its average value is zero.

This result was used by a clever young student who, on receipt of the first electricity bill, telephoned the company and said: 'What's this? Certainly on one half-cycle, you gave me electricity, but on the next half-cycle, I gave it all back. I have not kept any of your electricity. Why do you want money?'

Example – Sine-Wave Average
We can see this using schematic ('Sine wave.asc') if we display the trace *I(I1)* or by a *.meas* statement. Both give *39.293 nA* unless we have turned off compression. The difference is that the trace does not need to be visible for the *.meas* to work.

Half-Cycle Average

The difficulty with taking the full cycle was that the positive and negative halves cancelled. But if we limit the integration to the first half-cycle and divide either by π or $T/2$, we have, in the first case:

$$Y_{av} = \frac{1}{\pi} \int_{\alpha=0}^{\pi} Y_{pk} \sin\alpha . d\alpha \tag{6.13}$$

from which we find the magnitude is:

$$Y_{av} = \frac{1}{\pi} Y_{pk} \left[-\cos\alpha \right]_{\alpha=0}^{\pi} = \frac{2Y_{pk}}{\pi} \tag{6.14}$$

whilst in terms of the period we have:

$$Y_{av} = \frac{2}{T} \int_{t=0}^{\frac{T}{2}} Y_{pk} \sin\omega t . dt \tag{6.15}$$

or:

$$Y_{av} = \frac{2}{T} Y_{pk} \frac{1}{\omega} \left[-\cos\omega t \right]_{t=0}^{\frac{T}{2}} = \frac{2Y_{pk}}{T\omega} \left[-\cos\left(\frac{\omega T}{2}\right) + 1 \right] \tag{6.16}$$

If we substitute $T = 2\pi/\omega$, Eq. 6.16 becomes:

$$Y_{av} = \frac{\omega Y_{pk}}{\pi\omega} \left[-\cos\left(\frac{\omega 2\pi}{2\omega}\right) + 1 \right] = \frac{2Y_{pk}}{\pi} [2] = \frac{2Y_{pk}}{\pi} \tag{6.17}$$

and in either case, the result depends only on the peak value and we have:

$$V_{av} = 0.637 V_{pk} \qquad I_{pk} = 0.637 I_{pk} \tag{6.18}$$

and very often a good enough approximation is 2/3 of the peak value. It is a useful measure of the output of rectifier circuits.

Example – Sinewave Half-Cycle Average
To see this, we first need to create the appropriate waveform. An easy way is to use a voltage source to create a sinusoid and then follow it with an arbitrary voltage source where:

$$V = if(V(s) > 0, V(s), 0)$$

in schematic ('Half-Cycle Average.asc') which ensures that its output will follow the sinewave Vs if it is positive; else the output is zero. We first set the analysis time to exactly one half-cycle *0.5 ms* with a *10 V$_{pk}$ 1 kHz* input and find *6.36618 V*. If, however, we set the time to *5 ms*, we correctly find *3.183 V* because it is averaging over the whole cycle. And if we set the time to only *4.5 ms*, we find *3.53659 V* because we do not have an integral number of cycles.

RMS Value

This always gives a positive answer since any real value squared can never be negative. For any waveform we take the following steps:

- Square the waveform at every point on the waveform – which means squaring the expression for the voltage or current and gives us *volts squared* or *amps squared* and is always positive.
- Integrate over one cycle to give *volts squared seconds* or *volts squared degrees* if we use angle rather than time and similarly *amps squared seconds* or *amps squared degrees* for current.
- Find the average by dividing by the period, and again we have volts squared or amps squared.
- Take the square root – which brings us back to volts or amps.

To be specific, let us take voltage, so by the first step we have:

$$v^2 = V_{pk}^2 (\sin \omega t)^2 \tag{6.19}$$

Next we integrate over the period *T*:

$$v^2 t = V_{pk}^2 \int_{t=0}^{T} (\sin \omega t)^2 .dt \tag{6.20}$$

using:

$$(\sin \omega t)^2 = \frac{1}{2}(1 - \cos 2\omega t) \tag{6.21}$$

Equation 6.20 becomes:

$$v^2 t = \frac{V_{pk}^2}{2} \int_{t=0}^{T} (1 - \cos 2\omega t) dt \tag{6.22}$$

Performing the integration we have:

$$v^2 t = \frac{V_{pk}^2}{2} \left[t - \frac{1}{2\omega} \sin 2\omega t \right]_{t=0}^{T} = \frac{V_{pk}^2}{2} T \tag{6.23}$$

and finally dividing by T and taking the square root:

$$V_{rms} = \frac{V_{pk}}{\sqrt{2}} \tag{6.24}$$

and similarly for the current. It is important to remember that this result is true only for a complete sine wave.

As an extension, suppose we add a DC offset to the sine wave to give $Vpk(sin\omega t) + DC$. If we square the voltage at every point we have:

$$v^2 = \frac{V_{pk}^2}{2} (sin\omega t)^2 + DC^2 + 2DC.V_{pk} \sin(\omega t) \tag{6.25}$$

and we can use the same substitution of Eq. 6.21 and integrate Eq. 6.25. The integral of the last term is $2DC.\omega V_{pk} cos(\omega t)$ and evaluates to $[1-1] = 0$ at the limits leaving us with

$$\left[\frac{V_{pk}^2}{2} (t - \frac{1}{2\omega} \sin 2\omega t) + DC^2 \right]_{t=0}^{T} \tag{6.26}$$

The sine term also evaluates to 0 so we are left with:

$$\frac{V_{pk}^2}{2} T + DC^2 T \tag{6.27}$$

We then divide by T and take the square root to find:

$$V_{RMS} = \sqrt{\frac{V_{pk}^2}{2} + DC^2} \tag{6.28}$$

that is, it is the square root of the sum of the two RMS values squared.

Example – Sinewave RMS Value

The schematic ('Sinewave RMS.asc') returns somewhere around *7.06885 A* instead of *0.70710 A*. Reduce the time step to *0.1u* or turn off compression and it returns *0.70710 A*.

If we add a DC offset of *10 V*, Eq. 6.28 indicates that we should expect an answer of $V_{RMS} = \sqrt{\frac{100}{2} + 100} = 12.2474 \ V$, and this is exactly what we find where the *.meas* statement adds the last digit. We get a different answer if we only make the transient run over *0.5 ms.*

Power

If a sinusoidal voltage is connected across a resistor, at any instance t we can write:

$$V_{pk} \sin(\omega t) = i.R \tag{6.29}$$

where lower-case letters are used to denote instantaneous values. The current is therefore also a sinusoid with instantaneous value:

$$i = \frac{V_{pk} \sin(\omega t)}{R} \tag{6.30}$$

And as the instantaneous power is: $p = v.i$ We use Eq. 6.30 to substitute for i:

$$p = \frac{V_{pk}^2 \sin(\omega t)^2}{R} \tag{6.31}$$

and we can expand this as:

$$p = \frac{V_{pk}^2}{2R}[1 - \cos(2\omega t)] \tag{6.32}$$

which shows that the instantaneous power consists of a steady component $\frac{V_{pk}^2}{2R}$ plus a sinusoidal component at twice the frequency of the supply. The result is always zero or positive since $cos(2\omega t)$ can never be greater than one. We can understand Eq. 6.32 by noting that the instantaneous power waveform runs from zero to $2Vpk^2/R$ so; unlike pure AC, it does not go equally positive and negative. But if we subtract $Vpk^2/2R$, it does. Hence this term is the DC offset.

The work done over a complete cycle is found by integrating Eq. 6.32 over the time for a complete cycle and is:

$$W = \frac{V_{pk}^2}{2R} \int_{t=0}^{2\pi/\omega} [1 - \cos(2\omega t)]dt \tag{6.33}$$

so:

$$W = \frac{V_{pk}^2}{2R}\left[t - \frac{1}{2\omega}\sin(2\omega t)\right]_{t=0}^{2\pi/\omega} \tag{6.34}$$

and making the substitutions:

$$W = \frac{V_{pk}^2}{2R} \frac{2\pi}{\omega} \tag{6.35}$$

and finally the average power over one cycle is found by dividing Eq. 6.35 by $1/T$. whence:

$$P_{av} = \frac{V_{pk}^2}{2R} \frac{2\pi}{\omega} \frac{\omega}{2\pi} \tag{6.36}$$

which reduces to:

$$P_{av} = \frac{V_{pk}^2}{2R} \tag{6.37}$$

and is independent of the frequency. The same result is obtained by using the angle rather than the time.

In terms of RMS values we have:

$$P_{av} = \frac{V_{rms}^2}{R} = I_{rms}^2 R \tag{6.38}$$

which are more useful measures and – incidentally – explain the bill from the electricity company. We should also note that from Eq. 6.32, the peak power is $P_{pk} = \frac{V_{pk}^2}{R}$. The result is different if we choose another waveform, such as a

square or a triangle. In many cases, the calculations are quite simple.

Example – Power Consumed by an Electric Fire
Let us see what we can find out about a 1 kW electric fire connected to the 240 V AC mains supply.

First, the peak voltage is $\sqrt{2}.240 = 340$ V. Next, the RMS current is $Irms = 1\ kW/ 240 = 4.17\ A$, so the peak current is $\sqrt{2}.4.17 = 5.9$ A, and, finally, the hot resistance of the heater element is $240/4.17 = 57.6\ \Omega$ although its cold resistance will usually be a lot less. This is schematic ('Electric Fire.asc'). We find the result by two measurement steps. First we find the RMS current

.meas Irms RMS I(R1)

then the power:

.meas Pfire PARAM Irms ∗ ∗2 ∗ 57.6

The result is *999.833 W*. Close enough.

Explorations 2

1. Open schematic ('RMS and Power.asc') and display the current in R1. Move the cursor above the trace itself and onto the label 'I(R1)' where the cursor changes to a pointing hand. Hold down the **Ctrl** key and left click on the label, and LTSpice will open a small window showing the average and RMS values. Note that the average is not quite zero. This is caused by rounding errors of the finite step times. Try smaller steps.

2. Try adding a **DC offset(V)** to the sine wave. The RMS value of the resultant is:
$$V_{rms} = \sqrt{\left(V_{1rms}^2 + V_{2rms}^2\right)}$$
where *V1* is the DC offset and *V2* the sine, and check the LTSpice result with your calculation.

3. Change the time setting so that the waveform is no longer an exact multiple of the frequency, and note that the values change. In particular, make the time period exactly one half-cycle, and check the simulation result for the average with Sect. 6.2.3.2. Derive an expression for the RMS value of a half-cycle and compare that with the simulation result.

6.2.4 Adding Sine Waves

In many electronic circuits we encounter two or more sine waves of the same frequency but not in phase. By this it is meant that one waveform *leads* the other by a certain *phase angle* or, to put it the other way round, one *lags* the other by a certain phase angle. We must exercise considerable caution when adding two sine waves when we want to find the result of:

$$V_1 \sin \omega t + V_2 \sin (\omega t + \alpha)$$

where α is the phase angle between them. It should be noted in passing that the phase angle was ignored in the previous analyses and the reader is invited to repeat them including α when it will be found that results are the same.

Resultant Waveform

We can find this by two methods: one, rather tedious, is to draw the two waveforms, taking account of the phase difference, and add them point by point; see, for example, Fig. 6.5 (which is actually from an LTSpice simulation) where it is the sum. This shows the waveforms and the plotting points.

The other is to use the idea of a rotating vector show in Fig. 6.6 only this time we have two vectors rotating anti-clockwise with a constant phase angle between them so we can draw their peak values and then the resultant waveform *Vpk(res)* is found from the vector sum of the two and is also rotating anti-clockwise at the same frequency. And – this is important – because it is a rotating vector, it will also trace

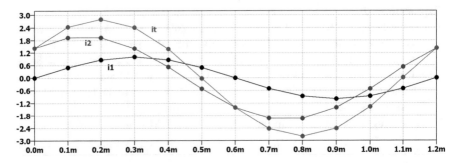

Fig. 6.5 Adding sinusoids

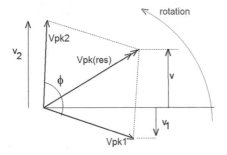

Fig. 6.6 Vector sum of two sine waves

out a sine wave where the protection on the vertical axis v is the instantaneous value just as before. But it is important to note that this only applies to the peak or RMS values which are constant.

In passing, we might note the projections of the two waveforms $v1$ and $v2$ on the y-axis. At the instant when the figure is drawn, $v1$ is negative and $v2$ is near its peak value. But imagine the vectors rotated by 90 degrees, and now $v1$ has a large positive value, whilst $v2$ is almost zero. This explains the strange dance of the amplitudes shown in Fig. 6.5.

Peak and RMS Value

If the two voltages are of the same frequency, we can redraw Fig. 6.6 where, for convenience, we draw $V1$ along the x-axis, Fig. 6.7. We then resolve $V2$ into its horizontal and vertical components to find the resultant as:

$$V_{pk(res)} = \sqrt{V_x^2 + V_y^2} \qquad (6.39)$$

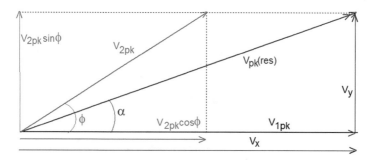

Fig. 6.7 RMS sum

The two components are:

$$V_x = V_2 \cos \phi + V_1 \quad V_y = V_2 \sin \phi \qquad (6.40)$$

substituting:

$$V_{pk(res)} = \sqrt{(V_2 \cos \phi + V_1)^2 + (V_2 \sin \phi)^2} \qquad (6.41)$$

expanding:

$$V_{pk(res)} = \sqrt{V_2^2 \sin \phi^2 + V_2^2 \cos \phi^2 + V_1^2 + 2V_1 V_2 \cos \phi} \qquad (6.42)$$

using: $\sin^2 + \cos^2 = 1$

$$V_{pk(res)} = \sqrt{V_2^2 + V_1^2 + 2V_1 V_2 \cos \phi} \qquad (6.43)$$

In the special case that the angle is zero:

$$V_{pk(res)} = \sqrt{(V_1 + V_2)^2} = V_1 + V_2 \qquad (6.44)$$

and if it is 90°:

$$V_{pk(res)} = \sqrt{(V_1^2 + V_2^2)} \qquad (6.45)$$

which is the hypotenuse of the right-angled triangle.

And in every case the RMS is just:

$$V_{RMS} = \frac{V_{pk(res)}}{\sqrt{2}} \tag{6.46}$$

and the phase angle α of the resultant relative to V_1 is:

$$\alpha = \tan^{-1} \frac{V_2 \sin \phi}{(V_2 \cos \phi + V_1)} \tag{6.47}$$

We can draw Fig. 6.7, using RMS values instead, and now the resultant is the RMS value of the sum, at the correct phase angle, but of course, this is not rotating.

Example – Adding Sinusoids

A sinusoid of *200 V_{pk}* is added to one of *100 V_{pk}* which leads it by 30° . Find the RMS value of the resultant waveform.

Using Eq. 6.43 we have:

$$V_{pk}(res) = \sqrt{100^2 + 200^2 + 2.200.100.\cos(30)} \quad \text{or}$$

$$V_{pk}(res) = \sqrt{1.10^4 + 4.10^4 + 4.10^4.0.866} = 291V_{pk} = 206V_{RMS}.$$

The simulation by schematic ('Adding Sinusoids.asc') returns *205.65 V*. As always, we may find the current in the resistor is in the wrong direction, and this can be resolved by rotating it by 180°.

Explorations 3

1. Use ('Adding sinewaves.asc') to try a number of different amplitudes and phases, and check the readings against calculation. In particular, replace *V1* with DC. Note that the result is only valid if there are an exact number of cycles shown.
2. Place two current generators in parallel, *B1* and *B2*, in ('Adding Sinusoidal Currents.asc'). These are arbitrary current generators of the same frequency and phase and having peak amplitudes of 10 mA and 5 mA, respectively. The currents are sinusoidal with a frequency of 1 kHz since $\omega = 2 \pi f$. Explore different settings.
3. In schematic ('Adding Sinewaves 1.asc'), the measurements are made at intervals of 30°. Note that the data points used by the simulation are closer. With **View->SPICE Error Log**, if we right click in the panel we can plot the stepped data. Figure 6.5 was created by plotting the stepped data from schematic ('Adding Sinewave 1.asc'). Explore.

6.2.5   Partial Sine-Wave

Thyristor circuits often delay the start of a waveform so that the load current and voltage are zero up to a delay angle. Now we no longer have a continuous wave, so we cannot integrate over the whole cycle, but must take each part separately, Fig. 6.8.

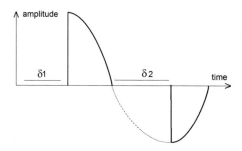

Fig. 6.8 Partial sinewave

Average Value

If we have the same delay on both positive and negative half-cycles, this will be zero. However, if there is an imbalance, the result can be positive or negative. Taking the general case of a delay angle of $\delta 1$ on the positive half-cycle and $\delta 2$ on the negative, the integral is:

$$V_{pk}\left(\int_{\alpha=\delta 1}^{\pi} \sin\alpha.d\alpha + \int_{\alpha=\pi+\delta 2}^{2\pi} \sin\alpha.d\alpha \right) \tag{6.48}$$

or:

$$V_{pk}(\cos\delta 1 - \cos\delta 2) \tag{6.49}$$

and the average is:

$$V_{av} = \frac{V_{pk}}{2\pi}(\cos\delta 1 - \cos\delta 2) \tag{6.50}$$

The practical significance of this is that the average value represents a DC which may have deleterious consequences if the circuit requires pure AC.

Example – Average of Partial Sine-Wave

A 240 V mains supply is delayed by 40° on the positive half-cycle and 45° on the negative. Find the average value.

From Eq. 6.50 we have: $V_{av} = \frac{240.\sqrt{2}}{2\pi}(0.766 - 0.707) = 3.2V$ and although this may seem trivial, it could be significant if this drove a current though an inductor since it could cause magnetic saturation.

This is illustrated in the schematic ('Partial Sinewave.asc'. For convenience, we define four parameters:

.param Vpk = 240∗sqrt(2) {to convert the 240 V RMS to the peak value}
.param twopi = 2∗pi {be careful! NOT *2pi*}

.param dt1 = 40/360   {the delay in radians of the first delay $\delta 1$ of 40°}
.param dt2 = 225/360   {ditto for the second delay $\delta 2$ of 225°}

The function for the arbitrary voltage source **B1** is:

$$V = if\,(time > dt1\,\&time < 0.5|time > dt2, Vpk * sin\,(twopi * time), 0)$$

The *if* will cause the output to follow the sinusoid if the time is greater than the first delay and less than the end of the half-cycle. The *&time<0.5* is important: without it the waveform will continue for the rest of the time. The vertical line I for 'or' enables the waveform again if the time is greater than the second delay.

We can alternatively expand the voltage as the first half-cycle plus the second:

$$V = if\,(time > dt1\,\&time < 0.5, Vpk * sin\,(twopi * time), 0)$$
$$+ if\,(time > dt2, Vpk * sin\,(twopi * time), 0)$$

After a run **Ctrl** and left click on the trace label V(out) shows an average of *3.1908 V* and an RMS of 230.27 V.

Half-Cycle Average

This is only meaningful if the delay angle is the same on both halves. Then we merely integrate to find:

$$V_{hcav} = \frac{V_{pk}}{\pi}(\cos\delta + 1) \quad \text{if} \quad \delta < 180° \quad \text{else} \quad V_{hcav} = \frac{V_{pk}}{\pi}(\cos\delta - 1) \quad (6.51)$$

which reduces to $2V_{pk}/\pi$ if the delay is zero – the same as for the full cycle.

Example – Half-Cycle Average of a Partial Sine Wave

Taking the previous values of *40°* and *45°* on the positive half-cycle and using Eq. 6.51 we have: $V_{hcav} = \frac{240\sqrt{2}}{\pi}(\cos(40) + 1) = 108.038(1.7669) = 190.80\ V$ and on the negative half-cycle it is: $108.036(\cos(225) - 1) = -184.429\ V$.

We can use ('Partial Sinewave.asc') again: the positive half-cycle value agrees exactly but the negative half-cycle returns $-184.435\ V$.

RMS

It is more usual to work in terms of delay angle δ in radians rather than delay time, so assuming the same delay on both halves of the cycle, we repeat the previous analysis leading to the first Eq. 6.23 for both halves but using angle α instead of ωt to give:

$$V_{rms}^2 = \frac{V_{pk}^2}{2} \left(\left[\alpha - \frac{1}{2} \sin 2\alpha \right]_{\alpha=\delta}^{\pi} + \left[\alpha - \frac{1}{2} \sin 2\alpha \right]_{\alpha=\delta+\pi}^{2\pi} \right) \qquad (6.52)$$

Because of the symmetry, and because we are squaring, negative terms become positive; we can handle the second term by subtracting π from it so that it runs from δ to π. Then:

Hence:

$$V_{rms} = \sqrt{\frac{1}{2\pi} \left(\frac{V_{pk}^2}{2} (2\pi - 2\delta) \right)} \qquad (6.53)$$

finally:

$$V_{rms} = \frac{V_{pk}}{\sqrt{2}} \sqrt{\frac{\pi - \delta}{\pi}} \qquad (6.54)$$

and if the delay is zero, this reduces to the full sine-wave figure. What is surprising at first is that if the delay is 90°, the RMS voltage (or current) is:

$$V_{rms90} = \frac{V_{pk}}{\sqrt{2}} \sqrt{\frac{1}{2}} \qquad (6.55)$$

In other words, it is not half the RMS value of the full sine wave, but the square root of half.

If the two halves do not have the same delay, equation 6.52 becomes:

$$V_{rms}^2 = \frac{V_{pk}^2}{2} \left(\left[\alpha - \frac{1}{2} \sin 2\alpha \right]_{\alpha=\delta 1}^{\pi} + \left[\alpha - \frac{1}{2} \sin s 2\alpha \right]_{\alpha=\delta 2}^{\pi} \right) \qquad (6.56)$$

$$V_{rms}^2 = \frac{V_{pk}^2}{2\pi} (2\pi - \delta 1 - \delta 2) \qquad (6.57)$$

$$V_{rm} = \frac{V_{pk}}{\sqrt{2\pi}} \sqrt{2\pi - \delta 1 - \delta 2} \qquad (6.58)$$

Example – RMS of Partial Sinusoid

We change the setting for **dt2** in schematic ('Partial Sinewave.asc'). to *220/360* giving a delay of 40° on both half-cycles and find the RMS is *232.01 V*.

Analytically from Eq. 6.54:

$$V_{RMS} = 240 \sqrt{\frac{\pi - 0.222222}{\pi}} = 240 \sqrt{\frac{2.9194}{\pi}} = 231.36V$$

which does not quite agree with the simulation.

If we keep the original delays of *40°* and *45°*, Eq. 6.58 is:

$$V_{RMS} = \frac{240}{\sqrt{2\pi}} \sqrt{6.2832 - 0.22222 - 0.25} \quad = 95.75 \times 2.411 = 230.84$$

whereas simulation gives *230.43 V* – an error of *0.18%* which is usually acceptable.
If we set $\delta 1 = 90°$ and $\delta 2 = 270°$, simulation and calculation both give *169.7 V*.
If we set $\delta 1 = 90°$ and $\delta 2 = 180°$, simulation and calculation both give *207.85 V*.

Example – .meas RMS, AVG
We can find this quite easily by:

.meas rms V RMS V(out) FROM 0 TO 1

where we find *230.429 V* for the original delay settings and *232.008 V* when they are
both *40°*, and these do not seem to be affected by the time step of the transient run
and are closer to the calculated values. Also the average voltage is now less than
0.6 nV.

Explorations 4
1. Open ('Partial Sinewave.asc') and explore different delay settings. Note carefully
    the effect of turning off compression and the time step. Compare the measured
    results with those obtained from the trace by **Ctrl** left click on *V(out)*.
2. Build a similar circuit with a current source.
3. Add a load resistor, and measure the load power for different delays. Check by
    calculation.

6.3   Rectangular Waves

If the waveform is rectangular (of which square waves are a special instance), the
analysis is much simpler. The waveform is not continuous, so we must treat it in two
pieces. During the period *t1*, we have *V−*, and at *t2* we have *V+* for the second, so the
average is:

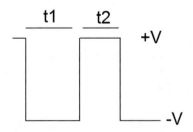

Figure   Rectangular wave

$$V_{av} = \frac{t_1 V- \; + t_2 V+}{t_1 + t_2} \tag{6.59}$$

The RMS is:

$$V_{RMS} = \sqrt{\frac{t_1 V-^2 \; + t_2 V+^2}{t_1 + t_2}} \tag{6.60}$$

In the special case that $V- = 0$, so that the wave is zero during period $t2$, the average value is:

$$V_{av} = \frac{t_2 V+}{t_1 + t_2} \tag{6.61}$$

and the RMS value is:

$$V_{RMS} = \sqrt{\frac{t_2 V+^2}{t_1 + t_2}} \tag{6.62}$$

and for a square wave with $t1 = t2$

$$V_{rms} = \frac{V+}{\sqrt{2}} \tag{6.63}$$

If the magnitude of the positive and negative voltages are identical, so that $V+ = V- = V$, the peak and RMS values are both equal to V and the average is zero.

Example – Rectangular Wave 'Trise' 'Tfall'
Open schematic ('Squarewave RMS.asc'). With the pulse settings of $Trise = Tfall = 0$, the sides are sloping. If we change to something like $Trise = Tfall = 1f$, they are vertical if we also make **Maximum Timestep** = 1u. It is easy to change the duty cycle and measure the RMS and average values of voltage and current.

Explorations 5
1. Open the schematic ('Squarewave RMS.asc'). A transient run can show the current and the instantaneous power and the current squared (which overlays the power). As before we can find the average and RMS values.
2. Change the ratio $t1{:}t2$ to $1{:}9$ by changing **Ton(s)** and **Tperiod(s)** but making the time an exact number of half-cycles (e.g. 1.5 ms with 1 kHz). Note that if the period is not an exact number of cycles, the average is not zero and the RMS voltage is not $V_{pk}/\sqrt{2}$.

3. Create an exact square wave with $t1 = t2$ and equal negative and positive magnitudes, i.e. $V- = V+$. Measure the average and RMS values, and compare to the calculated ones.
4. A rectangular wave has an average current of 2 A and $t1{:}t2 = 1{:}15$. Calculate the RMS and peak currents, and check with LTSpice.

6.4 Triangular Waves

This is another common waveform. We can create repetitive waves using the *PULSE* option of a voltage or current source and adjusting the two voltage settings and the *Trise, Tfall* and *Ton* times to create a sawtooth or ramp going positive and negative. Otherwise a *PWL* waveform enables just one waveform to be created.

6.4.1 Average Value

Unless either the rise or fall times are zero, these must be handled in two parts if the waveform starts from zero volts or zero current, Fig. 6.9(a). We shall use voltage in the following analysis, but it also applies to current.

As the area of a triangle is 1/2 *base * height,* the total area under the waveform in volt seconds is:

$$Area = \frac{1}{2}\left(V_p t_1 + V_p t_2\right) \qquad (6.64)$$

and as the period *Tper* is just $t1 + t2$, we have:

$$Area = \frac{1}{2} V_p T_{per} \qquad (6.65)$$

and if we divide by the period, the average is:

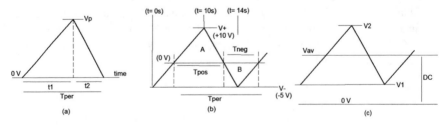

Fig. 6.9 Triangular wave average value

$$V_{av} = \frac{1}{2} V_p \qquad (6.66)$$

However, the picture becomes a little more complicated if the waveform goes positive and negative, Fig. 6.9(b), where the numerical values are for Example – Triangular Wave.

We can use Eq. 6.65(b) for both triangles A and B to find the total area, taking account of the sign, and we have:

$$Area = \frac{1}{2} V_+ T_{pos} - \frac{1}{2} V_- T_{neg} \qquad (6.67)$$

If $V+ = V-$ the average is zero.

6.4.2 RMS Value

In an unguarded moment, one can go astray visualising the waveform: squaring the voltage at every point does not give a bigger triangle, but a square law curve. Taking the waveform in parts again, during the first portion, we have:

$$v(t) = V_p \frac{t}{t_1} \qquad (6.68)$$

so at any point, the voltage squared is:

$$v^2(t) = \left(V_p \frac{t}{t_1} \right)^2 \qquad (6.69)$$

and integrating:

$$v^2 t = \frac{1}{3} V_P^2 \frac{t^3}{t_1^2} \qquad (6.70)$$

After putting in the limit:

$$v^2 t = \frac{1}{3} V_p^2 t_1 \qquad (6.71)$$

For the period t_2, we can recognize that the waveform is the same shape but the slope is reversed. It simplifies the mathematics if we take the origin at the termination of the waveform and work back to the end of the first portion, in effect reversing the x-axis. Then we can simply substitute in Eq. 6.71 to write:

$$v^2 t = \frac{1}{3} V_p^2 t_2 \tag{6.72}$$

To find the RMS value, we add Eqs. 6.71 and 6.72, divide by the total time and take the square root so:

$$V_{RMS} = \sqrt{\frac{1}{3} \frac{V_p^2 (t_1 + t_2)}{t_1 + t_2}} = \frac{1}{\sqrt{3}} V_P \tag{6.73}$$

which is independent of the period and of the slopes.

It the waveform is not unipolar nor symmetrical about zero volts, we can handle it in parts, Fig. 6.9(c), by splitting it into the DC component and a triangular AC component. Therefore the DC component is the average leaving a triangular wave equally balanced about the average as in (b).

Now we need to add the *volts squared time* for each part. For the DC the peak, average and RMS are all the same: V_{av}. For the triangle the peak is $(V2 - V1)/2$. Thus we have:

$$v^2 t = V_{av}^2 T_{per} + \frac{1}{3} \frac{(V_2 - V_1)^2}{4} T_{per} \tag{6.74}$$

whence the RMS is:

$$V_{rms} = \sqrt{V_{av}^2 + \frac{1}{12}(V_2 - V_1)^2} \tag{6.75}$$

Example – Triangular Wave
Using the values from Fig. 6.9b, we have from Eq. 6.73: $V_{RMS} = 1/3(10 - (-5)) = 5$ V, and from Eq. 6.67, we can find the average voltage, but first we need to find the times when the graph crosses zero volts, and these are $t_{+1} = 10$ s/$3 = 3.333$ s, and the second time is $t_{+2} = 10 + (2*4)/3 = 12.667$ s. Thus, the positive area is:

$$\frac{1}{2} (10\ V(t_{+2} - t_{+1}) = \frac{1}{2} (10\ V\ (2.667 - 3.333)) = 46.67\ Vs$$

and the negative area is:

$$\frac{1}{2}((14 - t_{+2}) + t_{+1}) = \frac{1}{2} (5\ V(1.333 + 3.333)) = 11.665\ Vs.$$

Hence the average is $V_{av} = (46.67 - 11.67)/14\ s = 2.5$ V.

If we add a positive offset of 3 V shown as **V2** in ('Triangular Wave.asc') to create a voltage that runs from $+3$ V at $t = 0$ s to $+15$ V at $t = 10s$ and falls to $+3$ V again at $t = 14$ s, we calculate the RMS value from Eq. 6.75 and it is: $V_{RMS} = \sqrt{(81 + 1/12 (15-3)^2)} = 9.643$ V and agrees with the simulation.

6.5   Other Waveforms

We can generalize this for any waveform that does not fit a simple analytic function, in particular, the exotic ones encountered in Power Electronics where we often find waveforms that are essentially discontinuous like those in Fig. 6.10.

We again find the *volts* or *current squared time* for each portion, but notice that in finding the average or RMS, we use the whole period, including the time when the waveform is zero as in the examples (a) and (b).

Example – Piecewise Waveform Average and RMS Voltages

We shall use the fanciful schematic ('Piecewise Waveform 2.asc') also shown as Fig. 6.11. Again we notice that to achieve a vertical edge, we cannot have exactly the same times, but must have a very small time increment. PWL(0 0 2 10 2.000000001 0 4 0 4.000000001 -5 6 -5 6.000000001 0 8 0 10 6 12 6 14 0) where after 2s, 4s, and 6s we add 1 ns for the next point.

Starting at the left we have:

Triangle area = 1/2 (10 V × 2 s) = 10 Vs
Rectangle area = (−5 V × 2 s) = − 10 Vs
Trapezoid area = (6 V × 2 s + 6 V × 2 s) = 24 Vs

Hence the average is *(10 − 10 + 24)/14 = 1.7143 V*

For the RMS we must find the total *volts-squared time:*
Three triangles using Eq. 6.70 = *1/3(200 + 144) = 114.67 V^2 s*
Two rectangles = *50 + 72 = 122 V^2 s*
Average = *(114.67 + 122)/14 s = 16.905 V^2*; thus V_{RMS} = *4.1115 V*, and this agrees with the simulation.

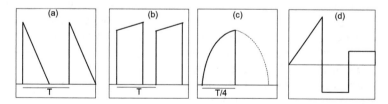

Fig. 6.10 Other waveforms

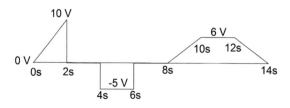

Fig. 6.11 Piecewise waveform

Just out of interest, we can find the RMS value from the component parts. One way is to integrate the individual voltages squared over their times as in the first three measurements and then finally to find the square root of their mean by adding them and dividing by the total time as the last measurement. The individual results are shown in brackets. With compression turned off, we find:

*.meas Vsqt1 INTEG V(out)**2 FROM 0 TO 2*   *(66.67)*
*.meas Vsqt2 INTEG V(out)**2 FROM 4 to 6*   *(50)*
*.meas Vsqt3 INTEG V(out)**2 FROM 8 to 14*   *(120)*
.meas VrmsTotal1 PARAM sqrt((Vsqt1 + Vsqt2 + Vsqt3)/14)   *(4.11154)*

A second way is to find the RMS value of each component by dividing its $V^2 t$ by the total time and taking the square root. Then we find the square root of the sum of the RMS of each component squared.

.meas Vrms1 = PARAM sqrt(Vsqt1 /14)   *(2.18218)*
.meas Vrms2 = PARAM sqrt(Vsqt2/14)   *(1.88982)*
.meas Vrms3 = PARAM sqrt(Vsqt3/14)   *(2.9277)*
*.meas VrmsTotal2 PARAM sqrt(Vrms1**2 + Vrms2**2 + Vrms3**2)*   *(4.11154)*

Explorations 6

1. Find the average and RMS values of examples (a) and (b) of Fig. 6.10(d). Choose your own time scales.
2. Set up a pulse train of rectangular waves with a duty cycle of 50% then increasing the amplitude as necessary, keep the same average voltage or current whilst reducing the duty cycle in stages to 1%, and notice how the peak and RMS values rise dramatically. This can be a significant problem in some circuits where the average current is low, but the peak and RMS values are excessive.
3. Fig. 6.10(c) is a bit difficult: the SINE option allows a delay to the start of the wave, but not the termination. Try using an *if* statement that the waveform is a sine wave if the time is less than *T/4* else it is zero.
4. Instead of a 1/4 cycle of the sine wave with the rest of the period at zero volts or current, try other fractions of a cycle.
5. In particular, try just the 'tip' of a sine wave, not starting at zero degrees, but 60° and stopping at 120°. Not easy, consider using a *B* source, not voltage.
6. Example Fig. 6.10(d) is an example of a piecewise waveform. Modify ('Piecewise Waveforms.asc') to create it.
7. Feel free to play with the *B* current and voltage sources.
8. Does the relationship $V_{RMS} = \sqrt{V_{RMS1}^2 + V_{RMS2}^2 + V_{RMS3}^2 + \ldots}$ hold good for any of these waveforms?

6.6 Other Forms of Trigonometrical Functions

At this point, this is 'nice-to-know' rather than essential, and it is not so straightforward to explore them in LTspice since, understandably really, it does not implement the square root of -1. Reading through the section **B. Arbitrary Behavioral Voltage or Current Sources** of the 'Help' file shows that *pow(x,y)* and exponentiation $**$ both return only the real part.

On the other hand, the following sections might, perhaps, give a deeper understanding of trigonometrical and exponential functions and the relationship between them.

6.6.1 Series Forms

These have been used by computers to calculate values, particularly of sines and cosines and their hyperbolic counterparts, although SPICE in all its forms might use more efficient algorithms. There are many websites that expound them in greater or lesser depth and with greater and lesser rigour and accessibility.

These forms are also applied to a wide range of functions of which several examples can be found on the web. As our scope is limited, we note only two essential conditions for using these series: one is that we must be able to differentiate the original function an infinite number of times, which is patently true of sines and cosines, and the other is that the series must converge to a limiting value. Here we anticipate Eq. 6.85 and note that the signs of the terms alternate, so by the Alternating Series Test (see https://en.wikipedia.org/wiki/Convergence_tests# Alternating_series_test) the series converges. We must also note that angles are in radians.

Maclaurin Series

So far we have happily used sine and cosine without bothering about possible other representation. However, we have seen that the Taylor Series enables us to expand a function as a series about a point *a*, and we write it again here using *n!* for *factorial n*:

$$f(x) = f(a) + \frac{f'(a)}{1!}(x-a) + \frac{f''(a)}{2!}(x-a)^2 + \frac{f'''(a)}{3!}(x-a)^3$$
$$+ \ldots . \frac{f^{(n)}(a)}{n!}(x-a)^n \tag{6.76}$$

We now turn to the Scottish mathematician Colin Maclaurin who, 300 years ago, modified the Taylor Series with $a = 0$ to give the Maclaurin Series:

$$f(x) = f(0) + \frac{f'(0)x}{1}! + \frac{f''(0)x^2}{2}! + \frac{f'''(0)x^3}{3}! + \ldots \frac{f^{(n)}(0)x^n}{n}! \qquad (6.77)$$

Using this we can derive series forms for trigonometric functions.

The Sine Series

Setting: $f(x) = \sin(x)$ in Eq. 6.77 when $x = 0$ we have:

$$f(0) = \sin(0) = 0 \qquad (6.78)$$

and the first derivative is:

$$f'(\sin(0)) = \cos(0) = 1 \qquad (6.79)$$

and for the second derivative, we differentiate the first derivative, $cos(0)$ in Eq. 6.79:

$$f''(\sin(0)) = f'(\cos(0)) = -\sin(0) = 0 \qquad (6.80)$$

and then for the third, we differentiate the second, and without setting out the whole chain of differentials, we have:

$$f'''(\sin(0)) = f''(-\sin(0)) = -1 \qquad (6.81)$$

and in like fashion the following terms are found by differentiating the term immediately before and for the next few terms we have:

$$f^4(\sin(0)) = \sin(0) = 0 \quad f^5(\sin 0) = \cos(0) = 1 \quad f^6(\sin(0)) = -\sin(0) = 0 \quad (6.82)$$

and we see a pattern emerging where all the even terms are zero and the sign of the odd terms alternate $1, -1$.

Returning to Eq. 6.77, the first term from Eq. 6.78 is $f(0) = 0$. The second term comes from Eq. 6.79 and is:

$$\frac{f'(0)}{1!}x^2 = \frac{f'(\sin(0))}{1!}x = \frac{1}{1!}x \qquad (6.83)$$

We have seen from Eq. 6.80 that the third term is 0, and from Eq. 6.81, the fourth term is:

$$\frac{f'''(0)}{3!}x^3 = \frac{f'''-\sin(0)}{3!}x^3 = \frac{-x^3}{3!} \qquad (6.84)$$

Now we substitute in Eq. 6.77 and continue in like manner:

$$\sin(x) = \frac{x}{1!} - \frac{x^3}{3!} + \frac{x^5}{5!} - \frac{x^7}{7!} = \ldots \tag{6.85}$$

which we can write compactly in sigma notation using *(2n-1)* to pick out the odd terms and $(-1)^{n+1}$ to get the correct sign:

$$\sin(x) = \Sigma_{n=1}^{\infty}(-1)^{n+1}\frac{x^{(2n-1)}}{(2n-1)!} \tag{6.86}$$

Example – Maclaurin Sine Expansion
This is schematic ('Sine Maclaurin.asc') where *B2* traces a sinusoid and *B1* implements the sine expansion. It is interesting to start with just the first one or two terms and note that the graph follows the sine wave for the start of the time then breaks away very steeply upwards or downwards as extra terms are added from the comment on the schematic. The period where it follows the sine wave increases as more terms are added. We might note that it will only follow the sine wave up to one complete revolution – afterwards it will break away again. Polynomial fitting to a function behaves in the same way of running off wildly outside the limits. However, we can note the symmetry of both sine and cosine so that we only need fit the series to the function to 90°·

The Cosine Series

We can use the same methodology starting with Eq. 6.77 and setting $f(x) = \cos(x)$ and then:

$$f(0) = \cos(0) = 1 \tag{6.87}$$

Differentiating gives:

$$f'(\cos(0)) = -\sin(0) = 0 \tag{6.88}$$
$$f''(\cos(0) = -\cos(0)) = -1 \tag{6.89}$$

and in like manner:

$$f'''(\cos(0)) = 0 \quad f^4(\cos(0)) = 1 \quad f^5(\cos(0)) = 0 \quad f^6(\cos(0))$$
$$= -1 \tag{6.90}$$

which gives a similar pattern only this time we have the even terms:

$$\cos(x) = 1 - \frac{x^2}{2!} + \frac{x^4}{4!} - \frac{x^6}{6!} = \dots \qquad (6.91)$$

and the sigma form is:

$$\cos(x) = 1 + \Sigma_{n=1}^{\infty}(-1)^{2n}\frac{x^{2n}}{2n!} \qquad (6.92)$$

Example – Maclaurin Cosine Expansion
This is ('Cosine Maclaurin.asc') and again shows that the series expansion stays closer to the actual cosine graph as more terms are added, and, once again, it is only valid for one complete cycle.

Explorations 7
1. Open ('Sine Maclaurin.asc') and note the divergence between the actual sine and the series. If $a = 1$ the .*meas* angles are correct, and if $a = 0.5$, they are divided by 2. It will be seen that with $a = 0.5$, so that .*meas m90 = 45°*, and with only the first three terms, the error is only *3.6e-5* showing that even with a limited series, the approximation is good enough for most engineering purposes.
2. Repeat using ('Cosine Maclaurin.asc'), and note the closeness of the series to the true cosine as the number of terms is increased. And as was said before, we should also note the symmetry of both sine and cosine so that if the error is acceptable up to 90°, we can derive the rest from it.

Applications

These series are mainly used for finding values of trigonometrical functions. Because of the difficulty of multiplying series to find, for example, $2cos(\theta)cos(\Phi)$, they are not of much use in generating trigonometrical identities which are handled far more easily using the exponential form of the functions.

6.6.2 Exponential Forms

These allow a more compact representation where both sines and cosines of the same angle are involved, especially in a Fourier series analysis which we shall cover in a later chapter. They also are a useful link to the hyperbolic functions.
   https://wstein.org/edu/winter06/20b/notes/html/node29.html

Real and Imaginary Numbers

In some ways this is a misleading description and arises from the fact that $\sqrt{-1}$ or any other multiple such as $\sqrt{-9}$ is not a 'real' number in the sense that $\sqrt{9} = 3$ and we can

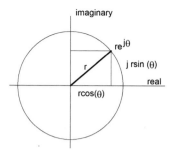

Fig. 6.12 Exponential angles

buy *3* bananas but we cannot buy √*-9* of them. We shall take up this theme again, but for now, in engineering, we can define √*-1* as an operator meaning 'rotate anti-clockwise by *90°* or *π/2* radians'. Mathematical texts use *i* for √-1; we shall use *j* to avoid confusion with the symbol *i* for current. The use of *j* leads directly to an alternative descriptions of trigonometrical functions.

Complex and Polar Form

We have seen that a sine wave can be constructed from an anti-clockwise rotating vector, Fig. 6.2. If we label the horizontal axis 'real' and the vertical 'imaginary' in Fig. 6.12, then the instantaneous voltage or current is:

$$re^{j\theta} = r \cos(\theta) + jr \sin(\theta) \qquad (6.93)$$

The History of e

But first we need to explore the constant *e*. It is said that this and *π* are the two most important constants in mathematics and engineering. We all know about the derivation of *π* starting with Archimedes (although the Babylonians and Egyptians had an approximate formula for it a thousand years earlier, https://www.pcworld.com/article/191389/a-brief-history-of-pi.html). However, *e* has a much more mundane history and derives from the work of Jacob Bernoulli.

The problem arose with the calculation of interest. There is no difficulty with simple interest where the sum accruing is just the fractional rate *r* multiplied by the number of years *n* so that the total repayable on a principal *P* is just *P(1 + rn)*.

Neither is there a serious difficulty calculating compound interest where the interest is added to the principal at the end of each year and becomes the new principal for the following year. Thus, at the end of the first year, the value is *P(1 + r)*, and this becomes the new principal at the start of the second year, so at the end of the second year the total is $P(1 + r)(1 + r)$ which in turn is the principal for the third year at the end of which the sum is $P(1 + r)(1 + r)(1 + r)$. We can write this as the formula for the final value F that:

$$F = P(1 + r)^n \qquad (6.94)$$

What taxed Bernoulli was what happens if the interest is not calculated annually, but at intervals during the year, for example, quarterly", or monthly or weekly. Now n becomes the number or periods in the year. Then to approximate to the original annual rate, we divide the annual rate by the number of periods, r/n, and Eq. 6.94 becomes:

$$F = P\left(1 + \frac{r}{n}\right)^n \qquad (6.95)$$

Example – Compound Interest

Suppose 1000 euros is invested at 5% for 1 year and the interest is calculated annually, then the final value will be $1000(1 + 0.05) = 1050$ euros. Now suppose the interest is calculated monthly and we have $1000(1 + 0.05/12)^{12} = 1051.16$ euros – a slight increase. But notice that this is a higher effective rate than 5%. And we can turn this round the other way: money lenders can 'improve' their rate by quoting the monthly rate where interest is charged every month. As an example, if we take a rate of 1% charged every month for 1000 euros over 1 year, the sum to be repaid is $1000(1 + 0.01)^{12} = 1126.8$ euros – an annual rate of 12.68%, not 12% although legislation in many countries has insisted that lenders also quote the true annual rate.

Continuous Interest and e

Bernoulli investigated what happened as the periods got shorter, and taking a rate of 1 to simplify matters, we want the limit of:

$$\lim n \to \infty \ \left(1 + \frac{1}{n}\right)^n \qquad (6.96)$$

where at the limit the interest is calculated continuously. We can try this on a hand calculator (which, unfortunately, Bernoulli did not have) and with $n = 1,000,000$, we find an answer of 2.7182805.

At first we might have thought that as n gets bigger and bigger, the bracket will tend to $(1 + 0) = 1$, but this is not so because at the same time, the exponential is increasing. The nicest demonstration is by using the binomial series, which is a special case of the Taylor series with $f(x) = (1+x)^n$, Then following Wikipedia writing the general term:

$$\binom{n}{k} = \frac{n!}{k!(n-k)!} \tag{6.97}$$

the series is:

$$\left(1+\frac{1}{n}\right)^n = 1 + \binom{n}{1}\frac{1}{n} + \binom{n}{2}\frac{1}{n^2} + \binom{n}{3}\frac{1}{n^3} + \dots \tag{6.98}$$

then where k starts at 1 the expansion of the kth term is:

$$\binom{n}{k}\frac{1}{n^k} = \frac{1}{k!}\frac{n(n-1)(n-2)\dots(n-k+1)}{n^k} \tag{6.99}$$

then as $n \to \infty$, we can discount subtracting $1,2,3$, etc. from the terms in n in numerator of Eq. 6.99 just to leave the ns and we have:

$$\binom{n}{k}\frac{1}{n^k} = \frac{1}{k!}\frac{n^k}{n^k} = \frac{1}{k!} \tag{6.100}$$

Substitute back in Eq. 6.98 and the first term is still 1 . The second term when $k = 1$ is $1/1!$ the third when $k = 2$ is $1/2!$ and so on leading to:

$$\left(1+\frac{1}{n}\right)^n = 1 + 1 + \frac{1}{2!} + \frac{1}{3!} + \frac{1}{4!} + \dots \tag{6.101}$$

where we note that the terms rapidly become smaller: $1 + 1 + \frac{1}{2} + \frac{1}{6} + \frac{1}{24} + \frac{1}{120} + \frac{1}{720} + \dots$

And the sum can be written as:

$$\left(1+\frac{1}{n}\right)^n = \sum_{n=0}^{\infty} \frac{1}{n!} \tag{6.102}$$

and we should note the change of limit from 1 to 0 because $0! = 1$ so the first four terms are:

$$\left(1+\frac{1}{n}\right)^n = \frac{1}{0!} + \frac{1}{1!} + \dots = 1 + 1 + \frac{1}{2} + \frac{1}{6}. \tag{6.103}$$

and we do not need the first factor 1 in Eq. 6.98.

Example – Compound Interest

The schematic ('Compound Interest.asc') returns the correct final value for Example – Compound Interest. If we insert the values of Eq. 6.96 with $n = 1e6$, we find the answer is e. However, the intermediate values are not correct: after 6 months, Eq. 6.95 should be $F = P\left(1+\frac{r}{n}\right)^{n/2}$. Rewrite the schematic to show intermediate values correctly.

Series Expansion of e

We can use the Maclaurin series with e^x. Then we know that $\frac{d(e^x)}{dx} = e^x$, that is, the differential is just the function itself, and so for the series about 0, we have: $\frac{d(e^0)}{dx} = 1$ and the same for all the higher derivatives, so putting the numbers in Eq. 6.77 gives:

$$e^x = 1 + x + \frac{x^2}{2!} + \frac{x^3}{3!} + \frac{x^4}{4!} + \dots \qquad (6.104)$$

From this it follows that:

$$e^0 = e = 1 + 1 + \frac{1}{2!} + \frac{1}{3!} + \frac{1}{4!} + \dots \qquad (6.105)$$

which can be written as:

$$e = \sum_{n=0}^{\infty} \frac{1}{n!} \qquad (6.106)$$

which is the series for e that converges to 2.7182......etc. and has been evaluated to over a thousand places. For those that are interested in such things, one website, https://www.mathsisfun.com/numbers/e-eulers-number.html, writes the first digits as 2.71828 1828 45 90 45 which may help if you want to remember it. This is the identical equation to 6.102 which are equivalent ways of expressing e . And Eq. 6.104 also explains the curious fact that the derivative of e^x is still e^x . Differentiate Eq. 6.104 and we have:

$$\frac{d(e^x)}{dx} = 0 + 1 + \frac{2x}{2!} + \frac{3x^2}{3!} + \frac{4x^3}{4} \qquad (6.107)$$

and simplifying:

$$\frac{d(e^x)}{dx} = 1 + x + \frac{x^2}{2!} + \frac{x^3}{3!} + \dots \qquad (6.108)$$

and remembering that this is an infinite series it does not matter that we have lost the original first term and we are back to the series for e^x.

Euler's Relationships

It may not seem that compound interest and trigonometry have much in common. Let us see. The starting point is Euler's relationships where the angles are in radians, not degrees, and are:

$$e^{j\theta} = \cos(\theta) + j\sin(\theta) \quad e^{-j\theta} = \cos(\theta) - j\sin(\theta) \qquad (6.109)$$

From this it follows by adding both together and dividing by 2:

$$\cos(\theta) = \frac{1}{2}\left(e^{j\theta} + e^{-j\theta}\right) \qquad (6.110)$$

and by subtracting the second from the first and dividing by 2j:

$$\sin(\theta) = \frac{1}{2j}\left(e^{j\theta} - e^{-j\theta}\right) \qquad (6.111)$$

and from this, we can see that if we differentiate a sine, we get a cosine, and the value of the cosine is the gradient of the sine, Fig. 6.13. At point a, the gradient of the sine is a maximum, and this results in the maximum of the cosine at $a1$. Likewise, at point b the gradient of the sine is less and so is $b1$, whilst at c the gradient is zero and so point $c1$ is zero. At d the magnitude of the gradient is the same as b but it is negative, and at e it is again zero.

We can prove Euler's relationships if we write $e^{j\theta}$ as a series:

$$e^{j\theta} = 1 + \theta^j + \frac{\left(\theta^j\right)^2}{2!} + \frac{\left(\theta^j\right)^3}{3!} + \frac{\left(\theta^j\right)^4}{4!} + \frac{\left(\theta^j\right)^5}{5!} + \cdots \qquad (6.112)$$

then remembering that:

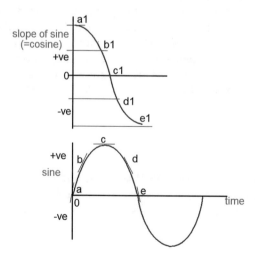

Fig. 6.13 Cosine as gradient of sine

$$j^0 = 1 \quad j^1 = j \quad j^2 = -1 \quad j^3 = j \quad j^4 = 1 \quad j^5 = j \qquad (6.113)$$

we have a repeating pattern which we can apply to Eq. 6.112 to give:

$$e^{i\theta} = 1 + i\theta - \frac{\theta^2}{2!} - i\frac{\theta^3}{3!} + \frac{\theta^4}{4!} + i\frac{\theta^5}{5!} - \frac{\theta^6}{6!} - i\frac{\theta^7}{7!} + \dots \qquad (6.114)$$

If we remove the odd terms from Eq. 6.114, we are left with:

$$j\theta - j\frac{\theta^3}{3!} + j\frac{\theta^5}{5!} - j\frac{\theta^7}{7!} + \dots \qquad (6.115)$$

which is just the sine series where we have replaced x by θ and multiplied by j. And the odd terms that we have removed are just the cosine series.

One special case of Eq. 6.114 which delights mathematicians is:

$$e^{j\pi} - 1 = 0 \qquad (6.116)$$

combining, as it does, three of the most important constants in mathematics and engineering.

Also at this point we can compare Eq. 6.109 with Eq. 6.93, and the difference is the scaling factor r, the radius of the circle. Thus we have the link between polar and exponential descriptions of angles.

Example – Euler's Relationship

We can partially illustrate this using the *hypot* function of an arbitrary voltage source where we make the *cos* function the real part and the *sine* at right angles as the imaginary part, schematic 'Cis,asc'. This works well with the two B sources used to generate $V=cos(2*pi*time)$ and $V=sin(2*pi*time+pi/2)$ where the *pi/2* creates the necessary phase shift and because the *hypot* is always positive $V=if(V(c)<0,hypot(V(c),V(s)),-hypot(V(c),V(s)))$ is needed to handle the negative half of the wave.

Natural Logarithms

Having discussed e this is a suitable place to say a little about natural logarithms. These are usually written $ln(x) = y$ or $log(x) = y$ where, by definition, $e^y = x$. We can find the differential of $ln(x)$ starting with $e^y = x$, and we differentiate this with respect to y:

$$\frac{d}{dx}(e^y) = \frac{d}{dx}(x) \text{ then } e^y\left(\frac{dy}{dx}\right) = 1 \qquad (6.117)$$

First substituting $e^y = x$ and then $ln(x) = y$ we have:

$$x\left(\frac{dy}{dx}\right) = 1 \quad \Rightarrow \quad \left(\frac{dy}{dx}\right) = \frac{1}{x} \quad \frac{d(\ln(x))}{dx} = \frac{1}{x} \qquad (6.118)$$

Example – Natural Logarithms

The schematic 'ln(x)asc 'draws the graph of *ln(time)* from *0 s* to *10 s*. Note that at *time = 1*, we have *ln(1) = e = 2.71829*, whilst at *time = −1*, we have *ln(−1) = 1/e = 0.367887*. At *time = 0*, we find *55.7549*. As *ln(x)* is only defined for *x > 0*, we need not worry too much about this. The only point is that to see the graph more clearly, we should set the y-axis to start from something like *−6 V* rather than *−60 V*.

We can also compare *ln(x)* to the tangent, taking point *(1,0)* where for small differences the tangent is close to the logarithm.

Trigonometrical Identities

Equations 6.109, 6.110 and 6.111 readily lend themselves to alternative derivations which are important in computing. The methodology is consistent in that each time we replace the function by its exponential representation and expand it. This can be simpler than the trigonometrical proofs. We shall now derive a few.

The Complex Conjugate

This is a very important topic that we shall return to in later chapters because it enables us to convert complex expressions into ones that are entirely real. Referring to Eq. 6.113 $j^2 = -1$, so if we multiply $(a + jb)$ by $(a - jb)$, we have:

$$(a + jb)(a - jb) = a^2 + b^2$$

and $(a - jb)$ is the *complex conjugate* of $(a + jb)$, and the multiplication removes the j leaving only the real products. Now we can attack some important trigonometrical identities.

$\text{Cos}(\theta)^2 + \text{Sin}(\theta)^2 = 1$

We note that $e^{j\theta}e^{-j\theta} = 1$ and that $e^{-j\theta}$ is the complex conjugate of $e^{j\theta}$ and then using Eq. 6.109:

$$e^{j\theta}e^{-j\theta} = (\cos(\theta) + j\sin(\theta))(\cos(\theta) - j\sin(\theta)) = \cos(\theta)^2 + \sin(\theta)^2 = 1$$

$Sin(2\theta) = 2 \sin(\theta)\cos(\theta)$

To create a multiple angle, we multiply the coefficients of *e*. Then:

$$\sin(2\theta) = \frac{e^{2j\theta} - e^{-2j\theta}}{2j} \tag{6.119}$$

and the expansion must involve a sine function because of the 2*j*. Let us try *sine(θ) x cosine(θ)*:

$$\left(\frac{e^{j\theta} - e^{-j\theta}}{2j}\right)\left(\frac{e^{j\theta} + e^{-j\theta}}{2}\right) = \frac{\left(e^{2j\theta} - e^{-2j\theta}\right)}{4i} \tag{6.120}$$

and if we multiply Eq. 6.120 by 2, we arrive at Eq. 6.119.

$Cos(2\theta) = Cos^2(\theta)\text{-}Sin^2(\theta)$

First we write *cos(2θ)*:

$$\cos(2\theta) = \frac{1}{2}\left(e^{-2j\theta} + e^{-2j\theta}\right) \tag{6.121}$$

Then we can expand the right-hand side in parts as:

$$\cos^2(\theta) = \frac{\left(e^{j\theta} + e^{-j\theta}\right)}{2}\frac{\left(e^{j\theta} + e^{-j\theta}\right)}{2} = \frac{1}{4}\left(e^{2j\theta} + e^{-2j\theta}\right) \tag{6.122}$$

$$\sin^2(\theta) = \frac{\left(e^{j\theta} - e^{-j\theta}\right)}{2}\frac{\left(e^{j\theta} + e^{-j\theta}\right)}{2} = \frac{1}{4}\left(e^{2j\theta} + e^{-2j\theta}\right) \tag{6.123}$$

and then subtracting:

$$\cos^2(\theta) - \sin^2(\theta) = \frac{1}{2}\left(e^{2j\theta} + e^{-2j\theta}\right) - \frac{1}{-4}\left(e^{2j\theta} + e^{-2j\theta}\right)$$

$$= \frac{1}{2}\left(e^{2j\theta} + e^{-2j\theta}\right) \tag{6.124}$$

which is what we wanted to show.

$2Cos(\theta)Cos(\Phi) = Cos(\theta\text{-}\Phi) + Cos(\theta + \Phi)$

We first write that:

$$2 \cos (\theta) \cos (\phi) = 2 \left(\frac{e^{j\theta} + e^{-j\theta}}{2} \right) \left(\frac{e^{j\phi} + e^{-j\phi}}{2} \right) \tag{6.125}$$

expanding Eq. 6.125

$$= \frac{1}{2} \left(e^{j(\theta+\phi)} + e^{j(\theta-\phi)} + e^{j(-\theta+\phi)} + e^{-j(\theta+\phi)} \right) \tag{6.126}$$

Then we have:

$$\cos (\theta - \phi) = \frac{1}{2} \left(e^{j(\theta-\phi)} + e^{-j(\theta-\phi)} \right) \tag{6.127}$$

which are the second and third terms of Eq. 6.125 and:

$$\cos (\theta + \phi) = \frac{1}{2} \left(e^{j(\theta+\phi)} + e^{-j(\theta+\phi)} \right) \tag{6.128}$$

are the first and fourth terms and thus we have proved Eq. 6.123.

Cos(θ + Φ) and Sin(θ + Φ)

These follow directly from Euler's Relationship. If we add both as:

$$\cos (\theta + \phi) + j\sin(\theta + \phi) = e^{j(\theta+\phi)} = e^{j\theta} e^{j\phi} \tag{6.129}$$

then using Eq. 6.109:

$$e^{j\theta} e^{j\phi} = (\cos (\theta) + j\sin(\theta))(\cos (\phi) + j\sin(\phi)) \tag{6.130}$$
$$= \cos (\theta) \cos (\phi) - \sin (\theta) \sin (\phi) + j(\cos (\theta) \sin (\phi) + \sin (\theta) \cos (\phi)) \tag{6.131}$$

Equation the real parts of Eqs. 6.129 and 6.131:

$$\cos (\theta + \phi) = \cos (\theta) \cos (\phi) - \sin (\theta) \sin (\phi) \tag{6.132}$$

and equating the imaginary:

$$\sin (\theta + \phi) = \cos (\theta) \sin (\phi) + \sin (\theta) \cos (\phi) \tag{6.133}$$

Explorations 8
1. Using schematic 'ln(x).asc' find *ln(x)* over a range of values. Selecting different values find the range about *ln(x)* over which the tangent differs by less than 1%.

2. Use a B voltage source to create the first few terms of $e = \sum_{n=0}^{\infty} \frac{1}{n!}$ and compare with an actual exponential, in particular, how many terms are needed for a given percentage error.
3. Expand some more of the double angle formulae given in https://en.wikipedia.org/wiki/List_of_trigonometric_identities, and check that you get the same result.
4. It is very tedious, but just about possible, to use the series forms of sine and cosine to test some of the trigonometrical identities, at least over part of a cycle.

6.7 '.four' Waveform Analysis

We very often need to know what frequencies are contained in a waveform. The reasons are manifold. One is to assess the electrical interference created by appliances (including computers); another is to measure the distortion created by a signal. The analysis gives a tabulation of harmonics provided we know what fundamental frequency to use. This is most important. It is therefore very useful if we are exciting the circuit with a known voltage or current source and is rather easier to interpret than the more comprehensive Fast Fourier Transform.

6.7.1 Application

It is invoked from the schematic and performed after an analysis where the results will be found in the **SPICE Error Log**. The essential is:

$$.four < frequency >< test\ point >$$

which by default finds the first nine harmonics where (correctly) the fundamental is counted as the first harmonic.

Example – .four of Square Wave
We can build a simple square wave source, schematic ('Fourier Squarewave.asc'). which consists simply of a voltage source **V1** and an earth. The source is declared as:

$$PULSE(0\ 10\ 1\ 1f\ 1f\ 1\ 2)$$

This is a waveform consisting of a *10 V* pulse *1 s* wide starting after *1 s* and a second starting after *2 s*. The analysis is:

$$.four\ \ 0.5\ \ V(out)$$

selecting a frequency of 0.5 Hz which is the fundamental of the square wave. The output can be accessed through **File→Open** and then select **All files** (∗.∗), and it can be found as a .TXT file.

Harmonic Number	Frequency [Hz]	Fourier Component	Normalized Component	Phase [degree]	Normalized Phase [deg]
1	5.000e-01	6.366e+00	1.000e+00	179.65°	0.00°
2	1.000e+00	9.330e-15	1.466e-15	85.02°	−94.63°
3	1.500e+00	2.122e+00	3.333e-01	178.95°	−0.70°
4	2.000e+00	8.255e-15	1.297e-15	81.55°	−98.10°
5	2.500e+00	1.273e+00	2.000e-01	178.24°	−1.41°
6	3.000e+00	9.071e-15	1.425e-15	86.22°	−93.42°
7	3.500e+00	9.092e-01	1.428e-01	177.54°	−2.11°
8	4.000e+00	8.233e-15	1.293e-15	81.07°	−98.58°
9	4.500e+00	7.070e-01	1.111e-01	176.84°	−2.81°

Total harmonic distortion: 42.873670%(48.004124%)

The interesting point here is that the amplitudes of the odd harmonics are in the sequence *1 1/3. 1/5, 1/7 . . . .* For example, taking the fifth harmonic, we have *1.273/ 6.366 = 1/5.* This is what we saw right at the start of the chapter.

Notice that the even harmonics are vanishingly small and at approximately 90°, whilst the odd ones are almost at 180°, but the phase decreases with frequency.

Adding More Harmonics and Cycles

Extra harmonics can be added by inserting the number after the frequency:

.four < frequency >< no. of harmonics >< test point >

By default the measurement is made over the last cycle of the waveform. This makes sense because very often it takes a few cycle for the waveform to settle down. However, we can specify an integer number of cycles working back from the last as:

.four < frequency >< no. of harmonics >< no. of cycles >< test point >

Entering more than the total does not flag an error, but the best way to cover the entire trace is to enter −1 for the no. of cycles.

Example – .four Number of Harmonics
It is interesting to see more harmonics of the square wave.

.four 0.5 21 V(out)

shows 21 from which two things are clear. The first is that the phase shift is slowly increasing; the second is that the contributions of the higher harmonics decrease slowly because it is a reciprocal relationship, and for an accurate representation, we need many harmonics.

The Effect of Not Using Exact Harmonics

A small frequency error, say 5%, will increase the even harmonics by two orders of magnitude, but as these are still very small, it is of no great importance; see the schematic ('Fourier Squarewave 1.asc'). The amplitude of the fundamental is not greatly changed, but there is an increasing error in magnitude and phase for higher harmonics. The total distortion remains about the same, but then, as that is nearly 50%, that is not very important.

So if an accurate analysis of the harmonic content is needed, it is essential that the analysis is made exactly on the fundamental frequency

Explorations 9
1. Create and analyse a triangular waveform.
2. Explore a rectangular waveform where the 'on' and 'off' times are not the same.
3. Create and explore a sawtooth waveform – a triangle with one side vertical.

6.8   Summary

This chapter has covered most of the essential mathematics needed to analyse waveforms. A little more will be added in the next chapter. What we have seen is:

- The *.meas* function is a more accurate way of finding values when specific conditions are met in the analysis and faster if the analysis conditions are changed.
- Any repetitive waveform can be reduced to the sum of a series of harmonics with appropriate coefficients.
- The average value of a sine wave is zero and its RMS value is *peak/√2*. The half-cycle average is *2peak/π*.
- The average and RMS values of rectangular waves depend on the duty cycle and DC offset, but for a square wave, the peak and RMS values are the same.
- A triangular wave, either all-positive or all-negative or bipolar, has an average of *peak/2* and an RMS of *peak/√3*.
- The RMS value of waveform with more than one component such as a DC offset or a piecewise waveform can be found from the *volts-* or *current-squared time* for all the parts, then adding them, and taking the square root.
- We can use the Maclaurin series to write sine and cosine as an infinite series.
- We can also derive a series for e as: $e = \sum_{n=0}^{\infty} \frac{1}{n!}$.
- Euler's relationships $e^{j\theta} = \cos(\theta) + j\sin(\theta)$ $e^{-j\theta} = \cos(\theta) - j\sin(\theta)$ enable sines and cosines to be expressed as exponentials, and hence trigonometrical relations such as $cos(2\theta) = cos^2(\theta) - sin^2(\theta)$ can be derived from them rather than using the series form.
- The *.four* analysis gives us the harmonic content of a signal provide we know the fundamental frequency.

Chapter 7
Capacitors

7.1 Introduction

Capacitance exists whenever two conductors are separated by an insulator. Strictly speaking, this is *mutual capacitance*, but the distinction need not bother us. At the very large scale, there is capacitance between the overhead wires of the high voltage electricity distribution system and ground; at the much smaller scale, there is capacitance between all the tracks on a printed-circuit board. And this is a nuisance.

Capacitors are used for four purposes. The first is to store charge to maintain the current in a circuit: the output of rectifiers in power supplies is lumpy, and a capacitor keeps the current flowing. The second use, which dates back to thermionic valves, is to block DC from passing from one circuit to the next: much of modern circuit design tries to use DC coupling without capacitors because they are relatively large and expensive. The third use is to remove AC where only the DC component is required as in the bias circuits of thermionic valves and transistors, and again modern design often uses different circuit configurations where capacitors are not needed. The fourth use is to shape the frequency response of a circuit. It can be argued that all the three previous uses are merely a subset of the fourth and it is true that in blocking DC, the capacitor also affects the low-frequency response of the circuit. These will be discussed later in this and the next two chapters.

7.2 Capacitors

A practical capacitor is made from two conducing surfaces separated by an insulator. To understand how it works, we take the simple case of two rectangular metal plates in a vacuum shown in Fig. 7.1 and apply a voltage between the two plates. This will create an electric field between the inner surfaces of the plates which will try to repel electrons from the left-hand plate and attract them to the right-hand one. We must not

© Springer Nature Switzerland AG 2020
C. May, *Passive Circuit Analysis with LTspice*®,
https://doi.org/10.1007/978-3-030-38304-6_7

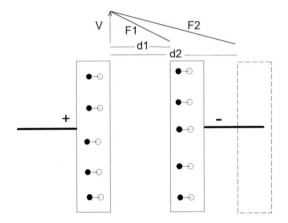

Fig. 7.1 Capacitor Concept

think of a wholesale flow of electrons so that they pile up on one plate of the capacitor and pour out of the other – remember that the drift velocity of electrons in a metal wire is very small, a few millimeters per second. And as there is no conducting path between the plates, there will be no conventional current.

Now we know that the electrons are randomly moving very rapidly, and in the absence of an electric field, their mean positions do not change. But under the influence of an electric field, there will be a small *displacement current* as electrons, shown as black circles, move from their mean equilibrium positions shown as empty circles. We can image the forces pushing them back as small springs, so they increase their potential energy when they are moved. And it is this potential energy that is stored in the capacitor.

From this it follows that if we increase the area of the plates, there will be more electrons to be displaced and the capacitance will increase with area. If we increase the spacing of the plates from *d1* to *d2* shown as a dashed rectangle and keeping the same voltage, the electric field will decrease inversely as the spacing from *E1* to *E2* and the displacement of the electrons will decrease accordingly, so that the capacitance is proportional to the inverse of the spacing of the plates.

A *dielectric* is an insulator that can be polarized, that is, that under the influence of an electric field, some charge carries will be displaced from their mean equilibrium positions as shown in Fig. 7.2. So if we interpose a dielectric between the plates of a capacitor, even more energy can be stored in the dielectric. Thus:

$$C = \epsilon_0 \epsilon_r \frac{A}{d} \tag{7.1}$$

where ε_0 is the *electric constant* of $\varepsilon_0 \approx 8.854 \times 10^{-12}$ F·m^{-1}) and ε_r is the *relative static permittivity*, often called the *dielectric constant* of the material which is 1.000 for a vacuum by definition and almost the same for air (actually 1.00058986) but is much larger for plastics, glass or mica. There are excellent Wikipedia articles on all of these topics.

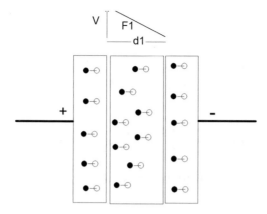

Fig. 7.2 Capacitor with Dielectric

7.2.1 Unit of Capacitance

This is the 'farad', defined as that capacitance where one coulomb of charge raises its potential by 1 volt. It is a large unit, and generally capacitors are measures in microfarads (μF) or smaller, although the new 'super capacitors' do have capacitances of the order of tens of farads.

7.2.2 Energy Stored in a Capacitor

This is of considerable importance because even very small capacitors can store enough energy to damage sensitive circuits. We suppose that the capacitor has charged to a voltage v and now add a small increment of charge dq. The work done in adding this is just $dq.v$. Hence, to find the total work done W in adding all the charge to the capacitor, we can integrate from zero charge up to the final charge q_f or:

$$W = \int_{q=0}^{q_f} v.dq \qquad (7.2)$$

but as the voltage is given by $v = q/C$, we can substitute in Eq. 7.2:

$$W = \frac{1}{C} \int_{q=0}^{q_f} q.dq \qquad (7.3)$$

and performing the integration gives:

$$W = \frac{1}{2C} q_f^2 \qquad (7.4)$$

and as $q_f = CV_f$ we can write Eq. 7.4 as:

$$W = \frac{1}{2}CV_f^2 \qquad (7.5)$$

Example – Energy Stored in a Capacitor

An electronic flash uses a xenon tube rated at *50 J, 330 V*. Using Eq. 7.5 we find we need a capacitor of:

$$50 = \frac{1}{2}C330^2 \quad C = \frac{100}{330^2} = \frac{100}{108900} \quad F = 0.918mF = 1mF$$

where we selected the next higher preferred value.

7.2.3 Capacitors in Parallel and in Series

If we place two or more capacitors in parallel, they will all have the same voltage across them, Fig. 7.3 is for two, and we can write:

$$V_{in} = \frac{Q_1}{C_1} = \frac{Q_2}{C_2} \qquad (7.6)$$

so the total charge is:

$$Q_t = Q_1 + Q_2 = V_{in}(C_1 + C_2) \qquad (7.7)$$

and the two capacitors can be replaced by one equivalent capacitor which is the sum of the two:

$$C_{eq} = C_1 + C_2 \qquad (7.8)$$

However, if we put capacitors in series, Fig. 7.4, the situation is a little more complicated.

First we can note that Kirchhoff's voltage law must be obeyed so:

$$V_1 = V_{C1} + V_{C2} \qquad (7.9)$$

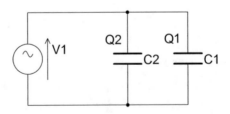

Fig. 7.3 Capacitors in Parallel

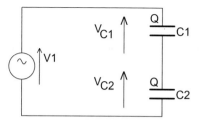

Fig. 7.4 Capacitors in Series

But before we apply a voltage, all four plates of the capacitors are at the same potential. Afterwards, the displacement current will result in a charge Q on the top plate of capacitor $C1$ and an equal and opposite charge on the lower plate, so the potential of the lower plate is no longer the same as the top plate of $C2$. Hence there will be another displacement current so that there is the same charge on each capacitor. Thus:

$$V_1 = \frac{Q}{C_1} + \frac{Q}{C_2} \tag{7.10}$$

or:

$$V_1 = Q\left(\frac{1}{C1} + \frac{1}{C2}\right) \tag{7.11}$$

and as $V/Q = C$ the equivalent capacitor now is:

$$\frac{1}{C_{eq}} = \frac{1}{C_1} + \frac{1}{C_2} \tag{7.12}$$

so, if we only have two capacitors:

$$\frac{1}{C_{eq}} = \frac{C_1 + C_2}{C_1 . C_2} \tag{7.13}$$

$$C_{eq} = \frac{C_1 C_2}{C_1 + C_2}$$

Compare this to two resistors in parallel.

Example – Capacitors in Series and Parallel

Given two capacitors, one *1 F* the other *2 F* and a *10 V DC* supply, If we place them in series the total capacitance is *2/3 F*. If we change the numbers in schematic ('Capacitors in Series.asc') to make $C2 = 2F$, we find its voltage is *3.33 V*. And if we move the mouse over it and hold down **Ctrl** so that the current clamp meter turns into a thermometer and right click, we find the power is plotted. This is a ramp to *1.1 MW* and we find the average is *222 kW*, but that is over the whole transient interval, and the energy is *11.1 J*. As this is an ideal capacitor, the power is entirely imaginary. We also find *22.2 J* for the *1F* capacitor.

If we place them in parallel, the capacitance is *3 μF* and using *1/2 CV²* we have *150 J*.

7.2.4   Capacitors in Series Voltage Ratings

It follows from the previous that if we have two identical capacitors, the voltage should divide equally between them. This is a common stratagem to deal with the situation where the voltage ratings of the individual capacitors is insufficient for the applied voltage. This, however, is not a safe practice since it ignores the leakage resistance of the capacitors which acts as a potential divider. So, for example, if it happens that one capacitor has a leakage resistance ten times the other, it will have to absorb almost all of the applied voltage. The solution is to connect resistors in parallel with the capacitors to even out the voltage across each As a rough guide they should draw ten times the leakage current.

Explorations 1
1. Use ('Capacitors in Series.asc') to examine the voltages. Note that the commands **.op** and **DC sweep** both set capacitors as open circuits; therefore we must use a **Transient** analysis. Moreover, we must check **Start external DC supply voltages at 0 V** (which will add the text **startup** to the transient command). Uncheck this box and it will be found that the voltage at node *2* is wrong.
2. Change the capacitor values so that they are not equal and measure voltages, charge and energy.
3. Try networks with three or more capacitors.

7.3   Capacitor Types

A good description of capacitor types can be found at https://www.electronics-tutorials.ws/capacitor/cap_2.html. The critical parameters are the equivalent series resistance (ESR) and equivalent series inductance (ESL) and the leakage resistance of electrolytic capacitors. The ESR is important because it determines the power dissipated by the capacitor as $P = ESR.I^2$. It also defines the rise time and hence the rate at which the capacitor can charge and discharge which limits its frequency response. The ESL is mainly of interest at high frequencies where the capacitor behaves like a damped tuned circuit.

The data sheet figures are not tightly controlled, and it is safest to take the worst-case values. A very good series of articles on capacitors by Ian Poole can be found starting at http://www.radio-electronics.com/info/data/capacitor/capacitor_types. php. Also of interest is http://conradhoffman.com/cap_measurements_100606.html.

7.3.1   Variable Capacitors

We can divide these into user controls and preset capacitors.

7.3.1.1   User Controls

These are widely used for tuning in radio stations on the amplitude modulated (AM) bands. For frequency-modulated and television frequencies of the order of a 100 MHz and above, the required capacitance is very small and varactor diodes are now used. These are smaller and lighter than traditional moving plate capacitors and, since the capacitance is varied by voltage, they are amenable to use with remote controls, whereas the traditional variable capacitor required manual intervention to turn the control knob.

AM radios require large capacitances, far beyond the range of varactor diodes. They often have two or three units ganged together, typically of 500 pF per section when the plates are fully meshed, each section consisting of a number of plates on a common spindle approximately shaped like a tapered 'D' so that the capacitance changes almost linearly with the spindle rotation. The fixed plates are generally earthed. Air is mainly used as a dielectric, so they are quite large units compared to fixed capacitors although thin mica sheets have been used on single units, making them much more compact. In an interesting article (http://g3rbj.co.uk/wp-content/uploads/2013/10/Measurements_of_Loss_in_Variable_Capacitors_issue_2.pdf), Alan Payne found the ESR of a single unit typically to be about 0.1 Ω and depended on frequency as:

$$R_{ESR} = R_f + \frac{a}{\left(fC^2\right)} + bf^{0.5} \tag{7.14}$$

7.3.1.2   Preset Capacitors

These are small cylindrical components mounted on a printed circuit board which are adjusted with a screwdriver, usually when the circuit is first set up. Component distributors such as Farnell offer a wide range of types.

The choice depends on the application. At one extreme are the very expensive products of Johanson costing tens of euros each where the capacitance is changed by the overlap of two concentric cylinders, one being moved by a lead screw allowing very precise adjustment, for example, https://uk.farnell.com/johanson-manufacturing/5201/cap-10pf-250v-can-smd/dp/1692432. At the other are simple small capacitors of a few picofarads where the plates form a V shape and the adjusting screw opens or closes the 'V' to reduce or increase their separation and capacitance.

For general use, we find small cylindrical units with multiple plates, where the overlap is adjusted by a screw, for example, https://uk.farnell.com/ vishay/bfc280811229/cap-150v-pp-through-hole/dp/1215709?MER=bn_level5_ 5NP_EngagementRecSingleItem_2, where the dielectric is plastic. The maximum capacitance is around 50 pF. A cutaway drawing of a preset capacitor, as well as illustrations of other types, can be found at http://binaryupdates.com/types-of-capac itor/#Trimmer_Capacitor.

7.3.2   Fixed Non-polar Capacitors

These can be inserted in the circuit either way round and there is no need to worry about reverse DC being applied provided the voltage rating is not exceeded. They are available in a wide ranges of dielectric types and with capacitances up to 10 F or so and working voltages to several kV (but not both together). They are used whenever possible because they approximate more closely to ideal capacitors than polar ones. We shall discuss them by the dielectric type.

7.3.2.1   Ceramic

These are the most widely used, often in the form of small rectangular blocks as surface-mount devices (SMD), but they can also be found as wire-ended components, often as a small disc. The Wikipedia article is comprehensive.

Three types of ceramic are in general use designated class 1, 2 and 3 (high K) with class 1 as the most stable but having the lowest capacitance per unit volume. It is difficult to generalize; hence Murata have a web tool (http://ds.murata.co.jp/ simsurfing/mlcc.html?lcid=en-us) giving graphs of the performance of their products. As a generalization:

- The tolerance is typically 20%.
- They are not stable against temperature or bias voltage with type 3 the worst showing a loss of 90% of capacity with a bias of 6 V and up to 60% over the temperature range. The Murata graphs for type 3 (http://ds.murata.co.jp/ simsurfing/mlcc.html?lcid=en-us) are V-shaped on log-log axes and show for type 3 a fall of impedance from 1 Ω at 100 kHz to 7 mΩ at 7 MHz rising again to 2 Ω at 1 GHz. This is due to the equivalent series inductance (ESL). Types 1 and 2 have a much better response with smaller changes.
- The maximum value is about 4 nF.

- They have low voltage rating, up to 6 V or so, and are best used without any bias, so they are not suitable for blocking DC.
- They have a slightly lower loss than silver mica which they generally replace.
- ESL. It is reported (https://pdfs.semanticscholar.org/d884/9416a13c24827650 eb58c4bd765653fa9aa9.pdf) that this is predominately due to the way the SMD device is placed on the PCB, that is, through holes, the mounting pads, track layout, etc. The capacitor itself contributes about 400 pH.

Murata have a library of the dynamic performances (https://www.murata.com/~/ media/webrenewal/tool/library/pspice/note_dynamic-model_mi101e.ashx? la=en-gb).

7.3.2.2   Silver Mica Capacitors

These are not so widely used today. They are made by silver-plating thin sheets of mica which can be made into a stack for higher values of capacitance. Leads are then attached at 0.1 in spacing to suit PCB layout and the whole encapsulated for protection. Their salient properties are:

- Close tolerance, 1% or so.
- High stability against temperature, frequency and time.
- Values from a few pF to 10,000 pF.
- Voltage ratings from 100 V to 400 V.
- Very small losses.
- But conradhoffman (see above) reported a very high dielectric adsorption of 0.82% for one specimen.

7.3.2.3   Film (Plastic) Dielectric Capacitors

Several plastics, polycarbonate, PTFE, polypropylene and others are used and have different properties. They can consist of a Swiss Roll of a very thin metallic foil on either side of the plastic file, or the metal can be deposited directly on the plastic. The roll is then enclosed in a protective sleeve and the ends sealed leaving axial leads, or the roll is flattened to a rectangle and put in a rectangular plastic case and sealed with radial leads on the 0.1 in pitch for PCBs. The salient features are:

- Tolerance from 1% to 10% depending on type.
- Positive and negative temperature coefficients depending on type with a change of a few per cent over the operating range.
- Changes of a few percent with frequency up to 100 kHz (see https://www.vishay. com/docs/26033/gentechinfofilm.pdf).
- Values from about 1 nF to 10 μF.
- Voltage rating to 400 V or more.
- Small to very small losses depending on type.

- Very low dielectric adsorption except polyester, and that is still only about 1/4 that of silver mica.
- Polycarbonate reportedly has a 'soakage' effect where the capacitor does not discharge completely and some more charge can be recovered if the capacitor is left for a little time.

7.3.3 Polar (Electrolytic) Capacitors

These encompass a large capacitance in a small volume. They are polar devices and will not withstand a reverse voltage.

7.3.3.1 Aluminium Electrolyte

These have long been used. A reverse voltage will dramatically increase the leakage current and power dissipation, thus leading to a high internal temperature to the point that the electrolyte boils. Modern capacitors either have a vent cut in the rubber gasket closing the end, or the gasket itself is only lightly held in place: in the first case, the steam can escape through the vent, and in the second, the gasket is blown out with some force. In thermionic valves circuits, the capacitors worked at quite high voltages and were solidly constructed so that an inadvertent reverse polarity could be very dangerous as the author can recall hearing one explode like a small bomb many years ago: the substantial aluminium can was fired the length of the laboratory with the contents of the capacitor streaming out behind and an acrid smell lingering afterwards.

An excellent discussion can be found at http://www.cde.com/resources/catalogs/AEappGUIDE.pdf. The salient properties are:

- Tolerance typically 20%.
- Leakage current increases strongly with temperature and voltage, rising from 10 µA to a few mA and is much higher than other types.
- ESR a little less than 1 Ω.
- Values from a few microfarads to over 1 farad.
- Voltage ratings to 400 V or more.
- Significant losses at frequencies more than a few kilohertz.

7.3.3.2 Tantalum Capacitors

These have a better frequency range and higher capacitance per unit volume but are limited in voltage rating and can easily be destroyed by a reverse voltage. The Wikipedia article is very comprehensive:

- Capacitance from 1 nF to 79 mF

- Working voltage to 125 V but need to be derated for improved lifetime
- Very sensitive to over-voltages and reverse voltages causing catastrophic failure
- Smaller than aluminium of the same capacitance and voltage
- Much smaller ESR than aluminium, typically less than 1 Ω but 5.5 Ω for 'wet' tantalum
- Very low leakage so can be used in sample-and-hold circuits
- Very good long-term capacitance stability compared to aluminium

7.3.3.3 Super Capacitors

These span a wide range of physical sizes from 8 mm diameter and 12 mm long to the size of vehicle batteries. The main characteristics are:

- Large capacitance to over 1000 F.
- The working voltage is less than 3 V but batteries of capacitors can be up to 160 V.
- Almost instantaneous charging/discharging and the large modules are ideal for supplying and adsorbing short busts of high energy, for example, in regenerative braking, or deploying aircraft escape slides if there is a power failure.
- Smaller units can be used in uninterruptible power supplies.
- Very many more charge/discharge cycles than rechargeable batteries.
- The small units mount like electrolytic capacitors.
- Very low internal resistance, typically of the order of 100 mΩ.
- Leakage current is higher than the above types, but still typically less than 1 mA.
- High power density, typically 7 kW/kg, many times greater than batteries.
- The lifetime depends on the temperature and voltage, ranging from about 1 year if used at full voltage and 65 °C to 10 years if derated to 2.2 V and a temperature of 25 °C.

7.3.4 SPICE AC Analysis

We mentioned this in a previous chapter; now it is time to look in more detail. A transient analysis enabled us to see the waveform by making repeat analyses point-by-point from the start time to finish time. Perforce this is a relatively slow process since many circuits contain non-linear components. In theory it would be possible to plot the frequency response from repeat transient analyses over the frequency range, but this would be painfully slow. Therefore SPICE includes an AC analysis which creates a plot of amplitude and phase against frequency.

To do this, SPICE first performs a DC analysis to find the quiescent point. Then it finds the tangents to the characteristics at this point. Then a sinusoidal signal is injected and the frequency is swept over the chosen range and the analysis performed point-by-point using these tangents. It is implicit that these extend infinitely in either direction, and it is not uncommon to find outputs of a kilovolt or two which would certainly have destroyed the real circuit.

Thus an AC will not reveal distortion nor clipping nor overload, but it will simulate very quickly and give the overall picture of the circuit's response. In effect, this makes the same assumptions as the manual methods to be described below and is the closest to them in terms of the output presentation.

7.3.4.1   LTspice AC Analysis

We have touched on this before, here we expand upon it. We need first to ensure that the voltage or current source has an AC value. Assuming a voltage source, right click on it to open **Independent Voltage Source – V1** and then set **AC Amplitude** to *1*. You can set a different value, for example, *10*, but then the analysis will show an asymptotic gain of 20 dB because the gain of the circuit is multiplied by the input.

The analysis is invoked by **Simulate- > Edit Simulation- > Cmd** from the main menu or by right clicking on the schematic then the **AC Analysis** tab.

The first option *Type of Sweep* is the way the measurements are made. For 'Octave' and 'Decade', these are the number of measurements per octave or per decade; for 'Linear' these are the grand total, whilst for 'List', a list of measurement frequencies must be supplied.

It pays to be generous with the *Number of points*... because the simulation runs very rapidly but LTspice will very often drastically reduce them if the frequency spread is very large.

The *Start frequency* and *Stop frequency* define the limits of the simulation. Again, it pays to be generous because the range can be edited on the waveform window.

7.3.4.2   Changing the Y-Axis

This can be done after the analysis. By moving the mouse into the left-hand y-axis region of the trace window, the cursor changes to a small vertical yellow ruler. Right click to open the **Left Vertical Axis –> Magnitude** dialogue. There are three choices for **Representation.**

Bode

This is most often used. We find three radio buttons that define the scale.

- **Linear** defaults to the maximum value at the top and usually 0 V at the origin. The three edit boxes to the left allow the range to be altered and the horizontal grid lines to be set.

- **Logarithmic** defaults to the range of the measurements with grid lines every decade. It is scaled relative to the exciting source – if that is *10 V* AC, the axis will start at *10* and decreases downwards. No units are given since there could be graphs of current and voltage.
- **Decibel** again defaults to the measurement range with the option to draw the grid lines at other dB levels. But be aware that with a small plotting pane, these may not be seen correctly.

The **Don't plot the magnitude button** suppresses the left-hand axis and magnitude traces. They can be restored by selecting another radio button.

We may similarly change the right-hand y-axis. Here we have the options of plotting the phase or the group delay which measures the time for each frequency to travel from input to output. The check box **Unravel branch wrap** is irrelevant here but ensures a smooth phase plot without a sudden vertical jump at 180° by re-scaling the axis.

Nyquist

This is not helpful at the moment. There is no choice of scale.

Cartesian

This divides the signal into a real part with the axis on the left, and the imaginary part with the axis on the right with an *i* dividing the unit name, for example, *50 miV*. There is no choice of scale. We shall use this very soon. The **Don't plot real component** leaves only the imaginary. On the right-hand axis, we have the option **Don't plot imaginary component**.

Note that these changes do not affect the underlying simulation data, only the presentation.

7.3.4.3 Changing the X-Axis

Move to the horizontal axis area and the cursor changes to a small horizontal yellow ruler. Right click to open the **Horizontal Axis** dialogue. The options are to change from a logarithmic to a linear scale and to change the limits of the plot. And again these changes do not affect the data.

We can adjust everything by right clicking in the trace window and selecting **View- > Manual Limits.**

7.3.4.4 Changing the View

Rather than adjusting the axes, we can choose an area for closer inspection. In every case, the cursor turns into a magnifying glass. From the main menu, chose **View- >** and then click:

- **Zoom Area** Hold down the left mouse button and drag to define the zoomed area.
- **Zoom Back** Each click on this reduces the vertical scale. It can only be undone by **Zoom to Fit** to restore the original traces.
- **Pan** Click in the upper half of the waveform panel to move the traces down and in the bottom half to move them up. This does not scale the traces.

There are few problem areas, the main one being of ensuring enough points are taken not to miss sharp resonances. Now, armed with AC analysis, we can embark on an exploration of capacitor models.

7.4 Capacitor Models

For non-critical, low-frequency, first attempts, a simple capacitor will suffice just to get a feel for how the circuit will behave. As a second stage, we can use the LTspice library and the LTspice model, again at low to medium frequencies. This, of course, begs the question of what is a medium frequency and here a rough guide is *10 Hz* for polar type else *10 MHz*. But this does not fully model the peculiar characteristics of some types, in particular, the dramatic reduction of capacitance with bias voltage of some ceramic capacitors. A clearer picture emerges running the various schematics presented below using the manufacturer's models. These are invaluable for accurate work at high frequencies.

7.4.1 The LTspice Model

The structure is shown in Fig. 7.5. It encompasses a popular model of a capacitor by the series circuit of *Rser,Lser,C*. To this it adds *Rpar* and Cpar and *RLshunt* across the inductor.

The essential description is:

$$C < name >< node\,1 >< node\,2 >< value >$$

which is usually filled in when the symbol is placed on the schematic.

However LTspice includes many additional parasitic elements in its model which are shown in the **Capacitor Cn** dialogue.

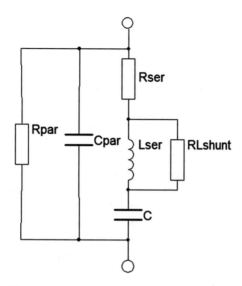

Fig. 7.5 LTspice Capacitor

Property	Comment
Capacitance (F)	Essential
Voltage rating (V)	Not acted on
RMS current rating (A)	Not acted on
Equiv. series resistance (Ω) (ESR)	Values are given for the capacitors that can be chosen by pressing the **Select Capacitor** button. It represents power loss.
Equiv. series inductance (H)	Due to lead inductance and the capacitor itself.
Equiv. parallel resistance (Ω) – Rpar	Leakage, especially electrolytic capacitors
Equiv. parallel capacitance – Cpar	Stray capacitance such as capacitance of PCB mounting tabs and through holes

If we click on the **Select Capacitor** button, we open the database where some of the previous parameters are listed. And if we then click **Quit and Edit Database**, we find the I_{RMS} current and **Lser** which, unfortunately, is empty for every entry.

These parameters are entered in the *Spice Line* in the **Component Attribute Editor.** The 'Help' page names another resistor *Rlshunt*, not shown in the dialogue, but can be entered in the *Spice Line*. However, this seems mysteriously to vanish from time to time and certainly is not acted upon.

It is better to include the parasitic effects in the capacitor rather than as explicit components because it reduces convergence problems and the circuit simulates faster. LTspice allows values to be entered directly, but they may not be qualified, unlike other versions of SPICE where linear and quadratic temperature coefficients are accepted.

7.4.1.1 Temperature Effects

A direct entry of, for example:

$$Lser = 1n * (1 - 0.001 * temp)$$

is not accepted, not for the capacitance itself nor the series resistance. However, these may be declared as parameters, and then there is flexibility in that powers, exponentials and most other functions are accepted.

Example – Capacitor Temperature Effects
The schematic ('Capacitor as Parameters.asc') defines **{Rser}** and **{Lser}**. The inductance has to be unrealistically large to see any temperature effects, and the resistor temperature effect should be disabled. The directive:

$$.\text{param Lser} = 1m * (1 + temp * 0.02)$$

shows a small increase in the capacitor current and a slight change in voltage. As it is now a series tuned circuit, if the series resistance $R1$ is too small, the circuit will exhibit dampted oscillations with a puse input., schematic ('Capacitor as Parameters 2.asc'). We can also change the parallel resistance, by, for example *.param Rpar = 1e5*(1-temp*0.01)* in schematic ('Capacitance Rpar tempco.asc').

7.4.1.2 Voltage Effects

Attempts to include voltage or frequency fail, schematic ('Capacitors as Parameters fail.asc'). This also applies to *Rser*. However, if we use the special variable x which is the voltage across the capacitor, we can make some headway. Indeed, we find that the possibilities are comparable to the arbitrary current and voltage sources.

7.4.2 Capacitor Losses

These can be significant at a few tens of kilohertz for polar types, but usually at megahertz for others.

7.4.2.1 Capacitor Dissipation Factor (DF)

The relationship is:

$$DF = \frac{ESR}{Xc} = 2\pi fC.ESR = \tan(\delta) \qquad (7.15)$$

where X_c is the *reactance* of the capacitor, $2\pi fC$ (see later).

7.4.2.2 Capacitor Self-Resonant Frequency f_r

The ESR and ESL form a series tuned circuit (to be discussed in a later chapter). For electrolytic capacitors, this can be around 10 kHz, but a few megahertz for others. The relationship is:

$$f_r = \frac{1}{2\pi\sqrt{ESL.C}} \tag{7.16}$$

7.4.2.3 Capacitor Q-Factor

This is the *quality factor* and is also something to be discussed later. For the present we just note that it is:

$$Q = \frac{2\pi f_r LSR}{ESR} \tag{7.17}$$

7.4.2.4 Capacitor Impedance

This is:

$$Z = \sqrt{ESR^2 + \left(\frac{1}{2\pi f C} - 2\pi f ESL\right)^2} \tag{7.18}$$

and at the resonant frequency, the impedance is just the ESR. This sets a maximum usable frequency for the capacitor, and it is safer to keep well below this.

7.4.2.5 Other Measures of Parasitic Properties

Instead of explicit value for ESR and ESL, other figures may be given (see conradhoffman above and http://www.atceramics.com/documents/notes/esrlosses_appnote.pdf).

7.4.2.6 Capacitor Loss Angle (δ)

This is measured at a specific frequency and is usually just a few degrees, Fig. 7.6, or its tangent is given. Otherwise we may find the phase given as a negative angle, usually more than $-80°$.

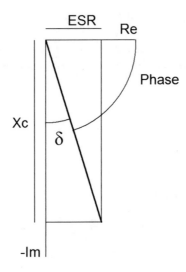

Fig. 7.6 Loss Angle

Explorations 2

1. Enter an arbitrary ESL data in a capacitor, and make an AC analysis to find the self-resonant frequency. Monitor impedance and current, schematic ('Self resonance.asc').
2. Add an ESR and note the reduction in Q as a wider resonance peak.
3. Use the websites mentioned above to download capacitor models and test them. In particular, download the Murata library.
4. Use the data from http://www.cde.com/resources/catalogs/AEappGUIDE.pdf to model an electrolytic capacitor and its temperature coefficient.

7.4.3 Capacitor as Charge

LTspice rather curiously allows a capacitor to be modelled by charge as:

$$\textbf{Cnnn} < \textbf{node1} >< \textbf{node2} >< \textbf{Q} = \textbf{value} >$$

where Q is the charge on the capacitor. This is a different way of thinking where the charge is placed directly on the capacitor, not by a voltage or current source, and this determines the voltage. It takes a little time to get used to it. It must be handled with some care. The outcome is that the capacitor can be given a charge in the absence of a current or voltage source.

However, the full capacitor model including series resistance, series inductance, parallel resistance and parallel capacitance is no longer available. This is important.

7.4.3.1   Charge = C.f(x)

This is the simplest where the charge is a linear function of the voltage across the capacitor x and the capacitance C which results in a constant capacitance. Other functions must be handled with some care to avoid the capacitance becoming zero at which point the simulation will effectively stop.

Example – Q = C.x
The statement:

$$Q = C.x$$

is a constant capacitance which we can test by the schematic ('Capacitor as Charge. asc') where we have a fixed capacitor C2 and for C1 $Q = 1u*x$. For both capacitors, we find the same constant 10 µA charging current and the same linear rise or fall in voltage. With a 1 kΩ resistor that equates to a 1 µF capacitor and with a final 10 V on it, the charge is 10 C, which agrees with Q = CV.

Example – Voltage Variable Filter
Anticipating Sect. 7.7.1.2, the 3 dB point for an RC low-pass filter is $f_3 = \frac{1}{2\pi RC}$, whence $C = \frac{1}{2\pi f_3 R}$. In schematic ('Voltage Variable LP Filter.asc') the charge on the capacitor is:

$$Q = 10n * x * k$$

where k is stepped as:

.step param k 0 10 2

From the resulting graphs, we find the 3 dB points, noting that when $k = 0$ the capacitor is zero, there is no 3 dB point. For the others, the results are measure with respect to the value at *100 Hz* which is virtually *0 dB*.
Measurement: 3db

step	v(c)	at
1	(-1.#INFdB,0°)	0
2	(-3.01099dB,-45.0045°)	7959
3	(-3.01304dB,-45.0181°)	3981.39
4	(-3.01647dB,-45.0407°)	2656.35
5	(-3.02126dB,-45.0722°)	1994.46
6	(-3.02741dB,-45.1127°)	1597.82

Taking the second entry, we calculate the capacitor to be *20 nF*. For the third, *40 nF*; for the fourth, *60 nF*; and so on showing that the capacitance increases linearly with the voltage across it.

If we change to a *2 ms* transient analysis with a *1 kHz* sine wave, we find a peak
current of *243.694* μA with $k = 3$, that is, $C = 40$ nF. Using: $Z =$

$$\sqrt{\frac{1}{2\pi 10^3 x40.10^{-9}} + 10^6} \quad I(V1) = 243.75 \mu A$$

Example – Q = C.x^2

The schematic ('Capacitor as Charge v2.asc') compares $Q = 1*x*x$ and $Q = 1*x$
where the linear relationship again creates a straight line graph. Both capacitors draw
the same current of *100 μA*. If we plot $V(c2)**0.5$, this falls on top of $Q = 1*x*x$.

However, changing the excitation to a sine wave **SINE(0 10 10)** causes problems
when the capacitor voltage reaches zero.

7.4.3.2 Charge = C.f(time)

A simple $Q = 1u*time$ results in a constant current. However, we can combine this
with the voltage *x* to create more interesting waveforms and also with trigonometric
functions.

Example – Q = 1u*x*(time*0.01)

This is schematic ('Capacitor as Charge V4.asc') which, with a sinusoidal excitation
at 1 kHz, shows that the voltage across the capacitor V(n001) is constant but the
current increases with the increasing capacitance.

Example – Q = sin(600*time)

This is the unlikely circuit ('Capacitor Oscillator.asc') consisting of a capacitor and
resistor in series, without any voltage or current source. First, we set the charge as
Q = 600*time which means the charge on the capacitor is changing at the rate of *600*
C/s or *600 A* so we find a voltage across the *100 Ω* resistor of a constant *60 kV* and
we have a capacitor as a current generator! To see this, we need to delay the start of
recording by *10 μs* or so using **.tran 0 0.1 10u** to avoid the starting transient.

If we change the capacitance to **Q = sin(600*time)**, a sinusoidal current flows in
the circuit creating a peak voltage of *60 kV*. If we change to **Q = sin(60*time)**, the
peak voltage declines to *6 kV*, and the frequency reduces by a factor of 10.

7.4.3.3 Charge = C.f(current)

Some of the more esoteric voltage functions might work better with currents.

Example – Q = f(current)

This is schematic ('Capacitor as Charge V3.asc'). We see the capacitor voltage
modulated by the sine wave and increasing at the same rate as the voltage ramp. We
should also note that, although the modulation is of constant amplitude, the capacitor
current increases, implying that its capacitance has decreased, and this we see from
the expression where the current is subtracted in:

$$Q = 1u * x * (1 - 0.1 * I(R2))$$

The capacitor voltage can never exceed the voltage $V1$, schematic ('Capacitor as Charge V3a.asc'), and hence we see the positive peaks are clipped at which time the capacitor current trace might break up. Changing to the Gear method improves it.

* * *

From these examples, we can assume that the other functions in the LTspice 'Help' file can be applied to the charge.

Explorations 3
Have fun exploring other combinations using x with power and trigonometric functions in various combinations.

7.4.4 Manufacturer's Capacitor Models

The SPICE capacitor in many cases has been superseded by more accurate models supplied by manufactures as libraries. These phenomenological models are based solely on measurements, or they may have some association with the construction. Usually, there is a base sub-circuit into which the numbers for the individual capacitors are inserted.

We should note that some of these models – the non-linear ones – may be valid only over a restricted range of voltages, somewhat larger than the normal working range, and that excessive inputs such as 1kV applied to the Murata MLCC will cause serious problems.

7.4.4.1 AVX Ceramic Capacitor Model

AVX have published models having two parallel arms each of a resistor and inductor in series and these in series with the ideal capacitor, Fig. 7.7. Samsung also have a model, Fig. 7.8. This has five inductors in series with parallel resistors and two capacitors. Both use fixed-value components, so temperature and voltage effects are not included.

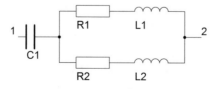

Fig. 7.7 AVX Ceramic Capacitor Model

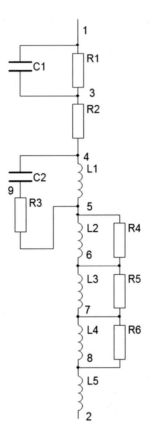

Fig. 7.8 Samsung Multilayer Ceramic

Multilayer Chip Ceramic (MLCC) Capacitor Models

These are different in that they are used at VHF and have series and parallel resonances. Microelectronics Inc. uses a model which is a slight variation on the LTspice one, schematic ('MLCC Microelectronics Model.asc'). On the other hand, Samsung adds a long RL ladder to the basic capacitor, Fig. 7.9, and some Taiyo Yuden models run to 42 passive components.

However, these models do not address the bias voltage dependence of the capacitance. Not surprisingly, really, manufacturers have chosen to encrypt the voltage effects in their models. The Murata models are based on their AC sub-circuits with current and voltage sources added. They can be downloaded but – oddly – each file can appear as a 'video clip'. The folder also contains inductors and MLCC sub-folders. Each contains pairs of the symbol files and model files. The simplest approach is not to split them by extension, but to make two new sub-folders *...lib/sub/Murata_caps* and *...lib/sym/Murata_inductors* and copy them to both. True, this does bloat the folders by duplicating everything, an extra 200 Mb. Then under **Tools→Control Panel→Sym & Lib Search Paths**, right click in each large

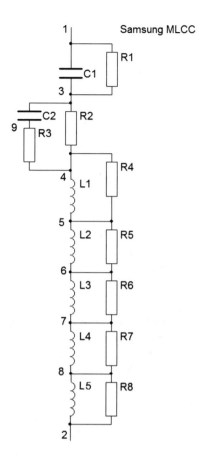

Fig. 7.9 Samsung MLCC Model

edit box and select **Browse** from the pop-up menu and add the relevant folders to each. It will be necessary to restart LTspice for the new folders to become accessible.

In an article (https://www.edn.com/design/components-and-packaging/4394275/ Quasistatic-Spice-model-targets-ceramic-capacitors-with-) by Hogo Coolens et al., it was found that a quadratic equation had a maximum error of 20% over 80% of the operating voltage range for a ceramic Y5V, which, in any case, is about the practical voltage limit. However, the schematic ('MLCC_model1.asc') of unknown origin extends this to $C(x) = \frac{a_1 + a_2 x}{1 + a_3 x + a_4 x^2}$ which is expanded into an exact and an approximate form. The circuit is Fig 7.10 showing the variable capacitor $C1$ in series with a resistor and inductor and with a parallel capacitor. It is a reduced version of the LTspice capacitor. This uses the specific LTspice ability to define a capacitor in terms of its charge. Other versions of SPICE allow a capacitor to be defined by linear and quadratic voltage coefficients.

Example – Murata MLCC Frequency Response

The schematic ('Murata MLCC.asc') has a capacitor GCD188R71E273KA01 which is *27* nF +/−10% and has a maximum 25 V bias. The simulation is at 50 °C stepping

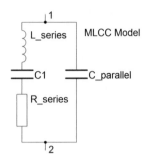

Fig. 7.10 MLCC Model

the voltage from 0 to 50 V in 10 V steps by changing the voltage source DC value to a parameter. This gives 3 dB points of 6442 Hz, 6704 Hz, 7805 Hz, 9921 Hz, 12,858 Hz and 16,737 Hz and shows that the modelling algorithm is still valid beyond the rated voltage.

The directive *.step temp − 55 125 25* fails because the capacitor defines its own *temperature*, but changing to *temperature = {t}* as a parameter and *.step t − 55 125 25* works. The *.meas* directive shows the change of 3 dB point with temperature with a 10 V bias. Taking the figures for *0 °C* and *25 °C*, we find a temperature coefficient of *33∗4/6607 = 2%*, (close enough to the published temperature coefficient, is 1.8% over its working voltage) but it increases rapidly above that, which we also see from the measurements.

At 25 °C the corresponding frequencies are 6328 Hz, 6588 Hz, 7683 Hz, 9572 Hz, 12,096 Hz and 15,206 Hz. The calculated upper 3 dB point for a 27 nF capacitor in series with a 10 kΩ resistor is $f_{3dB} = \frac{1}{2\pi x 10k\Omega x 27nF} = 5.89 kHz$ which is lower than any of the readings but adding 10% tolerance, and it is 6.48 kHz.

Example – Murata MLCC Distortion

If we use a sinusoidal input, schematic ('Murata MLCC Waveform.asc'), we can step the DC bias as we make a transient run and see that there is no distortion of the voltage and the capacitor and supply voltages fall on top of each other. However, there is distortion and a phase shift of the capacitor current. There appears to be very little power dissipation, even with a 100 V sinusoid, it is only 2 µW. A 1 kV input is problematic!

Example – Interdigitated Ceramic AVX Model

The schematic ('AVX Ceramic Cap Model.asc') uses Fig. 7.7 and shows the series resonance dip but not the parallel. The 'Low Inductance' type has a deeper resonance.

Example – Samsung MLCC No Bias

A Samsung 100 pF MLCC is modelled in schematic ('MLCC Samsung Model.asc'). The graph is similar to Microelectronics except that the multiple inductors widen the series resonance, and it has a sharper parallel resonance at *30 GHz*. A quick calculation shows that this is due to *C2L1*. The difference is sharper in the phase plot where the Samsung graph is smoother.

Example – MLCC with Bias
The schematic ('MLCC_model1.asc') has two models, **MCLL** and **MCLL1.** It will be found that traces for all the capacitor currents and voltages agree closely. Schematic ('MLCC Test.asc') shows a single resonance. There is a slight change of frequency of about *15 MHz* with a bias voltage change from *1 V* to *10 V*. Note that we cannot add parasitic properties to the capacitor defined by charge, so we add discrete components.

7.4.4.2 SPICE Polar Capacitor Models

The models for tantalum capacitors are more complex than aluminium electrolytic types. And none take account of reverse polarity.

Nichicon Aluminium Electrolytic Capacitors

The main part of the Nichicon model is an inductor, resistor and capacitor in series, Fig. 7.11. Branching off this are two RC networks representing some 12% of the capacitance and modelling the decay of capacitance with frequency. It is pointed out in the interesting paper (*https://www.digikey.fr/.../CDE_ImprovedSpiceModels.p..*) by Sam G Parler that the capacitor behaves like a lossy co-axial distributed RC element – this explains the structure of some commercial models. However, he also

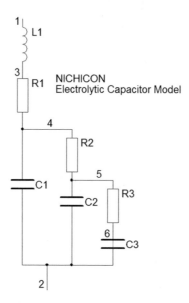

Fig. 7.11 Electrolytic Capacitor Model

says that the skin effect, oxide loss and Zener effect, if taken above the rated voltage, can be added – but so far these are not incorporated in commercial models.

Tantalum Capacitors

Kermet extend the Nichicon ladder to 5RC sections (*www.kemet.com/.../2010-03%20CARTS%20-%20Spice*). Samsung has created an extraordinary model of eight RC sections in series followed by a small resistor and inductor in series and then four LR sections in series with a very small capacitor in parallel with the final section, Fig. 7.12. The figure omits the repeated sections, but they can be seen in the simulation. This model of a 22 μF capacitor is a good example of capacitors in

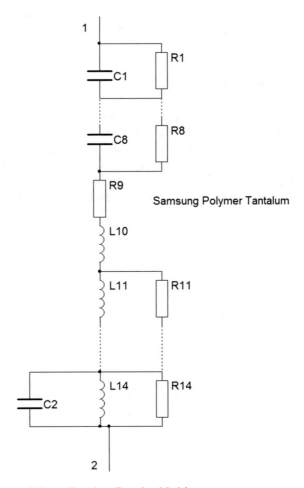

Fig. 7.12 Samsung Polymer Tantalum Capacitor Model

series. The capacitors C1...C7 are all hundreds of microfarads, and the final one, C8, is 9.6 μF.

Example – Ceramic Capacitor Model
Schematic 'Ceramic Cap Model.asc' is for an AVX *10 nF* ceramic capacitor. It can be seen that there is no significant divergence between it and an ideal capacitor *C2* up to about *10 MHz*. Using a logarithmic y-axis and plotting $V(c2)$-$V(c)$, the difference is about *2 μV* at *10 kHz* rising steadily to *440 μV* at *1 MHz* with a *1 V* input. The impedance falls to a minimum at just over *200 MHz*, and the phase plot shows that the capacitor has now become inductive.

Example – Electrolytic Capacitor Model and ESR
The schematic ('Electrolytic Cap. Model.asc') of a Nichicon type PCF0E221MCL4GB.V_100 is taken from their published SPICE models. The model and an ideal capacitor *C4* are very close to *10 kHz*, and at *100 kHz* the impedance of the capacitor is almost constant. If we draw the voltage at node *4*, it is almost that of a ideal capacitor.

   The data sheet gives *ESR = 20 mΩ*. In fact, *15 mΩ* is an almost perfect match. We see the difference at about *2 MHz* where the capacitor rapidly becomes inductive. As a rough measure, at *4.7 MHz*, the phase is *45°*, and the impedance of the inductor is *15 mΩ* which is slightly more than *R1*. We can discount the impedance of *C1*. We might ask who would use an electrolytic capacitor at such high frequencies. The answer is that the output of switch-mode power supplies is almost a square wave at a frequency of perhaps *100 kHz* and so has harmonics in the MHz region.

   The data sheet gives a maximum leakage of *100 μA*, but this is not modelled.

Example – Tantalum Capacitor Model
The schematic is ('Samsung Polymer Tant.asc'). Up to about *10 kHz*, a simple capacitor is a fair model, *C9* with voltage *V(cl)*, and as far as *40 MHz* or so, a capacitor with *ESR = 150 mΩ* is a reasonable fit, voltage *V(cesr)*. The *300 mΩ* is the maximum at *100 kHz*, so we are at liberty to adjust it.

7.5   Time Response of a Capacitor

This form of circuit occurs many times in electronics; indeed, it can be regarded as a fundamental circuit whose careful study will be repaid time and time again. We shall take the analysis step by step and view the result using LTspice. It may be asked why we do not just consider the response of a capacitor alone, but it should become clear that this is impossible as every circuit will contain some resistance. We will start by supposing we connect an initially uncharged capacitor alternately to two voltages representing the ON and OFF levels of the pulse, Fig. 7.13a where the off voltage is zero.

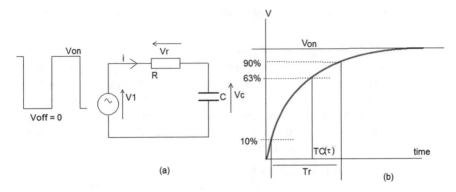

Fig. 7.13 Capacitor Charging

7.5.1 Capacitor Charging

Traversing the circuit clockwise, at any instant, KVL must apply. The voltage across
a capacitor is $V = Q/C$ where Q is the charge so at any instance we can write:

$$V_{ON} = i.R + \frac{Q}{C} \tag{7.19}$$

where I is the instantaneous current. But the charge on the capacitor is just the time
integral of the current, and if we make no assumption about its variation with time,
we write:

$$V_{ON} = i.R + \frac{1}{C} \int i.dt \tag{7.20}$$

where we have both the current and its integral.

Now for this equation to be dimensionally consistent, the integral of the current
must also be a current. It cannot be current squared or current cubed or any other
function of current. There are two common functions that fulfil this requirement, one
is a sinusoid because the integral of $sine(at)$ is $1/a.cos(at)$, ignoring the constant of
integration. The other is an exponential since the integral of e^{at} is $\frac{1}{a}e^{at}$.

As the input is a steady voltage, there is no good reason to suppose that the current
will oscillate; rather we should expect it to decay because as the capacitor charges,
the voltage across the resistor will fall and hence the current. So we shall try the
exponential solution. And as we know the current must decrease in time, this will be
of the form:

$$i = I_0 e^{-at} \tag{7.21}$$

where I_0 and a are constants and the exponent is negative, not positive. Both constants are needed because I_0 is the initial current at time $t = 0$ ($e^0 = 1$) and a is a scaling factor. Making this substitution in 7.20, we have:

$$iR + \frac{I_0}{C} \int e^{-at} dt = V_{ON} \tag{7.22}$$

On making the integration, we have:

$$iR - \frac{I_0 e^{-at}}{aC} + b = V_{ON} \tag{7.23}$$

where b is the constant of integration and this leaves three constants to be found.

After an infinite time, the exponential e^{-at} will have reduced to zero, and the current will also be zero, so from Eq. 7.23 we have $b = V_{ON}$. And at $t = 0$ the current is I_0 and making the substitutions for b and i, Eq. 7.23 becomes:

$$I_0 R - \frac{I_0}{aC} + V_{ON} = V_{ON} \tag{7.24}$$

whence $R = 1/aC$ and hence $a = 1/RC$. So returning finally to Eq. 7.21, we have:

$$i = I_0 e^{\frac{-t}{RC}} \tag{7.25}$$

which predicts that the current will start at some initial value I_0 and fall exponentially to zero. On the other hand, we would expect the capacitor voltage to rise exponentially to its final value as the capacitor charges up. The capacitor voltage at any time is:

$$v_c = \frac{1}{C} \int i.dt \tag{7.26}$$

and if we substitute for i from Eq. 7.25 we have:

$$v_c = \frac{1}{C} I_0 \int e^{\frac{-t}{RC}}.dt \tag{7.27}$$

and after integration this becomes:

$$v_c = \frac{-RC}{C} I_0 e^{\frac{-t}{RC}} + k \tag{7.28}$$

where k is some constant. Substituting $I_0 = V/R$ we have:

$$v_c = -V_{ON} e^{\frac{-t}{RC}} + k \tag{7.29}$$

and as $v = 0$ at time $t = 0$ we find $k = Von$. Thus, finally, the capacitor voltage is:

$$v_c = V_{ON}\left(1 - e^{\frac{-t}{RC}}\right) \tag{7.30}$$

which is an exponential starting from zero and tending to an asymptotic value of Von. We might compare this to $v = V_{ON} e^{\frac{t}{RC}}$ which also is a rising exponential, but the asymptote is infinite. And remembering that Kirchhoff's Laws still apply, the voltage across the resistor will decrease exponentially.

7.5.1.1   The Time Constant, Rise Time and Fall Time

Notice that at time $t = RC$, the current will have fallen to e^{-1} or 37% of its initial value and the voltage will have risen to $(1-e^{-1})$ or 63% of its final value. This product RC is the *time constant* (τ) because (remembering dimensional analysis) it has the dimensions of time. There is no fundamental significance about it; it is just that the sums are easy to do and it gives us a quick estimate of the performance of a circuit. To a good enough approximation, after one time constant, the current will have fallen to one third of its initial value, whilst the voltage will have risen to two thirds of its final value. A useful rule of thumb is that steady state is reached after five time constants.

Whilst we could use the time constant to characterize the circuit, a more usual measure is the *rise time*. This is the time for the signal to go from 10% to 90% of its value, Fig. 7.13b. These figures are chosen because we have seen that the final value of the voltage is not clear, whereas the 90% point is easy to measure; and if we have several CR circuits in cascade, the exact start of the voltage rise is also unclear. Substituting in Eq. 7.30 for $v_c = 0.1Von$:

$$0.1 = \left(1 - e^{\frac{-t}{RC}}\right) \tag{7.31}$$

then:

$$\ln(0.9) = \frac{t}{RC} \tag{7.32}$$

and the time to the 10% point is *0.1 RC* and the time to the 90% point is *2.3 RC*, whence the rise time t_r is:

$$t_r = 2.2\, RC \tag{7.33}$$

and similarly at the end of the pulse the fall time t_f is:

$$t_f = 2.2\,RC \tag{7.34}$$

Example – Time Constant, Rise and Fall Times

Schematic ('Capacitor Charge Discharge.asc') has measure statements that return *tr* = 0.21966 *ms* *tc* = 0.99966 *ms*. By calculation *tr* = 2.2*RC* = 0.22 *ms* and *tc* = 0.1000 *ms*. However, if we turn off data compression, *tr* = 0.21972 and *tc* = 0.1000 *ms*.

7.5.2 Capacitor Discharge

This, of course, only makes sense if we have first charged the capacitor. In effect, the circuit is that of Fig. 7.13a where, when *Voff* = 0 the voltage source is replaced by a short circuit so:

$$v_c = iR \tag{7.35}$$

where v_c is the capacitor voltage, and we again look for an exponential solution but this time with the capacitor current falling to zero. The charge on the capacitor at any time t is found from:

$$q(t) = Q_i - \int idt \tag{7.36}$$

where Q_i is the initial charge. Substituting $i = I_0 e^{-at}$ as before and integrating, we have:

$$q(t) = Q_i + \frac{I_0}{a}e^{-at} + k \tag{7.37}$$

where k is a constant and the initial charge on the capacitor is $Q_i = CV_{ON}$ where V_{ON} is the initial voltage. The capacitor voltage at any time is $q(t)/C$ and is equal to the voltage across the resistor:

$$\frac{q(t)}{C} = iR \tag{7.38}$$

Dividing Eq. 7.37 by C and substituting in Eq. 7.38:

$$V_{ON} - \left(\frac{I_0}{aC}e^{-at} + \frac{k}{C}\right) = iR \tag{7.39}$$

Now as the time goes to infinity, the capacitor voltage goes to zero and so does the current. Hence $k = CV_{ON}$ and Eq. 7.39 reduces to:

$$\frac{I_0}{aC} = iR \tag{7.40}$$

and as $I_0 = V_{ON}/R$ we find that $a = 1/RC$ as before. So, if we return to Eq. 7.39, after substituting for I_o and k, we have that the capacitor voltage is:

$$v_c = V_{ON}e^{\frac{-t}{RC}} \tag{7.41}$$

which is a decaying exponential from an initial value of V_{ON}.

The capacitor current follows very easily from Eq. 7.41, since the same voltage appears across the resistor and the same current flows through both; thus:

$$i_c = \frac{V_{ON}}{R}e^{\frac{-t}{RC}} \tag{7.42}$$

again showing a decaying exponential.

It is instructive to compare the two cases of charging and discharging. In the former, the capacitor voltage and current are effectively mirror images; in the latter they follow the same shape.

Explorations 4
1. Open ('Capacitor Charge Discharge.asc'). This is just the same old RC circuit but with a 1 ms square wave excitation.
2. Adjust the sweep to end at 2 ms. Calculate the time constant and rise time of the voltage. You should find the rise time is about 217 µs. Check the fall time. Plot the current waveform and note that the initial current is V/R, the final current is zero and the final capacitor voltage is V. Plot the voltage across the resistor as $V(n001)$-$V(n02)$ and check that Kirchhoff's Voltage Law holds good.
3. A rule of thumb is that the current has decayed to zero or the voltage has risen to its final value after five time constants. Test this from the graphs and by direct calculation.
4. Repeat the previous for some other resistor and capacitor values.

7.5.3   Sag

The top of the pulse is distorted and has a *sag* which gets bigger as the frequency is reduced. In effect, we are seeing the discharge of the capacitor. It can be reduced by increasing the capacitor or the resistor. From the previous section, this is a measure of the low-frequency performance of the circuit. A useful figure is the time for a 10% sag. Then from Eq. 7.31, we have $RC = 10\,t$ where t is the time to the 10% drop.

Example – Capacitor Initial Sag

The schematic ('Sag.asc') has a $1\,\mu F$ capacitor in series with a $1k\Omega$ resistor and the output taken from the junction. With an initial pulse of $1\,V$ amplitude and $1\,ms$ duration, the voltage has fallen to $370\,mV$ at the end of the pulse. Substituting in Eq. 7.41:

$$\ln(0.370) = \frac{-1ms}{R_1 C_1} \text{ whence } R_1 C_1 = 1.01 \times 10^{-3}$$ which is correct. We might note that rise and fall times of $1\,ps$ produce vertical edges of the pulse, whereas $1\,fs$ leaves them sloping.

7.5.3.1   Pulse Train Response

We have a little conundrum: let us suppose we start with a large uncharged capacitor in series with a large resistor, schematic ('Capacitor Pulse Train (a).asc'), and apply a step input of 10 V. The left-hand plate of the capacitor will rise to 10 V and therefore so will the right-hand plate. At the end of the pulse, the voltage of the left-hand plate will fall to zero and so will that of the right-hand plate. And this will continue with the subsequent pulses, and the voltage across the resistor will be a train of pulses from 0 V to 10 V which, of course, has a DC component. How, then, does a capacitor block DC? The answer is that we have failed to take into account the discharge of the capacitor during the pulse. This is neither visible nor measurable over 10 ms, but after 100 ms, we see a drop of a few microvolts in the peak and a couple of microvolts negative when the pulse is removed.

The second schematic with a small capacitor and resistor in conjunction with Fig. 7.14 makes the situation clearer where the shape of the capacitor voltage and current waveforms are coincidental.

We see that at the end of the pulse at time a, the input voltage falls to zero. However, the capacitor, whose voltage is shown in red, has not discharged completely, only to voltage V_y and the 10 V negative step to zero volts subtracts from this so the capacitor voltage becomes negative with voltage V_z where $V_z = V_{in} - V_y$. The capacitor now starts to charge but at a slower rate because the voltage across it is only V_z to voltage V_q, so on the next pulse at time b, the capacitor voltage only rises to V_r where $V_r = V_{in} - V_q$, and thus we see a steady decline in the positive

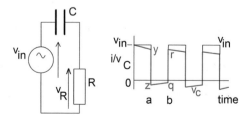

Fig. 7.14 Capacitor Pulse train

capacitor voltage and a steady increase in the negative voltage until the two are equal and the average is zero.

Example – Capacitor Pulse Train Sag

We extend the simulation schematic ('Sag.asc') to *25 ms* and just show the last *5 ms* when the positive peak is *731 mV* and the voltage at the end of the pulse is *269 mV*; thus we have *ln(0.368)* which again returns $R_1 C_1 = 1e\text{-}3$.

Explorations 5

1. Change the square wave to a pulse in the above schematic and increase the frequency to 100 kHz or open ('RC Circuit – Pulse Excitation.asc.') The circuit has nearly reached steady state at the end (100 µs). Vary the duty cycle from 0.1 to 0.9 and note the envelope of the capacitor voltage. What relationship is there between the asymptotic capacitor voltage and the duty factor? Could this circuit be used to convert a pulse train to almost DC? How large a capacitor would be needed and how long would it take to reach steady state? A good starting point is a capacitor of 1 µF and a time of 5 ms – but do not forget to increase the **Ncycles** of the voltage source or leave it empty.
2. Interchange the capacitor and resistor. Explain the shape of the waveform (hint: remember we are now looking at the voltage across the resistor and 'on' time of the waveform is much less than the time constant).
3. Using ('Waveform Synthesis.asc') of chapter 'AC' as a base, string together six or seven sine wave sources to approximate a square wave with a fundamental of 1 kHz at 10 V, so they should be 3 kHz 3.333 V, 5 kHz 2 V, 7 kHz 1.43 V, 9 kHz 1.11 V and 11 kHz 0.91 V. Now add a phase shift of 30° to the fundamental and note the large sag. Estimate the phase shift for a 10% sag and hence the frequency for a CR circuit to give this percentage sag. (This is making the approximation that the higher harmonics are unshifted which, of course, is not true, but the best that can be done without a lot of maths.)
4. String together four RC circuits and stimulate them with a pulse waveform. A commonly used approximation is that the total rise time t_r is given by $t_r = 1.1\sqrt{t_{r1}^2 + t_{r2}^2 + \dots}$. Test its validity.

7.5.4 Average Voltage

A simple RC circuit can be used to find the average of a waveform.

Example – Pulse Train Average

The schematic 'RC Average.asc' has a *10 V* pulse input with a duty cycle of *10%* applied to a resistor of *1 kΩ* in series with *10 µF*. The measured output across the capacitor is *1.00447 V*. With a duty cycle of *30%*, the capacitor voltage is *3.01037 V*.

We should note that we need a short step time of *0.1 μs* to resolve the trace correctly and compression turned off.

The measured rise time of *22 ms* is independent of the duty cycle and is just that of the RC circuit.

7.5.4.1   Amplitude Modulation Smoothing

Amplitude modulation (AM) is widely used for radio transmission. The signal is generated by multiplying the high frequency carrier sinusoid by the much lower frequency audio signal, which we shall also take as a sinusoid. The equation is:

$$V_{AM} = (A_c + A_a \sin (f_a)) \sin (f_c) \qquad (7.43)$$

where A_c is the carrier amplitude, A_a is the audio amplitude and f_c f_a are the respective frequencies. It is usual practice to make the audio amplitude no more than 80% of the carrier.

To recover the audio signal, the modulated waveform is rectified to leave only the positive half-cycles. But as these are very lumpy, it is usual to pass them through an RC circuit to generate a smooth output. There is, however, a difficulty: the audio signal might span a wide frequency range, for example, from *50 Hz* to *5 kHz*. If we choose the RC time constant to smooth the 50 Hz, it will not accurately follow the 5 kHz signal, Fig. 7.15, because the time constant is too long. This is clearly seen in the following example. It is a nice decision just how long to make it.

Example – AM Demodulation
Schematic 'AM Demodulation.asc' uses arbitrary voltage source *B2* to create a modulated wave:

$$\mathbf{V} = (\mathbf{1} + \mathbf{0.8} * \mathbf{sin}\,(\mathbf{1e3} * \mathbf{time})) * \mathbf{sin}\,(\mathbf{1e5} * \mathbf{time})$$

where, from Eq. 7.43, the modulation is *80%*. This is passed to arbitrary voltage source *B1* which cuts the negative half by:

$$\mathbf{V} = \mathbf{if}\,(\mathbf{V(am)} > \mathbf{0}, \mathbf{V(am)}, \mathbf{0})$$

so that if *V(am)* is positive, the output voltage follows it; else the output voltage is zero. In effect, this replaces the usual diode. If we make *C1 = 100 nF* with *R1 = 1 kΩ*, the smoothing is not very good. We can check the distortion using the *.four* command

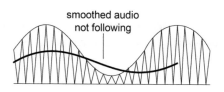

smoothed audio
not following

Fig. 7.15 AM Smoothing

where the frequency is *159.155 Hz*, and it is very low at *0.0005%*. However, if we increase the frequency to *1e4* and change the *.four* limits accordingly, we see *V(c)* no longer correctly follows the envelope, and the distortion has jumped to *20%*.

Example – AM Modulation
A not very practical schematic is 'Capacitor AM Modulator,asc' which does at least show the amplitude of the sine wave increasing with the modulating ramp. We can switch to a sinusoid instead of the ramp and increase the amplitude of *V2*.

7.5.4.2 Pulse Width Modulation Smoothing

This modulation does not rely on the height of the waveform, but on the duty cycle of constant height pulses at a fixed rate. The easiest way to test this is to use the LTC6992–2. This is limited to a duty cycle from 5% to 95% for a 0.05 V to 0.95 V *MOD* input. In fact, any of the LTC6992 series can be used here, schematic ('PWM. asc').

Example – PWM Smoothing
We use a sinusoid with an offset to ensure it is always positive:

$$SINE(0.45\ 0.45\ 1k)$$

The audio output in schematic 'PWM.asc' shows considerable distortion because the sampling rate is not fast enough. Increase *R3* and reduce *R4* to set a higher rate and the output is much improved.

7.6 Frequency Response of a Capacitor

We shall now use an AC stimulus. LTspice first finds the operating point and then takes the slope of all the parameters, so creating a linear circuit. And so far, this is exactly what we have used: capacitors whose value does not change, unaffected by current or voltage.

7.6.1 Voltages and Currents

Let us return to the simple circuit of a voltage source in series with two resistors – our old friend, the potential divider. Kirchhoff's Laws apply to every circuit. Thus we have:

$$v_1 = i.R_1 + i.R_2 \qquad (7.44)$$

where lowercase letters have been used to show these are instantaneous values. If we just focus on *R1* and replace v_i by a sine wave we have:

$$V_{pk} \sin (\omega t) = i.R_1 \qquad (7.45)$$

Now we can write:

$$R_1 = \frac{V_{pk} \sin (\omega t)}{i} \qquad (7.46)$$

from which it follows that:

$$i = I_{pk} \sin (\omega t) \qquad (7.47)$$

and the current and voltage are in phase.

Now let us replace *R2* by a capacitor *C1*, Fig. 7.16, and we have:

$$v_1 = v_r + v_c \qquad (7.48)$$

Now changing the resistor for a capacitor will not change the relationship between the current and voltage in resistor *R1*. The current and votage are always, exactly in phase for a pure resistance. Certainly the peak value may change, but current and voltage will still be in phase, and therefore the current will still be sinusoidal. So substituting in Eq. 7.48:

$$V_{pk} \sin (\omega t) = I_{pk} \sin (\omega t) R_1 + v_{C1} \qquad (7.49)$$

But the voltage on the capacitor is given by $v_C = \frac{q}{C} = \frac{1}{C} \int i.dt$ where q is the charge on the capacitor. And as the same current flows through all the components of a series circuit, we can write Eq. 7.49 as:

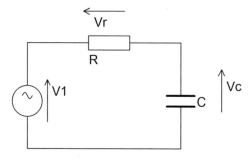

Fig. 7.16 Sinusoidal excitation of RC

$$V_{pk} \sin{(\omega t)} = I_{pk} \sin{(\omega t)} R_1 + \frac{1}{C_1} \int I_{pk} \sin{(\omega t)} dt \qquad (7.50)$$

After performing the integration we have:

$$V_{pk} \sin{(\omega t)} = I_{pk} \sin{(\omega t)} R_1 + \frac{1}{\omega C_1} I_{pk} \cos{(\omega t)} \qquad (7.51)$$

As this is a series circuit, Kirchhoff's Laws tell us that the instantaneous current must be same in every component. So the interpretation of Eq. 7.51 must be that the voltage across the capacitor lags the current by 90° because at time 0 the voltage is zero but the current is Ipk. Thus for a pure capacitance, the current leads the voltage by 90°.

The question now is how to calculate the current at any time. The mathematically minded will know that $cos(\omega t)$ can be written as $isin(\omega t)$ where i is the square root of minus one, $i = \sqrt{-1}$, although engineers usually prefer to use j. Here j is an operator meaning 'rotate anti-clockwise through 90 degrees'. So, going back to Fig. 6.2, in the previous chapter, we can reinterpret the left-hand part of the very first diagram by noting that the vector O-D is the vertical axis and so represents vector O-A rotated anti-clockwise through 90°. Thus we can rename it as the j or $imaginary$ axis, abbreviated to 'Im', Fig. 7.17, whilst the horizontal axis represents the 'real' part, abbreviated to 'Re'. And as the current leads the voltage, we can replace $cos(\omega t)$ by $-jsin(\omega t)$ on the negative axis, so Eq. 7.51 finally becomes:

$$V_{pk} \sin{\omega t} = I_{pk} \sin{\omega t} R - j\frac{I_{pk}}{\omega C} \sin{\omega t} \qquad (7.52)$$

or:

$$V_{pk} = I_{pk} R - j\frac{I_{pk}}{\omega C} \qquad (7.53)$$

As the RMS voltage is just $V_{rms} = V_{pk}/\sqrt{2}$ and $I_{rms} = I_{pk}/\sqrt{2}$, Eq. 7.53 is valid for RMS values as well.

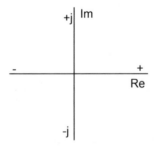

Fig. 7.17 Complex Plane

Example – Frequency Response of an RC Circuit
This is schematic ('Series RC Circuit.asc'). The 3 dB point is *1.59 kHz* using a *Bode* plot. Changing to *Cartesian* and using the cursor, we find both real and imaginary parts are equal at *1.594 kHz* and are *500 mV*.

If we make a transient run with a sinusoidal input of 1.59 kHz, we find a *45°* phase shift between input and output.

Example - Sine Train Response
Schematic ('Sine Train.asc') shows that the final conditions are established in the first cycle of the input waveform.

Explorations 6
1. Open ('Series RC Circuit.asc'). It takes about 1 ms for the circuit to reach steady state (check by extending the transient run time). Note that the capacitor and resistor currents are identical. You will have to delete one of the traces to see this. Note that the capacitor current leads its voltage V(n002) by 0.25 ms or 90°. Notice that as the voltages are not in phase, the instantaneous values do not add directly. Note also the relationship between the input voltage V(n001) and the capacitor voltage.
2. Interchange the resistor and the capacitor and repeat.
3. Change the start time of the waveform window to 1 ms so one complete cycle is shown. Note the RMS voltages and currents.

7.6.1.1 Reactance and Impedance of a Capacitor

Example – 'Frequency Response of an RC Circuit' should have found an RMS current of 3.76 mA and an RMS input voltage of 7.07 V. This corresponds to a resistance of 1.88 kΩ, whereas the resistor is only 1 kΩ. It seems, therefore, that the capacitor is behaving in some way like a resistor.

If we gather terms in Eq. 7.53, we have:

$$V_{pk} = I_{pk}\left(R - \frac{j}{\omega C}\right) \tag{7.54}$$

which has certain similarities to Ohm's Law. The difference is the $j/\omega C$ term. Dimensionally, it is the same as resistance and is measured in ohms. But because the energy is stored and not dissipated, it is called *reactance* and defined as:

$$X_C = \frac{1}{\omega C} \qquad \text{or}: \quad X_C = \frac{1}{2\pi f C} \tag{7.55}$$

And it gets smaller as the frequency increases. The *impedance* of a capacitor is similar:

$$Z_C = \frac{-j}{\omega C} \text{ or } \frac{1}{j\omega C} \tag{7.56}$$

which includes the j to remind us that we cannot just add it arithmetically to a resistor but must rotate the phase by $-90°$.

7.6.2 Manual Circuit Analysis

There are two ways of attacking circuits which include resistive and reactive elements. The first, and perhaps more illuminating, is to use the complex plane. The second (and most usual) is to recognize that we are dealing with complex numbers and use the complex conjugate to reduce the expressions to real numbers. This we will do soon.

7.6.2.1 Using the Complex Plane

We need the total circuit impedance. For this we can develop the concept begun in Fig. 7.17. As the resistor voltage is in phase with the supply voltage, we draw its resistance along the real axis with magnitude R in the impedance diagram of Fig. 7.18. The capacitor voltage lags the supply by $90°$, so we draw its impedance along the negative imaginary axis with magnitude $1/\omega C$. And the resultant of these two phasors is Z where:

$$Z = \sqrt{R^2 + \frac{1}{(\omega C)^2}} \tag{7.57}$$

However, we can also draw a voltage diagram on the same axis because the voltages are in the same directions as the impedances, and now the lengths of the phasors correspond to the voltages. Hence we have Fig. 7.19. The peak voltage across the resistor is in phase with the input so that is drawn along the real axis, and

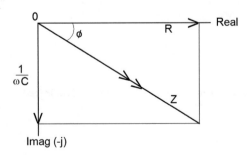

Fig. 7.18 Impedance Diagram

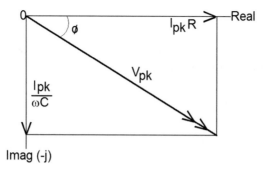

Fig. 7.19 Voltage Diagram

the capacitor voltage is drawn along the negative imaginary axis. The sum of these from the figure must equal the peak input voltage or:

$$V_{pk} = I_{pk}\sqrt{R^2 + \frac{1}{(\omega C)^2}} \qquad (7.58)$$

or we can use Z from the previous figure and Eq. 7.54 and:

$$V_{pk} = I_{pk}Z \qquad (7.59)$$

The same is true for the RMS voltage, and of course there is no corresponding current diagram because it is in the same direction for both the resistor and the capacitor as we saw in Explorations 6.

It is easy to get in a muddle if one does not take care. The capacitor voltage leads the current by 90°. This is always true. Similarly the voltage across a resistor is always in phase with the current. But when we consider the circuit as a whole, Fig. 7.19 'Voltage Diagram', shows us that the voltage across the resistor leads the supply voltage by an angle ϕ given by:

$$\varphi = \tan^{-1}\left(\frac{1}{\omega CR}\right) \qquad (7.60)$$

and the supply voltage Vpk leads the capacitor voltage by $90° - \phi$, not 90 °.

7.6.2.2   Using Complex Numbers

In finding the resultant impedance or voltage, we used Pythagoras' theorem that:

$$z^2 = x^2 + y^2 \qquad (7.61)$$

so given a number of the form $z = a + jb$, we can arrive at the form of Eq. 7.61 if we multiply it by $(a - jb)$ since:

$$(a + jb)(a - jb) = a^2 + b^2 \tag{7.62}$$

where $(a - jb)$ is the *complex conjugate* which in this case is $R + \frac{j}{wC}$. Then we take the square root to find the modulus:

$$V_{pk} = I_{pk}\sqrt{\left(R - \frac{j}{\omega C}\right)\left(R + \frac{j}{\omega C}\right)} \tag{7.63}$$

so:

$$V_{pk} = I_{pk}\sqrt{\left(R^2 + \frac{1}{\omega^2 C^2}\right)} \tag{7.64}$$

and this is the same as before.

Notice, however, that manual methods have only enabled us to calculate RMS values and phase angles. Unlike a SPICE transient analysis, we cannot see the actual waveforms. These methods have implicitly assumed that the waveforms are all sinusoids and the component values are constant. If for any reason they are not, the previous manual analysis will not detect it, for example, if a resistor or capacitor varies with voltage.

Example – RC Impedance

The schematic 'RC Impedance.asc' has an RC series circuit excited at first by a *1 V* sinusoid of *159 Hz*. We should note that unless the time is an exact multiple of the period, the average currents and voltages are not zero. The quickest remedy is to make the run over many cycles. We find the RMS current is *494 μA* and the RMS output voltage is *493 mV* compared to an input RMS voltage of *700 mV*. By use of the above equations, we find that these numbers agree with theory. And if we zoom the trace to show just a couple of cycles, we find the phase angle is correct.

And it is interesting to repeat at higher and lower frequencies.

7.7 Frequency Response of Series RC Circuits

We have noted the amplitude and phase of the voltage across a capacitor at some arbitrary frequency ω. But we also need to describe the behaviour of the circuit over a range of frequencies. It will be seen in later chapters that knowledge of the frequency response is important for many circuits: in audio amplifiers, poor low- or high-frequency response leads to loss of bass or treble, in pulse amplifiers poor frequency response causes the pulse shape to be distorted, and the phase shifts (see below) can lead to instability of feedback circuits.

7.7.1 Manual Analysis

The essentials have been covered above. To find the output voltage, we note that Fig. 7.16 is a complex potential divider where the output voltage v_{out} is the voltage across the capacitor v_c so:

$$A_v \equiv \frac{v_{out}}{v_{in}} = \frac{\frac{1}{j\omega C}}{\frac{1}{j\omega C} + R} \tag{7.65}$$

where we use A_v for the voltage gain. Multiplying by $j\omega C$ we have:

$$A_v = \frac{v_{out}}{v_{in}} = \frac{1}{1 + j\omega CR} \tag{7.66}$$

and the denominator is now in the form $(a + jb)$. So at any frequency, the magnitude of the voltage gain is:

$$|A_v| = \frac{1}{\sqrt{1 + (\omega RC)^2}} \tag{7.67}$$

and at low frequencies, this tends to unity, whilst at high frequencies it tends to zero. So the 'gain' is actually a loss. The phase angle between input and output is:

$$\varphi = \tan^{-1}\left(\frac{1}{\omega CR}\right) \tag{7.68}$$

which tends to 0° at low frequency and 90° at high frequencies. We can see this graphically by calling up ('Series RC Circuit Frequency Response.asc'). At low frequencies $1/\omega C$ is very large, and most of the input voltage appears across the capacitor, Fig. 7.20. But at high frequencies the impedance of the capacitor is very small, and most of the voltage is across the resistor, Fig. 7.21. However, as the

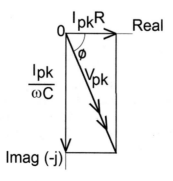

Fig. 7.20 Low Frequency Phasor Diag

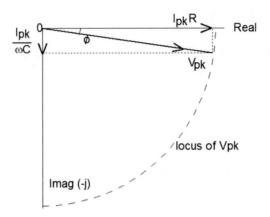

Fig. 7.21 High Frequency Phasor Diag

input voltage is constant, the locus of its tip is a quadrant of a circle moving from
$-90°$ to $0°$.

On the other hand, if we take the voltage across the resistor, Eq. 7.65 becomes:

$$A_v = \frac{R}{\frac{1}{j\omega C} + R} \tag{7.69}$$

which reduces to:

$$A_v = \frac{1}{1 - \frac{j}{\omega CR}} \tag{7.70}$$

so the magnitude is:

$$|A_v| = \frac{1}{\sqrt{1 + \frac{1}{(\omega CR)^2}}} \tag{7.71}$$

and the phase angle is:

$$\varphi = \tan^{-1}\left(\frac{1}{\omega CR}\right) \tag{7.72}$$

and now the ratio tends to zero at low frequency and unity at high, whilst the phase
angle moves from $90°$ to $0°$, the reverse of the previous. This, however, only gives us
the asymptotic values; it does not tell us about the shape of the response in between.
So it is helpful to plot the voltage gain against frequency. In many cases, we use the
decibel to measure the gain.

Example – Complex Potential Divider

We can use the schematic ('Complex Pot Div.asc') with an AC input, and we switch to a logarithmic left-hand trace scale. We find at low frequencies the voltage gain is almost unity up to about *50 Hz*. The trace then curves smoothly to become at about *300 Hz* a downward straight line. However, the phase shift becomes significant even at *10 Hz* and still has not moved to *90°* at *10 kHz*. If we take measurements of gain and phase, we find they agree with the above theory.

If we swap the resistor and capacitor, we find the graph laterally inverted with the voltage gain rising to a final steady value.

7.7.1.1 The Decibel

The decibel is a unit of either absolute or relative power. According to the Wikipedia article, its precursor was created at Bell Labs Inc. in the USA and was used to measure power loss in telegraph and telephone wires. And because telephone wires could run from a few meters to hundreds of kilometers, to avoid large numbers, the unit is logarithmic and defined as $10\log_{10}(P_1/P_2)$ where P_1 and P_2 are two powers.

As an absolute unit, there are two definitions. For measuring acoustic noise, the reference of 0 dB in air is a pressure of 20 micropascals (μP). This is about the limit of hearing in the most sensitive region of the human ear, which is from 1 kHz to 5 kHz. On this scale, a domestic refrigerator generates 20 dB of noise, shops and city traffic 60 dB and chain saws, lawnmowers and the like around 110 dB, and the threshold of pain is often given as 150 dB, but this is subjective and some sources place it as low as 120 dB.

The other absolute unit, the dBm, defines 0 dBm as the power dissipated by a 600 Ω resistor connected to a source of 0.775 V, so it dissipates 1 mW and is much used by audio engineers. It has the dimensions of power. The Wikipedia article is very informative.

We can also use the decibel as a measure of relative power; we first note that the power gain of a circuit is the ratio of the output and input powers or:

$$G_P = \frac{P_{out}}{P_{in}} \tag{7.73}$$

Because the ratio can vary over several decades, it is often convenient to take logarithmic rather than linear ratios and the power gain can be defined in decibels as:

$$G_{P(dB)} = 10\log_{10}\left(\frac{P_{out}}{P_{in}}\right) \tag{7.74}$$

which is a pure number.

We can extend this by supposing that we measure the input and output powers by the voltage developed across a certain resistor R:

$$G_P = \frac{V_{rmsout}^2 R}{V_{rmsin}^2 R} \tag{7.75}$$

and the resistance will cancel. This leads to the concept of measuring voltage ratios in decibels:

$$A_v = 10\log_{10}\left(\frac{V_{rmsout}^2}{V_{rmsin}^2}\right) \tag{7.76}$$

or generally:

$$A_v = 20\log_{20}\left(\frac{V_1}{V_2}\right) \tag{7.77}$$

where the voltages are peak or RMS. Thus we have a factor of 10 measuring power but 20 measuring voltage.

7.7.1.2 The 3 dB Point

When $\omega RC = 1$ Eq. 7.66 becomes $1/(1 + j)$, the magnitude $1/\sqrt{2}$ and the phase angle $45°$. There is no fundamental significance to this; it just happens to be an easy calculation to make. Substituting in Eq. 7.77, we have:

$$A_v = 20\log_{20}\left(\frac{1}{\sqrt{2}}\right) = -3.01 dB \tag{7.78}$$

and we ignore the 0.01 so this is the *three dB point* also called the *breakpoint* or the *turning point* which occurs at a *corner frequency, cut-off frequency* or *3 dB frequency* of:

$$\omega_c = \frac{1}{RC} \quad \text{or}: \quad f_c = \frac{1}{2\pi RC} \tag{7.79}$$

and it depends only on the RC product, not the individual values. This is worth noting because it often gives the circuit designer some freedom in choosing values to meet a given cut-off frequency.

Example – 3 dB Point
We can use schematic ('RC Impedance.asc') to measure the 3 dB point from the amplitude of the waves. With the values as set and a *159 Hz* input, the ratio of output/ input is *700 mV/1 V* = *−3.1 dB*. If we change the frequency slightly, say to *100 Hz*, we find *678 mV/1 V* and the attenuation is *−3.38 dB*. If we keep the RC product constant, we find other combinations give the same result.

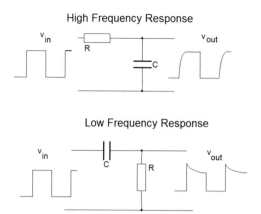

Fig. 7.22 Square wave Response

7.7.1.3 Relation Between Frequency Response and Time Response

Old hands at circuit testing will be well aware that square waves can quickly give us essential information about a circuit's performance without making a frequency response over several decades of frequency. Here is how it works.

We can substitute the time constant τ in Eq. 7.79:

$$f_c = \frac{1}{2\pi T} = \frac{0.16}{T} \tag{7.80}$$

or, using the rise time:

$$f_c = \frac{0.35}{t_r} \tag{7.81}$$

and this allows us to translate between the frequency and time domains when we measure the voltage across the capacitor. So a quick assessment of the high-frequency response of a circuit can be made with a square wave input, Fig. 7.22. For example, if we measure a rise time of 1 µs, the upper 3 dB point is 350 kHz.

In a similar manner, the time to a sag of 10% – which is about the least we can measure accurately – can give us the lower 3 dB point. For example, if it is 10 ms, the lower 3 dB point is 35 Hz. But, of course, unless we fiddle with the square wave frequency, it is unlikely that we will be lucky enough to have a sag of exactly 10%. So if we go back to Eq. 7.41, in the general case:

$$\frac{V_c}{V_{ON}} = e^{\frac{-t}{RC}} \tag{7.82}$$

if we define $sag = \frac{V_c}{V_{ON}}$ and take natural logarithms and use $RC = \frac{1}{2\pi f_c}$ we have:

$$\ln{(sag)} = \frac{-t}{RC} = -t2\pi f_c \tag{7.83}$$

we substitute in Eq. 7.83 and finally:

$$f_c = \frac{\ln{(sag)}}{-2\pi t} \tag{7.84}$$

However, we must also remember that at the 3 dB frequency there is a phase shift of 45°, and this may be significant.

Example – Square Wave Testing

We can use the schematic ('Bode.asc'). We often need an amplifier to work over a frequency range of at least *100:1*. So we need to change *C1 = 10 nF, C2 = 100 nF*. And cascade the first two stages by making *V = V(out)* for *B2* and change to a square wave for *V1*. If we measure *V(out2)*, the rise time is approximately 24 *μs*, and then Eq. 7.86 gives a 3 dB point of *14.6 kHz*, whereas it is actually *16 kHz*.

We measure the sag as *133 mV/933 mV* in *2 ms*, and then from Eq. 7.89, we have the cut-off frequency − *1.95/(2 π 2 ms) = 155 Hz* which is very close to the correct *159 Hz*.

7.7.1.4 Bode Plot

Thus far we have found the asymptotic response of a circuit and the 3 dB points. To view the response over a range of frequencies, a common method of presenting the frequency response is the Bode plot because it is easily constructed manually and gives a clear picture of the response of the circuit. Taking Eq. 7.67 and replacing *RC* by *1/ω_c*:

$$A_v = \frac{1}{\sqrt{1 + \frac{\omega^2}{\omega_c^2}}} \tag{7.85}$$

and the phase is:

$$\varphi = \tan^{-1}\left(\frac{\omega}{\omega_c}\right) \tag{7.86}$$

or in dB's:

$$A_v = 20\log_{10}\left(1 + \frac{\omega^2}{\omega_c^2}\right)^{-\frac{1}{2}} \tag{7.87}$$

We should note that it is the ratio ω/ω_c that is important, so these are universal equations that apply to every circuit of this form. And as they are relative to the cut-off frequency, if we know that we will find the same change of gain with frequency whatever the actual cut-off value.

We can try differentiating Eq. 7.87 to find how A_v changes with frequency, but the result is not very illuminating, and it is better to make numerical substitutions over a spectrum. So we tabulate the voltage gain and phase as ratios, not absolute frequencies, to arrive at the following:

ω/ω_c	A_v in dB's	Phase
0.01	−0.00	0.57
0.03	−0.00	1.72
0.1	−0.04	5.71
0.3	−0.37	16.70
1	−3.01	45.00
3	−10	71.57
10	−20.04	84.29
30	−29.55	88.09
100	−40.00	89.43

From this we plot the voltage gain in dBs on the vertical axis and the logarithm of the frequency on the horizontal axis. The table shows that the voltage gain is substantially constant at almost zero dB's up to $\omega/\omega_c = 0.3$. Above $\omega/\omega_c = 3$ it falls at a rate of −20 dB/decade of frequency or −6 dB/octave where 'octave' is the musician's 'octave' of a doubling of the frequency.

The great advantage of the Bode plot is that the overall gain of cascaded stages is given by:

$$A_v = A_{v1} \times A_{v2} \times A_{v2} \times \dots \dots \qquad (7.88)$$

whereas using decibels, we merely add:

$$A_v = A_{v1} + A_{v2} + A_{v3} + \dots \qquad (7.89)$$

Thus the overall response of a circuit can be found by adding the Bode plots for each frequency-sensitive network, as we shall see shortly.

7.7.1.5 Manual Construction

A quick construction is to draw a horizontal line from the *0* dB point *P* on the voltage gain axis to the 3 dB frequency *Q* in Fig. 7.23 and from there to draw another straight line *QR* with a slope of −20 dB/decade. This will be in error by no more than 3 dB, and the biggest error is at the 3 dB point itself. We can make the plot more accurate by joining the points $\omega = 0.1\omega_c$ (point *S*) and $\omega = 10\omega_c$ (point *T*) with a smooth

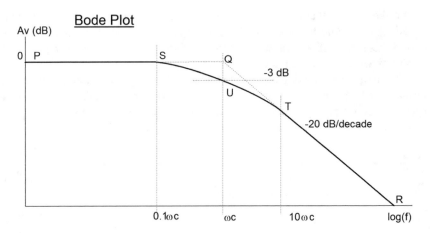

Fig. 7.23 Bode Plot

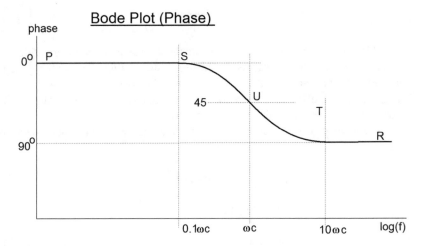

Fig. 7.24 Bode Plot Phase

curve through the 3 dB point U. Carefully done, this will be in error by less than 0.1 dB.

The phase can be constructed similarly by a horizontal line at $0°$ (Fig. 7.24) and a smooth curve from $\omega = 0.1\omega_c$ (point S) through $45°$ at the corner frequency U to a phase shift of $90°$ at $\omega = 10\omega_c$ (point T) followed by another horizontal straight line to R. This should not be in error by more than $10°$.

If we interchange the capacitor and resistor, we get a result that is a left-to-right reflection.

In the following chapter, we shall put these methods to good use as we explore circuits using resistors and capacitors.

Example – Bode Plot

The schematic ('Bode.asc') consists of four RC circuits in series. Arbitrary behavioural voltage sources are used to transfer the output voltage of the previous stage to the present. In this way there is no loading of one stage on another and each can be analysed separately by setting their inputs to V(in). So we find the first stage has an upper 3 dB point at 159 Hz, the second has a lower 3 dB point at 157 kHz, the third has an upper 3 dB point at 15.9 kHz, and the last has a lower 3 dB point at 1.58 MHz. We can cascade the stages by setting the inputs of the Bs to be the output of the previous stage.

Explorations 7

1. Make an AC analysis of ('Series RC Circuit.asc') to find the 3 dB point by measuring the voltage across the capacitor. Compare this to the time constant.
2. Measure all the voltages and currents.
3. Change the component values and repeat.
4. Run ('Series RC Circuit1.asc') and repeat. Compare the results with theory.
5. Try the schematic ('Bode.asc') with different values and swapping the components around.

7.8   Summary

This chapter has taken us through the SPICE model of the capacitor and the response of a capacitor to pulse and sinusoidal waveforms. From that we have found:

- The LTspice capacitor incorporates parasitic resistive and inductive elements which it is computationally more efficient to include in the capacitor rather than as separate entities.
- The energy stored in a capacitor is $\frac{1}{2} CV^2$ J.
- Capacitors in parallel add as $C = C1 + C2 + C3 + \dots$.
- Capacitors in series add their reciprocals $1/C = 1/C1 + 1/C2 + 1/C3 + \dots$ but in the case of two capacitors $C = C1C2/(C1 + C2)$.
- A capacitor does not dissipate power but impedes the flow of current. The impedance is $Xc = 1/2\pi fC$.
- Non-polar capacitors range from a few picofarads to a few farads. The capacitance of MLCC types falls dramatically with bias voltage.
- Polar capacitors range from a few microfarads to over a hundred farads and are only usable up to a few kilohertz.
- Three useful measures of the time response of a capacitor in series with a resistor are the *time constant* which is the time to 2/3 of the final value and the *rise time* and *fall time* which are the times from 10% to 90% and are 2.2∗*time constant.*
- The current in a capacitor lags the voltage by 90° but the lag of the current in an RC circuit depends also on the resistance.

- An RC circuit can settle to the average value of a unipolar input.
- We may use the decibel to measure voltage gain as *20 log(V1/V2)* and the *3 dB point* occurs when $R = 1/2\pi fC$.
- The 3 dB point is related to the rise time by *fc = 0.35/rise time*.
- A Bode plot has a y-axis of decibels and an x-axis of log frequency to plot the frequency response of a circuit.
- A square wave test can give the lower and upper 3 dB points of a circuit using $f_{lower} = \frac{\ln(sag)}{2\pi t_s}$ $f_{higher} = \frac{0.35}{t_r}$ where t_s is the time interval for measuring the sag.
- LTspice has flexible scaling for the trace window axes which can be logarithmic or linear, phase angle or group delay.

Chapter 8
RC Circuits

8.1 Introduction

Now it is time to analyse the behaviour of capacitor circuits starting with two bridges for measuring capacitance – the De Sauty bridge and the Schering bridge which can also measure the losses of a capacitor. Next we shall examine how we can make an instrument probe – typically using a length of coaxial cable – so that it has a flat frequency response. Then we shall explore relaxation oscillators, and then pass lightly over RIAA filters before exploring some guitar 'tone stacks' and two classic hi-fi tone controls where there is good reason to think that some of the designs were made, not by rigorous circuit analysis but by an initial outline idea followed by much bench testing.

Resistors always generate thermal noise. This is a bedevilling problem if the signal strength is low so we shall examine how noise is created and how it is modelled in LTspice. Finally we shall explore RC ladder networks, first using the Elmore delay method and then the LTspice lossless RC line. These models can also be applied to thermal modelling of semiconductor devices where the Foster and Cauer models are widely used. And we can also model the thermal performance of buildings.

Whilst most of the values quoted in the examples were obtained by using the cursor on the graphs, better accuracy can be obtained by measure statements.

8.2 Simple Capacitor-Resistor Circuits

We can now apply the analytical methods developed in the previous chapter and SPICE to a couple of useful combinations of capacitors and resistors. Much of this is an exercise in j-notation.

© Springer Nature Switzerland AG 2020
C. May, *Passive Circuit Analysis with LTspice*®,
https://doi.org/10.1007/978-3-030-38304-6_8

8.2.1 De Sauty Capacitance Bridge

There are many forms of capacitance bridges, and the one shown, Fig. 8.1, is not the best, and modern commercial instruments use better circuits to reduce errors. However, this is a simple exercise in circuit analysis.

8.2.1.1 Analysis

Figure 8.1 shows a Wheatstone bridge arrangement with two resistors replaced by capacitors. The balance condition is that the voltage between points 1 and 2 is zero. The arm C1R2 is a potential divider so we write:

$$V1 = V_{in} \frac{\frac{1}{j\omega C1}}{R1 + \frac{1}{j\omega C1}} \tag{8.1}$$

which reduces to:

$$V1 = V_{in} \frac{1}{1 + j\omega C1 R1} \tag{8.2}$$

and similarly:

$$V2 = V_{in} \frac{1}{1 + j\omega C2 R2} \tag{8.3}$$

at balance, V1 = V2 so:

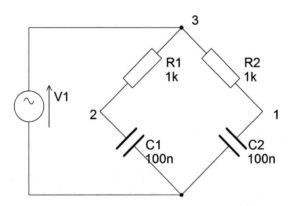

Fig. 8.1 De Sauty Capacitance bridge

$$\frac{1}{1 + j\omega C1R1} = \frac{1}{1 + j\omega C2R2} \tag{8.4}$$

from which we have: $C1R1 = C2R2$ or, if $C2$ is the unknown:

$$C2 = C1R1/R2 \tag{8.5}$$

and $R1/R2$ again are the ratio arms so that if we have a precision variable capacitor for $C1$, we can measure any capacitance by suitable choice for the ratio arms. This bridge will not measure the ESR of the capacitor.

Example – De Sauty Bridge
This is schematic ('Capacitance Bridge.asc'). By stepping C_2, a 0.1% difference is detectable creating an out-of-balance voltage of *0.4 mV* with a *1 V* excitation.

Explorations 1
1. Build the bridge of Fig. 8.1 and set the voltage source to 1 kHz. Make a transient run over 5 ms. Measure the voltages at nodes 1 and 2 and show that they are in phase and of the same amplitude. Even if you expand the voltage axis to +1fV to −1fV the difference is still zero.
2. Measure the voltage at node 3. It is larger than the voltage at node 2 and shifted in phase. Measure the phase difference converting the time difference between the peaks into an angle. Check your answer by calculation.
3. In theory, the balance is unaffected by the frequency of the source if it is an ideal capacitor. Test this by an AC analysis from 100 Hz to 1 MHz. In like manner, the output should not be affected by the excitation voltage. Test this by sweeping the generator's voltage from 1 mV to 100 V. What happens if we add ESR and ESL?
4. A question of considerable importance with a bridge is 'How sharp is the balance condition?'. We can answer that by a transient analysis whilst sweeping C2 from 50 nF to 150 nF in 100 steps and probing the voltage between nodes 1 and 2. The result is a pretty pattern of sine waves. Reduce the spread of values from 99 nF to 101 nF, and it will be seen that a difference of 0.1 nF gives a voltage of about 0.5 mV – enough to be easily detectable. Even with a difference of 10 pF, the detector voltage is 45 μV.
5. If C2 is slightly out of balance (make it 100.1 nF), how does the error voltage change with the excitation frequency?
6. Will the bridge work if we swap *R1* and *C2*? If so, is there any advantage.

8.2.2 Schering Bridge

This is of some antiquity and nowadays there are more sophisticated instruments. It is more comprehensive than the De Sauty bridge because it can also determine the

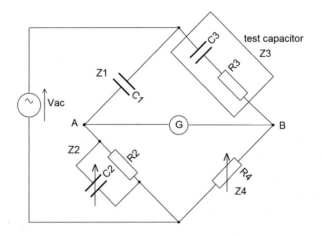

Fig. 8.2 Schering bridge

series resistance of the capacitor. The circuit is 'Schering bridge'. It can also measure the properties of insulators in the form of a thin slab held between two flat plates so as to form a capacitor. To measure losses it can use a high-voltage excitation. The circuit is Fig. 8.2 where Z_3 represents the capacitor under test.

At balance the voltages A and B are equal:

$$\frac{Z_1}{Z_2} = \frac{Z_3}{Z_4} \tag{8.6}$$

It is easier to handle this as a product:

$$Z_1 Z_4 = Z_2 Z_3 \tag{8.7}$$

where:

$$Z_1 = \frac{1}{j\omega C_1} \tag{8.8}$$

$$Z_2 = \frac{R_2 \frac{1}{j\omega C_2}}{R_2 + \frac{1}{j\omega C_2}} \quad \text{or} \quad Z_2 = \frac{R_2}{1 + j\omega C_2 R_2} \tag{8.9}$$

$$Z_3 = R_3 + \frac{1}{j\omega C_3} \tag{8.10}$$

$$\text{and } Z_4 + R_4 \tag{8.11}$$

Hence Eq. 8.7 becomes:

$$\left(\frac{1}{j\omega C_1}\right) R_4 = \left(\frac{R_2}{1 + j\omega C_2 R_2}\right)\left(R_3 + \frac{1}{j\omega C_3}\right) \tag{8.12}$$

Multiplying equation 8.12 by $(1 + j\omega C_2 R_2)$:

$$\frac{R_4}{j\omega C_1}(1 + j\omega C_2 R_2) = R_2(R_3 + \frac{1}{j\omega C_3}) \tag{8.13}$$

expanding:

$$\frac{R_4}{j\omega C_1} + \frac{R_4 C_2 R_2}{C_1} = R_2 R_3 + \frac{R_2}{j\omega C_3} \tag{8.14}$$

Equating the imaginary parts the capacitor is:

$$C_3 = C_1\frac{R_2}{R_4} \quad R_3 = R_4\frac{C_2}{C_1} \tag{8.15}$$

We balance the bridge by making R_2 switchable in decades whilst C_2 and R_4 are variable which puts a practical upper limit on C_2 of around 100 pF because it also needs to be calibrated and the limit is about 1% accuracy. This is not too onerous because it only appears in the resistance equation, but then the ESR is not a close tolerance parameter. Capacitors of up to a microfarad or so can be used for C_1. These must be of good quality, highly accurate and stable because it is the reference for measuring the unknown capacitor. The balance is independent of frequency and the amplitude of the exciting wave form.

8.2.2.1   Loss Angle (δ) and Dissipation Factor

These can be calculated using the bridge at balance.
   The loss angle is

$$\tan(\delta) = \frac{R_3}{X_3} = \omega C_3 R_3 \tag{8.16}$$

which is the dissipation factor so substituting from Eq. 8.15:

$$DF = \omega\left(C_1\frac{R_2}{R_4}\right)\left(R_4\frac{C_2}{C_1}\right) = \omega C_2 R_2 \tag{8.17}$$

And we can conveniently measure it with these two calibrated variable components.

Example – Schering Bridge
This is schematic ('Schering Bridge.asc'). It is a good idea to drop half the input voltage across the upper half of the bridge. Therefore with a *1 kHz* excitation and a

notional *10 nF* capacitor being measured, its impedance is 159 *kΩ*, and we can ignore its ESR so we can try $R_2 = 150\ k\Omega$. The bridge balance is not very sensitive because a 1% error of making $C_2 = 99\ pF$ instead of *100 pF* only creates an out-of-balance voltage of around *1 μV* with a *1 V* excitation.

Explorations 2
1. Repeat Explorations 1 using the Schering bridge. In particular, note the interaction of the balance condition.
2. We may ask if these results are realistic. For example, suppose one capacitor has a leakage resistance of 100 MΩ (rather on the low side for a good quality component), and we will find a small voltage between nodes 1 and 2. And the test leads joining C2 could have a resistance which can be included by inserting a small resistor in series with the capacitor. Finally, the resistors will have some tolerance, perhaps as low as 0.01%. All these factors (not to mention the fact that the resistors and the test leads could be inductive) will affect the accuracy. Explore their relative significance, and hence try and devise ways of reducing the errors. Also, but rather difficult, assign a value to the ESL and note the effect.

8.2.3   The Compensated Potential Divider

The problem is this: we wish to attenuate the input signal to a circuit or instrument. The obvious solution is a potential divider, but now we encounter the difficulty of the input capacitance of whatever it is that we are connecting to. Let us suppose that we want to attenuate the input signal of an oscilloscope by a factor of ten. The circuit ('Uncompensated Probe.asc') is typical, showing an oscilloscope with an input resistance of 1 megohm in parallel with a capacitance of 30 pF. It is reproduced here as Fig. 8.3.

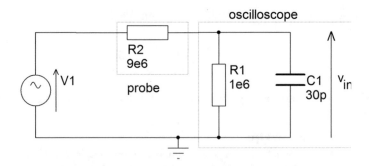

Fig. 8.3 Uncompensated Probe

8.2.3.1 Analysis

The analysis is simple and yields:

$$\frac{V_{in}}{V_1} = \frac{R_2}{R_1 + R_2 + j\omega CR_1R_2} \tag{8.18}$$

from which we see that as the frequency tends to zero, the attenuation tends to:

$$\frac{V_{in}}{V_1} = \frac{R_2}{R_1 + R_2} \tag{8.19}$$

which is what we require. However, Eq. 8.18 shows us that at high frequencies, the input tends to zero – which is not what we want. And if we take the 3 dB point as a measure of the useful upper frequency limit of the circuit, we find that:

$$\omega_c = \frac{R_1 + R_2}{CR_1R_2} \tag{8.20}$$

similar to a simple CR circuit except that the two resistors appear to be in parallel. In fact, one way of handling this circuit is to remove the capacitor and make the Thevenin equivalent of the two resistors and then that is clearly seen. Thus, taking the numbers from the circuit, we expect an upper 3 dB point at around 5.8 kHz – which is far too low for most purposes. One solution is to add another capacitor, Fig. 8.4. This is a complex potential divider. If we denote the two RC parallel combinations as Z1 and Z2 and we ignore Rs for the moment, we have:

$$v_{in} = v_s \frac{Z2}{Z1 + Z2} \tag{8.21}$$

where:

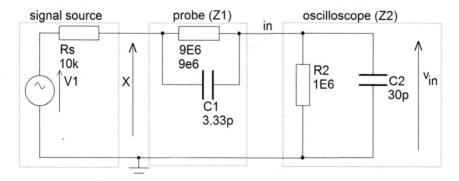

Fig. 8.4 CompProbe

$$Z1 = \frac{R_1}{1 + j\omega C_1 R_1} \qquad (8.22)$$

and similarly:

$$Z2 = \frac{R_2}{1 + j\omega C_2 R_2} \qquad (8.23)$$

substituting in Eq. 8.21:

$$v_{in} = v_s \frac{\frac{R_2}{1+j\omega C_2 R_2}}{\frac{R_1}{1+j\omega C_1 R_1} + \frac{R_2}{1+j\omega C_2 R_2}} \qquad (8.24)$$

from which we find:

$$v_{in} = v_s \frac{R_2}{R_2 + R_1 \frac{(1+j\omega C_2 R_2)}{(1+j\omega C_1 R_1)}} \qquad (8.25)$$

Now if the input is to be independent of frequency, we require C1R1 = C2R2. Hence, with the values given, we require the capacitor C2 to be 3.33 pF.

However, although the frequency response of the probe may be flat (see Explorations below), the impedance of the network is not constant and so neither is its loading on the circuit under test. This can lead to spurious results.

Let us include the Thevenin equivalent of the signal source resistance Rs which we have so far ignored in Fig. 8.4. There are two ways of handling this: one is to recalculate the potential divider including Rs in the upper arm so:

$$Z1 = R_s + \frac{R_1}{1 + j\omega C_1} \qquad (8.26)$$

and then to continue as before; the other is to consider the original circuit as an external load Z_t consisting of Z1 in series with Z2 connected to the signal source:

$$Z_t = \frac{R_1}{1 + j\omega C_1 R_1} + \frac{R_2}{1 + j\omega C_2 R_2} \qquad (8.27)$$

and if the compensation has been made, so that $C_1 R_1 = C_2 R_2$ this reduces to:

$$Z_t = \frac{R_1 + R_2}{1 + j\omega C_1 R_1} \quad \text{or} \quad Z_t = \frac{R_1 + R_2}{1 + j\omega C_2 R_2} \qquad (8.28)$$

In the first case, using Eq. 8.26, the analysis is not quite as straightforward as before; in the second case, we know the compensated probe has a flat response, so we need to only find the 3 dB point at X.

Example – Compensated Probe

Let us suppose the output resistance of the source to be 10 kilohms. Then taking Eq. 8.28 and manipulating the resulting equation, the voltage at X can be found:

$$v_x = v_s \frac{\frac{R_1+R_2}{1+j\omega C_2 R_2}}{R_s + \frac{R_1+R_2}{1+j\omega C_2 R_2}} \tag{8.29}$$

$$v_x = v_s \frac{R_1 + R_2}{R_s + R_1 + R_2 + j\omega C_2 R_2 R_s} \tag{8.30}$$

and the 3 dB point is at:

$$f_3 = \frac{R_1 + R_2 + R_s}{2\pi C_2 R_2 R_s} \tag{8.31}$$

or *5.31 MHz*. If we change to a *Cartesian* y-axis, we find both real and imaginary parts of the output are *49.9 mV* at *5.31 MHz* using the cursor, and clearly it is very sensitive to the value of *Rs*. This brings out the important point that the probe itself may have a flat response but the signal source resistance can significantly degrade it. And we should also note if we again use a *Cartesian* y-axis and plot *V(vp)/I(Rs)*, we find real and imaginary parts are equal to *5 MΩ* at only *5.3 kHz*, and at *100 kHz* the real part is only *28 kΩ* whilst the imaginary part is *528 kΩ* and the phase is *87°*. This agrees with calculation. A transient run with *1 μs* pulses shows some rounding with a rise time of *66 ns* which agrees with the two capacitors charging in series through the *10 kΩ* source resistance.

Explorations 3

1. Build a compensated probe – or use ('Compensated Probe.asc'). Run a frequency response with Rs = 1 Ω and the response is flat at −20 dB, which is correct for an attenuation of 1:10 and the upper 3 dB point is at about 55 GHz.
2. Reduce the signal source resistor to 1 nΩ and note that the response is flat to about 10e19 Hz. Do you think this is realistic? If so, build the circuit and try shining a torch on the input.
3. With Rs = 1 Ω, increase C2 to 4 pF and note the reduced attenuation above 10 kHz. Likewise, if C1 is increased to 40 pF, the attenuation above 10 kHz is increased.
4. Try a square wave input but set the rise and fall times to 1 ps, not zero.
5. Increase Rs to 10 kΩ and the upper 3 dB point is 5.8 MHz. Can this be improved? Also note the input impedance as *V(s)/I(Rs)*.
6. Make the internal resistance of the source 1e20 Ω or something very big and plot the impedance of the probe by *V(n002)/I(V1)* by clicking in the y-axis area then select **Bode** from the list and click **Logarithmic**. The low-frequency impedance is 10 MΩ falling to 1 MΩ at about 58 kHz. Check by calculation.

The attenuator probe is generally at the end of about 1 m of coaxial cable which can add another 100 pF to the input capacitance of the oscilloscope. It is still possible to compensate the probe, but now the input impedance will be much smaller. Bearing all this in mind, it would be nice if we could design a better probe that the user can adjust in case the cable capacitance is not exact.

8.2.4   RIAA Filters

Now that long-playing (LP) records ('vinyl') are making a comeback, it is appropriate to include the Recording Industry Association of America specification for recording and playback filters (RIAA). The need for a filter is due to the characteristics of the record. The lower sound level limit for an LP is set by the size of the atoms and molecules in the groove walls, https://www.st-andrews.ac.uk/~jcgl/ Scots_Guide/iandm/part12/page2.html, which defines the minimum significant excursion of the stylus since it must be larger than the random excusions due to the irregularities in the walls, whilst the upper is set by the 80 micron maximum lateral excursion of the groove else if it were more the tracks would have to be further apart and the total playing time less; additionally, the lateral acceleration of the stylus would be increased and hence the force and wear on the grooves.

These limits give a practical range of some 70 dB: interestingly, this is about the same as CDs. So both are capable of recording the full dynamic range of a symphony orchestra of about 60 dB. Which sounds better? There is a lengthy discussion at https://www.analogplanet.com/content/does-vinyl-have-wider-dynamic-range-cds-heres-some-math, but this is irrelevant to our present concern.

Our ears are highly non-linear and are most sensitive at around *1 kHz* but very insensitive at low frequencies. The equal loudness contours in https://en.wikipedia. org/wiki/Equal-loudness_contour show that we need an increase of some *30 dB* in sound pressure level at *50 Hz* compared to *1 kHz* for the same perveived loudness. It is manifestly impossible to encompass this in the restricted lateral movement of the stylus, so record makers have agreed to reduce the amplitude of bass notes on the LP meaning that they must be boosted on playback. They also agreed to boost the treble during recording, so that during playback when the treble is cut, the noise created by dust, pops and scratches would be attenuated.

The specification gave three time constants (not frequencies) of 75 µs (corresponding to 2122 Hz), 318 µs (500 Hz) and 3180 µs (50 Hz). The schematic ('Inverse RIAA Filter.asc') is used during recording, and a filter such as the schematic ('RIAA Filter.asc') is included in the playback preamplifier. In theory, passing a signal though both in series should give a flat response. However, there are variations on the filter, http://crude.axing.se/projriaa2.html, but a deviation of 1 or 2 dB is scarcely audible and, besides, the listener usually adjusts the sound by the tone controls.

It is interesting that the published circuits do not have accompanying theory. This could be because they were prototyped 'by God and by guesswork' in the pre-SPICE days or – more likely now – they were simulated in SPICE.

Example – RIAA Filter

Stepping the capacitor values in schematic ('RIAA Filter.asc') shows that *C1* has little effect and just blocks DC whereas *C2* mainly affects the curve between 60 Hz and 1 kHz and *C3* is effective above 1 kHz. And *R3* slightly alters the attenuation above 20 kHz but dramatically changes the phase shift.

We can make an approximate analysis by assuming *C1* is a short-circuit and taking *R1 C2 R2* as a potential divider which has a 3 dB point at around 67 Hz. At a higher frequency, we may take *R1 C3* as another potential divider, ignore in *R3,* and we find a breakpoint at about 2 kHz.

This circuit is also a good candidate for turning into a sub-circuit. This is ('RIAA SubCkt Test.asc') where the circuit is derived directly from ('RIAA Filter.asc') and is:

```
* RIAA Filter
.subckt RIAA in 0 out 0
C1 N002 in 470n Rser=0.1
C2 N003 out 22n Rser=0.1
C3 N004 N003 7.2n Rser=0.1
R1 out N002 105k
R2 0 N003 11k
R3 0 N004 390
R4 0 out 2Meg
.ends
```

which is a two-port network. The symbol is just a rectangle with two input pins on the left, *in* and *0*, and two on the right *out* and *0*. The file is *RIAA Filter.txt.*

Example – Laplace RIAA Filter

We can also implement the filters directly in LTspice using the Laplace function, https://www.instructables.com/id/RIAA-Equalization-with-analog-electronics/, as schematic ('RIAA Laplace Vinyl Out.asc') which includes another breakpoint at *3.18 μs.* The function is:

$$\text{Laplace} = (((s * T0 + 1) * (s * T2 + 1)/(s * T1 + 1) * (s * T3 + 1)))/10$$

where *s* replaces *jω* and *T0..T3* are the time constants. Those in the numerator are high-pass breakpoints and those in the denominator low-pass. The inverse filter is constructed by inverting the function:

$$\text{Laplace} = (((s * T1 + 1) * (s * T3 + 1)/(s * T0 + 1) * (s * T2 + 1))) * 10$$

which is schematic ('RIAA Laplace Vinyl In.asc') and the graph is the inverse of the previous.

In both cases we see the phase changing, but if we select **Group Delay** for the right-hand y-axis, we find that it is a constant *0 ms* above *200 Hz.* This is more important.

If we cascade the two filters, schematic ('RIAA Laplace Vinyl In Out.asc'), the combination is flat to *10 kHz* but shows a treble boost of *1.3 dB* at *20 kHz*. This is not significant since we cannot easily differentiate such a small change and also this is the limit of hearing of young adults and beyond the audible range of those approaching early middle-age (i.e., over 60). The treble boost increases to *14 dB* at *100 kHz* which may be of interest to a passing bat.

An additional point from this is that, for simulation, we do not always need to build the full circuit but can use the transfer function. We shall develop this theme in the next chapter.

8.2.5 Relaxation Oscillator and 'sw' Component

This is a simple circuit that can be used to flash a light, or control the speed of the windscreen wipers on a car, or something similar by making the resistor *R1* in Fig. 8.5 variable.

The circuit uses the voltage controlled switch *Sw1* which would usually have a very low-resistance R_{on} and a very high-resistance R_{off}. This is not a switch to be found at the local electronics shop, but a versatile single-pole on-off device with 'on' and 'off' being variable and requires a *.model* card to describe it.

The way the circuit works is that the capacitor *C1* charges up to the cut-in voltage of the switch, *Vc* in Fig. 8.6, which then closes to R_{on} as shown by the dashed line in Fig. 8.5, and the capacitor partially discharges until the voltage reaches the drop-out voltage *Vd* of the switch) and then it turns off, switches back to R_{off}, and the capacitor starts to recharge again. We should note that R_{on} forms a potential divided with *R1* which can be significant if R_{on} is of comparable value.

8.2.5.1 Charging Time

LTspice defines two voltages, the threshold voltage *Vt* and the hysteresis voltage *Vh*. Contrary to general usage, their behaviour in this application is that $V_d = Vt - Vh$ and $V_c = Vt + Vh$.

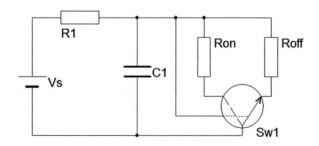

Fig. 8.5 Relaxation oscillator

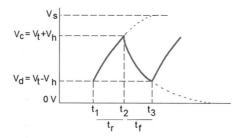

Fig. 8.6 Relaxation Oscillator Waveform

If we assume R_{on} is very small and R_{off} is very large, then after the initial pulse, the voltage across the switch will have fallen to the holding voltage V_h at time t_1. From here it will rise exponentially towards the supply voltage V_s until it reaches the strike voltage V_{st} at time t_2. During this period, the charging voltage is $V_s - V_d$ so the equation for the voltage is:

$$v_c(t) = V_d + (V_s - V_d)\left(1 - e^{\frac{-t}{R1C1}}\right) \tag{8.32}$$

which simplifies to:

$$v_c(t) = V_s - (V_s - V_d)e^{\frac{-t}{R1C1}} \tag{8.33}$$

Time $t_1 = 0$ is the starting time so:

$$v(t_1) = V_d \tag{8.34}$$

At time t_2 when the capacitor voltage is V_c, we have:

$$V_c = V_s - (V_s - V_d)\left(1 - e^{\frac{-t_2}{R1C1}}\right) \tag{8.35}$$

$$1 - \left(\frac{V_c - V_s}{V_d - V_s}\right) = e^{\frac{-t_2}{R1C1}} \tag{8.36}$$

Then taking natural logarithms:

$$t_2 = \ln\left(\frac{V_s - V_c}{V_s - V_d}\right)R1C1 \tag{8.37}$$

Example – Relaxation Oscillator

The schematic ('Relaxation Oscillator.asc') uses a voltage controlled switch with the following specification: *.model aSw SW(R_{on} = 10 R_{off} = 1e8 Vt = 8 Vh = 2)* and a DC supply of *12 V*, a capacitor of *100 nF* and a resistor of *10 kΩ*. The cut-in voltage

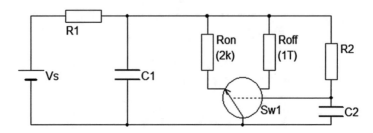

Fig. 8.7 Improved Relaxation Oscillator

is $V_c = Vt + Vh = 10$ V, and the drop-out voltage is $V_d = Vt - Vh = 6$ V, and the output therefore oscillates between 6 V and 10 V. Thus we find from Eq. 8.37 that $t_2 = ln((12-10)/(12-6)) = ln(2/6)10^{-3} = 1.01$ ms. By measurement with the cursor, the value is 1.04 ms.

Setting $R_{on} = 1\Omega$ has no effect on the charging time but making $R_{on} = 0.01\Omega$ causes problems with the discharge time. We can understand this because the time constant is 100 nF x $0.01\Omega = 10^{-9}$s and we need very short time steps, 10 ns or less, to resolve the discharge current.

Improved Relaxation Oscillator
The solution is to add a second time constant composed of R2C2 so that the discharge now is controlled, not by $C1R_{on}$ but by this second time constant, Fig. 8.7, where the switch is shown in the discharge position and we can dispense with C1 and the difficulty of very short discharge times goes away. Notice this changes the waveform and we no longer have a sawtooth but rectangular current and voltage pulses which fall to zero, not Vh. The solution is to add Vser, and now the pulse only falls to that value, schematic ('Improved Relaxation Osc.asc'). However, there is still a sawtooth waveform at the switch control terminal.

8.2.5.2   Using the LTspice Neon Bulb

The Fig. 8.8 is a partial expansion of the LTspice sub-circuit found on the 'Misc' sheet but omitting the second anti-parallel switch where the device can be used with either polarity. The full sub-circuit listing is:

```
* Copyright © Linear Technology Corp. 1998, 1999, 2000, 2001, 2002, 2003.
All rights * *reserved.
.subckt neonbulb 1 2
S1 1 2 2 N001 G
S2 2 1 N001 2 G
R1 1 N001 100Meg
C1 N001 2 {Tau/100Meg}
.model G SW (Ron={Zon} Roff=1T Vt={.5*(Vstrike+Vhold)} Vh={.5*
(Vstrike-Vhold)} Vser={Vhold-Ihold*Zon})
.param Vstrike=100 Vhold=50 Zon=2K Ihold=200u Tau=100u
.ends neonbulb
```

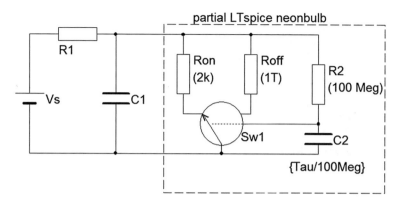

Fig. 8.8 Neon Bulb

The sub-circuit parametrises the values so that alternative bulbs can be used. The parameters *Vstrike = 100* and *Vhold = 50* means that it will turn on at *100 V (Vt + Vh)* and turn off at *50 V (Vt−Vh)*, but this is overridden by *Vser*. This is defined in terms of a holding current *Ihold* which is the minimum current to ensure that the gas remains ionized and is a data sheet parameter of the bulb, typically *200 μA*.

The idea is to make *R2C2* slightly larger than *C1Ron*, but if it is made too large, there is a dead time before *C1* starts to charge again as we see in the following Example. However, as before, this overrides the drop-out voltage and the capacitor *C1* will discharge completely. This can be solved as before by adding a series voltage to the switch **Vser** when, provided *C2R2 > C1Ron* it will only discharge to *Vser*.

Example – Neon Bulb Oscillator
The internal capacitor *C1* is surprisingly large, several orders of magnitude larger than before, typically *10 mF*, and will cause the lamp to oscillate even without an external capacitor, schematic ('Neon Osc.asc'), at about *1.4 kHz.*. We can reduce the frequency by adding an external capacitor *C1* in schematic 'Neon Osc.v2' and now the current is a sawtooth.

8.2.5.3 Discharge Time

To find the discharge time in Fig. 8.6, if R_{on} is small, at time t_2 the capacitor starts to discharge towards *0 V* with a starting voltage of V_c and continues to discharge until its voltage reaches V_h. The voltage during this period is:

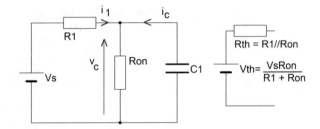

Fig. 8.9 Relaxation oscillator discharge

$$v_c(t) = V_c e^{\frac{-t}{RC}} \tag{8.38}$$

However, if R_{on} is significant, the picture changes, Fig. 8.9.
We can turn this into its Thevenin form, shown on the right. We now have:

$$i_1 = \frac{(V_s - v_c)}{R_1} \tag{8.39}$$

$$v_c(t) = (i_1 + i_c)R_{on} \tag{8.40}$$

Substituting for i_1 in Eq. 8.40:

$$v_c = \left[\frac{(V_s - v_c)}{R_1} + i_c \right] R_{on} \tag{8.41}$$

Expanding:

$$v_c = V_s \frac{R_{on}}{R_1} - v_c \frac{R_{on}}{R_1} + i_c R_{on} \tag{8.42}$$

Collecting terms:

$$v_c \left[1 + \frac{R_{on}}{R_1} \right] = V_s \frac{R_{on}}{R_1} + i_c R_{on} \tag{8.43}$$

So cross-multiplying:

$$v_c = V_s \left[\frac{R_{on}}{R_1} \frac{R_1}{R_1 + R_{on}} \right] + i_c R_{on} \left(\frac{R_1}{R_1 + R_{on}} \right) \tag{8.44}$$

$$v_c = V_s \frac{R_{on}}{R_1 + R_{on}} + i_c \frac{R_1 R_{on}}{R_1 + R_{on}} \tag{8.45}$$

And we have arrived at a Thevenin description with the capacitor voltage being V_s divided by the potential divider of the two resistors and the current flowing through the two resistors in parallel, Fig. 8.9.
From this we see that the capacitor cannot discharge to a lower voltage than:

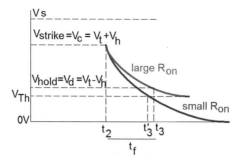

Fig. 8.10 Relaxation Oscillator Capacitor discharge

$$V_{min} = V_s \frac{R_{on}}{R_1 + R_{on}} \tag{8.46}$$

And if we increase R_{on} so that $V_{min} > V_h$, it will not oscillate at all.

At the instant that the switch closes to R_{on}, the voltage across the capacitor is $V_{strike} - V_{Th}$, Fig. 8.10 and the capacitor discharges towards V_{Th} not zero, but stops at V_d at time t_3. So for the decay time t_f: by comparison with eqn. 8.38:

$$v(c)t = (V_{strike} - V_{Th})e^{\frac{-t_2}{R_{Th}C}} \tag{8.47}$$

Example – Relaxation Oscillator Calculations

If we make $R_{on} = 5\ k$ in schematic ('Relaxation Oscillator.asc'), we have $V_{Th} = 4\ V$, $R_{th} = 3.33\ k$ and Eq. 8.37 becomes $ln(2/6) = -t_2 /(3.33\ k*100n) = 0.37\ ms$ and agrees with the measured value.

Taking $R_{on} = 7\ k$, we have $R_{Th} = 4.12\ k$ $V_{Th} = 4.94\ V$; hence $ln(1.06/5.06) = -t_2 /(4.12\ k*100n) = 0.644\ ms$. and it measures the same.

Finally if $R_{on} = 9.5\ k$, then $R_{Th} = 4.87\ k$ $V_{Th} = 5.85\ V$ and $t_2 = 1.607\ ms$, and by measurement it is *1.60 ms,* it being difficult to assess the minimum.

Explorations 4

1. Open ('Relaxation Oscillator.asc') and try different values. Check that the maximum value of R_{on} for oscillations is 10 kΩ.
2. Open ('Neon Osc2.asc') and explore. Note that if *Tau* is large, the pulse repetition rate is set by the internal time constant of the bulb and can result in a long dead time, for example, with *Tau = 100u* and a capacitor of *10 pF* on the schematic.
3. Try using a Current Controlled Switch instead.
4. Test ('Neon Osc.asc') with an AC input.

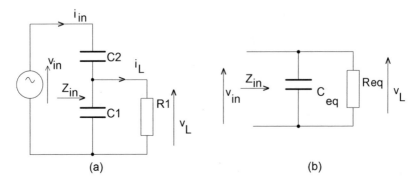

Fig. 8.11 Tapped capacitor

8.2.6   Tapped Capacitor Impedance Matching

This is an innocuous-looking circuit, Fig. 8.11a, of only three components yet needs careful handling. The treatment here follows that of R V Patron at www.zen22142. zen.co.uk/Analysis/analysis.pdf. The idea is to match the source impedance to the load without using a transformer. If we directly connect a high impedance source to a low impedance load, it is clear that there will be considerable wasted power.

8.2.6.1   Input Admittance

The impedance of the circuit is:

$$\frac{v_{in}}{i_{in}} = \frac{1}{j\omega C_2} + \frac{R\frac{1}{j\omega C_1}}{R + \frac{1}{j\omega C1}} = \frac{1}{j\omega C_2} + \frac{R}{1 + j\omega C_1 R} \tag{8.48}$$

$$Z_{in} = \frac{1 + j\omega C_1 R + j\omega C_2}{i\omega C_2(1 + j\omega C_I R)} \tag{8.49}$$

After expanding and inverting, the input admittance is:

$$Y_{in} = \frac{j\omega C_2(1 + j\omega C_1 R)}{1 + j\omega(C_1 + C_2)R} \tag{8.50}$$

We expand the numerator and multiply throughout by the complex conjugate of the denominator to make the denominator real:

$$Y_{in} = \frac{(1 - j\omega(C_1 + C_2)R)(j\omega C_2 - \omega^2 C_1 C_2 R)}{1 + (\omega(C_1 + C_2)R)^2} \tag{8.51}$$

$$Y_{in} = \frac{j\omega C_2 + \omega^2 C_2(C_1 + C_2)R - \omega^2 C_1 C_2 R + j\omega^2 C_1 C_2(C_1 + C_2)R^2}{1 + (\omega(C_1 + C_2)R)^2} \tag{8.52}$$

We can safely ignore the factor 1 in the denominator, and as $\omega \gg 1$ we may discount the first term in the numerator because the others contain ω^2. Also the terms in $\omega^2 C_1 C_2 R$ cancel, so we can divide the admittance into real and imaginary parts:

$$Y_{in} = \frac{C_2^2}{(C_1 + C_2)^2 R} + \frac{j\omega C_1 C_2}{C_1 + C_2} \tag{8.53}$$

This means that, looking from the input side, we can replace the original circuit by an equivalent resistor representing the real part of Eq. 8.53:

$$R_{eq} = \left[\frac{C_1 + C_2}{C_2}\right]^2 R \quad R_{eq} = \left[1 + \frac{C_1}{C_2}\right]^2 R \tag{8.54}$$

and as R_{eq} is just a real number, this indicates that the load voltage will be in phase with the input. To this is added a capacitor representing the imaginary part, Fig. 8.11b:

$$C_{eq} = \frac{C_1 C_2}{C_1 + C_2} \tag{8.55}$$

Equation 8.54 enables us to find the ratio of the capacitors for a given impedance, but not their value.

8.2.6.2   Voltage Gain

$$A_V = \frac{v_L}{v_{in}} = \frac{\frac{R}{1 + j\omega C_1 R}}{\frac{1}{j\omega C_2} + \frac{R}{1 + j\omega C_1 R}} = \frac{\frac{R}{1 + j\omega C_1 R}}{\frac{1 + j\omega C_1 R}{j\omega C_2} + R} \tag{8.56}$$

In most cases the factor $j\omega C_1 R$ is not much greater than 1. But as an approximation we can ignore the 1 to find:

$$A_v = \frac{1}{\frac{C_1}{C_2} + 1} \quad A_v = \frac{C_2}{C_1 + C_2} \tag{8.57}$$

8.2.6.3   Q-Factor

This is defined as:

$$Q_E = \omega \, C_{eq} \, R_{eq} \tag{8.58}$$

If we substitute from Eqs. 8.54 and 8.55, we have:

$$Q_E = \omega \frac{C_1 C_2}{(C_1 + C_2)} \left[\frac{C_1 + C_2}{C_2} \right]^2 R \quad Q_E = \omega \frac{(C_1 + C_2)}{C_2} C_1 R \tag{8.59}$$

And from this and Eq. 8.54, we can find the capacitor values.

Example – Tapped Capacitor
The schematic ('Tapped Capacitor.asc') uses the values from R V Patron. If the transient run time is not an exact multiple of the period we will get false readings. This can be mitigated by making a run over *20* cycles or more and we may need to delay the start of the readings until the circuit reaches steady state.

If we connect the source directly to the load *R1*, we find it dissipates *934 nW* with a current of *137 µA*. This agrees with the source impedance of *5 kΩ*.

When we connect the tapped capacitor the load power falls to only *10.8 nW* and the source power is *12.5 nW*. However, if we add the inductor, schematic ('Tapped Capacitor with L.asc'), the load voltage peaks at about *45 mV* – much higher than the *1 mV* before – and the load power is *17.7 µW* which is virtually the same as the source power. We might compare the load RMS current of *595 µA* and load voltage of *29.7 mV* to directly connect with the source.

8.3   Passive Tone Controls

Most circuits are based on resistors and capacitors, not inductors, because capacitors are smaller and cheaper than an inductor of comparable impedance and can be made to closer tolerance. As we shall see in the next chapters, inductors, being made from wire, always dissipate power.

What follows are examples of *reverse engineering* where we work backwards to analyse circuits others have designed. This is not bad, for it gives an insight into circuit design – but it cannot replace the creative intuition that invented the circuit in the first place. Some of the websites referred to are examples of approximate analysis. This also is a valuable skill to acquire since it gives a feel of what the circuit should do without resorting to the full detailed mathematics.

Tone controls come in two classes. The first, and with the greater variety, are the 'tone stacks' incorporated in guitar amplifiers, so called because the variable controls are often wired in series as a stack, see, for example, the 'Fender' in Duncan's 'Tone Stack Calculator' at http://www.duncanamps.com/tsc/download.html which

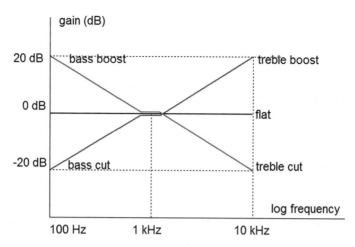

Fig. 8.12 Bow Tie Response

contains a simulator that can be downloaded (free) and several examples. More examples can be found at:

https://www.ampbooks.com/mobile/classic-circuits/bassman-tonestack-intro/
https://www.ampbooks.com/mobile/classic-circuits/cp103-tonestack-analysis/
http://members.optusnet.com.au/batwingsoup/builds/tone.htm
https://www.ampbooks.com/mobile/classic-circuits/gretsch-chet-atkins-tone-stack/

The site also has some interesting examples of classic guitar circuits.

Guitar tone stacks are part of the sound creation system of guitar, amplifier and loudspeaker and are often designed to allow the amplitude of the bass, middle and treble parts of the audio spectrum to be varied dramatically.

The second class are tone controls used in radios and hi-fi systems. These are part of the sound reproduction system rather than sound creation, and their purpose is to compensate for any deficiencies in the frequency response of the system. The ideal is the 'bow tie' response, Fig. 8.12, where bass and treble can be cut or boosted independently and there is a setting – usually with the controls in the mid position – where the response is flat or as flat as possible. The extreme boost/cut settings are shown in red and blue. The numbers in the figure are typical.

Example – Bass and Treble Cut

The simplest, and earliest, tone controls were just bass and treble cut. These were nothing more than cascaded low-pass and high-pass filters, ('Simple Bass and Treble Cut.asc'), so they had asymptotic slopes of 20 dB/decade. As variable resistors are cheaper and available over a wider range than variable capacitors, in almost every case, the capacitors are fixed and the resistors variable. In this instance we see the 3 dB point steadily reducing to lower frequencies, but we should note that this is not a simple case of the graph sliding to the left but rather that the graphs diverge from a common hinge-point of about *200 Hz.*

The schematic ('Bass Treble Cut.asc') is an alternative where we see that the degree of bass attenuation depends upon *R2* and is mainly effective below *500 Hz.* However, the treble cut *R1* shunts the signal source and so is only effective if we have a substantial source impedance.

8.3.1   The Big Muff

This is the simplest guitar tone stack, designed in 1969 as part of a system for electric guitars. It is shown here as Fig. 8.13. You can find the full circuit and analysis at

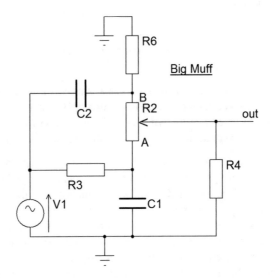

Fig. 8.13 The Big Muff

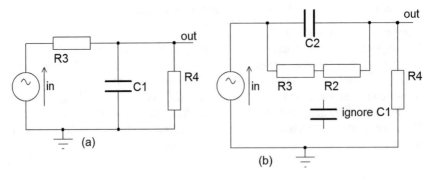

Fig. 8.14 Big Muff analysis

https://www.electrosmash.com/big-muff-pi-analysis, but we can make approximations if the potentiometer slider is at either end of its track, Fig. 8.14.

First, assume it is at *A*. As a quick estimate, we can take *R4* in parallel with *C1* and that combination in series with *R3* (Fig. 8.14a) – like the uncompensated probe – and we then use Eq. 8.20 with suitable change of component identifications to find the 3 dB point is 568 Hz. Likewise, if the slider is at *B* (Fig. 8.14b), we have a high-pass filter of *C2* in parallel with *R2* + *R3* all in series with *R4,* and we find the 3 dB point is at almost 2.1 kHz – which is just over 10% too low. Try varying the ratio *R1/R2* keeping the total constant and ensuring neither is zero, and you should get the response published by Duncan. The schematic is ('Big Muff Tone Control.asc'), and note that the potentiometer has been replaced by two resistors.

8.3.2   The James

This was published by E.J. James in 1949 and was the precursor of the circuit published by Peter Baxandall and is sometimes incorrectly called the 'passive Baxandall tone control'. It is shown as Fig. 8.15 with typical values. It suffered from an insertion loss of around −20 dB. We can get a hint of that from the circuit where we see that at least *R1* and *R4* are going to attenuate the output. It was designed for radios and record players at that time, so it had a flat response with the two controls at mid-position. However, relative to that 20 dB loss, bass and treble could be boosted as well as cut.

The analysis is complicated rather than profound and uses only the analytical tools developed so far. The basis is to consider the case when both controls are at maximum boost as Fig. 8.16. The capacitor *C1* is short-circuited and can be ignored.

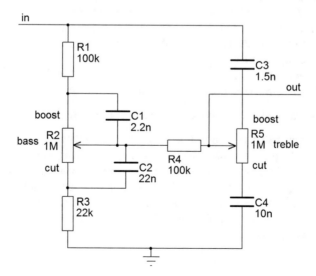

Fig. 8.15 The James

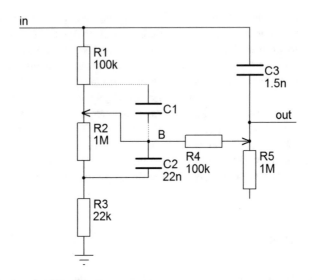

Fig. 8.16 The James at Max Boost

To estimate the relative importance of $C2$ and $R2$, we can note that the 3 dB point of the two in parallel is about 7 Hz, so at higher frequencies, the impedance of the capacitor is less than the resistance of $R2$ and we can ignore the resistor. So approximately the bass at point B consists of the potential divider of:

$$\frac{R_1}{R_1 + R_3 + \frac{1}{j\omega C_2}} \qquad (8.60)$$

and the 3 dB point is at 60 Hz. Turning to the output circuit, we may ignore $R5$ and assume there is no external load applied to the output. The impedance of $C3$ only falls to about 1 kΩ at 1 kHz, so at lower frequencies, we may ignore it and so the bass contribution to the output is just that of B. We then consider the case where the controls are set to maximum cut and repeat the analysis.

The full analysis can be found at: http://makearadio.com/tech/tone.htm. This substitutes the Laplace operator s for $j2\pi f$. The analysis can also be found at:

https://www.ampbooks.com/mobile/amp-technology/james-tonestack-analysis/

whilst a simpler explanation can be found at: http://www.learnabout-electronics. org/Amplifiers/amplifiers42.php which does not use the same component values as ('James Tone Control.asc') – indeed there are several variations on the component values, usually preserving the 10:1 ratio between the capacitors and the resistors.

More variations can be found at http://www.angelfire.com/electronic/ funwithtubes/Amp-Tone-A.html where the circuit is placed between stages of valve (tube) amplifiers.

8.3.3   Baxandall Tone Control

This circuit rendered every other tone control obsolete and is still, in one form or another, in wide use. A copy of the original article published in 'Wireless World' in 1952 can be found at http://www.learnabout-electronics.org/Downloads/NegativeFeedbackTone.pdf and is the best explanation of how the circuit works.

   The circuit is very similar to the James. The crucial difference is that the output of the network *vc* in schematic ('Baxandall Tone Control.asc') is amplified by *E1* which has a voltage gain of 100 (in the original circuit, this was a pentode valve here simulated by a voltage controlled voltage source) whose output is returned to the network, out of phase with the input, through the capacitor *C4*. It was thus incorporated in a *feedback loop*.

Example – Baxandall Tone Control
The schematic is shown with variable treble boost/cut. In effect, this creates a potentiometer for the bass and could be replaced by the one we created in chapter 4. With an AC input, we see a smooth bass response hinging at about 2 kHz. It is worth also making the treble circuit variable. If we apply a pulse input and adjust the setting of **TsliderPos** from **0.01** to **0.99** over a few runs, we find it takes a few cycles for the output to reach steady state. We can see the change in amplitude and also the rounding of the rise and fall and, at **0.99**, the sag.

Explorations 5
1. Explore as many of the circuits as you have a mind to. In particular, if tone control potentiometers are split into two resistors, each part can be made a parameter rather than an explicit value and swept. This has already been done in some of the circuits; alternatively, the potentiometer sub-circuit can be used. Removing one or two components can be helpful in understanding how the circuit works to isolate, for example, the bass or treble sections.
2. Notice that the James and Baxandall tone controls for audio amplifiers have a 'bow tie' response where bass and treble can both be cut or boosted by about 20 dB. In the case of the Baxandall, the common hinge point is at about 1 kHz. It can be argued that there should be a lower-frequency bass hinge and a higher-frequency treble hinge instead of just the one to avoid interaction. Try it.
3. Note carefully the effect of changing component values. Especially note the effect of reducing the gain of *E1* of the Baxandall circuit.
4. Measure the rise and fall times and the sag for the Baxandall tone control, changing the pulse settings if necessary, and compare them to the frequency response.
5. Explore the ESeries tone stack ('E Series.asc').
6. Most tone controls have asymptotic slopes of −20 dB/decade. Design one that has −40 dB/decade.
7. Design and test a tone control allowing bass, middle and treble adjustment.

8.4   Noise

This is a topic of some importance, especially with low signal levels. It consists of unwanted signals introduced into the system.

8.4.1   Noise Sources

There are three possible sources: (a) thermal or Johnson noise, (b) shot or partition noise applicable only to thermionic devices or semiconductor junctions and (c) flicker or 'one over f' noise prevalent at low frequencies in thermionic and semiconductor devices. To this we could also add *excess noise* where a component generates more noise that the previous three categories indicate. This has been attributed variously to impurities and crystal lattice defects causing phonon scattering.

8.4.1.1   Johnson (Thermal) Noise

For simple resistors, we are concerned with thermal noise which is ubiquitous and is caused by the random thermal vibrations of the electrons in a conductor. This was first reported by J. B. Johnson in 1927, and H. Nyquist derived the equation in 1928 using quantum mechanics and statistical thermodynamics.

We can find the derivation of Eq. 8.63 at many internet sites. However, the derivation of the full equation is more complicated. One approach regards the noise as two resistors and a transmission line, www.pas.rochester.edu/~dmw/ast203/.../ Lect_20.pdf, where we find the mean square noise voltage is:

$$v_n^2(f) = 4R \frac{hf}{e^{\frac{hf}{kT}} - 1} \Delta f \qquad (8.61)$$

where h is *Plank's constant* and k is *Boltzmann's constant* $= 1.38 \times 10^{-23}$ *J/K*, T is the absolute temperature, R is the resistance and Δf is the bandwidth. The importance of this equation is that it allows us to integrate to find the total noise power of a resistor from zero frequency to infinity. It also shows us that the noise density is constant up to very high frequencies, 10^{12} *Hz* (in the infra-red) and then it falls rapidly.

But for electronic applications, $\frac{h \Delta f}{kT} \ll 1$, and we can expand the denominator using Taylor's series as: $e^{\frac{hf}{kT}} = 1 + \frac{hf}{kT} + \ldots$ and just use the first two terms.

then $\frac{1}{e^{\frac{hf}{kT}}} = \frac{kT}{hf}$ and Eq. 8.61 becomes:

$$4Rhf\Delta f \frac{kT}{hf} \Rightarrow v_n^2 = 4RkT\Delta f \tag{8.62}$$

and Eq. 8.62 leads to the familiar RMS noise voltage v_n:

$$v_n = \sqrt{4kTR\Delta f} \tag{8.63}$$

The derivation of Eq. 8.62 and further explanation can be found at: https://www.physics.queensu.ca/~phys352/lect04.pdf as well as the Wikipedia article.

At a temperature of 300 K, $4kT$ is 1.656×10^{-21}, so Eq. 8.63 is:

$$v_n = \sqrt{1.656 \times 10^{-21} R\Delta f} \tag{8.64}$$

Excess Noise

Data sheets show that carbon film resistors have a significant excess noise due the current, and for a 1 kΩ resistor, it is typically 0.02 μV/V over the working bandwidth compared to 40 μV for the thermal noise. And a rather ancient report by P. D. Smith published by Texas Instruments in 1961 pointed out that current noise can be significant at low frequencies, typically less than 50 kHz, and this follows a *1/f* relationship. These effects are not included in the LTspice resistor noise which only implements thermal noise. Nevertheless, assuming that the theoretical relationship is obeyed, we can make some useful examinations.

We might note that LTspice allows us to define a resistor as noiseless by adding that word to its description.

Adding Voltage and Current Noise

We saw in 'non-linear resistors' that we could include self-heating or temperature coefficients, and we could attribute the current or voltage noise to that. But the noise analysis ignores any DC or AC setting of the source, or stepping it, so, sadly, it is not possible this way. However, we may first find the temperature rise then make a noise analysis with that value, see 'Example – Temperature, Frequency and Johnson Noise'.

8.4.1.2 Adding Noise Contributions

These add as the square root of the sum of the squares if they are not correlated (which is true). That is:

$$v_{\text{total}} = \sqrt{v_1^2 + v_2^2 + v_3^2 + \ldots} \tag{8.65}$$

$$\text{and } v_{\text{total}}^2 = v_1^2 + v_2^2 + v_3^2 + \ldots \tag{8.66}$$

for voltage and the same for current.

8.4.2   LTspice and Noise

Noise is a special analysis which requires a noiseless current or voltage source to
which the noise is referred. The DC or AC settings of the source are ignored. It was
pointed out in http://www.dsplog.com/2012/03/25/thermal-noise-awgn/ following
'*Johnson-Nyquist Noise*', *Turner C. S., Jan 2007* that the Johnson spectral noise
density was flat to some 1e12 Hz and then fell rapidly as we saw in Eq. 8.61. The
LTspice noise shows no fall.

8.4.2.1   Noise Settings and Measurements

The circuit is built in the usual fashion except that for the analysis we open the *Noise*
tab in the **Edit Simulation Command** dialogue. The first two edit boxes are specific
to noise analysis; the remaining four retain their usual functions for an AC analysis.

Output and V(onoise)
The noise measurement is made between two nodes, for example, *V(1,2)*. If the
second node is omitted, it is assumed to be *0*, Fig. 8.17. On the trace panel, it will be
referred to as **V(onoise)** rather than the two nodes and is always available.

Input and V(inoise), inoise
This is the name of the noiseless source, for example, *I2*. Either one is available,
depending on the type of source, if the input and output nodes are not identical. If the
noiseless source is a voltage, it is *V(inoise)*; if it is a current source, then it is *inoise*.
And note the difference in notation for current.

Gain
This is the ratio of either *V(onoise)/V(inoise)* which is a pure number or *V(onoise)/*
inoise in which case it has the dimensions of resistance. Thus *V(inoise) = V(onoise)/*
gain. It is available if the two measurements are also available.

Example – Temperature, Frequency and Johnson Noise
The schematic ('Resistor Noise 1.asc') has a current source in parallel with a resistor
of *1 kΩ*. The analysis is:

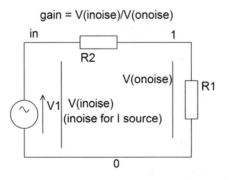

Fig. 8.17 Noise Definition

.noise V(1, 0) I1 dec 100 1 1e6

which runs from *1 Hz* to *1 MHz* with *100* steps per decade. From this one analysis, we glean two useful pieces of information. The first is that the output shows a constant noise voltage over the whole spectrum – as we expect. The second is that noise voltage increases with temperature as we see on the graph, and it is quantified by the *.meas* statement which returns the noise voltage at *10 Hz*.

.meas NOISE nv FIND V(onoise) AT 10

Note that we must select **NOISE** as the **Applicable Analysis** and the **Measured Quantity** is not the voltage across the nodes but the alias name **V(onoise)**. We make the measurement at *10 H*.

Plotting Stepped Measurements
If we step a measurement, rather than do the sums ourselves on the data, and if we right click in the **Error Log** in the small pop-up menu, we are offered the option of **plot step'ed meas data**. If we click this, the measurements are shown as a graph.

Example – Plotting Stepped Data
The schematic ('Resistor Noise 1.asc') steps the temperature in *10 K* intervals, and we therefore find 11 noise voltages. If we plot them, we find a straight-line graph from *temp = 0 V = 3.88 nV* to *temp = 100 V = 4.54 nV*. This is the noise voltage per root Hertz and agrees with Eq. 8.75.

If we assign a temperature coefficient to a resistor, schematic ('Resistor Noise 3.asc'), it is acted upon, and now the plot of the stepped data is no longer straight.

In passing, we can note that changing the DC or AC settings of *I1* has no effect and also that changing the frequency has no effect. We can also note that if we exchange the current source for a voltage source, there is no noise because the voltage source has zero internal impedance and short-circuits the resistor.

Example – Noise 'Gain' and RMS Noise
The schematic ('Resistor Noise 2 .asc') has a current source and two *1 kΩ* resistors in series forming a potential divider. After a run, if we hold down **Ctrl** and left click the trace names, we find a total noise current of *40.714 nA* for **inoise**. As the analysis it to *100 MHz* this is *4.07 pA/Hz$^{1/2}$* And **V(onoise)** is *40.714 μV* corresponding to the *inoise* current in *1 kΩ*.

The *gain* is a constant *1 kΩ* in agreement with *V(noise)/inoise*. It makes no difference if *R2* is noiseless or not.

If, however, we start with a voltage source, then the resistors form a potential divider, schematic ('Resistor Noise 4.asc') and *V(onoise)* is that due to *R1* and *V (inoise)* is the noise that the voltage source must generate to give *V(onoise)* with the potential divider of *1/11* formed by the two resistors.

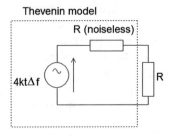

Fig. 8.18 Noise Thevenin model

Example – Noise from Multiple Resistors

Schematic ('Resistor Noise 5.asc') has three *1 kΩ* resistors in series with a current source. The output voltage noise is *7.0518 μV* which agrees with Eq. 8.63. The noise current is *7.0518 μV/3 kΩ = 2.3506 nA*. We might notice that ('Noise.asc') has the same noise voltage for a single resistor and for the equivalent resistor of three resistors in parallel.

Resistor Noise Power

Equation 8.76 gives the noise voltage, so squaring that and dividing by the resistance will give the power if the resistor is short-circuited:

$$P_n = \frac{\left(\sqrt{4kTR\,\Delta f}\right)^2}{R} \quad = 4\,kT\,\Delta f \tag{8.67}$$

and is independent of the resistance. But as that noise voltage is not accessible to us, a better approach is to create a Thevenin model of a noiseless resistor in series with the noise voltage generator. For maximum power transfer, we attach an identical resistor, Fig. 8.18. As only half the voltage appears across each resistor, the power now is:

$$P_n = kT\,\Delta f \tag{8.68}$$

and this is usually expressed in *dBm* where *1 mW = 0 dBm*. For room temperature of *300 K* and a bandwidth of *1 Hz*, we write Eq. 8.67 as:

$$P_n = 10\log_{10}(1.38\text{e-}23 \times 300 \times \text{e}3) = 173.83\ dBm/Hz \tag{8.69}$$

where the *e3* converts to *mW*.

The Wikipedia article quotes typical figures for several examples such as WLAN 802.11 channels.

8.4.3 Noise Generator

There are two possibilities: one is to use the *white* option with an arbitrary source, and the other is to create a voltage source with the appropriate voltage.

8.4.3.1 Using *white* Arbitrary Source

The simplest setting is schematic ('Noise Generator 2.asc'). It is important to set a short time step. With the setting:

$$V = white(time * 1000)$$

and a **10 us** time step an FFT shows a flat distribution up about 300 Hz. However, with no minimum time step, it is only flat to about 100 Hz. The peak amplitude is +/ − *500 mV*. This can be reduced by multiplying by a small power of ten, for example, *1e6* gives *67.8 nV RMS*, but at the same time, that reduces the frequency range which requires a large power of ten to compensate:

$$V = white(time * 1e10) * 1e - 6$$

This also is flat to about *300 Hz*.

We might remember that white noise creates a new value if the term in brackets changes. As time is linear, the values are created at equal intervals and the noise density is constant. But we can – with some difficulty – try powers of time so that the noise density is not linear. And we shall also need to use a small time step.

We should also note that we can only use this with a transient run.

8.4.3.2 Using a Voltage or Current Source

We can use a voltage or current generator with the noise voltage derived from Eq. 8.63 and run an AC analysis. Alternatively, to make a transient run, we can use an arbitrary voltage or current source and assign that the appropriate value with a variable frequency f where kb is Boltzmann's Constant and we have a *1 kΩ* resistance schematic ('Noise Generator 1.asc').

$$V = sqrt(4 * kb * \{temp\} * 1k * f$$

then:

.step param f 1 10k 1k

will make 10 runs and we find the last gives *400 nV* at *temp = 300 K*. This gives constant values rather than a series of noise voltages.

8.4.4   Noise Reduction

The obvious way is to reduce the resistance. Otherwise, reduce the bandwidth to exclude unwanted signals. Careful choice of resistor type to reduce excess noise is important with wire wound resistors being the best – but they are also inductive and not available at high resistance values. It is also important to exclude DC if possible. The following are some useful references:

https://www.analog.com/en/analog-dialogue/raqs/raq-issue-25.html
https://www.analogictips.com/tips-electrical-noise-reduction-faq/
http://www.ecircuitcenter.com/Circuits/Noise/Noise_Analysis/res_noise.htm
http://www.engr.usask.ca/classes/EE/323/notes_2005/chapter8.pdf

Capacitor Noise
First, let us note that a capacitor does not generate noise; it is just a convenient way of expressing the noise from a filter. What we need is an ideal square filter that will pass the same noise power as the actual filter, Fig. 8.19, where the two areas a are equal. We need to integrate the noise density of the RC filter. As there is no low-frequency cut-off, the integration is from zero to infinity. We quote the result from www.ee.bgu.ac.il/.../Copy%20of%20_6.%20Noise%20t that:

$$BW = 0.5\pi f_{3dB} \quad = 0.5\pi \frac{1}{2\pi RC} \quad = \frac{1}{4RC} \tag{8.70}$$

which can be derived from Eq. 8.61 by integrating to find the total noise and as this BW is the Δf in Eq. 8.67 then:

$$v_n^2 = 4kTR\frac{1}{4RC} = \frac{kT}{C} \tag{8.71}$$

showing that the noise depends only on the capacitor and temperature.

1/f Noise
This is proportional to the DC current density and not the resistance. Quoting from the above website:

$$i_n^2(f) = \frac{K_f^m I^m}{f^n} \quad 1 < m < 3 \quad 1 < n < 3 \tag{8.72}$$

where K_f is a constant that depends upon the material.

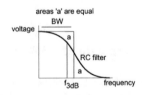

Fig. 8.19 Noise bandwidth

Example – Noise Reduction

A simple low-pass filter, schematic ('Noise Reduction.asc'), reduces the total RMS output noise from *57.578 μV* to *2.0337 μV*. This assumes an ideal capacitor with no equivalent resistance. If we remove the resistor in parallel with the capacitor, the total RMS input noise is *40.7 μV* and the output noise is *2.0347 μV*. The *3 dB* point is *159 kHz*, and there we find the noise has fallen by *1/√2*.

Explorations 6

1. Explore different resistor values in ('Resistor Noise 1,2,3.asc').
2. Create some resistor networks such as ('Resistor Noise 4.asc') and measure the output noise across different terminals. The easy way to check the result is to remember that noise is proportional to the square root of the resistance, so we can use 1 kΩ as a reference and then taking the resistor in kilohms; *noise* = √R × *4.07 nV Hz $^{1/2}$*.
3. Explore noise reduction using capacitors from the LTspice database so the ESR is included.

8.5   RC Delay Lines

An RC delay line is potentially formed whenever two lengths of conducting material are separated by an insulator. Any linear sequence of resistors and capacitors where one side of the capacitors is grounded and there are no closed loops will also form a delay line.

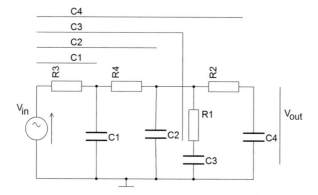

Fig. 8.20 Elmore delay

8.5.1   *Elmore Delay*

The analysis of such structures, Fig. 8.20, was studied by W. C. Elmore in 1948. According to https://www.google.com/search?client=firefox-b&q=elmore+delay, the formal method of finding the delay due to each capacitor is to earth V_{in}, apply a current across the capacitor and measure V_{out} then $R_{eff} = V_{out}/I$. An approximate method is to find the sum of the resistors from any capacitor back to the source multiplied by the capacitor. Thus, for the Fig. 8.20, the total delay is:

$$D_t = C_1 R_3 + C_2(R_3 + R_4) + C_3(R_3 + R_4 + R_1) + C_4(R_3 + R_4 + R_2) \qquad (8.73)$$

This is a rather pessimistic figure for the delay.

8.5.1.1   Elmore Delay for RC Line

The previous tree structure is used for IC interconnections, but there is an easier method if we are dealing with a simple line, Fig. 8.21, where the resistance and capacitance are the distributed resistance and capacitance and not discrete components.

We divide the wire into n equal sections so that $n = L/dL$ and thereby divide the resistance and capacitance into n equal sections of dR and dC. Applying Eq. 8.73 we have:

$$D = dRdC(1 + 2 + 3 + \ldots n) \qquad (8.74)$$

this is an arithmetic progression so its sum is:

$$D = dRdC \frac{(1+n)}{2} n \qquad (8.75)$$

and if n is very large we may ignore the distinction between n and $(n+1)$ and this becomes:

$$D = \frac{(dR.n)(dL.n)}{2} = \frac{RC}{2} \qquad (8.76)$$

and as each of the bracketed terms is proportional to L, then $D \propto L^2$.

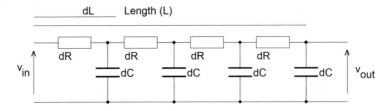

Fig. 8.21 Elmore wire

We may compare Eq. 8.76 to a simple RC circuit:

Delay	RC circuit	Elmore delay
$0 \rightarrow 50\%$	0.69 RC	0.38 RC
$0 \rightarrow 63\%$ (time constant)	RC	0.5RC
$10\% \rightarrow 90\%$ (rise time)	2.2 RC	0.9 RC

And the Elmore delay is half that of the simple RC circuit. The output voltage can be found using the Laplace transform, https://engineering.purdue.edu/~chengkok/ee695K/lec3a.pdf, and is:

$$V_{out}(s) = \frac{1}{s \cosh \sqrt{sRC}} \quad \cosh(x) = \frac{e^x - e^x}{2} = 1 + \frac{x^2}{2!} + \frac{x^4}{4!} + \frac{x^6}{6!} + \dots \quad (8.77)$$

An improved method of calculating the delay can be found at https://www.sciencedirect.com/science/article/pii/S0895717709002866. In the main, the Elmore delay is used for interconnections on integrated circuits where the metallization is not just point-to-point but can have extensive branching.

8.5.2   The Uniform RC Line

We mention it here but will explore it in more detail in the chapter 'Transmission Lines' The LTspice 'Help' file says that these are intended for interconnections on ICs but rarely used. And we shall later see that they can be used for thermal modelling. Also they can model the delays and distortions caused by the tracks of a printed circuit board. Once we understand that, the unusual syntax falls into place. Typically the tracks will all be of the same conductive material (usually copper), and on the same substrate, only their lengths and widths will differ. And this is reflected in the syntax.

There is a considerable improvement in the rise time of a distributed component line compared to a lumped RC line, Fig. 8.22. We shall also find that the LTspice URC line differs greatly from lumped RC lines, especially at frequencies higher than the 3 dB point.

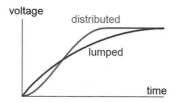

Fig. 8.22 Rise time comparison

8.5.2.1 Creating an Instance

For this we need to place a model statement on the schematic and one or more instances of the URC. At its simplest we just press *s* on the schematic and type:

.model URC URC()

That will use the default resistance and capacitance of *1 kΩ* and *1.5e-15 F* of which the resistance is usually too big by an order of magnitude and the capacitance too small by three or four orders of magnitude. Therefore we are more likely to need:

.model URC URC(Rperl = 100 Cperl = 10n)

The capacitance and resistance are *per length* where length is in metres and is the same for every instance.

8.5.2.2 Placing an Instance

There are two assemblies differing only in the number of zigzags in the resistor. They are found in the **Misc** folder near the end. Having placed it on the schematic, we connect one end of the resistor to a voltage or current source and the other end usually to a resistor. The capacitor plate is almost always connected to ground.

We must then open the **Component Attribute Editor** and add a length in **Value2** as *L = n:* there must be a value. We may optionally add *N* which determines how many sections there are in the line. If none is given, LTspice will make an appropriate choice, but we shall see the number is not critical.

Example - RC Delay Line We place an URC on the schematic ('RC Delay Line. asc') and make its length a parameter which we step by **.step param L list 1 5 10**. With *L = 1* this shows a low frequency attenuation of *26.4 dB* when terminated by 50 Ω which agrees with *50/1050*. Likewise with a *L = 5* we find an attenuation of *50/5050 = 40 dB* and with *L = 10* we have *50/10050 = 45 dB*.

The upper 3 dB point of the input impedance *Z(in)* falls dramatically with *N*, being 380 MHz, 16 MHz and 4 MHz, accompanied by significant phase shifts. This is not the same as a simple RC circuit, see Example - The Length Parameter.

8.5.2.3 Creating Different Instances

If the PCB tracks are not all the same width, so that the resistance and capacitance per unit length differ, we can create as many models as are necessary by defining new models with the appropriate parameters:

.**model MyUrc1 URC(Rperl = 200 Cperl = 1n)**

.**model MyUrc2 URC(Rperl = 50 Cperl = 1u)**

and so on. The difference now is in editing it where **URC** is replaced by **MyUrc1**, etc.

8.5.2.4 The URC Parameters

We have no access to the internal structure of the line. However, we may deduce something about it from measurements.

Example – The Length Parameter
The schematic 'URC Length.asc' steps the length from *1* to *10* in unit steps with a load resistor matching the unit resistance. From the measurements with an AC input, we see that the low-frequency attenuation exactly agrees with a potential divider. We can use the first set of measurements in dBs at *0.01 Hz* or we can convert the y-axis to linear and use the cursor to read the values. For example, with $L = 1$ we have $V(out) = 100\,\Omega/200\,\Omega \times 1\,V = 500\,mV = -6.02\,dB$ and at $L = 5$ we have $V(out) = 1/6 = 167\,mV = -15.56\,dB$, and $-20\,dB$ at step *9* and so on. This confirms that the resistance increases linearly with length.

This graph also gives us the upper 3 dB points. As the number of lengths increases, the upper 3 dB point tends to $f_{3\,dB} \propto \frac{1}{L^2}$ which implies a constant capacitance per unit length. If we plot the stepped data on a linear x-axis and logarithmic y-axis, they are nearly straight lines. To resolve the input voltage, we need some source resistance.

Example – URC Rise Time
This too shows a dramatic change with just one length, schematic ('URC Risetime. asc'). The graph there appears to be a straight line, but extending the time will resolve it. What is interesting is that the rise time is approximately *10 μs* and does not change much for $N > 1$. And it does not change greatly for different load resistors up to *1 M Ω* – only the amplitude increases. We should note that the rise time is the time-constant of the circuit, not the *2.2RC* we should expect for a single RC. Compare this with a pulse excitation of schematic ('URC Length.asc').

Example – Propagation Constant
This is a global constant and must be included in the .*model* statement if we wish to change it from the default of *2*, schematic ('URC Prop Const.asc') and must be at least 2. It has a marginal effect on the shape of the pulse, and also the delay, which becomes clearer if we zoom in.

Example – URC Compared to Distributed
The schematic ('URC v Distributed.asc') shows the dramatic difference between the two. It might be interesting to add a lumped single RC filter.

Explorations 7
1. Create a simple RC tree and apply the Elmore delay to calculate the delay to each node. Compare theory and measurements.
2. Set up a discrete component RC delay line and calculate the Elmore delay. Compare with mismeasurements and how the difference depends upon the length of the line – that is the number of sections.
3. More difficult, but create a connection tree to model the interconnections on an IC. There are a few web sites that have examples, for example, 'science direct' mentioned above.
4. Explore the URC delay line and compare it to the Elmore results of an RC ladder.

8.6 Thermal Modelling

Thermal modelling has an obvious application in electronics in calculating temperatures, in particular, the junction temperature of semiconductor devices where even a brief excess temperature can cause catastrophic failure. But the same thermal models can be applied to a wider field, particularly the thermal performance of buildings. And we shall see that these models consist of resistors and capacitors and voltage and current sources.

8.6.1 Heat Transfer Mechanisms

There are three: radiation, convection and conduction. Their relative importance depends upon the temperature, very strongly in the case of radiation.

8.6.1.1 Radiation

This is significant chiefly at temperature above 2000 K and so can be discounted in electronic circuits. It is described by the Stefan-Boltzmann Law which states that it is proportional to the fourth power of the temperature. Radiation requires no physical medium to transfer the energy, which is an electromagnetic wave, part of the spectrum extending to γ-rays. The Wikipedia article is very informative. However, it can be significant in thermal models of buildings where insolation is important.

8.6.1.2 Convection

Convection transfers heat from a hot body to a surrounding fluid, often air, but it can be oil or anything else, by the movement of the fluid over the hot surface. We may distinguish between *natural convection* where the movement of the fluid depends upon thermal gradients and *forced convection* where it depends upon fans or pumps. The calculations rely on mass flow of the liquid and its specific heat capacity. However, it is usual to assume that the flow is sufficient to maintain a constant temperature of the hot body and so avoid some rather awkward calculations. So, for example, we assume a constant temperature for the external surface of a heat-sink. The important point is that if the flow of the cooling fluid is restricted, in the case of electronic equipment, either by obscuring air vents on the enclosure or by the failure of a cooling fan, the surface temperature of the heat-sink will rise with potentially disastrous results.

8.6.1.3 Conduction

Conduction is the mechanism whereby heat is transferred from a hot surface to a cold one, usually through a solid medium, by way of the thermal gradient created by the kinetic energy of the components of the body. It can be described by the diffusion equation although that is not necessarily helpful. In the general case, Fig. 8.23, the heat can flow laterally as well as downwards. Fortunately in electronics, the height h is usually much smaller than the area A and it becomes a one-dimensional problem where we ignore lateral heat flow.

Conduction can be modelled by a passive circuit of resistors and capacitors. The resistors model the ability of the heat to travel from the hot surface to the cold and are analogous to electrical resistance except that it is usual to work with the *thermal conductivity* κ so that $R_{th} = \frac{1}{\kappa}\frac{L}{A}$ where L is the length in the direction of heat flow and A is the cross-section area, Fig. 8.24. It follows, then, by analogy with Ohm's Law, that the flow of heat is:

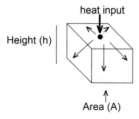

Fig. 8.23 Heat conduction in 3 Dimensions

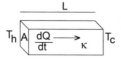

Fig. 8.24 Heat conduction in 1 Dimension

$$\frac{dQ}{dt} = \frac{T_h - T_c}{R_{th}} = (T_h - T_c)\kappa\frac{L}{A} \qquad (8.78)$$

Therefore we only need to concern ourselves about keeping the cold surface cold and ensuring the hot does not get too hot and not to worry about how the heat is removed from the cold end.

However, we must also take into account the thermal capacity of the conducting medium. This is found from the *specific heat capacity s* of the material which is the number of joules required to raise the temperature of 1 kg by 1 K. Thus the energy dQ needed to raise a mass m by dT degrees is given by $dQ = msdT$ which we can write as $dQ = C\,dT$ where C is the thermal capacity in *J/K* and this is directly comparable to the change of charge on a capacitor where $dQ = C\,dV$. The temperature T_c does not immediately change when heat is applied to the hot end because there is a thermal time-constant modelled by a capacitor. Hence we expect to model thermal properties by an RC ladder representing the distributed thermal resistance and thermal capacitance.

8.6.2 Semiconductor Thermal Models

A simple model is just a resistor and a capacitor as we saw when we modelled thermistors. That is sufficient if we are dealing with a steady heat input but is not adequate for one of the main applications of thermal modelling, which is to find the junction temperature of a semiconductor device where the heat generated is not constant but often is in short pulses and then Eq. 8.78 is too simplistic. Consider the typical cooling situation of Fig. 8.25. This is a stylized sketch of a TO-220 device mounted on a heat-sink but insulated from it. This illustrates two important points. The first is that we have three thermal resistances in series. The second is that it is clear that heat can flow laterally as well as downwards in the heat-sink shown in the figure.

Taking the thermal resistances in turn, the first is that from the junction to the case of the semiconductor, R_{jc}, over which we have no control and is set by the construction of the device. The second is the thermal resistance between the case and the heat-sink, R_{cs}. At the microscopic level, the surfaces of the semiconductor case and the heat-sink are not flat, so to improve thermal contact, it is usual to liberally apply thermal grease – often based on zinc oxide, ZnO – to one surface. However, as the case of most semiconductors is 'live', it is generally necessary to use

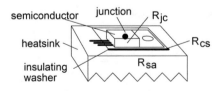

Fig. 8.25 Semiconductor cooling

a greased insulating washer between the case and the heat-sink which is usually earthed. This thermal resistance is under our control in the sense that by using thermal grease and careful assembly, we can ensure that it is not excessive. The final, and often the most important, thermal resistance is that between the heat-sink and ambient temperature, R_{sa}. Here we encounter lateral heat flow across the heat-sink so we are dealing with complex conduction paths as well as convection. However, these are all engrossed in the value of R_{sa} supplied by the heat-sink manufacturer who may supply different values, or a graph, to describe the behaviour with forced convection rather than free air. Thus we can use the simple Eq. 8.79 for the total thermal resistance:

$$R_{th} = R_{jc} + R_{cs} + R_{sa} \qquad (8.79)$$

However, we also need to add capacitors to model the delay between heat being applied to one face and the temperature of the other rising. This applies especially to the heat sink where the thermal time constant can even be tens of seconds but also to the much shorter time constants of R_{jc} and R_{cs}. For this there are two popular models.

8.6.2.1 Foster Model

It is pointed out in several websites, for example, Infineon Technologies articles such as   https://studylib.net/doc/18731091/thermal-equivalent-circuit-models   that   the parameters of this model can be fitted using measurements on the cooling curve of a device and in particular of an Insulated Gate Bipolar Transistor (IGBT).

The Foster model consists of a series of resistors and capacitors in parallel, Fig. 8.26. It is also known as the *partial fraction model*. Unfortunately, the internal nodes of the model have no relationship to the physical device; only the very first is meaningful.

The thermal impedance is given by the equation:

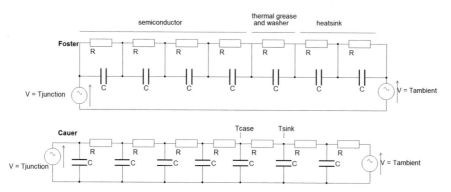

Fig. 8.26 Foster and Cauer Models

$$Z_{th} = \sum_{i=1}^{n} R_n \left(1 - e^{\frac{t}{t_n}} \right) \tag{8.80}$$

where t is the time and $t_n = R_n C_n$, the n th time-constant, where we usually have four but there can be more, for example, NXP semiconductors use seven. This is a closed form and therefore amenable to curve-fitting techniques. Both Infineon and Vishay do so for the thermal data of their power IGBTs and MOSFETs. The junction temperature $T_j(t)$ then is:

$$T_j(t) = P(t)Z_{th}(t) + T_{case}(t) \tag{8.81}$$

where $P(t)$ is the power as a function of time, $Z_{th}(t)$ is the thermal impedance of Eq. 8.80 and $T_{case}(t)$ is the temperature of the case, also called the mounting base.

The thermal grease and insulating washer can be modelled by a second stage in series and the heat-sink by one or two more. The difficulty is that the heat generated at the junction flows immediately to the heat-sink with no delay. As the thermal time-constant of heat-sinks is tens of seconds compared to a second or so from junction to case, this is not too great a problem.

'.net' Statement

We shall use this extensively in the next chapter; here we note that it is a convenient way of finding the impedance of a circuit. The syntax is:

.net < output node >< name of reference source >
< optional input resistance >< optional output resistance >

The minimum description is just the reference source when the measurement will be made on the whole circuit. And as the circuits we shall deal with here terminate in a voltage source which has zero impedance, we need not worry about the default terminating impedance so the simple directive:

.net I1

or similar will be found.

Example – Foster Model by Vishay
Vishay, in Application Note AN609, shows a four-stage model from junction to case of a power MOSFET type S17390DP. The input pulse is modelled by a switching waveform described as:

Time (ms)	0	0.1	0.99	10	20	2.1	5	5.1	10
P(W)	0	100	100	0.1	0.1	20	20	0	0

This is saved in a file and is used in a continuously repeating loop in some circuits:

$$\textit{PWL repeat forever} \left(\textit{file} = "\textit{Vishay SI7390DP Foster.txt}" \right) \textit{endrepeat}$$

Schematic ('Vishay SI7390DP Foster.asc') has a terminating voltage *V1* representing the temperature of 25 °C. The temperature graph for the first 10 ms agrees with application note AN609 being just the first input pulse. However, if we extend the time, we soon find the device overheating as the junction temperature soars well above 160 °C. This is not surprising: the problem is the small size of the device alone, without a heat-sink, which severely limits its ability to dissipate heat by convection to the surrounding air and hence the thermal resistances are high, in particular *R4,* the thermal resistance of *50.6 Ω* from the case to air.

To measure the input impedance, we need only convert the transient run into a comment and activate the AC analysis. It is helpful to convert the x-axis to logarithmic and the y-axis to linear in schematic ('Vishay SI7390DP Foster Rth. asc') and we see a low-frequency impedance of *70 Ω,* which agrees with the data sheet.

If we add a heat-sink in schematic ('Vishay Si7390DP Foster Expanded.asc'), the four-stage RC components relate purely to the MOSFET R_{jc} where – somehow or other, and it is not easy – the case is kept at 25 °C, and now we see there is no overheating and the thermal resistance is only *3.2 Ω* because, by assuming a constant case temperature, we have removed the very high thermal resistance of the case to air.

We might notice that LTspice often interposes many data points between those in the table above and most of these are removed when a run is made resulting in a string of messages in the **SPICE Error Log** of 'Removing PWL point (x,y)'.

Example – Foster Model Thermal Impedance

We use schematic 'Vishay Si7390DP Foster Rth.asc' where we use:

$$\textit{.net } V(a) \textit{ V2 Rout} = 1e12$$

to measure between the output voltage at *V(a)* and the source *V2*. In this case, setting *Rout = 1e12* is not needed because the circuit is terminated by a voltage source which is a short circuit. An AC run:

$$\textit{.ac lin 100 0.1m 1}$$

shows that the impedance at what is effectively DC is 70 Ω in agreement with the data sheet.

Example – Foster Model by NXP

The thermal model for a BUK7Y7RB is schematic ('NXP BUK7Y7RB-40E Foster. asc'). This has a complex input pulse modelling real-life power dissipation in a

switching   application,   https://assets.nexperia.com/documents/application-note/
AN11261.pdf which is used to create a text file called in by:

$$PWL\ file = {}^{''}NXP\ BUK7Y7RE\text{-}40E\ Foster.txt^{''}$$

With this *1 A* pulse, the thermal profile can be measured and compared to the data
sheet. Using a transient run of *10 ms*, we can see the first power pulse.

8.6.2.2   The Cauer Model

This is also known as the *continued fraction model*. The internal nodes now can
accurately represent the thermal conditions at the node. Topologically it is the same
as the URC. The difficulty is in finding the component values since, unlike the Foster
model, they do not relate to measurements made on the device and must be derived
by curve fitting to the graphs of thermal behaviour.

Example – Cauer Model by Vishay
Vishay in Application Note AN609, https://www.vishay.com/doc?73554, use four
RC stages to model heat transfer. In schematic ('Vishay SI7390DP Cauer.asc') we
use their data for heat transfer from junction to ambient, and it shows the junction
temperature climbing to more than the maximum allowed of *150 °C* and therefore is
not a safe design. The input pulse is from the same file used to excite the Foster
model.
    There are also values for heat transfer from junction to case. These are used in
schematic ('Vishay SI7390DP Cauer Expanded.asc') where we have added a
notional thermal washer and grease *C4R4*, and a heat-sink *C5R5*, and still the
junction temperature rises above the limit on an extended transient run unless we
cool the heat sink to *0 °C* – which adds a margin of safety. If we apply a short as
shown in the circuit, the measured thermal resistance junction to case is just within
the data sheet limits.

8.6.2.3   The URC Model

The LTspice URC can also be used for thermal modelling with the advantage that a
line with many sections is contained in one model. Analog Power http://www.
analogpowerinc.com/applications.html has a model for an SO-8 device, schematic
('Thermal Model Using URC.asc'). The URCs model:

- Vertical heat flow from the die (URC_DIE)
- Heat flow from the die attach (URC_DA)
- Vertical heat flow in the copper of the lead frame under the die (URC_LF_Vert)
- Horizontal heat in the lead frame (URC_LF_Horiz)
- Heat flow in the PCB (URC_PCB)

    Using a ramp input to *1 A DC* enables the thermal resistance to be read off directly
from the y-axis of voltage; these are the *.meas* statements from *1 ms to 500 s*.

8.6.3   Thermal Models of Buildings

Steady-state models have been used by engineers for decades using heat transfer values through walls, floors and ceilings to size the air-conditioning plant and boiler to the building. However, it is also of interest to know how long it will take for the building to reach a given temperature or to cool down.

This is possible using resistors, capacitors and voltage and current sources to model the building. It can get complicated, unsworks.unsw.edu.au/fapi/datastream/ unsworks:10641/SOURCE01 by Gerard Parnis. Modelling relies on accurate data for building materials, and it must be remembered that the simple thermal resistance of things like glass and plasterboard is not enough and we need the *U-value* which takes into account the layer of stagnant air adjacent to the surface. The model may have resistors in parallel, Fig. 8.27 of a semi-detached house, representing heat transfer through walls, windows and doors, and resistors in series to model the wall composition of, perhaps, plasterboard, insulation and brick outer layer. Likewise heat transfer upwards will pass through the ceiling, loft insulation, the air in the loft void and the tiles or whatever roof covering is used. This models the steady-state, but to find the time to heat or cool we need to add the thermal mass of the building modelled by capacitors.

The model must also take into account the fact that the sink temperatures may not be constant, certainly the air temperature *Tair* will not, nor need the room temperature next door *Troom* be consonant. And the model should also include insolation, wind speed and the use of a reflective foil facings on insulating batts to reflect infrared radiation back into the roof and to handle heating and cooling rather than just steady state, and the thermal capacity of the materials. So some of the structures may best be modelled by a URC.

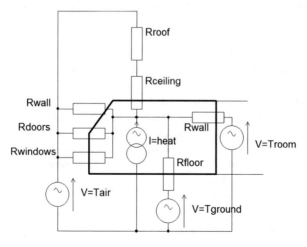

Fig. 8.27 House thermal model

The Master's thesis of 2012 by Gerard Parnis referred above is a very useful resource.

Explorations 8

1. Using manufacturer's data, create Foster models for some semiconductors and compare the simulations with the published graphs.
2. Compare these with Cauer models – where available, and measure the temperatures at the interfaces, for example, between case and heat sink.
3. Compare the distributed thermal models with *4* or *7* stages with the use of a URC for each stage. Is it easier to create and does it give a better fit?
4. Explore the above models with different heat pulse inputs.

8.7   Summary

This chapter has taken us through some common uses of simple RC circuits. We have seen:

- Real capacitors have an ESR, and it is better to use non-polar types if possible.
- A Wheatstone bridge with two capacitors replacing resistors can measure capacitance but a Schering Bridge also measures ESR.
- The LTspice capacitor cannot have its parasitic resistance and inductance directly tolerated or given a temperature coefficient – only if they are declared as parameters.
- LTspice accepts a description of a capacitor as charge, and then any arbitrary function can be applied.
- An approximate hand-waving analysis of guitar 'tone stacks' and hi-fi tone controls will give a useful guide to the operation of the circuit.
- Resistor noise is a problem, and they generate excess noise above the thermal noise, especially carbon film ones.
- Noise may be reduced by small resistor values and limiting the bandwidth.
- The Elmore delay gives a good approximation to the performance of distributed RC ladders, especially where the components do not all have the same value.
- The LTspice URC is an alternative and gives markedly different results.
- To model one-dimensional heat transfer, which is valid for semiconductor devices, the Foster model can be used, usually of four RC stages, where the values are found by curve fitting and have no relationship to the properties.
- The Cauer model can also be used, and the thermal resistances and capacitances can model the physical device.
- The URC can also be used.
- Networks of resistors, capacitors and voltage and current sources can model the thermal performance of a building.

Chapter 9
Second-Order RC Filters

9.1 Introduction

In previous chapters, we have analysed circuits using j notation. This works well, and we shall use it again. However, there are cases where a different approach is more fruitful. Second-order filters are a case in point where we shall start with two cascaded RC networks. It will be found that their response can be reduced to a fraction with a quadratic equation in the denominator and either 1, s or s^2 as the numerator. LTspice allows us to enter these directly in the simulation without having to draw the circuit first. This we shall explore and compare with the actual circuit.

It is also useful to draw the Bode plot and we shall see that, provided the roots of the equation are real, this can be done quite easily with sufficient accuracy for most purposes and we can extend the plots to three or four breakpoints.

From there we move on to notch filters where the analysis is different, in the case of the Bridged-T, because the numerator roots are complex, and in the case of the Twin-T because the response has a zero in the numerator. But first we need to understand the use of s notation for these circuits.

9.2 The Laplace 's' Function

LTspice offers this powerful option for the transfer function of arbitrary behaviour voltage or current sources B, voltage-controlled voltage sources E or voltage-controlled current sources G. It is invoked by using the **Value** field such as:

$$Laplace = s/(0.1 * s^2 + 1001.1 * s + 1e4)$$

and noting the circumflex ($^$) used for a power although two asterisks ($**$) are also accepted. The s is not case-sensitive so S is accepted. To find the frequency response,

© Springer Nature Switzerland AG 2020
C. May, *Passive Circuit Analysis with LTspice*®,
https://doi.org/10.1007/978-3-030-38304-6_9

LTspice uses $s = j2\pi f$; for the time response, LTspice uses the impulse response of the Fourier transform of the frequency response. There is no constraint on the form of the transfer function which we can write as a polynomial or as factors where both are illustrated in Eq. 9.1:

$$H(s) = \frac{4s^2 + 12s + 20}{(s+4)(s-5)} \tag{9.1}$$

9.2.1   Comparison of ω and s Transfer Functions

We should recall from Sect. 7.7 of Chap. 7 'Capacitors' that the transfer function of a simple circuit of a voltage source in series with a resistor and a capacitor with the output taken across the capacitor can be written as:

$$H(\omega) = \frac{1}{j\omega CR + 1} \tag{9.2}$$

And we know that this has a breakpoint at $\omega = \frac{1}{RC}$ because then Eq. 9.2 becomes:

$$H(\omega) = \frac{1}{j+1} \tag{9.3}$$

and the voltage gain is $-3\ dB$ and the phase angle is $45°$. If we now substitute $s = j\omega$, we have:

$$H(s) = \frac{1}{sCR + 1} \tag{9.4}$$

and when $s = j/(CR)$ Eq. 9.4 is identical to Eq. 9.3. So both forms yield the same breakpoint. And we also know that for Eq. 9.2 below this frequency the voltage gain is almost constant at $0\ dB$ and above it the voltage gain falls at the rate of $-20\ dB/$ octave, and therefore we shall find the same behaviour with Eq. 9.4.

   If we swap the resistor and capacitor and take the output across the resistor, the transfer function is:

$$H(\omega) = \frac{R}{R + \frac{1}{j\omega C}} \quad \text{or} \quad H(\omega) = \frac{j\omega CR}{1 + j\omega CR} \tag{9.5}$$

and again we have the same breakpoint but the frequency response is not the same. This time at low frequencies, we expect the voltage gain to be zero, whilst above the breakpoint, we can ignore the factor 1 in the denominator and the voltage gain is 1 or $0\ dB$. On writing $s = j\omega$, we have Eq. 9.6:

$$H(s) = \frac{sCR}{1 + sCR} \tag{9.6}$$

and if we make the same substitution $s = j/(CR)$, we find an identical frequency response.

9.2.2 Poles and Zeros

We have belaboured the point in the previous section because the attentive reader might have noticed that if we make $s = -4$ in Eq. 9.1, the denominator of $H(s)$ becomes infinite and this is called a *pole*. And if we have:

$$H(s) = \frac{s + 6}{10} \tag{9.7}$$

then if $s = -6$, the output will be 0 and, naturally, this is called a *zero*.

These, however, are something of snares and delusions, because we must remember that $s = \sigma + j\omega$ and for passive components $\sigma = 0$ and we only have $s = j\omega$, so for the zero, we require $s = -j6$, and likewise for the pole we require a frequency $s = -j4$, and these are negative frequencies which are hard to come by in the real world. So although the transfer function can theoretically become zero or infinite, with real passive components in real circuits, it does not happen. And because of the j we should note that the magnitudes of $s + a$ and $s - a$ are the same.

This is not to decry the use of poles and zeros nor negative frequencies: in the esoteric world of digital signal processing and control systems analysis they are vital. But for us lesser mortals, it is enough just to use the positive values of s to give us the breakpoints. We shall have a little more to say about this in Chap. 13 'Passive Filters'.

The very important point, though, is that contrary to expectations, a pole creates a decrease in the gain of $-20\ dB/decade$, not an increase. Conversely, a zero results in an increase of $20\ dB/decade$.

Example – Single Pole
We start with a simple CR circuit schematic ('Laplace RC.asc') which is nothing more than an AC voltage source in series with a resistor and a capacitor. Using Laplace we substitute $s = j\omega$ and it becomes:

$$H(s) = \frac{1}{sR1C1 + 1} \tag{9.8}$$

Inserting the numbers from the schematic, we have:

$$H(s) = \frac{1}{6s + 1} \tag{9.9}$$

giving a breakpoint at $s = 1/6$ rad/s or $f = 1/(12\pi) = 26.5$ mHz. It can be seen that the Laplace response coincides with that of the circuit itself and the gain falls above the breakpoint.

The schematic ('Laplace CR.asc') swaps the capacitor and resistor giving:

$$H(s) = \frac{sC1R1}{1 + sC1R1} \tag{9.10}$$

Inserting the values from the circuit, one form of the transfer function is:

$$H(s) = \frac{1}{1 + \frac{1}{6s}} \tag{9.11}$$

resulting in the same breakpoint and once more the two coincide. But this time the gain is rising at *20 dB/decade* before the breakpoint which subtracts -20 *dB/decade* from the slope so that it becomes horizontal.

Example – Single Zero

We may profitably ask what happens if the numerator involves s. A simple example is schematic ('Laplace Zero.asc') where we have:

$$H(s) = \frac{s + 5}{5} \tag{9.12}$$

and now if $s = -5$, the numerator is *0* and this point is a *zero*. But this does not mean that the output is zero; rather, that at a positive frequency of $f = 5/(2\pi) = 796$ *mHz*, we have a breakpoint but this time it is upwards. So now, contrary to expectations, a zero is an upward breakpoint with a slope of 20 dB/decade. The attentive reader might notice that there is no corresponding circuit: that is left as an exercise.

9.2.3   Types of Poles

Poles only originate in the denominator. We can write a general Laplace function as:

$$H(s) = \frac{f(s)}{s(s + a)} \tag{9.13}$$

where $f(s)$ is a constant value multiplied by a polynomial, or a sequence of factors, and is irrelevant for the following discussion.

When we later come to create a Bode plot, we need to consider two possibilities.

9.2.3.1 Pole = (s + a)

A Bode plot, perforce, cannot extend to zero frequency. However, we have seen above that if the expression includes a constant a then that defines a breakpoint at some frequency greater than zero. And we know that the low frequency plot is a horizontal line from the DC gain A_0. So to start the plot we draw a horizontal line from the frequency at the origin to the breakpoint, Fig. 9.1.

Example – Pole = s + a
If we take the schematic ('Laplace RC.asc'), the plot starts at *100 μHz*, and we found the breakpoint at *26.5 mHz*. We calculate the DC voltage gain to be *0 dB*, and so we draw a horizontal line of *0 dB* from the origin to the breakpoint and then a straight line downwards at *−20 dB/decade*.

9.2.3.2 Pole = s

This is essentially the same but without the constant a, the point being that there is no horizontal component before the breakpoint, and we are faced with a downward slope of *−20 dB/decade*, Fig. 9.2. There are now two options: the easiest is to find the next higher breakpoint *BP2* and draw a line from it to the frequency at the origin

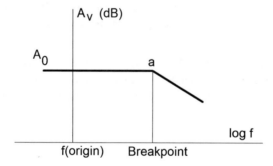

Fig. 9.1 Pole (s + a)

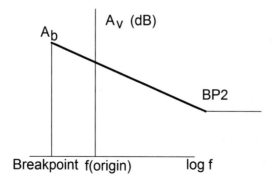

Fig. 9.2 Pole s

with an upward slope of *20 dB/decade.* The alternative is to calculate the voltage
gain at some convienient frequency point A_b such as 1 Hz by: $H(s) = 1/s$ or to find
the frequency for a gain of 0 dB.

Example – Pole = s
If we substitute *Laplace = 1/(s)* in schematic ('Laplace RC.asc') and use the starting
frequency of *100 μHz*, we find $1/s = 1/(2\pi10^{-4}) = 1591$, and this corresponds to a
gain of *64 dB,* which is what is measured.

9.2.4   Types of Zeros

This is the converse of the poles where we have a function of the form:

$$H(s) = \frac{s(s+a)}{f(s)} \qquad (9.14)$$

9.2.4.1   Zero = s + a

This predicts a breakpoint at $s = a$ with a constant gain at frequencies less than this
and an upward slope of *20 dB/decade* at higher frequencies.

Example – Zero = s + a
If we take the schematic ('Laplace Zero.asc'), this has *Laplace = (s + 5)/5,* and when
$s = 0$, the voltage gain is *0 dB* and the breakpoint is at $s = 5$ or $f = 5/(5\pi) = 796\ mHz$.
It is left for the reader to construct a physical circuit for this example.

9.2.4.2   Zero = s

This also exists in Eq. 9.13 and corresponds to an upward slope of *20 dB/decade*
with no breakpoint. We find the value at the origin in exactly the same way as above.

Example – Zero = s
The starting frequency is *1 mHz,* so we find $s = 2\pi10^{-3}$ whence $H(s) = 6.28 \times
10^{-3} = -44\ dB.$

9.2.5   Voltage Gain

For a passive circuit, this can never exceed *0 dB.* We can find the DC voltage gain by
setting $s = 0$. For any other frequency, we must reduce the expression to:

$$H(s) = \frac{A + Bs}{C + Ds} \qquad (9.15)$$

and remember that the s terms are in quadrature. Many times the numerator will just be a constant.

9.2.6 Laplace Numerator

These are simple equations with no denominator in the schematic ('Laplace Basic HP.asc.') This perhaps perverse description of 'numerator' is in preparation for the later section where we will have both numerator and denominator. Our purpose here is to predict the shape of the plot and calculate a few values in preparation for the Bode plot later on.

Example – Laplace(a + s)
The output $V(l1)$ is from the simple expression $E = Laplace(a)$ where a is a constant. This creates a constant value of a times the control voltage and does not depend upon frequency. In this case we have $Laplace = 10$ which multiplies the 1 V input by 10, and we see a horizontal line at 10 V or 20 dB for $V(l1)$.

The output $V(l2)$ with $Laplace = s$ means the output is $s = j2\pi f$ – the input AC voltage – and depends directly upon frequency and has an upward slope of 20 $dB/decade$ and a constant phase angle of $90°$.

The trace $V(l3)$ with $Laplace = (50 + s)$ is similar in concept to Eq. 9.5 with a denominator of 1. Therefore we find a horizontal section from the y-axis to the breakpoint at $s = -5$ above which it sits on top of $V(l2)$.

Example – Laplace(a∗s)
The traces $V(l4)$ with $Laplace = 5*s$ and $V(l5)$ with $Laplace = 10*s$ are straight lines with upward slopes of 20 $dB/decade$ similar to $V(2)$ and in agreement with $s = j2\pi f$.

If we take the left-hand side of the trace panel where $f = 1 \mu Hz$ and set the scale to *Logarithmic*, we find for $V(l4)$ that 5 $s = 31.4$ μV in agreement with $(5\ s) \times (2\pi) \times (1\ \mu Hz)$ and for $V(l5)$ at 1 μHz it is 63.3 μV, again in agreement. And the phase is a constant $90°$ because there is no real part to the expression, only the imaginary j. This is unlike a simple RC circuit where the magnitude and phase change with frequency but is totally in agreement with $s = j2\pi f$. However, adding a real constant b, $Laplace = a*s + b$ creates a breakpoint and phase shifts because we now have real and imaginary parts to the transfer function.

Example – Laplace(a∗s^n)
The traces $V(l6)$, $V(l7)$, $V(l8)$ and $V(l9)$ have powers of s where we can see that the slope increases by 20 $dB/decade$ and the phase by $90°$ for each power. Of especial note is $V(l8)$ with $Laplace = (50 + 10*s^3)$ where the graph is virtually horizontal at 34 dB or 50 V corresponding to the 50 in the Laplace. In point of fact, there is a

miniscule change of less than *0.01 dB* due to the cubic term which can be shown by using the cursor to measure from *1 mHz* to *100 mHz* where it will be found that the numbers agree with direct substitution for *s* in the Laplace. The point being that taking the cube of a small number less than *1* is an even smaller number that is insignificant compared to *50*. The value rises to *37 dB* at the upward breakpoint of *274 mHz*. At this point we have $50 = 10\,s^3$ from which $s = 1.71$ *rad/s* in close agreement with the frequency *274 mHz* $\times\,2\times\pi = 1.72$ *rad/s*. We must remember that $s^3 = (j\omega)^3 = j\omega^3$ and so is imaginary. Therefore the modulus is $\sqrt{(50^2 + 50^2)} = 70.7$. If we zoom in and change to a logarithmic y-axis, that is what we read. For higher frequencies the trace follows exactly that of *V(l7)*.

Trace *V(l9)* with **Laplace = (50 + 60∗s^3)** is similar to *V(l8)*, but the breakpoint is at *150 mHz*. This agrees with $50 = 60\,s^3$ and then $s = 0.941$ *r/s* which is also *150 mHz*. And because we have *s^3*, the slope is *60 dB/decade*, the same as *V(l8)*. So we see that a constant term controls the gain, whilst the coefficient of *s* controls the breakpoint, and the power of *s* controls the slope.

9.2.6.1   Laplace Denominator

For this we take the reciprocal of the previous in schematic ('Laplace Basic LP.asc').

Example – Laplace(1/(a + s))
This time for *V(l1)*, we have **Laplace = 1/10**, and this divides the control voltage by *1/10* so *V(l1)* = −20 *dB* or *0.1* V. In the case of *V(l2)*, we have **Laplace = 1/s**, and this divides the control voltage by *jω* leaving the control voltage itself as *0 dB* or *1 V*. The trace *V(l3)* falls on top of *V(l2)* above *20 Hz* because again **Laplace = 1/ (50 + s)** adds a breakpoint so its phase is not a constant *0°*. This is consistent with what we found before.

Example – Laplace(1/(a∗s))
The first example *V(l4)* is **Laplace = 1/(5∗s)** where we find the trace of *V(l4)* sloping downwards at −20 *dB/decade* and we find *0 dB = 31.8 mHz*. This agrees with $1/(5*s) = 1/(5 \times 2 \times \pi \times 31.8\ mHz) = \log_{10}(1) = 0$. The phase angle is a constant −90° because now *j* is in the denominator, equivalent to −*j* in the numerator.

For *V(l5)* we have **Laplace = 1/(10∗s)** which also has the same slope and phase shift as *V(l4)* but with a vertical shift of −6 *dB*. This time *0 dB* occurs at *15.9 mHz* or half the previous value in accord with the *6 dB* loss.

Example – Laplace(1/(a∗s^n))
These are almost mirror images of the numerators, again with slopes of *20 dB/decade* for each power of *s*. Plots *V(l8)* and *V(l9)* have a horizontal portion due to the constant *50*.

The first is *V(l6)* with **Laplace = 1/(s∗s)** which shows a slope of −40 *dB/decade* and a constant phase shift of *180°*. The *0 dB* point is at *159 mHz* which agrees with $1/s^2 = 1/(2 \times \pi \times 0.159)^2 = 0.999 = 0\ dB$.

The second, $V(l7)$, with **Laplace = 1/ (10∗s^3)** increases the slope to -60 dB/ decade and has 0 dB = 73.9 mHz. Inserting this value, we have $1/(10 \times (2 \times \pi \times 0.0739)^3) = 0.999 = 0$ dB as before.

The third, $V(l8)$, with **Laplace = 1/ (50 + 10∗s^3)** has the effect of introducing a breakpoint of some 270 mHz after which it follows the $V(l7)$ trace. Below that frequency the output is asymptotically -34 dB corresponding to the 50 in the Laplace. We can see this by moving the cursor to 1 mHz and reading 33.979 dB. This is the exact figure for 50. If we move to 100 mHz, the figure is $50 + 60 \times (0.1/ (2 \times \pi))3 = 33.98$ dB. The breakpoint is identical to the case where this Laplace function was in the denominator.

The final trace is V(l9) with **Laplace = 1/ (50 + 60∗s^3)**, and this has the same asymptotic horizontal -34 dB but moves the breakpoint back to 150 mHz, which agrees with $V(l9)$ when the same factor was in the numerator.

We might note in passing the enormous voltages that appear in these simulations.

9.2.6.2   Laplace Fractions

We divide these into three classes. They can be found at schematic ('Laplace Basic Fraction.asc'). Because we factorize the expression, there are no powers of s – only plain s. The constant terms differ by 1000 to give a clear separation between the breakpoints.

Example – Laplace(s/(a∗s))
The presence of the s term in the numerator means that there is no horizontal portion from the origin. The first trace, $V(l1)$, is **Laplace = s/10** and is simply s divided by 10, so we find an upward slope of 20 dB/decade and a constant phase angle of $90°$. We find 0 dB = 1.58 Hz and the voltage is 0.1 V, whereas previously, without the division by 10, it had been 1.0 V.

Trace $V(l2)$ with **Laplace = s/(5 + s)** has a pole at $s = -5$ rad/s and below that a slope of -20 dB/decade. The asymptotic high-frequency voltage gain is 1 or 0 dB. The trace $V(l3)$ has **Laplace = s/(50 + s)** which therefore shifts the breakpoint to a frequency ten times higher than before.

Trace **Laplace = s/(5∗s)** for $V(l4)$ has a constant voltage gain of -14 dB from the factor 5 and a phase angle of $0°$ because the expression simplifies to **Laplace = 1/5**.

Example – Laplace((a + s)/s)
The two traces $V(l5)$ and $V(l6)$ show an initial downward slope because of the pole s. We can recast $V(l5)$, which is **Laplace = (5∗s + 10)/(s)** as **Laplace = (5 + 10/s)**, and then it is clear that at low frequencies where $s \ll 1$, we can ignore the 5 and we are back to $1/s$ only this time multiplied by 10 so we expect a downward slope of -20 dB/decade. However, from the original expression, when $s = 2$ then $10/s = 5$ and this is at 318 mHz. At this point, again remembering that $s = j\omega$, the Laplace is $5 + j5$ and the modulus is 7.07 which we see on the simulation.

At higher frequencies, the gain rapidly tends to $5 = 13.9$ dB – which is what we find. We can see that by substituting 1 Hz in the Laplace (and remembering the s terms is imaginary) and then the gain is $5 + j1.59 = 5.24 = 14.4$ dB and with 10 Hz, it is $5+ j0.159 = 13.98\ dB$. The phase changes from $-90°$ at low frequencies to $0°$ at high frequencies.

Taking trace $V(l6)$ of **Laplace** $= (s + 1)*(s + 1000)/s$, we expand it to **Laplace** $= (s^2 + 1001s + 1000)/s = s + 1001 + 1000/s$. This is a little more complex. As before, at very low frequencies, we are left only with the final term $1000/s$, and the result can be hundreds of thousands. For example, at $1\ mHz$ we have $158\ kV$. By direct calculation at that frequency, we have $1000/s = 159\ kV$.

Compare this to $V(l5)$ and the gain is 100 times greater $= 40\ dB$, and this is what we see. Now as the frequency increases, from the original Laplace, the first breakpoint will be when $s = 1$ or $158\ mHz$ – which is what we find. As this is a zero, it is an upward breakpoint. Above that frequency, the contribution from $1000/s$ will rapidly decrease, so whilst $s << 1001$ the gain will be constant at $1001 = 60\ dB$. Eventually, the gain will increase starting at $s = 1001$ when the s term will predominate which we correctly measure at approximately $159\ Hz$, and this is another upward breakpoint. This time the phase shifts from the initial $-90°$ to $90°$.

Example – Laplace((s + a)/(s + b)(s + c))
The last two examples are essentially the inverse of the previous two. Taking $V(l7)$, this is **Laplace** $= s/((s + 1)(s + 1000))$. We expand this and divide by s to get **Laplace** $= 1/(s + 1001 + 1000/s)$. At low frequencies this tends to $s/1000$ which is a $20\ dB/decade$ slope up to $s = 1000 = 158\ mHz$ where there is a downward breakpoint and the trace becomes horizontal at a constant $1/1001 = -60\ dB$ until $s = 1$ then we have a second downward breakpoint.

On the other hand, $V(l8)$ is **Laplace** $= (s + 1)/((s + 1)(s + 1000))$ which divides to **Laplace** $= 1/(s + 1000)$, and at low frequencies this tends to a constant $1/1000 = -60\ dB$ until the upper breakpoint at $158\ Hz$ when $s = 1000$. This is effectively the same breakpoint for $V(l7)$ above which both fall at $-20\ dB/decade$. If we zoom in there is a difference between the two of $0.009\ dB$.

Now, armed with these examples, we can tackle some filters. But first we shall introduce the Two-Port Network which enables us to make additional useful measurements on the filters such as their input and output impedances. This can be important because these can change dramatically between their pass and stop bands.

Explorations 1
1. Notice that the shapes of the graphs is the same whether we have s or $-s$. Thus $Laplace = (5*s + 10)/(s)$ is the same as $Laplace = (5*s-10)/(s)$ except for the phase.
2. Create and test a few more examples.

9.3 Two-Port Networks

Many circuits can be reduced to two input terminals and two output terminals and hence are described as a two-port network, Fig. 9.3. The ports are electrically independent with no signal path between them, although often one terminal is

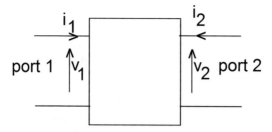

Fig. 9.3 Two-Port Network

common to both input and output and is usually the ground connection, but this is not a necessary condition. In effect, these are extensions of the Thevenin and Norton equivalent circuits.

Each port is described by an equation consisting of an impedance or conductance – which could be complex – with a current or voltage source controlled by the signal in the other port. This is the means by which a signal at port 1 is scaled and passed to port 2 and vice versa. The crucial point is that if a signal is applied to port 1, unless there is a voltage or current source in port 2 controlled by the voltage or current in port 1, there will be no signal in port 2, and conversely, if there is no voltage or current source in port 1 controlled by the voltage or current in port 2, there will be no reverse transfer of the signal to port 1. We must remember that these voltage and current sources are entirely fictitious and are merely a means of transferring a signal from one port to the other.

The network must be linear so non-linearities must be removed by finding the DC operating point and then taking the slopes of the characteristics at the operating point. This is what LTspice does. Otherwise there is no restriction on the composition or complexity of the network. The various analyses return measurements made on the network ports without knowledge of the internal structure of the network. Each analysis is different: the most useful here perhaps are the z-parameters but keep the others in mind.

In the schematic ('H-Parameter Example.asc'), the components are all resistive so we have ignored the phase differences of zero.

9.3.1 The '.net' Directive

We may productively make several measurements upon the network using the '.net' directive. This is a powerful tool for measuring many properties of the network and is applicable only to AC analysis. We here expand on the brief description in the previous chapter.

It is generally better in the simulations to change the y-axis to *linear* rather than the default *dB*

.net <source>
The is the simplest form. It is particularly useful if the circuit is terminated by a short-circuit. The source is used only as a reference point to make the measurement, and its AC value is ignored. The command will return $Z(in)$ and $Y(in)$ for the whole circuit.

Example – .net <source>
This is the circuit in schematic ('net Examples1.asc'). Only the input impedance Z *(in)* and admittance $Y(in)$ are available. The input impedance is *1.1 kΩ*, and, on careful inspection, a dotted trace can also be seen showing that the phase angle is *0°*. If we show *Yin(I1)*, we find *1/1.1 kΩ* also at *0°*. Swapping the current source for a voltage source returns the same answer. We can also change the AC setting to, say, *100 V* and get the same answer.

.net <Output Port><Source><Output Resistance>
If the output port is defined by a single node such as *V(2)*, then the ground *node 0* is implicitly used for the other node; otherwise we may supply two nodes as *V(2,7)*. This time more measurements are available to us which we shall discuss a little later.

Example – .net<output><source><Rout>
If we just use *.net V(out) I1* in schematic ('net Examples2.asc') without defining an output resistance, LTspice will apply the default of *1 Ω* across the terminals *out* and *0* so that it is in parallel with *R2*. Then if we make *R2 = 1 Ω*, we correctly find *Zin(I1)* is *100.5 Ω*. However, with the original *R2 = 1 kΩ*, the output voltage is *1 kV* showing that the default *1 Ω* is ignored during the AC analysis. But if we supply an output resistance, for instance, *.net V(out) I1 Rout = 1e20*, the default is no longer used and, in this case, the impedance is *1.1 kΩ*.

If we have a terminating voltage source, its *Rser* appears in parallel with the output resistance, schematic ('net Examples3.asc').

.net <Port><Source>
This measures the impedance seen from the output port, and for this we need at least both the port and the source: *.net V(out) V1* in which case a default *Rin* of *1 Ω* will be used. This can be overridden by an explicit value or, if there is a voltage source at the input, its *Rser*.

Example – .net<port><source>
This is schematic ('net Examples4.asc') where we find *Zout(V1) = 50.25 Ω* which is the parallel sum of *R2* and *R1* with the default *1 Ω*:

$$Zout = \frac{100 \times 101}{201} = 50.249\Omega$$

The easiest test is to make *Rser = 1 Ω*, and then *Zout* is again *50.249 Ω* showing it has overridden the default rather than being in parallel with it.

If we set *Rin = 100 Ω* and *Rser = 1e6 Ω*, we find *Zout = 66.667 Ω* showing that *Rser* is ignored.

9.3.2 H-Parameters

These are 'hybrid' parameters because their dimensions are not all the same and are widely used in bipolar transistor circuits, but not limited to them. We often, quite naturally, refer to the 'input' and 'output' ports, and, indeed, that is very commonly their functions. But this is not a necessary distinction, and we shall refer to them only as ports *1* and *2*. The ordering of the numbering is significant and is:

$$H(< \text{destination port} >< \text{source port} >)$$

so *H(21)* is a change in port 2 caused by port 1. The four parameters are described by two equations:

$$v_1 = H(11)i_1 + H(12)v_2 \tag{9.16}$$

This is the voltage in port *1* caused by the resistance *H(11)* and the current i_1 in that port, plus the voltage generated in port *1* by the voltage v_2 in port 2 and shown by a voltage source *H(12)*. This does not mean that there is an actual voltage source there that – if we were clever – we could cut out. It is simply a means of graphically representing the behaviour of the circuit. The second equation is:

$$i_2 = H(21)i_1 + H(22)v_2 \tag{9.17}$$

This follows a similar logic and is the current generated in port *2* by the current in port *1* plus the current drawn by the conductance *H(22)* in port 2.

From these two equations, we can construct an equivalent circuit, Fig. 9.4. It will be seen that the measurements require us to place AC open- or circuits across the ports. In practice, an AC short can be a large capacitor; in simulations, often we can just replace an existing resistor by a very small one such as $1 n\Omega$ or a very large one such as $1e100\Omega$ and never mind if these are not off-the-shelf components. But first we define the parameters.

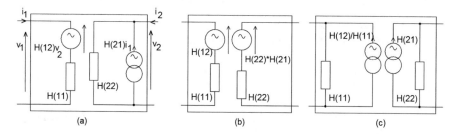

Fig. 9.4 H-Parameters

Port 1 Impedance H(11)
This is defined as the ratio of the current in port *1* to the voltage across port *1* so to measure it we need a voltage source across the input terminal

$$H(11) = \frac{v_1}{i_1} \quad (v_2 = 0) \tag{9.18}$$

with no voltage across the output terminals v_2. This is important because Eq. 9.16 shows that a voltage in port 2 will create a voltage in port *1* so, to avoid confusion, we make sure there is no voltage at port 2. If we are dealing with AC, we simply connect a large capacitor across port 2. The point here is that if we are making measurements on a circuit with a DC supply, it is not a good idea to short-circuit the output to ground. However, in the following example, we only have AC excitation so short-circuits are acceptable.

The parameter has the dimensions of impedance, but it often reduces to a simple resistance.

Reverse Voltage Transfer H(12)
A voltage applied to port 2 can produce a voltage at port *1*. The parameter is the ratio of the voltage generated across port *1* due to a voltage across port 2.:

$$H(12) = \frac{v_1}{v_2} \quad (i_1 = 0) \tag{9.19}$$

where we set the current in port *1* to zero to avoid complications from H(11). It is a dimensionless number. In effect, we are reversing the network by applying a signal at port 2 instead of port 1 and measuring the conditions in port 1.

Forward Current Transfer H(21)
This is the ratio of the current generated in port 2 by a current in port *1*. The definition is:

$$H(21) = \frac{i_2}{i_1} \quad (v_2 = 0) \tag{9.20}$$

with no signal voltage at port 2. This also is a dimensionless number.

Port 2 Conductance H(22)
This is ratio of the current in port 2 to the voltage across port 2:

$$H(22) = \frac{i_2}{v_2} \quad (i_1 = 0) \tag{9.21}$$

This has the dimensions of a conductance. To avoid interaction with H(21), we set the current in port *1* to zero.

9.3.2.1 H-Parameter Equivalent Circuit

We can create a lumped parameter equivalent circuit as Fig. 9.4. We can recognize this as a Thevenin port 1 and a Norton port *2* shown in Fig. 9.4a. But we may also convert it into both Thevenin ports (Fig. 9.4b) or both Norton ports (Fig. 9.4c) or a Norton input an Thevenin output.

This equivalent circuit is useful if the parameters do not depend upon frequency. If they do, either we need the parameters at a particular frequency or we need to create parameters which reflect the frequency response. Trying to do this from the traces is difficult; otherwise we need access to the internal circuit to create what could be quite complicated Thevenin and Norton equivalents.

The benefit of this and other equivalent circuit is that they enable us to make transient analyses on a much simpler circuit than the original where it is far easier to see the effects of source and load impedances. But it must always be remembered that these are strictly linear models and will not show distortion or clipping.

Example – H-Parameter Model
The middle circuit of schematic ('H-Parameter Model.asc') consists of four resistors which constitute the original circuit. From the h-parameter definitions, we need to short-circuit the output port 2 to measure *H(11)* and *H(21)*. However, that means there is no output node and we are restricted to *.net V1,* and as there are no h-parameter terms available, we can only measure *Zin(V1) = 2.2 kΩ* which agrees with the calculation of *(1 k + 600/500) = 2.2 kΩ.*

However, if we leave port 2 open-circuit with *R7 = 1e20* or something similar, we can use *.net V(out) V1 Rout = 1e12*, and now all the h-parameters are available and LTspice will apply short-circuits as necessary. The result is *H(11) 2.22 kΩ, H(21 = H(12) = 400 m* and *H(22) = 1/450μS = 2.22 kΩ.*

We can check these results by effectively short-circuiting port 2 by making *R7 = 1nΩ* and measuring the voltages and currents and we find *H(11) = V(in)/I (V1) = 2.22 kΩ* and *H(21) = I(R7)/I(V1) = 400 m.*

If we turn to the left-hand circuit, it is a good idea first to convert the previous *.net* statement from the middle circuit to a comment and uncomment the *.net* directive of this circuit to avoid a superfluity of parameters that can be plotted. Here we have a voltage source applied to the output, and we find *H(22) = I(V3)/V(out1) = 1/450 μΩ* and *H(12) = V(in1)/V(out1) = 400 m.* All of these values can be checked by direct circuit analysis and agree with the above results.

The h-parameter equivalent circuit is on the right. And it is worth noting that our entirely passive circuit has now acquired two (fictitious) generators *E1* and *F1.* *H(11)* is a resistor of *2.222 kΩ* in series with a voltage-controlled voltage source *E1* representing *H(12)* and having a value of *400 m.* The output port 2 consists of a resistor of *2.222 kΩ* representing *H(22)* rather than a conductance of *450 μS.* Finally, the current transfer *H(21)* is realized by a voltage-controlled current source *F1* with a gain of *400 m.* Using a transient run we can measure currents and voltage and their ratios to find two of the h-parameters. For the other two we replace *R13* in the right-hand circuit by a voltage source and remove V2 and compare the results with the left-hand circuit, we again find they are identical.

We may well ask what has been gained by replacing four resistors with a somewhat more complex circuit. The chief answer is that it is far easier to calculate the effects of changing the external load or the input source impedance. This circuit is deliberately simple to make it easy to compare the h-parameter results with direct calculation, but life becomes more difficult with extended circuits: please feel free to add a dozen more resistors and a handful of capacitors to prove this. We should also note that we can create the equivalent circuit from the prescribed measurements on an actual circuit without having access to its internal structure.

9.3.3 Z-Parameters

Unlike the h-parameters, these are all impedances. The equations are:

$$v_1 = Z(11)i_1 + Z(12)i_2 \qquad\qquad (9.22)$$

$$v_2 = Z(21)i_1 + Z(22)i_2 \qquad\qquad (9.23)$$

and we can again construct an equivalent circuit similar to the one for h-parameters, Fig. 9.5. We again make measurements on the ports but this time the other port is open-circuit.

Port 1 Impedance Z(11)
This is defined as the impedance measured at port *1* with zero current in port *2* to avoid contributions from Z(12):

$$Z(11) = \frac{v_1}{i_1}(i_2 = 0) \qquad\qquad (9.24)$$

It is also known as the *driving point impedance*.

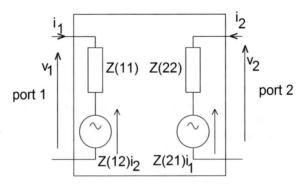

Fig. 9.5 Z-Parameters

Port 1 Transfer Impedance Z(12)
This is the impedance reflected from port 2 into port 1. It is measured with no current in port 1 to avoid contributions from Z(11):

$$Z(12) = \frac{v_1}{i_2} (i_1 = 0) \qquad (9.25)$$

Port 2 Transfer Impedance Z(21)
This is similarly defined as Z(21) and is the impedance of port 1 reflected into port 2. This time the current in port 2 is zero to avoid any effects from Z(12):

$$Z(21) = \frac{v_2}{i_1} (i_2 = 0) \qquad (9.26)$$

Port 2 Impedance Z(11)
This is defined by the current and voltage in port 2 with no current in port 1, again to avoid possible errors due Z(21):

$$Z(22) = \frac{v_2}{i_2} (i_1 = 0) \qquad (9.27)$$

and this also is a *driving point impedance*.

Example – Z parameters
The schematic ('Z Parameter Example.asc') uses the same component values as the ('H Parameter Example.asc') and *.net V(out) V2* which leaves the default values of *Rin = Rout = 1* Ω. Once again LTspice sets up the correct conditions for the measurements if we have an open-circuit output, and we find Z(11) = 2.55 kΩ, Z (12) = 889 Ω Z(21) = 889 Ω, and Z(22) = 2.222 kΩ, where Z(11) and Z(22) correctly have the same values as the h-parameter model.

To realize an equivalent circuit, we can use current sources so that H1 = Z(12)*I (out2) and H2 = Z(21)*I(in2) and with R7 = R8 = 1 kΩ shown in the right-hand circuit where we find the input and output currents for both circuits coincide with a transient run. There is a very slight difference in the output voltages with an AC run, about half a millivolt and is due to rounding errors in component values.

Z(in), Z(out)
These are not part of the model but measured across the terminals of the port with the specified terminating resistances. By default these are both *1* Ω. We can change them by adding to the *.net* statement **Rin = <value> Rout = <value>.**

Example – Z(in)
With the default *Rin = Rout = 1* Ω, we can ignore R7 in schematic ('Z-Parameter Example.asc'), and we find $Z(in) = 1k + \frac{(2k \times 3k)}{6k} = 2.2k\Omega$ and this is what we measure.

Example – Z(out)
Similarly with the default values looking back into the circuit from port 2, we have
$R1//R2$ and in series with $R3$ then:

$$Z(out) = \frac{(3.667k \times 4k)}{3k + 4k} = 1.91k\Omega$$

If we change the *.net* statement to: *.net V(out) V1 Rout = 1 k.*We now have
$R4/1\ k = 0.8\ k\Omega$ and so $Z(out) = 1k + \frac{2 \times 3.8k}{5.8k} = 2.31k\Omega$ which is what we measure.

9.3.4 Y-Parameters

These are the converse of the z-parameters and are admittances. The equations are:

$$i_1 = Y(11)v_1 + Y(12)v_2 \tag{9.28}$$

$$i_2 = Y(21)v_1 + Y(22)v_2 \tag{9.29}$$

and we can again construct an equivalent circuit, Fig. 9.6 using voltage-controlled
current sources. As these are admittances, they add in parallel, not series. We can
note that the parameters depend only on voltages and not upon the source or load
currents.

Port I Admittance Y(11)
This is defined as:

$$Y(11) = \frac{i_1}{v_1} \quad (v_2 = 0) \tag{9.30}$$

where again we stipulate that the port 2 voltage is zero to avoid a contribution from
$Y(12)$. In other words, this is the short-circuit port 1 admittance.

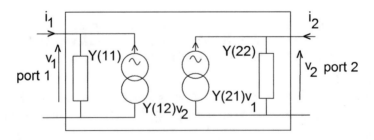

Fig. 9.6 Y-Parameters

Port 1 Transfer Admittance Y(12)

Now we have:

$$Y(12) = \frac{i_1}{v_2} \quad (v_1 = 0) \tag{9.31}$$

And if the port 1 voltage is zero, then Y(11) is zero so we measure Y(12) alone.

Port 2 Transfer Admittance Y(21)

This is the converse of the previous and is:

$$Y(21) = \frac{i_2}{v_1} \quad (v_2 = 0) \tag{9.32}$$

This time the port 2 voltage is zero so that Y(22) is also zero.

Port 2 Admittance Y(22)

This is:

$$Y(22) = \frac{I_2}{v_2} \quad (v_1 = 0) \tag{9.33}$$

And with no voltage at port 1, we simply have the conductance.

Example – Y-Parameters

We use schematic 'Y-Parameter Example' to model the original circuit used for the h- and z-parameter models. The resistors $R1$ and $R2$ are declared as parameters so that we can calculate their conductances. This time we use voltage-controlled current sources to model the transfers.

With $R7 = 1e100\ \Omega$, we measure the parameters as $Y(11) = 454.54\mu S$, which corresponds to a resistance of $2.2\ k\Omega$, and $Y(22) = 522.70\mu S$ which corresponds to $1.913\ k\Omega$. The other two reading are $Y(12) = 181.8\mu S$ and $Y(21) = 181.8\mu S$.

If we set *.net* measurements for both circuit, we find a very small difference between the parameters, typically much less than 1%, due to rounding errors in the component values.

If we make a transient run with a *1 V 1 kHz* input (remembering this is peak voltage so the RMS is *0.707* V), we find that the RMS input current is *276.7 μA* and the RMS output voltage is *245.96 mV*. These agree with the previous measurements. From this we find the input impedance is $0.707/276.7\ \mu = 2.55\ k\Omega$. If we set $Y(12) = Y(21) = 0$, the RMS input current rises to *454 μA*, and then the input impedance is *2.2 kΩ*. From this we see that the difference is due to the current from *G1*.

If we set $R3 = R7 = 1\ k\Omega$ and make a transient run, we again find the two input currents and the two output currents coincide and a minute difference in the voltages.

Y(in) and Y(out)

These follow the same pattern as Z(in) and Z(out) and suppose terminating admittances of 1 μS.

Example – Y(in) and Y(out)
The schematic ('Y-Parameter Example.asc') has $Y(in) = 454.51$ μ and Y *(out)* $= 522.67$ μ. These are very slight differences compared to $Y(11)$ and $Y(22)$ and are due to the default load and source admittances. If we add a $1e6$ Ω resistor in parallel with $R2$ (which is $Y(22)$), the current in port *1* increases to 276.81 μA because of the reduced port *2* voltage which has fallen from 245.96 mV to 245.49 mV.

9.3.5   Scattering Parameters

These are mainly of interest at RF and microwave frequencies. They are mentioned here for completeness. They do not use short-circuits and open-circuits to terminate the ports to determine the parameters but matched loads which are easier to realize at high frequencies. Also the equations are in terms of power rather than voltages and currents. The Wikipedia article offers a readable explanation starting at 'Two-Port S-Parameters'. The equations are:

$$b_1 = S(11)a_1 + S(12)a_2 \qquad (9.34)$$

$$b_2 = S(21)a_1 + S(22)a_2 \qquad (9.35)$$

which relate to the incident power a and the reflected power b. Figure 9.7 shows that the incident power may result in power transferred to port *2* or reflected from port *1* itself. And similarly for power incident upon port 2:
   We define voltage waves $a_1 = v_1^+$ $a_2 = v_2^+$ $b_1 = v_1^-$ $b_2 + v_2^-$

Port 1 Voltage Reflection Coefficient S(11)
If port *2* is terminated in the characteristic impedance of the system:

$$S(11) = \frac{b_1}{a_1} = \frac{v_1^-}{v_1^+} \qquad (9.36)$$

This is complex and can be shown on a polar diagram or a Smith chart.

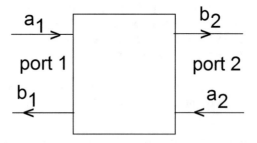

Fig. 9.7 S-Parameters

Reverse Voltage Gain S(12)
And also if port *2* is terminated by the characteristic impedance of the system:

$$S(12) = \frac{b_1}{a_2} = \frac{v_1^-}{v_2^+} \qquad (9.37)$$

Forward Voltage Gain S(21)
If port *1* is terminated by the characteristic impedance of the system:

$$S(21) = \frac{b_2}{a_1} = \frac{v_2^-}{v_1^+} \qquad (9.38)$$

This is the complex voltage gain and is the ratio of the reflected power wave over the incident power wave at port *1*. If both the impedances are identical, it will equal the voltage gain.

Port 2 Voltage Reflection Coefficient
Similarly if port *1* is terminated by the characteristic impedance of the system:

$$S(22) = \frac{b_2}{a_2} = \frac{v_2^-}{v_2^+} \qquad (9.39)$$

Example – S-Parameters
Measured on the schematic ('H-Parameters Example.asc'), we have *S(11) = 999.09 m, S(21) = S(12) = 363.3 μ S(22) = 998.95 m.*

Explorations 2
1. The schematic ('2-Port Network.asc') contains three trivial circuits. Measure them using all the parameter types.
2. Set *.net* analyses for both circuits in schematics ('H-Parameter Example.asc') and ('Z-Parameter Example.asc'), and compare the parameters of the original circuit and the parametrized circuit.
3. Check the response of all the circuits to different source and load resistances.
4. Change values in the schematic ('H-Parameters Example.asc'), and compare measured and calculated values.

9.4 Second-Order RC Cascade Networks

We saw that a single resistor in series with a capacitor could serve as a low-pass filter with an asymptotic slope of 20 dB per decade or a high-pass filter with the same asymptotic slope. So frequencies ten times higher or lower than the cut-off could still

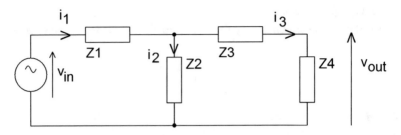

Fig. 9.8 Second Order RC Filter

make a considerable contribution to the output. One answer is to add a second RC circuit. As we can use the same configuration for low pass, high pass and band pass, we shall use a generalized circuit, Fig. 9.8.

9.4.1 General Analysis

We shall give the analysis in full, starting with the input loop:

$$v_{in} = i_1 Z_1 + i_2 Z_2 \qquad (9.40)$$

for the output we have:

$$v_{out} = i_3 Z_4 \qquad (9.41)$$

and also:

$$i_2 Z_2 = i_3 (Z_3 + Z_4) \qquad (9.42)$$

whence:

$$i_2 = i_3 \frac{(Z_3 + Z_4)}{Z_2} \qquad (9.43)$$

then with:

$$i_1 = i_2 + i_3 \qquad (9.44)$$

we have enough equations. Substituting in Eq. 9.40 for i_1:

$$v_{in} = i_2 (Z_1 + Z_2) + i_3 Z_1 \qquad (9.45)$$

Substituting for i_2 from Eq. 9.43:

$$v_{in} = i_3 \left[\frac{(Z_3 + Z_4)(Z_1 + Z_2) + Z_1 Z_2}{Z_2} \right] \tag{9.46}$$

then:

$$\frac{v_{out}}{v_{in}} = \frac{Z_2 Z_4}{Z_1 Z_3 + Z_2 Z_3 + Z_1 Z_4 + Z_2 Z_4 + Z_1 Z_2} \tag{9.47}$$

9.4.1.1 An Alternative Approach

In this particular case, we can remove $Z4$ and create a Thevenin equivalent of the rest. We find:

$$v_{Th} = v_{in} \frac{Z_2}{Z_1 + Z_2} \qquad Z_{Th} = Z_3 + \frac{X_1 Z_2}{Z_1 + Z_2} \qquad = \frac{Z_1 Z_3 + Z_2 Z_3 + Z_1 Z_2}{Z_1 + Z_2}$$

then:

$$v_{out} = v_{Th} \frac{Z_4}{Z_{Th} + Z_4}$$

$v_{out} = \frac{Z_2}{Z_1 + Z_2} \cdot \frac{Z_4}{Z_4 + \frac{Z_1 Z_3 + Z_2 Z_3 + Z_1 Z_2}{Z_1 + Z_2}}$ and this expands to Eq. 9.47.

It pays to be flexible in these matters.

9.4.1.2 The Transfer Function

The transfer function is simply the ratio of *output* over *input,* usually voltages, but can be currents, often denoted by *H*. The general *transfer function* (the ratio of output over input) for a second-order network is a quadratic equation of the form:

$$\frac{v_{out}}{v_{in}} = H(s) = \frac{as^2 + bs + c}{As^2 + Bs + C} \tag{9.48}$$

This is a *biquadratic filter.* LTspice, however, only allows us to find the DC transfer function so it is of no help here. We can enter the transfer function in any form – as factors, as a quadratic (or higher) equation or as a collection of terms such as **Laplace = s + 2∗s + 10 + s∗5** so our sole concern is in writing the transfer function in terms of *s*. It is more useful to write it as factors so Eq. 9.48 becomes:

$$\frac{v_{\text{out}}}{v_{\text{in}}} = H(s) = \frac{(s + z_1)(s + z_2)}{(s + p_1)(s + p_2)} \tag{9.49}$$

Now we can see that when $s = -z_1$ or $s = -z_2$, the output will be zero and if $s = -p_1$ or $s = -p_2$, the denominator will be zero and the function goes to infinity but with real practical passive components, this never happens. As the equation involves the square of s, it is a second-order system.

There are five classes of filter (excluding 'all-pass'). An ideal *low-pass* filter would only let frequencies up to f1 pass, Fig. 9.9, whilst a *band-pass filter* would only pass frequencies between *f2* and *f3* and a *high-pass filter* would only pass frequencies greater than *f4*. A *band-stop* filter passes all frequencies except those between *f2* and *f3*. There is also an extreme case – the notch filter – which ideally rejects just one frequency. In analogue circuit, these abrupt stop and start frequencies are impossible, and we have to accept a more or less gentle transition.

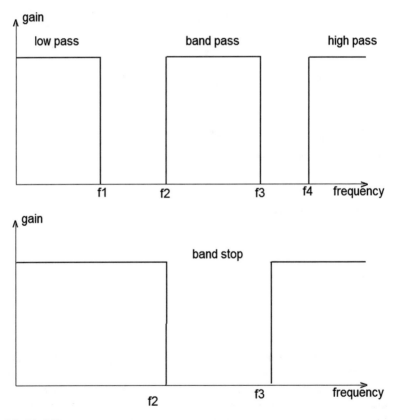

Fig. 9.9 Ideal filter

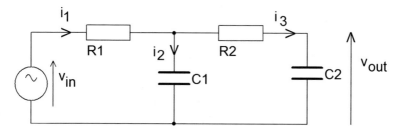

Fig. 9.10 Low-Pass Filter

9.4.2   Low-Pass Filter

We can now substitute component types from Fig. 9.10 in Eq. 9.50:

$$\frac{v_{out}}{v_{in}} = \frac{X_{C1}X_{C2}}{R_1R_2 + X_{C1}R_2 + R_1X_{C2} + X_{C2}X_{C1} + R_1X_{C1}} \qquad (9.50)$$

writing s for $j\omega$ and grouping terms yields:

$$\frac{v_{out}}{v_{in}} = \frac{\frac{1}{s^2C_1C_2}}{R_1R_2 + \frac{R_1}{sC_2} + \frac{R_1}{sC_1} + \frac{R_2}{sC_2} + \frac{1}{s^2C_1C_2}} \qquad (9.51)$$

multiplying by $s^2\,C1C2$ gives:

$$\frac{v_{out}}{v_{in}} = \frac{1}{s^2C_1C_2R_1R_2 + s(C_1R_1 + C_2R_1 + C_2R_2) + 1} \qquad (9.52)$$

we define:

$$\Pi = C_1C_2R_1R_2 \quad \Sigma = C_1R_1 + C_2R_1 + C_2R_2 \qquad (9.53)$$

and Σ is interesting because it omits the last possible term of *C1R2*. This is significant because if we interchange the resistors or capacitors, we will have a slightly different sum, and although the asymptotic values are the same, the shape of the curve between them will be different.

Rewriting Eq. 9.52 using Eq. 9.53 gives:

$$\frac{v_{out}}{v_{in}} = \frac{1}{s^2\Pi + s\Sigma + 1} \qquad (9.54)$$

To find the poles, we need to solve the quadratic equation of the denominator with just s^2 – not multiplied by Π, so now we divide the denominator of Eq. 9.20 by Π and it becomes:

$$s^2 + s\left(\frac{\Sigma}{\Pi}\right) + \frac{1}{\Pi} \tag{9.55}$$

Using the formula for the solution of a quadratic equation, the poles p_1 and p_2 are:

$$p_1, p_2 = \frac{-\left(\frac{\Sigma}{\Pi}\right) \pm \sqrt{\left(\left(\frac{\Sigma}{\Pi}\right)^2 - \frac{4}{\Pi}\right)}}{2} \tag{9.56}$$

which simplifies to:

$$p_1, p_2 = \frac{-1}{2\Pi}\left[\Sigma \pm \sqrt{(\Sigma^2 - 4\Pi)}\right] \tag{9.57}$$

and the transfer function is:

$$H(s) = \frac{1}{(s + p_1)(s + p_2)} \tag{9.58}$$

And the poles give us the breakpoints. Three possible situations arise:

- **The two poles are real and different** – which is the general case.
- **They are real and identical** – in which case the poles coincide and we find the −6 dB point, not the −3 dB point.
- **The poles are a complex conjugate pair** if the square root is negative.

So we must test Eq. 9.57:

$$\text{root test} = (C_1 R_1 + C_2 R_1 + C_2 R_2)^2 - 4C_1 C_2 R_1 R_2 \tag{9.59}$$

squaring and collecting terms:

$$(C_1 R_1)^2 + (C_2 R_1)^2 + (R_2 C_2)^2 + 2C_1 C_2 R_1^2 + 2C_2^2 R_1 R_2 - 2C_1 C_2 R_1 R_2 \tag{9.60}$$

and Expression 9.60 must be positive for real roots.

Example – Low-Pass Filter

This is not to design a filter but, given a set of values, to investigate the response. If we use $R1 = R2 = 10\ k\Omega$ $C1 = 1\ \mu F$ and $C2 = 10\ nF$, we find $\Sigma = 10^{-2} + 2 \times 10^{-4} = 1.02.10^{-2}$ and $\Pi = 10^{-6}$. If we approximate $\Sigma = 10^{-2}$, a swift and dirty analysis from Eq. 9.57 is:

$$p_1, p_2 = -5 \times 10^5 \left(10^{-2} \pm \sqrt{\left(10^{-4} - 4 \times 10^{-6}\right)}\right)$$

where we first ignore the *4Π* term of -4×10^{-6} as being much smaller than Σ^2 which gives one root of *1e4* and the other of zero. However, if we then reintroduce the $\sqrt{4 \times 10^{-6}}$, we find a second root at *100* which gives *(s + 100)(s + 1e4)*. A proper analysis is:

$$\frac{-1}{2.0\text{e-}6}\left[1.02\text{e-}4 \pm \sqrt{(1.02\text{e-}2)^2 - 4.0\text{e-}6}\right] = -5.0\text{e} + 5 \times [0.0102 \pm 0.010002]$$

and then the denominator is *(s + 99)(s + 10101)*. This is not a great difference, but now the Laplace falls exactly on the circuit response. However, *1/((s + 99) (s + 10101))* shows a low frequency loss of *−120 dB* because at DC we have *H (s) = 1/(99 × 10101)* and we need to add *1e6* to the numerator to correct this. This is the low-pass filter in ('Laplace LF Pass Example.asc').

It is interesting to compare the result with the trace of *LP2* where the capacitors have been interchanged. The section from *10 Hz* to *10 kHz* differs slightly.

9.4.2.1 Filter Design

The biggest problem is keeping track of the powers of ten. The analysis is conventional. We expand Eq. 9.58:

$$H(s) = \frac{1}{s^2 + s(p_1 + p_2) + p_1 p_2} \tag{9.61}$$

Then by comparing the denominator with Eq. 9.21:

$$\frac{\Sigma}{\Pi} = p_1 + p_2 \qquad \frac{1}{\Pi} = p_1 p_2 \tag{9.62}$$

Example – Low-Pass Filter Design

Suppose we want breakpoints at 20 rad/s and 2000 rad/s. From Eq. 9.62, we find $\Pi = 2.5 \times 10^{-5}$ and $\Sigma/\Pi = 2020$ from which $\Sigma = 5.05 \times 10^{-2}$. As we only have two values to find and four variables at our disposal, we can freely choose two of them. As close tolerance capacitors of non-standard values are more difficult to find and more expensive than resistors, let us try *C1* = *C2* = *1 μF*, then from Π, we find that the product of the resistors is *R1R2* = 2.5×10^7 *Ω*. And thus *R2* = 2.5×10^7 / *R1*.

We can now substitute for Σ to give $5.05 \times 10^{-4} = 10^{-6}$ *(2R1 + R2)*. On substituting for *R2* and solving, we find one solution that *R1* = *24.8 kΩ* and *R2* = *1.01 kΩ*. These are the values in ('Low-Pass Design Example.asc'). And note that the order of the resistors is important, swapping them gives a different response as we saw above. And there is a second solution to the quadratic equation for the resistors that can be tried.

Example – Low-Pass Filter with Identical Breakpoints

Suppose we want a filter with both breakpoints at the same frequency. We now have only one condition to satisfy which from Eq. 9.21 is:

$$2p = \frac{\Sigma}{\Pi} \quad p^2 = \frac{1}{\Pi} \tag{9.63}$$

We might naively suppose that we could make the CR products the same. Then $\Pi = (CR)^2$ from which $p = 1/CR$. However, in this case $\Sigma = 3CR$, then from Eq. 9.63. $2p = 3/CR$ or $p = 3/2CR$ and these are incompatible. The difficulty lies with Σ. If that can be made to approximate to $2\ RC$, we can proceed and we can do that by keeping the same time constants but with radically different values, for example, $R1 = R$, $R2 = 100R$, $C1 = C$ and $C2 = C/100$. Then following the order of Eq. 9.19, we have $\Sigma = (CR + 0.01CR + CR)$. This is the strategy used in the schematic ('TwoTau.asc'). This is equivalent to making the input impedance of the second RC circuit much higher than the output impedance of the first, so we can get an approximate answer by taking each CR product separately – which is the case here because the second resistor $R2$ is 100 times the first and the second capacitor only 1/100 of the first so it has little loading on the first section.

If, as here, we make the two time constants the same, then the two breakpoints will almost coincide, for example, ('TwoTau.asc'), where we use the transfer function:

$$H(s) = \frac{1}{(1 + sCR)^2} \tag{9.64}$$

If we substitute the component values from **LTspiceXVII->examples->Educational** ('TwoTau.asc'), we have $C1C2R1R2 = 10^{-6}$, $C1R1 = C2R2 = 10^{-3}$ and $C2R1 = 10^{-5}$, and on discarding $C2R1$ as contributing only 1% of Σ, we have the second implementation of the filter, output B, using **Laplace = 1/(1 + 0.001*s)**2.**

A plot of the circuit output A and the Laplace output B look the same but expand the plot, and a very slight difference can be seen because the $0.01RC$ term was ignored in Σ. Restore it and from Eq. 9.20 the Laplace is:

$$Laplace = \frac{1}{10^{-6}s^2 + 0.00201s + 1}$$

and the Laplace trace exactly fits over the circuit trace.

We can calculate the poles from Eq. 9.57, and we find the frequencies differ by about 4% because we had ignored the loading of the second stage, but, still, they are quite close together.

Explorations 3

1. Plot the frequency response of ('Low-Pass RC Filter.asc'). This has identical time constants. Measure the 6 dB frequency and compare it with the calculated one.

Compare this with the Laplace response, schematic ('Low-Pass RC Filter Laplace.asc'). The importance of the seemingly insignificant s∗0.0003 is only visible if we zoom in. Note the fall in input impedance above the breakpoint.

2. Multiply the values of C2 and R2 by 10 in ('Low-Pass RC Filter.asc'), and note the breakpoints no longer coincide. As they are a decade apart, to a first approximation, we can deal with them separately using the two time constants R1C1 and R2C2. Measure Vm to confirm this and check by calculation.
3. Interchange the capacitors in the low-pass filter in ('Laplace Example.asc'). And note that the response has changed because the product term is different.
4. Change the input to a square wave of 2 s duration, and measure the rise time and fall times. Compare these to the frequency response.
5. Increase the waveform to a prf of 1 kHz note that it takes some time for steady state to be reached.
6. Note any differences between the real circuit and the Laplace.
7. Repeat 4,5 and 6 with a sine wave input.
8. Design and test a few other filters using different values.

9.4.3 High-Pass Filter

We again cascade stages in Fig. 9.11 but with the resistors and capacitors interchanged, so now we substitute $C1 = Z1$ $R1 = Z2$ $C2 = Z3$ $R2 = Z4$ in Eq. 9.50 to find:

$$\frac{v_{out}}{v_{in}} = \frac{R_1 R_2}{\frac{1}{s^2 C_1 C_2} + \frac{R_1}{sC_2} + \frac{R_2}{sC_1} + R_1 R_2 + \frac{R_1}{sC_1}} \tag{9.65}$$

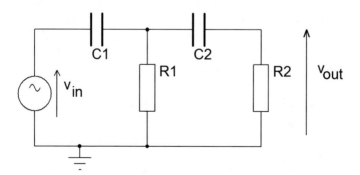

Fig. 9.11 High-Pass Filter

Multiply by s^2 $C1C2$ as before:

$$H(s) = \frac{s^2 R_1 R_2 C_1 C_2}{1 + s(R_1 C_1 + R_2 C_2 + R_1 C_2) + s^2 R_1 R_2 C_1 C_2} \qquad (9.66)$$

and dividing by Π and rearranging:

$$H(s) = \frac{s^2}{s^2 + s\left(\frac{\Sigma}{\Pi}\right) + \frac{1}{\Pi}} \qquad (9.67)$$

The only difference from the low-pass case is an s^2, so again if the square root is positive, the poles are real.

Example – High-Pass Filter
Suppose we make $C1 = C2 = 100\,nF\ R1 = R2 = 2k\Omega$ to give $\Sigma = 6\ 10^4\ \Pi = 4.10^{-8}$ and substitute in Eq. 9.67 to find:

$$H(s) = s^2 + s1.5.10^4 + 2.5.10^7 \qquad (9.68)$$

and we can insert this directly into the 'Laplace' statement, schematic ('High-Pass RC Filters.asc'). Or we can find the factors:

$$H(s) = \frac{s^2}{\left(s + 1.309.10^4\right)\left(s + 0.191.10^4\right)} \qquad (9.69)$$

and find the breakpoints at *2.08 kHz* and *304 Hz*. These are quite close together, so it is not easy to check the *304 Hz*, and the phase shift of *135°* is easier to see. And it is also interesting to note that although the time constants are identical, the breakpoints are not as we found with the low-pass case. And again we can find an approximate solution if the input impedance of the second stage is much larger than the output impedance of the first so that we can ignore its loading on the first stage. The low-frequency slope is *40 dB/decade* as we would expect. The AC responses of the two circuits agree within a few microvolts. The pulse responses both show a steady fall of ON and OFF levels from 0 V and 1V towards a DC level of 0 V, but the Laplace response does not have vertical transitions.

Explorations 3
1. Open and run ('High-Pass RC Filters.asc'). Note the *z-parameters* and identify them with the circuit analysis.
2. Change the values to $C1 = 1\ \mu F\ R2 = 200\ \Omega$ and test. Then interchange the resistor and capacitors and test.
3. Design and test filters for other breakpoints, identical or different.
4. Test using a square wave with prf = *5 kHz*. Notice that it takes some time for the output to settle and there is a considerable sag which gets worse at lower frequencies. Note how far the Laplace output diverges from the real circuit.

9.4.4 *Band-Pass Filter*

Cascading low-pass and high-pass sections gives a band-pass filter. We can construct these in a several ways using series or parallel sections. Three popular circuits now follow.

9.4.4.1 Cascaded High Pass and Low Pass

Cascading low-pass and high-pass stages gives a band-pass filter. This is shown in Fig. 9.12. If the capacitors and resistors are identical, it is almost a *notch filter* which passes one frequency, whereas if the breakpoints are well separated, we get a flat pass band where the attenuation is almost zero. It is typical of the traditional interstage coupling between circuits where resistor $R2$ could be the output resistance of a signal source or a previous stage in the circuit, and $R1$ represents the input resistance of the next stage. The capacitor $C2$ prevents any DC component from the signal source $V1$ being passed to the next stage, and capacitor $C1$ is either a physical capacitor to attenuate the high-frequency response or the input capacitance of the next stage.

As this is not the same topography as ('TwoTau.asc'), we can attack it as a complex potential divider:

$$\frac{v_{\text{out}}}{v_{\text{in}}} = \frac{Z_1}{Z_1 + Z_2} = \frac{\frac{R_1}{1+sC_1R_1}}{\frac{R_1}{1+sC_1R_1} + R_2 + \frac{1}{sC_2}}$$

$$= \frac{R_1}{R_1 + R_2(1 + sC_1R_1) + \frac{1}{sC_2}(1 + sC_1R_1)} \qquad (9.70)$$

If we divide by R_1 and multiply by s, we arrive at the Laplace form of:

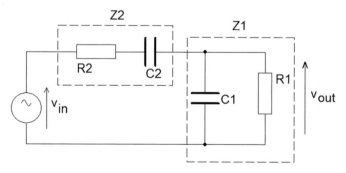

Fig. 9.12 Band-Pass Filter V1

$$Laplace = \frac{s}{s^2 C_1 R_1 + s\left(1 + \frac{R_2}{R_1} + \frac{C_1}{C_2}\right) + \frac{1}{C_2 R_1}} \qquad (9.71)$$

Example – Band-Pass Filter V1

If we insert the values from schematic ('Band-Pass Filter V1.asc') in Eq. 9.71, we find:

$$\textbf{Laplace} = s/(1e\text{-}5 * s * *2 + 2.01 * s + 1e3)$$

and this sits exactly on top of the filter output. The phase changes smoothly from $90°$ to $-90°$ with a loss of $-6.06\ dB$ at $1.592\ kHz$ where the phase is $0°$. At that frequency, the imaginary parts must cancel, so from Eq. 9.70, $s^2 = 1/C_1 C_2 R_1 R_2$ and $s = 10^4 = 2\pi f$ and $f = 1.591\ kHz$ in close enough agreement. However, we again find a divergence with a pulse excitation.

9.4.4.2   Cascaded Two 'L' Filters

Figure 9.13 is an alternative. If the capacitor values differ by at least two powers of ten for example with $C1 = 10C2$, we can simplify the analysis by regarding $C2$ as an open-circuit at low frequency, and with no external load, we are just left with $C1$:

$$v_{out} = \frac{R1}{R1 + \frac{1}{j\omega C1}} \qquad (9.72)$$

showing us that there will be a 3 dB point at $\omega = \frac{1}{R1C1}$.

Similarly, at high frequencies, we can consider $C1$ to be a short-circuit so R1 is a parallel load across the source and we have a potential divider:

$$v_{out} = \frac{\frac{1}{j\omega C2}}{R2 + \frac{1}{j\omega C2}} \qquad (9.73)$$

which reduces to:

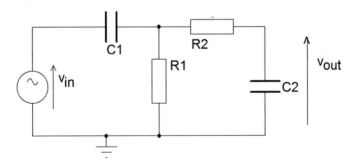

Fig. 9.13 Band-Pass Filter V2

$$v_{out} = \frac{1}{1 + j\omega C2 R2)} \tag{9.74}$$

and this will have a high frequency 3 dB point when $\omega = \frac{1}{C2 R2}$.

However, this has the same topography as Fig. 9.8, so in the general case, we can resort to Eq. 9.47 to find:

$$\frac{v_{out}}{v_{in}} = \frac{\frac{R_1}{sC_2}}{\frac{R_2}{sC_1} + R_1 R_2 + \frac{1}{s^2 C_1 C_2} + \frac{R_1}{sC_2} + \frac{R_1}{sC_1}} \tag{9.75}$$

We multiply by $s^2 C2/R1$:

$$H(s) = \frac{s}{\frac{sR_2 C_2}{R_1 C_1} + s^2 C_2 R_2 + \frac{1}{R_1 C_1} + s + \frac{sC_2}{C_1}} \tag{9.76}$$

collect terms and the result is the same format as Eq. 9.69 but with s in the numerator.

$$H(s) = \frac{s}{s^2 C_2 R_2 + s\left(1 + \frac{R_2 C_2}{R_1 C_1} + \frac{C_2}{C_1}\right) + \frac{1}{R_1 C_1}} \tag{9.77}$$

Example – Band-Pass Filter V2

Taking the numbers from schematic ('Band-Pass Filter V2.asc') which are $C1 = 1\ \mu F$, $C2 = 10\ nF$ and $R1 = R2 = 1\ k\Omega$ and applying them to Eq. 9.77, we have:

$$H(s) = \frac{s}{10^{-5} s^2 + s(1 + 0.02) + 10^3} \tag{9.78}$$

and we might note the large range of powers of ten. And we might also note that the *0.02* in the s term is significant. The measured flat gain is *0.172 dB*. By zooming the trace, *1.6 kHz* appears to be the minimum corresponding to $s = 10^4$. Then we must remember that $s^2 = -(2\pi f)^2$ with the minus sign because $s = j\omega$. Then Eq. 9.78 becomes:

$$H(s) = \frac{10^4}{-10^{-5} \times 10^8 + 1.02 \times 10^4 + 10^3} = \frac{10^4}{1.02 \times 10^4} = -0.1723 dB$$

and we are left only with the s terms in numerator and denominator and the js cancel, and at that frequency, the phase shift is zero as the trace shows. We can check the breakpoints by direct substitution of *154.88 Hz* and *16,454 Hz* in Eq. 9.78 or by using Σ and Π as before.

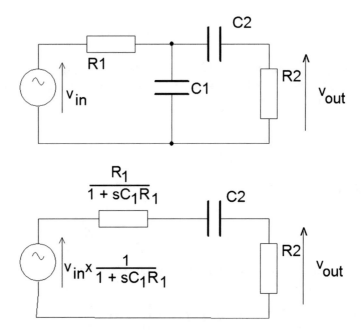

Fig. 9.14 Band-Pass Filter V3

9.4.4.3 Band-Pass Filter V3

This interchanges the capacitors and resistors of schematic ('Band-Pass Filter V2. asc'), Fig. 9.14, but the response is not symmetrical. We can again use Eq. 9.47, but, just for a change, we shall analyse it by turning R_1, C_1 and V_1 into their Thevenin equivalent, and then we have:

$$H(s) = \frac{R_2}{\left(\frac{R_1}{1+sC_1R_1} + \frac{1}{sC_2} + R_2\right)} \frac{1}{(1+sC_1R_1)} \tag{9.79}$$

and this expands to:

$$Laplace = \frac{s}{s^2 C_1 R_1 + s\left(\frac{R_1}{R_2} + 1 + \frac{R_1 C_1}{R_2 C_2}\right) + \frac{1}{C_2 R_2}} \tag{9.80}$$

which is very similar to Eq. 9.77.

Example – Band-Pass Filter V3
This is schematic ('Band-Pass Filter V3.asc'). Using the values from the schematic, we have:

$$\mathbf{Laplace = s/\left(0.1 * s^2 + 1001.1 * s + 1e4\right)}$$

and this agrees with the actual circuit. If, out of curiosity, we plot the difference between the traces, **V(p)-V(l2)**, we find the difference is more than $-300\ dB$ which is down at the noise level. It may be difficult to resolve this trace. A transient run with a *50 Hz* square wave shows spikes of just over *0.6 mV* on the leading and trailing edges and rounding of the waveform, but these differences are only seen at high magnification.

Explorations 4

1. Plot the response of the ('Band-Pass RC Filter V1.asc'). Compare this to the response of ('Band-Pass RC Filter V2.asc.') In particular, note the shape of the phase plot. Change the capacitors values and explore. Calculate the 3 dB points and compare them to your measured values.
2. Excite the circuit with a square wave of *50 kHz* prf. The outputs are heavily distorted but the Laplace settles to the true *+/− 180 mV* faster. Change the prf and measure the rise and fall times, and compare them to the frequency response.
3. Increase *R2* in ('Band-Pass RC Filter V1.asc') and *R1* in ('Band-Pass Filter RC V2.asc.') by a factor of 10. Note the changed response.
4. Design and test a band-pass filter to have 3 dB points at 10 Hz and 100 kHz.
5. Stimulate ('Band-Pass RC Filter V2.asc') with a 1 kHz square wave. Measure the rise time, and hence show that it agrees with the upper 3 dB point of 15 kHz. It will be found that the sag of the pulse is about 63% and so in close agreement with a lower 3 dB point of 160 Hz. Such heavy distortion would not be acceptable for good-quality sound. Reduce the lower 3 dB point to 16 Hz and repeat.
6. Repeat 4 with ('Band-Pass RC Filter V1.asc').
7. Interchange the series and parallel sections in Fig. 9.12 (Band-Pass Filter V1).
8. So far, we have assumed ideal capacitors: but on the **Capacitor** dialogue, when we click **Select Capacitor**, we shall see that they have series resistances of up to some 100 mΩ. This will affect their high-frequency response. Explore some high-pass filters to 100 MHz or more.
9. Repeat 6 adding ESL or capacitor sub-circuits or models from Murata and elsewhere.

9.5   LTspice "Laplace"

Having used this in a fairly simple manner where the numerator was either $1, s$ or s^2, we now need to explore it further. And it is worthwhile spending a little time understanding the basics of the Laplace Transform.

9.5.1   The Laplace Transform

This transforms the time domain to the frequency domain and is widely used in the solution of differential equations in many branches of science and engineering. The formal definition is:

$$F(s) = \int\limits_{t=0}^{\infty} f(t)e^{-st}dt \qquad (9.81)$$

where $F(s)$ is the Laplace Transform and s is a complex frequency, $s = \sigma + j\omega$ with σ in Nepers. So the transform is the integral of the signal multiplied by e^{-st} and is only applicable to positive time starting at zero. A concise explanation of the transform of some common functions can be found at https://web.stanford.edu/~boyd/ee102/laplace.pdf, and the videos at https://www.khanacademy.org/math/differential-equations/laplace-transform/modal/v/laplace-transform-1 from the first and onwards explain it carefully in detail. There are also many websites that go into this in great detail.

And (of course) having transformed everything and done the simulation, we afterwards need the reverse transform – which LTspice does before it presents the results. After we enter the Laplace equation, LTspice does it all for us – it transforms the input signal, processes the transfer function and does the inverse transform. So, apart from the esoteric delay introduced in ('TwoTau.asc') we can proceed with the usual input signals and not bother our heads any further.

9.5.2 Laplace Transform Examples

Just for practice, we shall try a few examples, including a time delay. It will be found that most of these are exercises in integration by parts, integrating exponentials and handling the exponential forms of trigonometrical functions.

9.5.2.1   Unit Step Input

At time zero the input rises instantaneously from zero to unity so $f(t) = 1$ if $t > 0$. The transform therefore is:

$$F(s) = \int\limits_{t=0}^{\infty} e^{-st}f(t)dt \qquad (9.82)$$

performing the integration we have:

$$F(s) = \frac{-1}{s} |e^{-st}| \Big|_{t=0}^{\infty} \qquad (9.83)$$

at $t = \infty$ the exponent is zero so the result is:

$$F(s) = \frac{1}{s} \tag{9.84}$$

9.5.2.2   Time Delay

Suppose we have a time-varying signal $f(t)$ and we delay it by D. The integration now starts at D, not zero, and the transform is:

$$G(t) = \int_{t=D}^{\infty} e^{-st} f(t-D) dt \tag{9.85}$$

writing τ for $t-D$, the integration now starts from zero and becomes:

$$G(t) = \int_{t=0}^{\infty} e^{-s(\tau+D)} f(\tau) d(t) \tag{9.86}$$

as e^{-sD} is constant we can take it out of the integration:

$$G(t) = e^{-sD} \int_{t=0}^{\infty} e^{-st} f(\tau) d(\tau) \tag{9.87}$$

The integration now is just the definition of the transform in Eq. 9.81 so finally:

$$G(t) = e^{-sD} F(s) \tag{9.88}$$

and multiplying the transformed function by e^{-sD} will delay it by time D.

9.5.2.3   Sine Wave

We write the wave in exponential form as:

$$F(t) = \int_{t=0}^{\infty} e^{-st} \frac{1}{2j} \left(e^{j\omega t} - e^{-j\omega t} \right) \tag{9.89}$$

Extracting the factor $1/2j$ and multiplying out the exponentials:

$$F(s) = \frac{1}{2j} \int_{t=0}^{\infty} (e^{(-s+j\omega)t} + e^{(-s+j\omega)t}) \qquad (9.90)$$

Performing the integration where the exponent evaluates to zero or *-1*:

$$F(s) = \frac{1}{2j} \left(\frac{1}{s - j\omega} + \frac{1}{s + j\omega} \right) \qquad (9.91)$$

Expanding:

$$F(s) = \frac{1}{2j} \left(\frac{(s + j\omega) + (s - j\omega)}{s^2 + \omega^2} \right) \qquad (9.92)$$

Finally as $\omega = s/j$, we have:

$$F(s) = \frac{\omega}{s^2 + \omega^2} \qquad (9.93)$$

9.5.2.4 A Decaying Exponential

The exponential is:

$$f(t) = Ae^{-\alpha t} \qquad (9.94)$$

and the transform is:

$$F(s) = \int_{t=0}^{\infty} Ae^{-\alpha t} e^{-st} dt \qquad (9.95)$$

Collecting terms:

$$F(s) = A \int_{t=0}^{\infty} e^{-(\alpha+s)t} dt \qquad (9.96)$$

$$F(s) = \frac{A}{s + \alpha} \qquad (9.97)$$

But let it be said again that LTspice will perform the transform of a signal and the inverse transform to recover the waveform.

9.6 Sketching the Bode Plot

In some ways this is totally fatuous since in a matter of 5 minutes, we can create the schematic and see the accurate results. However, in the unlikely event of not having a computer to hand, we shall see how to do it.

We have covered much of the ground earlier in the introduction where we saw how poles and zeros created breakpoints. Also, we have drawn Bode plots before in the simple case of RC filters. We can extend it to cover many circuits whose response can be expressed as a fraction with both numerator and denominator as products of factors like the second-order case of Eq. 9.49 which we repeat here as Eq. 9.98:

$$\frac{v_{out}}{v_{in}} = H(s) = k \frac{(s + z_1)(s + z_2)}{(s + p_1)(s + p_2)} \tag{9.98}$$

where k is a possible multiplier. And we should note that the AC setting is passed to k, so it is a good idea to set the input level to 1 V AC in the simulations: if we set it, for example, to $5\ V$, we will find the output is $14\ dB$ too high.

If numerator or denominator is given as quadratic or higher equations $as^2 + bs + c$, we write it as factors.

There are then three approaches to drawing the plot; one, explained by Swarthmore College, *https://lpsa.swarthmore.edu › Bode › BodeExamples*, is to plot each component and then add them to get the final plot; the second is to draw the plot section by section from the left, which is what we shall do; the third is to calculate the output at each pole and zero, mark it on the plot and join them with straight lines. This last is the most accurate but takes a lot longer.

9.6.1 Magnitude

We handle this in stages, first by finding the initial value.

9.6.1.1 Initial Value

If we start by setting $s = 0$, this will give us the DC gain. Unfortunately, the Bode plot uses a logarithmic x-axis for frequency so it never reaches zero. Instead we calculate the voltage gain in dBs at the starting frequency and mark that on the left-hand y-axis.

Example – Bode Plot Initial Value
To take three examples:

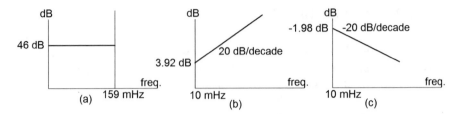

Fig. 9.15 Bode Initial

$$(a) \quad H(s) = \frac{(s+5000)(s+20)}{(s+1)(s+500)} \quad (b) \quad H(s) = \frac{s(s+500)}{(s+20)} \quad (c) \quad H(s)$$
$$= \frac{s+50}{s(s+1000)}$$

We can find the DC voltage gain by setting $s = 0$, then *(a)* has a DC gain of $20 \log_{10} (1e6/500) = 46 \ dB$, for *(b)* it is zero and for *(c)* it is infinite.

Suppose we start the plot at $f = 10 \ mHz$ or $0.0628 \ rad/s$. Then for *(a)* we still have *46 dB* because the plot is horizontal. A quick estimate for *(b)* is to discount s in the brackets as being very small compared to the constants, and then we have $H(s) = 2\pi \times 0.01 \times 25 = 3.92 \ dB$, and doing the same for (c), we find $-1.98 \ dB$. These differ by less than 1% from the measured values of schematic ('Bode Examples abc.asc') and are shown in Fig. 9.15.

9.6.1.2 Remaining Values Using Slopes

We now find the next higher pole or zero which is the next smallest number added to s and draw a vertical line at that frequency. We now draw a line from the initial value with the correct slope to meet the new vertical. If the next breakpoint is a zero, there is an upward change of slope of *20 dB/decade*; if it is a pole, there is a downward change of slope of $-20 \ dB/decade$. In general the change of slope is $20 \times n$ *dB/decade* where n is the number of coincident breakpoints and the sign is determined by whether it is a pole or zero.

Example – Plotting Values Using Slopes

We take the poles and zeros in ascending order of the added constant. Example *(a)* is Fig. 9.16 with the frequencies inserted. It has a pole at $s = -1$ which is *159 mHz*, point *(1)*. As we have just seen, the voltage gain here is virtually the same as the DC value. However, this pole results in a downward breakpoint until $s = -20$ at *3.18 Hz*, point *(2)*, when it is cancelled by the zero and the plot is horizontal until the pole at $s = -500$ at *79.6 Hz (3)* then it breaks downwards again, and finally becomes horizontal because of the zero at $s = -5000$ which is *796 Hz (4)*. Note that only the starting voltage gain is calculated; the remaining values are found by the intersection of the slopes with the uprights.

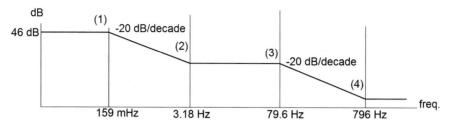

Fig. 9.16 Bode Example(a)

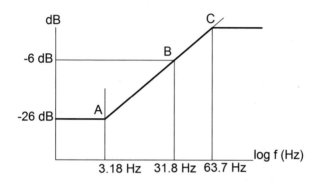

Fig. 9.17 Bode Example 1

Example *(b)* starts with an upward slope of *20 dB/decade*, but the pole at $s = -20$ reduces it by the same amount so the plot is horizontal until the zero at $s = -500$ when it breaks upwards again.

Example *(c)* starts with a downward slope which is cancelled by the upward breakpoint of the zero at $s = -50$, so its plot is horizontal until the pole at $s = -1000$ when slopes downwards at -20 *dB/decade*.

This method creates line segment plots. The true plot of these examples is available as schematic ('Bode Example abc.asc').

Examples – Line Segment Bode Magnitude Plots

1. Laplace $= (s + 20)/(s + 400)$: $A_0 = -26$ dB. We draw a horizontal line at $A_0 = -26$ dB as far as the pole at *4.18 Hz* (20 rad/s – point A), and then it slopes upwards at 20 dB/decade. We can fix the slope by a second point at A_0 by adding 20 dB and a frequency ten times greater, *318 Hz*,(point B) which is *(−6 dB,200 rad/s)*. Then at *63.7 Hz* we draw a horizontal line (point C). This is Fig. 9.17.

2. Laplace $= 10*(s + 300)*(s + 600)/(s + 20)$; $A_0 = 99$ dB. This starts with a horizontal line up to *3.18 Hz*, and then it falls at -20 dB/decade until *47.7 Hz* when we add 20 dB/decade to give a horizontal line. At *95.5 Hz*, we find a second zero so now the line slopes upwards at 20 dB/decade. Figure 9.18.

3. Laplace $= (s + 10)/(s + 400)^2$; $A_0 = -84$ dB. The plot is horizontal to *1.59 Hz* and then slopes upwards at 20 dB/decade to *63.7 Hz* when there are two zeros giving a total of -20 dB/decade downward slope. This is Fig. 9.19.

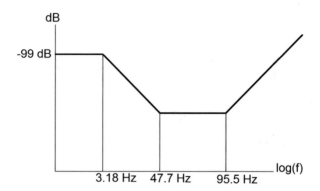

Fig. 9.18 Bode Example 2

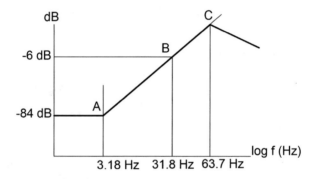

Fig. 9.19 Bode Example 3

These have all been drawn as line segments. And in many cases, this is sufficient, and it is quicker to use LTspice to draw the smooth curve. However, if we want to practice, here is a fanciful transfer function as an example:

Example – Smooth Bode Magnitude Plot
We sketch the Bode plot of:

$$Laplace = 1.6 * (s + 1)^2 * (s + 1e5) / \left((s + 400)^2 \right)$$

where we have added $A_i = 1.6$. This is Fig. 9.20. First we find the low-frequency asymptote as 0 dB and draw a horizontal line to 1 rad/s. Here is a double zero, so we draw a line upwards at 40 dB/decade to meet the line at 400 rad/s. This is point A where we encounter the double pole so now the line becomes horizontal again until 1e5 rad/s where there is another zero and the upward slope is 20 dB/decade. This is drawn as line segments, and note that we cannot predict the amplitude apart from the start as we only have the breakpoint frequencies. But we can always calculate it.

We can smooth the plot by joining the point at one tenth of the breakpoint frequency and ten times the breakpoint frequency and passing through the point $n3$ dB where n is the number of coincident breakpoints. In fact, it is often difficult to sketch accurately if the breakpoints are at all close together. This is schematic ('Laplace Test Jig.asc').

Explorations 5

1. Plot the magnitude of Laplace = s*(s + 2000)/(s + 40); $A_0 = 34\ dB$
2. Plot the magnitude of Laplace = (s + 10)*(s + 2000)/((s + 40)*(s + 400)); $A_0 = 1.9\ dB$
3. Plot the magnitude of Laplace =50*(s + 200)/((s + 4)*(s + 400)); $A_0 = 16\ dB$

9.6.1.3 Plotting the Phase

With a single breakpoint, we know that the relative phase will start at 0° and, at the 3 dB point, it will be +45° for a pole and − 45° for a zero. The asymptotic phase will be +90° for a pole and − 90° for a zero. These will double for two identical poles or zeros. So in this case of Fig. 9.20 the phase starts at 0°, Fig. 9.21, and will be 90° at

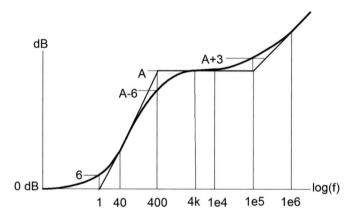

Fig. 9.20 Bode Plot Example Smoothed

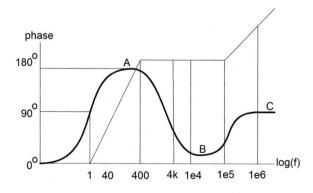

Fig. 9.21 Bode Example Smoothed Phase

1 rad/s tending to 180° at 400 rad/s at point *A*. Unfortunately, this conflicts with the zero at 400 rad/s which will cause negative phase shifts, so in actual fact the phase at *A* is only 170°. This time the phase should return to zero at *B* because of the two coincidental zeros but, because of the conflict with the pole at 1e4 rad/s, it only falls to 10°. The final phase shift of 90° at *C* is correct and can be deduced by adding all the asymptotic phase shifts.

Explorations 6
1. Run ('TwoTau.asc') and compare the first two results. Increase the numerator to 2 and notice the amplitude doubles. Try other numbers.
2. Use ('Laplace test Jig.asc') to run some of the examples, (you can copy from Explorations 5 and paste in **Enter new value for E1**) and make up more of your own.
3. It is also worthwhile and instructive to create a few circuits, derive their transfer functions and simulate them as circuits and as Laplace functions.

9.7 Band-Stop Filters

From the exploration of Bode plots, we know that zeros mean an upward breakpoint and poles mean a downward one. Thus, with a careful choice of components, we should be able to get the inverse of a band-pass filter.

9.7.1 A Simple Band-Stop Filter

This will have the form of Eq. 9.49:

$$\frac{v_{out}}{v_{in}} = H(s) = \frac{(s + z_1)(s + z_2)}{(s + p_1)(s + p_2)} \tag{9.99}$$

where if $z_1 z_2 = p_1 p_2$, the low-pass and high-pass products are the same giving unity gain below and above the stop band.

Example – Band-Stop Filter
One possibility is **Laplace = (s + 10)*(s + 5000)/((s + 0.1)*(s + 5e5))** which has $A_0 = 10*5000/(0.1*5.10^5) = 1$ and so the graph starts horizontally at 0 dB. There is a pole at 0.1 rad/s making the graph slope down at 20 dB/decade until 10 rad/s when there is a zero creating an upward breakpoint cancelling the previous zero and the graph is horizontal. At 5 krad/s there is another zero and the graph slopes upwards again until the final pole at 5.10^5 where the graph again levels out, ('Laplace Bandstop Filter.asc'). If we multiply by a constant, say by 10, we shift the whole

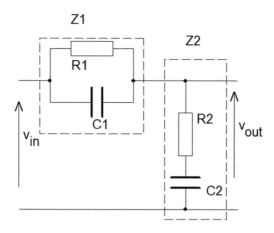

Fig. 9.22 Band-Stop Filter

graph upwards, in this case by 26 dB $= (20 \times \log(20))$. If we do not make the pole products equal to the zero products, the low-frequency gain will not equal the high-frequency gain.

How to build the filter is another matter. If we start with the Laplace polynomial, we see that we need a quadratic expression in both numerator and denominator. Now so far, with high-, low- and band-pass filters, the highest numerator has been s^2, so we can be sure that simple linear RC combinations will not work. On the other hand, a circuit like a complex potential divider has factors of the form *(sa + b)* in both numerator and denominator, so, by careful choice of components, when we multiply out, we will have a quadratic in the numerator. We can be guided in this by supposing that we have an upper arm that is essentially a high-pass filter and a lower arm that is a low-pass filter and the interaction between the two will create the band-stop characteristic. Thus we start hopefully with Fig. 9.22.

We can immediately set the ratio of two parameters by noting that in the stop band, C_2 will effectively be a short-circuit and C_1 an open-circuit so the attenuation is given by $A = R_2 /(R_1 + R_2)$.

Our starting point is an upper parallel arm Z_1 and a series lower arm Z_2.

$$Z_1 = \frac{R_1 \frac{1}{sC_1}}{\frac{1}{sC_1} + R_1} \Rightarrow Z_1 = \frac{R_1}{1 + sC_1 R_1} \tag{9.100}$$

$$Z_2 = \frac{1}{sC_2} + R_2 \Rightarrow Z_2 = \frac{1 + sC_2 R_2}{sC_2} \tag{9.101}$$

For the potential divided we have:

$$H(s) = \frac{Z_2}{Z_1 + Z_2} \tag{9.102}$$

$$H(s) = \frac{\frac{1+sC_2R_2}{sC_2}}{\frac{R_1}{1+sC_1R_1} + \frac{1+sC_2R_2}{sC_2}} = \frac{(1 + sC_1R_1)(1 + sC_2R_2)}{sC_2R_1 + (1 + sC_1R_1)(1 + sC_2R_2)} \qquad (9.103)$$

Equation 9.103 is not in standard form, but it shows that we have two zeros at:

$$f_1 = \frac{1}{2\pi C_1 R_1} \quad f_2 = \frac{1}{2\pi C_2 R_2} \qquad (9.104)$$

Equation 9.103 also shows that, unlike the example above, the poles and zeros are not independent and, if it were not for the factor $sC_2\,R_1$, would be coincidental.

Example – Band-Stop Filter
The schematic ('Band-Stop Filter.asc.') uses $R_1 = 99\ k\Omega$, $C_1 = 100\ pF$, $R_2 = 1\ k\Omega$ and $C_2 = 50\ \mu F$ to give a stop band attenuation of -40 dB. The zeros calculated from Eq. 9.104 are $1.6\ kHz$ and $16\ kHz$ which agree with measurements using the cursor. Changing the capacitors slides the left or right sections of the graph horizontally but not the slope.

9.7.2 The Bridged-T

As originally created in 1940 by W. N. Tuttle, this was a very sharp notch filter designed to reject one specific frequency. However, by a judicious – or injudicious – choice of components, it can be made to preform in a surprising number of ways. We shall first make a general analysis making no prescription about component types or circuit behaviour.

9.7.2.1 General Analysis

This can be found at https://scholarworks.montana.edu/xmlui/bitstream/handle/1/4951/3176210014.261.pdf;sequence=1 being a very ancient thesis by N Choudhury. The paper by Sulzer https://tf.nist.gov/general/pdf/2526.pdf is more concise.

We start on the assumption that the input is a low impedance source and the output is an open-circuit. We also will use general impedances for the elements, leaving the choice of resistor, capacitor or inductor to later.

There are a number of ways of attacking the circuit drawn in Fig. 9.23. One is to note that, going from the input to node A, impedances 2 and 4 in series are in parallel with impedance 1. As the output is open-circuit, the input current must flow through $Z3$. This leads to:

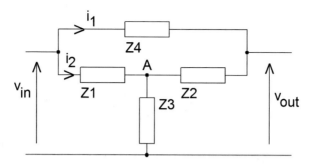

Fig. 9.23 Bridged-T

$$i_1(Z_4 + Z_2) = i_2 Z_1 \tag{9.105}$$

whence:

$$i_1 = i_2 \frac{Z_1}{Z_2 + Z_4} \tag{9.106}$$

We also have that:

$$v_{in} = i_2 Z_1 + (i_1 + i_2) Z_3 \tag{9.107}$$

and:

$$v_{out} = i_1 Z_2 + (i_1 + i_2) Z_3 \tag{9.108}$$

substituting for i_1 in Eq. 9.107:

$$v_{in} = i_2 \left((Z_1 + Z_3) + \frac{Z_1 Z_3}{Z_2 + Z_4} \right) \tag{9.109}$$

or:

$$v_{in} = i_2 \frac{Z_1 Z_2 + Z_1 Z_4 + Z_2 Z_3 + Z_3 Z_4 + Z_1 Z_3}{Z_2 + Z_4} \tag{9.110}$$

substituting for i_1 in Eq. 9.108:

$$v_{out} = i_2 \left(\frac{Z_1}{Z_2 + Z_4} (Z_2 + Z_3) + Z_3 \right) \tag{9.111}$$

or:

$$v_{\text{out}} = i_2 \left(\frac{Z_1 Z_2 + Z_1 Z_3 + Z_2 Z_3 + Z_3 Z_4}{Z_2 + Z_4} \right) \qquad (9.112)$$

finally:

$$\frac{v_{\text{out}}}{v_{\text{in}}} = \frac{Z_1 Z_2 + Z_1 Z_3 + Z_2 Z_3 + Z_3 Z_4}{Z_1 Z_2 + Z_1 Z_3 + Z_2 Z_3 + Z_3 Z_4 + Z_1 Z_4} \qquad (9.113)$$

It may be noticed that the difference between numerator and denominator is the inclusion of the term *Z1Z4* in the denominator.

This is the analysis making no assumptions about component types or values. N. Choudhury used resistors, capacitors and inductors, so we are free to use whatever components we fancy and continue the analysis to Eq. 9.114 where we insert the numbers in the LTspice Laplace.

9.7.2.2 Notch Filter ω_0

One version of this circuit uses capacitors for *Z1* and *Z2* and resistors for *Z3* and *Z4* in Eq. 9.113. (A second version is to interchange them.) And for a notch filter, Sulzer uses *C1* = *C2* = *C* and *R3* = *R/a*, *R4* = *aR*. Making these substitutions, Eq. 9.113 becomes:

$$H(s) = \frac{\frac{1}{sC}\frac{1}{sC} + \frac{R}{asC} + \frac{R}{asC} + R^2}{\frac{1}{sC}\frac{1}{sC} + \frac{R}{asC} + \frac{R}{asC} + R^2 + \frac{aR}{sC}} \qquad (9.114)$$

Collecting terms:

$$H(s) = \frac{\frac{1}{s^2 C^2} + \frac{2R}{asC} + R^2}{\frac{1}{s^2 C^2} + \left(\frac{2}{a} + a\right)\frac{R}{sC} + R^2} \qquad (9.115)$$

Now we can turn this into the standard form by multiplying by s^2/R^2:

$$H(s) = \frac{s^2 + \frac{s2}{aRC} + \frac{1}{R^2 C^2}}{s^2 + s\left(\frac{2}{a} + a\right)\frac{1}{RC} + \frac{1}{R^2 C^2}} \qquad (9.116)$$

We then test to see if the roots of the numerator are positive by examining the (b^2 − 4 *ac*) factor for solving the quadratic equation which is:

$$\frac{4}{(aRC)^2} - \frac{4}{(RC)^2} \qquad (9.117)$$

and as for any useful notch we have $a > 1$ the square root of the numerator is negative and there are no real zeros and we cannot draw a Bode plot even though the roots of the denominator are positive. Instead, we return to Eq. 9.115 and multiply throughout by sC/R, and then it becomes:

$$H(s) = \frac{\frac{1}{sRC} + \frac{2}{a} + sCR}{\frac{1}{sRC} + \left(\frac{2}{a} + a\right) + sCR} \qquad (9.118)$$

and substituting $\omega_0 = 1/CR$ whence $s = j/RC$ Eq. 9.118 is:

$$H(s)_{min} = \frac{-j\omega_0 + \frac{2}{a} + j\omega_0}{-j\omega_0 + \frac{2}{a} + a - j\omega_0} = \frac{\frac{2}{a}}{\left(\frac{2}{a} + a\right)} \qquad (9.119)$$

which is the minimum transfer function and does not involve j so there is no phase shift.

We note in passing that in general:

$$\omega_0 = \frac{1}{\sqrt{R_1 R_2 C_1 C_2}} \qquad (9.120)$$

but in this case, with this choice of resistor ratios, the frequency does not change with a.

9.7.2.3 Notch Depth

In this particular case, it is delightfully simple. We have it from Eq. 9.119:

$$H(s)_{min} = \frac{\frac{2}{a}}{\frac{2}{a} + a} = \frac{2}{2 + a^2} \qquad (9.121)$$

and if $a \gg 1$ Eq. 9.121 approximates to:

$$H(s)_{min} \approx \frac{1}{a^2} \qquad (9.122)$$

And the notch increases sharply with the ratio of the resistors. We should also note that as the frequency tends to infinity, the sCR term predominates in Eq. 9.118 and $H(s) = 1$ or 0 dB. Similarly, as the frequency tends to zero, the $1/sCR$ term predominates and again $H(s) = 1$. So the notch depth is absolute, being relative to 0 dB.

There are formulae for the general case of different resistor values which do not always seem to agree with the previous analysis.

9.7.2.4   Q-Factor

We define this as the two points where the modulus of the transfer function is 3 dB more than the minimum at notch ω_0. First we find the modulus by replacing s with $j\omega$.

Now if we expand the equations as they stand, we soon find ourselves with very unwieldy expressions; so instead we follow Sulzer and define:

$$u = \frac{\omega}{\omega_0} - \frac{\omega_0}{\omega} \qquad (9.123)$$

and use this to rewrite Eq. 9.118 as:

$$\frac{\frac{2}{a} + ju}{\left(\frac{2}{a} + a\right) + ju} \qquad (9.124)$$

Then we multiply by the *complex conjugates* to get real numbers, without expanding $(2/a + a)$. The 3 dB point is where the response is $\sqrt{2}$ times the minimum or $\sqrt{2} \times \left(\frac{2}{a} + a^2\right)$ which we leave squared here:

$$\frac{\frac{4}{a^2} + u^2}{\left(\frac{2}{a} + a\right)^2 + u^2} = 2\left(\frac{2}{2 + a^2}\right)^2 \qquad (9.125)$$

this reduces to:

$$u^2 = \frac{4}{a^2} \frac{(2 + a^2)^2}{(2 + a^2)^2 - 8} \qquad (9.126)$$

and finally:

$$Q = \frac{a}{2}\sqrt{1 - \frac{8}{(2 + a^2)^2}} \qquad (9.127)$$

and if $a \gg 1$:

$$Q \approx \frac{a}{2} \qquad (9.128)$$

And a similar argument holds for the other form of the circuit with resistors and capacitors interchanged. So we can flatten the notch but not turn it into a 'bath-tub' band-stop response.

Example – Bridged-T Filter

This is schematic ('Bridged-T.asc'). From the trace panel, we can see that the notch frequency does not depend upon a, but we could add a measure statement to prove it. If we measure the minimum voltage for ten values of a, we see the notch depth increasing with a, and if we plot the stepped data, we find an approximately linear fall with a y-axis in dB or logarithmic and a linear x-axis in the value of a. We may discount the very small variations in phase which are thousandths of a degree.

We may take various measures of the notch width, and all show a decrease with increasing a.

9.7.3 Twin-T Filter

This also is the work of Tuttle. The circuit is Fig. 9.24 where we have made the necessary assumptions about the relationships between the component values. The clearest analysis – from which the following is taken – can be found at: https://www.millersville.edu/physics/experiments/111/index.php.

9.7.3.1 Filter ω_0

If we take loop 1, we have:

$$v_{\text{in}} - i_1 R - i_2 \frac{X_c}{2} = 0 \qquad (9.129)$$

if we take loop 2, we have:

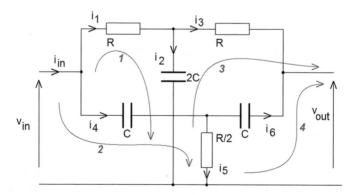

Fig. 9.24 Twin-T

$$v_{in} - i_4 X_C - i_5 \frac{R}{2} = 0 \tag{9.130}$$

Turning to the output, we have from loop *3*:

$$v_{out} + i_3 R - i_2 \frac{X_c}{2} = 0 \tag{9.131}$$

and from loop *4*:

$$v_{out} + i_6 X_c - i_5 \frac{R}{2} = 0 \tag{9.132}$$

Also from KIL, we have:

$$i_1 = i_2 + i_3 \quad i_4 = i_5 + i_6 \quad i_6 = -i_3 \tag{9.133}$$

If we substitute for *i1*, *i4* and *i5* in the above equations, we find:

$$v_{in} = i_2 \left(R + \frac{X_c}{2} \right) + i_3 R \tag{9.134}$$

$$v_{in} = i_2 \left(X_c + \frac{R}{2} \right) - i_3 X_c \tag{9.135}$$

$$v_{out} = i_3 X_c + i_2 \frac{R}{2} \tag{9.136}$$

Equating Eqs. 9.134 and 9.135, we have:

$$i_3 = i_2 \frac{(X_c - R)}{2(R + X_c)} \tag{9.137}$$

After substituting for *i3* in Eqs. 9.135 and 9.136 and dividing:

$$\frac{v_{out}}{v_{in}} = \frac{\frac{(X_c - R) X_c}{2(X_c + R)} + \frac{R}{2}}{X_c + \frac{R}{2} + X_c \frac{(X_c - R)}{(X_c + R)}} \tag{9.138}$$

after multiplying throughout by *2(Xc + R)*, we arrive at:

$$\frac{v_{out}}{v_{in}} = \frac{X_C^2 + R^2}{X_c^2 + 4RX_c + R^2} \tag{9.139}$$

remembering that $Xc = 1/j\omega C$, the numerator becomes zero when $f_0 = 1/2\pi RC$ with zero phase shift. This is in contrast to the Bridged-T where we only had a minimum of *2/a*.

9.7.3.2 Notch Depth

This again is delightfully simple. Equation 9.139 shows that the output should be zero at ω_0. However, LTspice reports $-88\ dB$. This is not due to limited data points nor to the limits of the iterations, *abstol, reltol* and *vntol*, and remains unexplained. However, the schematic ('notch.asc') in the next section does have a slightly deeper notch.

But because of the infinite notch depth, the Q-factor is meaningless.

9.7.3.3 Time Response

We usually consider only the steady state of these filters. However, if we set an AC signal of 50 Hz and make a transient run, we will find a large initial voltage surge which only settles to a sine wave after some 120 ms. If we measure the distortion, it is 2.16%, and a FFT analysis yields a single peak at exactly 50 Hz and at -86 dB.

Example – 50 Hz Twin-T
This is schematic ('Twin-T 50 Hz.asc') where a number of options have been tried to increase the depth of the notch (which should be infinite).

9.7.4 Other Notch Filters

Four more examples can be found at **LTspiceXVII->examples->Educational** as ('notch.asc'). They all have the same configuration, but interestingly, only two component values change. The approach we shall use is to note that the circuit looks like two Πs – or more exactly – the lower part is actually an inverted Δ, Fig. 9.25.

The original circuit is shown on the left at '(a)' where *Z1, Z2* and *Z4* can be considered as part of a bridged-T where the original *Z3* is replaced by the network of *Z5, Z6* and *Z7*. We now use the Delta-Star transformation to turn *Z5, Z6* and *Z7* into

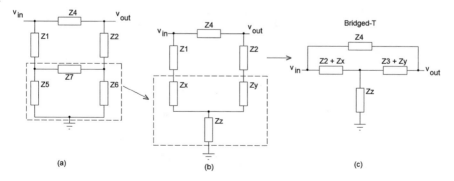

Fig. 9.25 Twin Pi Filter Alt

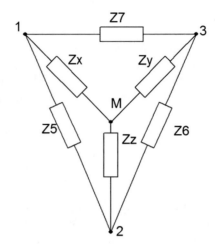

Fig. 9.26 Delta-Star

Zx, Zy and Zz in figure '(b)' showing that Zx and Zy are in series with $Z1$ and $Z2$ respectively so that finally we can rearrange this in '(c)' to show it as a Bridged-T network. And if we have a large A4 pad and nothing better to do for an hour or two, we can solve the following equations. We shall stop short of this final step and just find the four impedances.

To derive the equations, the original Delta-Star figure from chapter 2 is redrawn as Fig. 9.26 using the component labels of Fig. 9.25 which we now replace by those of the first example of schematic 'notch.asc', keeping the parameter w. These are:

$$Z4 = 6R \quad Z1 = Z2 = Z7 = C \quad Z5 = wR \quad Z6 = (1 - wR)$$

First we note that the same denominator applies to all the terms of the transformation (as we found in chapter 2), so we define:

$$T = Zx + Zy + Zz = \frac{1}{sC} + wR + (1 - w)R \quad = \frac{1}{sC} + R \qquad (9.140)$$

We then have:

$$Zx = \frac{Z5.Z7}{T} = \frac{wR}{sCT} \qquad (9.141)$$

$$Zy = \frac{Z6.Z7}{T} = \frac{(1 - wR)}{sCT} \qquad (9.142)$$

$$Zz = \frac{Z5.Z6}{T} = \frac{wR(1 - wR)}{T} \qquad (9.143)$$

Then the impedances for the Bridged-T are:

$$Z1 = \frac{1}{sC} + \frac{wR}{sCT} \quad = \frac{1}{sC}\left(1 + \frac{wR}{T}\right) \tag{9.144}$$

$$Z2 = \frac{1}{sC} + \frac{(1-wR)}{sCT} \quad = \frac{1}{sC}\left(1 + \frac{(1-wR)}{T}\right) \tag{9.145}$$

$$Z3 = \frac{wR(1-wR)}{T} = \frac{wR - w^2R^2}{T} \tag{9.146}$$

$$Z4 = 6R \tag{9.147}$$

It now becomes very messy. The product terms that we need are:

$$Z1Z2 = \frac{1}{(sC)^2}\left(1 + \frac{wR}{T}\right)\left(1 + \frac{(1-wR)}{T}\right) = \frac{1}{(sC)^2}\left(1 + \frac{(1-wR)}{T} + \frac{wR}{T} + \frac{wR(1-wR)}{T^2}\right)$$

$$= \frac{1}{(sC)^2 T}\left(T + 1 + wR(1-wR)\right)$$

$$\tag{9.148}$$

$$Z1Z3 = \frac{1}{sCT}\left(1 + \frac{wR}{T}\right)\left(wR - w^2R^2\right)$$

$$= \frac{wR}{sCT}\left(1 - wR + \frac{wR}{T} - \frac{w^2R^2}{T}\right) \tag{9.149}$$

$$Z2Z3 = \frac{1}{sCT}\left(1 + \frac{(1-wR)}{T}\right)\left(wR - W^2R^2\right)$$

$$= \frac{1}{sCT}\left(wR - w + \frac{(1-wR)wR}{T} - \frac{(1-wr)(w^2R^2)}{T}\right) \tag{9.150}$$

$$Z3Z4 = \frac{1}{T}\left(wR - w^2R_2\right)6R \tag{9.151}$$

$$Z1Z4 = \frac{1}{sC}\left(1 + \frac{wR}{T}\right)6R \tag{9.152}$$

And all we have to do now is to plug these into Eq. 9.113. Or we could just solve the circuit as it stands, Fig. 9.27, using the analysis tools we have developed in previous chapters. Fig 9.27 is a starting point.

If we run the example circuits, we see they have the same response as the Bridged-T, and we might well ask why the complication of the added components. The answer is the very great ease with which we can tune the circuit. It may be recalled that with the Bridged-T attenuator we had to adjust two resistors in opposition so that as one increased, the other decreased, and this required very close matching. But now the two resistors $Z5$ and $Z6$ have a constant total resistance so they can be just one variable potentiometer with the slider connected to ground, very neat Figs. 9.28.

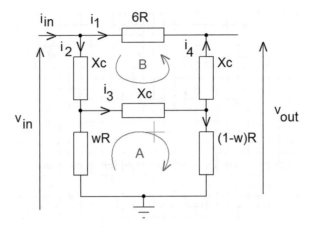

Fig. 9.27 Twin Pi Filter

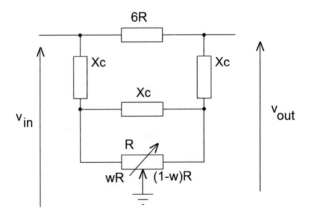

Fig. 9.28 Tunable twin-pi filter

Explorations 7

1. Note the effect of the different values of *R1* in 'Bridged-T.asc'. The notch frequency (of course) will change. Recalculate the capacitor value to restore it.
2. Add a load resistor of 100 kΩ across the output. The low-frequency attenuation is increased because of the potential divider formed by *R3* and the load, but at high frequencies, the two series capacitors offer very little impedance.
3. Repeat the analysis for two identical resistors. Change *C2* in ('Bridged-T V2. asc') and note the change in notch depth and compare to theory.
4. Explore ('Twin-T 50 Hz.asc'), and note that the internal node voltages at *c* and *r* do not presage the sharp notch.
5. Explore the 'examples' filter discussed above.

9.8 Summary

This chapter started with an introduction to the Laplace 's' and then used it extensively to explore second-order RC circuits where we have found:

- As a simple equation, we see that a constant term controls the gain, the coefficient of s the breakpoint and the power of s the slope.
- We may use Kirchhoff's Laws to find the transfer function as a fraction of two quadratic equations which we solve to find the roots.
- If we express the equations as factors, the numerator gives the zeros and the denominator the poles.
- The .net directive is used with an AC analysis and returns h-, z-, y- and s-parameters of the circuit.
- It also returns the port impedances of Z(in) and Z(out) where terminating resistors of $1\,\Omega$ are assumed, and port admittances of Y(in) and Y(out) admittances of $1\,\mu S$ are assumed, but these can be changed in the .net directive.
- For second-order low-, high-, and band-pass filters, we can reduce the numerator to $1, s^2$ or s, respectively, and not a polynomial and the roots of the denominator are the breakpoints.
- We can draw the Bode plot magnitude by erecting verticals on the graph at the breakpoints and drawing line segments between them with an upward slope of 20 dB/decade for poles and a downward slope of -20 dB/decade for zeros.
- For more accuracy, use the 3 dB points and the line segments to draw a smooth curve.
- The asymptotic low-frequency phase is $0°$. For each pole, it increases by $+45°$ at the 3 dB point and has a high-frequency asymptote of $90°$. The phase of each zero decreases by $-45°$ at the 3 dB point and tends asymptotically to $-90°$.
- LTspice accepts the transfer function as an input in the form ***Laplace*** $= (s + z_1)*$ $(s + z_2)/((s + p_1)* (s + p_2))$.
- The Bridged-T filter is a notch filter whose notch depth approximates to $1/a^2$ where $a \gg 1$ and is the ratio of the two resistors
- The Twin-T filter also is a notch filter with a theoretical infinite notch depth.

Chapter 10
Transmission Lines

10.1 Introduction

These are often parasitic effects, prevalent in connecting cables in RF circuits, and can even extend to audio frequencies. We may fruitfully distinguish between lines consisting only of capacitance and resistance which will attenuate the signal and those essentially consisting of inductance and capacitance where the signal is not attenuated but can be reflected from the ends of the cable. To this we may also add resistance and then the signal will also be attenuated.

A transmission line is automatically formed by any two adjacent parallel conductors. It may be the flat cable used in data transmission in computers, or a coaxial cable, or a twisted pair, or a strip-line, or the cable connecting an amplifier to a loudspeaker, or an overhead power line. At one extreme we have power distribution at hundreds of kV over hundreds of km. At the other extreme, we have the strip-line connections on a PCB. In theory, then, all interconnections are transmission lines. However, we only need treat them as such if the transmission path length is greater than 1/10 of the wavelength of the signal, or if the rise time of the signal is less than twice the delay time. With this in mind we can get a feel for significant lengths. The speed of light in air is near enough 3×10^8 m/s. Thus the wavelength of *50 Hz* power transmission is $(3 \times 10^8)/50 = 6000$ km. But in a coaxial cable, the speed of light is only about 2/3 of this or 2×10^8 m/s. So given a *10 MHz* signal (a period of *0.1 μs*), the wavelength is 20 m. And at *10 GHz*, the wavelength in air is *3 cm*. Hence the interconnecting metallization of an integrated circuit is treated as an Elmore RC delay line and not as a transmission line.

LTspice has three varieties, the internals of which are not open to investigation. The first is an RC line, the second is an LC line, and the third is an RLC line.

© Springer Nature Switzerland AG 2020
C. May, *Passive Circuit Analysis with LTspice®*,
https://doi.org/10.1007/978-3-030-38304-6_10

10.2   Uniform RC 'URC' Line

Although we have discussed this before, we expand upon it here and compare it to a discrete RC line and also add the reverse-biased diode parameters. It consists of 'L' sections of series resistors and shunt capacitors. The ends can be connected either way round, and the capacitors are usually connected to ground. We may alternatively replace the capacitances by revere-biased diodes.

There is no delay of the pulse as a whole, only the increased rise time through the various sections. The comment in the **Select Component Symbol** dialogue is that this is intended to model the interconnections in an IC, but we can equally well use it for tracks on a PCB if we ignore any inductive effects.

The symbol is found in the *Misc* folder. There are two versions differing only in the number of zigzags of the resistor.

10.2.1   Syntax

This is slightly unusual. It requires a model card and also a value for the length and optionally for the number of sections which defaults to one. Thus we have on the schematic *.model MyURC URC()*. This is the basic form. Then when we add an instance on the schematic in the netlist, we will have:

> < **Instance Name** >< **node in** >< **node out** >< **common** >< **model name**
> **> L =< length in metres >**

And notice that the length of the line is not in the *model* card. A typical instance would be:

U1 N001 out 0 MyURC L = 1

where *U1* is the default name for the first instance, *out* is the name of the far end of the line, the capacitors are connected to ground by the *0*, and the length of the line is 1 m. This netlist entry is automatically created when we place an instance on the schematic and add the **Spice Model** entry **MyURC** and the **Value** field **L = 1** in the **Component Attribute Editor** dialogue. Some examples follow in the next few paragraphs.

10.2.2   Parameters

Using it as a passive component, six parameters are at our disposal. The last two shown in the LTspice 'Help' are only applicable if we replace the capacitors by reverse-biased diodes. The first four, *Fmax, K, Rperl* and *Cperl*, are added to the model statement; the last two, *L* and *N*, are added to one of the **Value** or **Spice Line** fields. With so many parameters to play with, KISS is essential and we should avoid changing several at once; in particular, set *N=1* rather than the unknown value that LTspice will choose if we do not define it. Also, we may be straining at a gnat because most of the parameters are only significant above the *3 dB* point.

10.2.2.1 Global Parameters

There are two which apply irrespective of whether we use diodes or not.

Fmax
This is *1 GHz* and is not the upper *3 dB* point. It appears to be the point at which $Z(\text{in})$ has fallen to a constant value and is not a parameter we can change and attempting to add it to the model statement results in an error.

K
This is declared as the *propagation constant* with a default of *2*. The segments of the line are graded geometrically towards the centre, and this parameter controls the rate. It must be greater than *1*.

Example – URC 'Fmax' and 'K'
An *ac* analysis of ('URC Line Default.asc') shows that *Fmax* seems to be the start of the straight line fall in output if $N = 1$ when there is no gradation of the segments and so only one graph. An *ac* analysis with $N > 1$ shows that increasing *K* reduces the steepness of the fall at high frequencies. In short, *N* and *K* have little effect below the *3 dB* point.

Example – Discrete URC
We can build a five-stage line with the default values as schematic ('URC 5 Discrete. asc.'). This exhibits similar behaviour with a input impedance of *1TΩ* but only up to *1 Hz*. The impedance of the capacitors in this region is far greater than the resistors, so effectively they are in parallel and all draw substantially the same current up to *10 GHz*.

Thus, for the input impedance at *1 Hz*, we have $Z(\text{in}) = \frac{1}{2\pi \times 5 \times 10^{-15}} = 3.18 \times 10^{13}\Omega$ which is what we measure. A point in passing, if we do not actually remove the load resistor, we must make it at least 10^{20} Ω; else it is comparable to the low frequency impedance of the capacitors.

Rperl, Cperl
These are the resistance in ohms and the capacitance in farads per unit length. As the length is in metres, it follows that these too should be per metre. The RC product affects the upper 3 dB point. The total resistance of the line is simply $R = Rperl \times L$.

L
The length must be specified in the **Value** field as **L = <length>**. It acts with the resistance per length to give the total resistance of the line.

N
It is the number of sections and is **Value N = <no. of sections>**. With one section the roll-off above the 3 dB point is -20 *dB/decade* and increases at the rate of -40 *dB* per unit increase of N, so with $N = 3$ we have $-20 + 2(-40) = -100$ *dB/decade*.

Example – URC Length
An effect which is not seen in the discrete equivalent is that above the breakpoint the input impedance *Zin(V1)* falls at *10 dB/decade* and at high frequencies becomes horizontal again at half its low frequency value, schematic ('URC Len.asc'). This occurs when the impedance of the capacitors falls to a low value. As we would expect, the breakpoint of the impedance is highest for a single length.

The upper 3 dB point also falls with increasing length where $f_{3dB} \propto \frac{1}{L^2}$ which is consistent with constant capacitance per unit length.

The discrete URC line with the $C = 50\,pF$ $L = 50\,\mu H$ exhibits similar behaviour, but the fall of impedance is -20 *dB/decade*, and the phase shift is $-90°$, whereas it is only $-45°$ with the URC.

If we apply a *10 ns* input pulse to both the URC and the discrete line, we find the voltage pulse at the nodes along the line starting from the input has a similar shape to the URC line with different lengths only more rounded.

Example – URC Line 'Rperl'
The *Z(in)* in schematic ('URC Rperl.asc') has *300 Ω* at low frequency which is correct for $L = 5$ of *Rperl = 50 Ω* and the matched load of *50 Ω*. The low frequency output voltage is $-15.5\,dB$ - which is correct for the potential divider *50 Ω/300 Ω*. It starts to fall off a decade or so before the *3 dB* point, but this depends on the component values. We can also find the asymptotic slope is *20 + 40 N dB/decade*.

If we step *N* with an *ac* run, we find the upper *3 dB* point is largely unchanged, but there is a marked increase in the shape of the graph if *N* is more than *1*.

A transient run shows a rise time of *8 ns*. This is much shorter than the Elmore delay and becomes shorter as *K* is increased. The pulse shape is almost unchanged with a line length of *1* but broadens if we increase it.

10.2.2.2 Diode Parameters

These only apply if *Isperl > 0* and are in addition to the previous parameters.

Isperl
This is the diode leakage current per length and defaults to zero. It is used in the diode equation (which we shall not bother our heads with). This does not change the frequency response as we see in schematic ('URC Diode V1.asc'), and unless it is made unrealistically large, or *L* is very large, it does not significantly increase the DC

current. As the *.op* directive changes capacitors to open-circuits, that does not help us here. The reverse bias voltage has a much greater effect than the changes to *K*, and the capacitance follows the diode equation:

$$C_j = \frac{CJO}{(1 - V)^M} \tag{10.1}$$

where *CJ0* is the zero bias capacitance, *V* is the voltage and *M* is a factor of about 0.5.

Rsperl
It is the series resistance of the diode and independent of *Rperl* and defaults to zero.

Explorations 1
1. Explore the effects of changing the parameters of the URC line, in particular note how something like *N* = 50 greatly increases the attenutation below *Fmax*. Confirm the rate of high frequency roll-off with N.
2. Add more sections to the schematic ('URC 5 Discrete.asc)'. This is easiest by clicking **F6** and then dragging a rectangle enclosing the original five sections. LTspice will give them new names, and they can be added to the end.
3. Save the five or ten section lines as sub-circuits and use them to create even longer lines. Of course, the RC segments are equal, but if we make up a line of 7 or more segments, we could change the values in each segment to be a geometrical progression towards the centre like the LTspice URC.
4. Carefully explore the performance of the discrete URC and compare it to the LTspice one.
5. Explore *Isperl* and *Rsperl* of the URC line and find a value for *M* in Eq. 10.1. Schematic ('URC Diode.asc') is a test jig where we have measured the *3 dB* point at different bias voltages and find *318 MHz* at zero bias voltage with the default 1 kΩ whence *CJO* = 1 fF, the default. Plotting the capacitance against voltage looks like it could be a reciprocal relationship. And (as we should expect) this does not change with *Isperl*.
6. Use the measurements from schematic ('URC Line Default.asc') to try and quantify the effects of *K* and *N*.

10.3   Transmission Lines

A scan through the websites shows that this is a topic of considerable complexity. However, on the basis of KISS, we will keep it as simple as possible and deal with it only insofar as the results are amenable to simulation in LTspice; for example LTspice iterates to a steady-state solution so temporal progression of a signal along the line can be ignored. Therefore we shall proceed with cavalier disregard of the subtle distinctions between ∂v and *dv*. This is not as reckless as it may seem

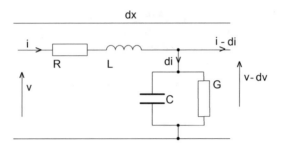

Fig. 10.1 Lossy line

because we shall be deriving equations from the equivalent circuit where *dv* is appropriate rather than from the 3-dimensional physical line.

For those who wish to dig deeper into things such as standing wave ratios and Smith Charts, an extremely comprehensive chapter with LTspice simulations can be found at *www.eas.uccs.edu/~mwickert/ece3110/lecture_notes/N3110_2.pdf*.

The signal travels as an electromagnetic wave. There are many possible *modes*, but in this case it is effectively the Transverse Electro-Magnetic (TEM) mode having orthogonal electric and magnetic fields. So now we here move from simple circuitry where it is implicitly assumed that a voltage at point *A* will immediately be seen at point *B*. Now we are dealing with wave propagation where there is a finite time between launching the wave and its arrival at the far end.

The besetting problem is that of representing a travelling wave by an electrical circuit: a problem that is compounded because the wave has two components so we can encounter effects that are counter-intuitive if we think in terms of an electric circuit. In the simulation the two are separated as a voltage and a current. It may help to visualize each as a transverse wave travelling along a length of string. On the other hand, a large G&T may be better.

10.3.1   *Equivalent Circuit*

Now we recall that a current in a wire creates a magnetic field around it, and this is the basis of inductance: if the current varies, so does the field and a back EMF is induced in the wire. Therefore we can model the magnetic component by a series inductor together with an optional resistor representing its DC resistance. Also there is capacitance between the two conductors which we can model by a shunt capacitor in parallel with an optional conductance to model the dielectric losses. We then have the model which consists of repeated sections of Fig. 10.1.

10.3.2   Analysis

We must here anticipate the next chapter to note that the impedance of an inductor is $Z_l = j\omega L$. If the section dx is infinitesimally short, we can write the equations as differentials and the change of voltage over the section dx is:

$$\frac{-dv(x)}{dx} = i(x)(j\omega L + R) \qquad (10.2)$$

and the current through the shunt components is:

$$\frac{-di(x)}{dx} = (j\omega C + G)v(x) \qquad (10.3)$$

If we differentiate Eq. 10.2 with respect to current, we have:

$$\frac{di(x)}{dx} = \frac{-d^2v(x)}{dx^2}\frac{1}{(j\omega L + R)} \qquad (10.4)$$

Rearranging Eq. 10.4 as the first equation below and then substituting for di/dx from Eq. 10.3 we have the second equation:

$$\frac{d^2v(x)}{dx^2} = -(j\omega L + R)\frac{di(x)}{dx} \quad \Rightarrow \quad \frac{d^2v(x)}{dx^2} = (j\omega L + R)(j\omega C + G)v(x) \qquad (10.5)$$

similarly rearranging Eq. 10.3 and then substituting for dv/dx from Eq. 10.2:

$$\frac{d^2i(x)}{dx^2} = -(j\omega C + G)\frac{dv(x)}{dx} \quad \Rightarrow \quad \frac{d^2i(x)}{dx^2} = (j\omega L + R)(j\omega C + G)i(x) \qquad (10.6)$$

These are known as the *telegrapher's equations* developed by Oliver Heaviside in the 1880s.

We should take careful note that this says nothing about what happens when we reach the end of the line. In fact, it is implicitly assumed that the line is infinitely long so that there are no reflections from the far end.

Propagation Constant
This tells us how the wave propagates along the line. The two bracketed terms in the right-hand Eqs. 10.5 and 10.6 are identical and are the *complex propagation constant* denoted by γ:

$$\gamma = \sqrt{(j\omega L + R)(j\omega C + G)} \qquad (10.7)$$

This is expanded as real and imaginary parts:

$$\gamma = \alpha + j\beta \tag{10.8}$$

then:

$$\alpha^2 = RG \quad \beta^2 = \omega^2 LC \tag{10.9}$$

The α is a resistance and represent the attenuation in Np/m or dB/m. The β term is the phase constant in rad/s. And from this we can see that there is always a β, but (naturally) if there is no resistance, the lines are lossless.

The One-Dimensional Wave Equation

This describes the current or voltage as a function of distance. We note that, from Eq. 10.7, $\gamma^2 = (j\omega L + R)(j\omega C + G)$ so we can write the second. Equations 10.5 and 10.6 as:

$$\frac{d^2v(x)}{dx^2} - \gamma^2 v(x) = 0 \tag{10.10}$$

$$\frac{d^2i(x)}{dx^2} - \gamma^2 i(x) = 0 \tag{10.11}$$

These are second-order differential equations. We require a solution which, when differentiated twice, has the same dimensions as the original. One function that will satisfy this requirement for Eq. 10.10 is $v = v_0^-\, e^{\gamma x}$ because on substitution we have $\gamma^2 v_0^- e^{\gamma x} - \gamma^2 v_0^- e^{\gamma x} = 0$. But we can equally well have $v = v_0^+\, e^{-\gamma x}$ to give $\gamma^2 v_0^+ e^{-\gamma x} - \gamma^2 v_0^+ e^{-\gamma x} = 0$ and the solution is the sum:

$$v(x) = v_0^+ e^{-\gamma x} + v_0^- e^{\gamma x} \tag{10.12}$$

where v_0^+ is a wave propagating in the x direction and v_0^- in the -x direction from $x=0$. We similarly find for the current:

$$i(x) = i_0^+ e^{-\gamma x} + i_0^- e^{\gamma x} \tag{10.13}$$

Characteristic Impedance

If we differentiate Eq. 10.12, we can replace $dv(x)/dx$ by Eq. 10.2 to give:

$$-\gamma v_0^+ e^{-\gamma x} + \gamma v_0^- e^{\gamma x} = (j\omega L + R)i(x) \tag{10.14}$$

We now replace $i(x)$ from Eq. 10.13:

$$-\gamma v_0^+ e^{-\gamma x} + \gamma v_0^- e^{\gamma x} = (j\omega L + R)i_0^+ e^{-\gamma x} + (j\omega L + R)i_0^- e^{\gamma x} \tag{10.15}$$

By definition, $Z_0 = \frac{v_0^+}{i_0^+} = \frac{-v_0^-}{i_0^-}$ where in the second equation we have a minus sign because this is a wave travelling in the reverse direction. Equating terms in $e^{-\gamma x}$ from Eq. 10.15 and dividing by the exponentials:

$$Z_0 = \frac{v_0^+}{i_0^+} = \frac{-(j\omega L + R)}{\gamma} = -\sqrt{\frac{j\omega L + R}{j\omega C + G}} \qquad (10.16)$$

Similarly we can equate the terms in $e^{\gamma x}$:

$$Z_0 = \frac{v_0^-}{i_0^-} = \frac{j\omega L + R}{\gamma} = \sqrt{\frac{j\omega L + R}{j\omega C + G}} \qquad (10.17)$$

This allows us to write the current in Eq. 10.13 in terms of voltage by using Z_0:

$$i(x) = \frac{v_0^+}{Z_0} e^{-\gamma x} - \frac{v_0^-}{Z_0} e^{\gamma x} \qquad (10.18)$$

The difference in sign between Eqs. 10.16 and 10.17 refers only to the direction of the current. We must take careful note that the characteristic impedance refers only to a wave travelling from start to finish of the line or, since the line is uniform, a wave from finish to start, in either case, one wave only with no reflections. This means that the line must be infinitely long or, usually, that it is terminated by its characteristic impedance so that there is no reflection. Under any other conditions, Eqs. 10.16 and 10.17 do not apply, and the impedance is a function of frequency.

Voltage Reflection Coefficient 'Γ'
This is defined as the ratio of the voltage reflected from the load to the voltage incident on the load. However, instead of solving this directly, we can arrive indirectly by terminating the line at the end l with a load Z_l. We can set $l=0$ and then all the exponentials are unity and they cancel. Then dividing Eq. 10.12 by Eq. 10.18 we have:

$$Z_l = \frac{v_l}{i_l} = \frac{v_l^+ + v_l^-}{v_l^+ - v_l^-} Z_0 \qquad (10.19)$$

We can rearrange this to give:

$$v_l^- = \frac{Z_l - Z_0}{Z_l + Z_0} v_l^+ \qquad (10.20)$$

then:

$$\Gamma = \frac{v_l^-}{v_l^+} = \frac{Z_l - Z_0}{Z_l + Z_0} \qquad (10.21)$$

From which we see that if the line is terminated in a matched load, there is no reflection. This can be very important in data transmission where the last thing we want is a signal travelling back up the line.

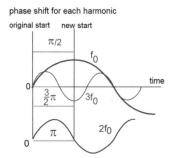

Fig. 10.2 Phase shift

Distortion

The conditions for no distortion are that the attenuation is independent of frequency so that all frequencies are equally attenuated and that the phase shift depends linearly on frequency. This second requirement can be seen in Fig. 10.2 where if the fundamental is shifted by $\pi/2$, the second harmonic must be shifted by π and the third by $3/2\pi$. From Eq. 10.16 this condition exists if $R/L = G/C$. We may also note from Eq. 10.9 that α does not depend upon frequency and β depends linearly on frequency.

Velocity of Propagation

This is how fast the wave progresses along the line, the *phase velocity*. That is $u_p = wavelength/period$ and is $u_p = 1/\sqrt{(LC)}$ from which we can derive:

$$T_d = \sqrt{LC} \quad C = \frac{T_d}{Z_0} \quad L = Z_0 T_d \tag{10.22}$$

where T_d is the *delay time* and does not depend on the resistance.

10.4 Lossless Transmission Line 'tline'

This component is found at the end of the opening component selection dialogue, just above the voltage source. It is a simple two-port model that is characterized by two parameters. It is symmetrical in that either end and can be used as the input. It models a single conductor and therefore supports only one transmission mode. If the port connections are one-to-one, the reflected wave is the same polarity as the incident wave; if the connections are crossed, the reflected wave is inverted. And although they are not marked on the symbol, the terminals are a,b,c,d and drawn on schematic ('Lossless Gamma v1.asc'). Their names are also shown in the Status Bar if the mouse hovers over the ends of the line after we have made a run.

We shall deal with the line under two separate heads: the first is transient response and the second frequency response. These are fundamentally different. In the transient response, we must not be led astray by imagining that we are watching the progression of a train of waves. No, we must rather think that we are sitting at the ends of the line and watching what appears at certain times. And these times are multiples of the delay time: we are not privy to what happens in between while the wave travels through the delay line.

Creating an Instance
This is quite straightforward. The only point to watch is when trying to connect an end directly to earth. To confirm that the earth is actually connected to the line, use **F8** to slide it a little (not **F7** – that will only slide the earth symbol itself), and note if there is a wire connection to the line.

10.4.1 Voltage Reflections

These are crucial to understanding what is happening. There are a number of useful animations to be found on the Web. There are three special conditions as well as a general mismatch condition which is none of the three.

10.4.1.1 Open-Circuit End

If the end of the line is open-circuit, from Eq. 10.20, the reflection coefficient is *1*, and a reflected wave is created whose amplitude adds to the incident wave at that point and at that time only. We might like to think of the reflected wave as climbing over the incident wave. The energy of the incident wave is transferred to the reflected wave which travels back down the transmission line to the start. We can use the analogy of a wave travelling down a rope to explain this, Fig. 10.3. The wave first

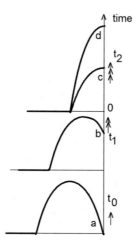

Fig. 10.3 Open-circuit voltage

arrives at time *t0* where *a* is the end of the rope which starts to move upwards. As the wave progresses, the end of the rope *b* moves upwards faster at time *t1* shown by two arrowheads. Finally, at time t_2, at the crest of the wave, the end *c* is moving at its maximum velocity, shown by three arrowheads, and will continue upwards loosing kinetic energy until it comes to an instantaneous stop at *d* where, assuming no losses, the distance *c-d* is the same as *0-c*. Thus we have a maximum deflection of the far end that is twice the original peak. The end will then fall back, converting potential energy into kinetic energy of the end so that, at point *c* again, the end of the rope has the same vertical velocity as before, but downwards instead of upwards, and the amplitude of the reflected wave is exactly that of the incident. It is here that we must be careful because the LTspice simulation shows the peak at the output with the wave twice the amplitude of the input: but we do not see the return wave starting back down the line.

That the wave must return back along the rope with the same amplitude as the original and not twice follows from the conservation of energy since none is expended by the movement of the end of the rope so the wave must be in the reverse direction and of the same amplitude. We can furthermore indulge in the physically impossible thought experiment of supposing the start of the rope were also free and the amplitude of the return wave were the double amplitude shown in the simulation. When the wave returned to the start, it would be reflected again with double the amplitude or four times the original, and this would continue as long as the wave bounced to and fro along the rope, doubling its amplitude at each rebound. This would create energy from nothing.

10.4.1.2   Matched Impedance

The analogy here is that the rope is joined to the end of another rope so that the energy continues to flow along that and there is no reflection.

10.4.1.3   Short-Circuit End

The analogy now is that the far end of the rope is fixed so that it cannot move, Fig. 10.4. When the travelling wave encounters the fixed end, the tension in the rope at *a* pulls it downwards. This tension increases at *b* and reaches a maximum at *c*.

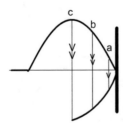

Fig. 10.4   Short-circuit voltage

Thus, to conserve energy, a wave of equal amplitude but inverted is created which again travels back to the start.

10.4.1.4 Mismatch

Here we have neither an open-circuit, nor a short-circuit, nor a matched load, but something in between. In every case there will be a partial reflection: if the mismatched load is greater than the matched load so that it tends to an open-circuit, the reflected wave will not be inverted but less than the incident wave. Likewise, if the mismatched load is less than the matched load, the reflected wave will be inverted and less than the incident wave.

10.4.2 Current Reflections

For a start, we can have current in the earth line, whereas we cannot have a voltage. Remember that the LC sections are infinitely short so in Fig. 10.5 at time t_0, the input pulse immediately drives a current i_0 through L_0C_0 so we shall see an equal and opposite current in the earth line. And as time progresses and the pulse moves along the line we shall see equal and opposite currents $t_1\ t_2 \ldots t_n$ created in sections $1,2,.. \ n$ then falling to zero after the current wave has passed.

10.4.2.1 Open-Circuit End

If the far end is open-circuit, there can be no current at that point. So the current flow proposed in Fig. 10.5 must result in a decaying currents along the line, which we cannot see, resulting in very small equal and opposite currents at the far end if the open-circuit is not perfect.

10.4.2.2 Matched Impedance

This time the incident current pulse is totally adsorbed by the load, and there is no reflected current.

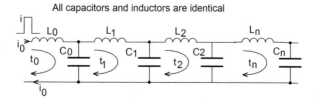

Fig. 10.5 Current wave

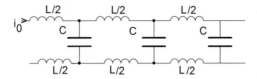

Fig. 10.6 To see the Voltage wave

10.4.2.3 Short-Circuit End

The two incident currents are inverted and reflected with the amplitude the same as the incident wave. Once again we must be careful to note that the currents shown by LTspice are the instantaneous superimposition of the incident and reflected waves at the ends of the line and not the reflected wave.

In passing, we can note that if we want to see the voltage wave in the second conductor, we can halve the inductors and place the other half in the second conductor, Fig. 10.6, but as the second conductor is usually earthed, the model and analysis are based on Fig. 10.1.

10.4.2.4 Parameters

The line is defined by two parameters entered in the **Value** field of the **Component Attribute Editor**.

Td
This is the delay before the signal reaches the far end and defaults to 50 ns. It effectively represents the length of the line. A useful figure for the propagation speed in co-axial cable is 20 cm/ns which is 2/3 of that for air (which is 1 ns/foot).

Z0
Is the characteristic impedance which defaults to 50 Ω. There is no restriction on its value, but we can see some interesting effects if we make it unrealistically large.

10.4.3 Single Mode Behaviour

We can use Eq. 10.21 to find the default parameters. As $Z_0 = 50 \; \Omega$ and $T_d = 50 \; ns$ then:

$$C = \frac{T_d}{Z_0} = 1nF \quad L = \frac{T_d^2}{C} = 2.5\mu H \qquad (10.23)$$

A good approximation to co-axial cables is $C = 100 \; pF/m \; L = 240 \; nH/m$. The default numbers from Eq. 10.23 are quite large and correspond to *10 m* of line.

It is important to remember that we have both voltage and current that can be reflected. This can be rather difficult to get to grips with at first. Bear in mind that voltage and current no longer go together. As we have seen, the amplitude of the reflected waves depends upon how well the termination matches the characteristic impedance: if it is a perfect match, there are no return voltage or current waves, and all the power is adsorbed by the termination; if it is open-circuit the return voltage wave amplitude will be the same as the input and in phase and the current essentially zero; and if it is terminated by a very small resistance approximating to a short-circuit, the return voltage wave will be almost zero and we have only current.

It is convenient to divide the examples into transient and AC analyses starting with transient.

10.4.3.1   Transient Analysis

We shall explore different terminations with current or voltage excitation.

Example – 'Tline' Open-Circuit, Voltage Input

This is schematic ('Lossless Gamma V1.asc'), Let us remind ourselves that the trace shows the conditions at the measuring points as time progresses: the x-axis is not a plot of the conditions along the length of the line. The far end is effectively an open-circuit between nodes b and d with $R1 = 1e100 \, \Omega$, or we may simply cut out this resistor. We apply an input voltage pulse of $1 \, V$ at time $t = 0 \, s$. The Fig. 10.5 shows that we should immediately find a current $I0$ flowing through capacitor $C0$ to create an equal and opposite current in c. As the line impedance is $50 \, \Omega$, we therefore should expect $Ia(T1)$ and $Ic(T1)$ each of 20 mA, which is exactly what we see during the time $t=0$ to $t=20 \, ms$.

As time progresses, the current will transfer to $C1, C2$, and so on and cease in $C1, C2$ behind the pulse as the pulse moves down the line. As the RC sections are assumed to be infinitely short, it is a question of no importance to ask how many of them are in the contiguous group carrying the current at any one time because we are not privy to what happens inside the line; all we need to know is that the pulse will travel along the line.

After $50 \, ns$ our interest turns to the far end of the line, and as the line is open-circuit between nodes b and d, there can be no current between them, so there is no external current. This is confirmed if we trace the currents $Ib(T1)$ and $Id(T1)$ which are both horizontal lines of 0 mA; but there will be current in the final internal capacitor Cn. Thus we find horizontal traces for $Ia(T1)$ and $Ic(T1)$ at $50 \, ns$, but internally there will be inverted current pulses each of amplitude $20 \, mA$ reflected back from the far end along the two lines, and these will arrive at the start after $100 \, ns$.

However, the input is terminated by the voltage source $V1$ which, with resistor $R2$, are effectively a short-circuit, so at $100 \, ns$ the current pulses are reflected in phase which, at the instant that the reflection takes place, means that we have the new waves reflected from the start at terminals a and c and sitting on top of the reflected waves from the far end so giving twice the current of *40 mA* and these pulses are inverted with respect to the original input. But do remember that this picture is valid only at that instance; the actual current pulse is still only *20 mA*. These reflections will continue ad infinitum since the line is lossless.

If we look at the voltages, we find a *1 V* initial input pulse and after *50 ns* a *2 V* pulse *V(out)* at node *d*. This is consonant with the *40 mA* current pulse and a *50 Ω* line. This voltage pulse is reflected back to the input at *100 ns* where, because it is effectively a short-circuit, there is no new reflected voltage pulse, but, as we have seen, there is an inverted current pulse at the input and this travels to the far end creating a new *2 V* pulse of opposite polarity at *150 ns*. And these reflections also will continue forever.

If we exchange the connections to the output terminals so that node *b* is earthed the picture is essentially the same only the first voltage reflection is inverted, not in phase.

Example – 'Tline' Open-Circuit Current Input

If we exchange the voltage source for a current source, schematic ('Lossless Gamma V2.asc'), the input also is an open-circuit so the voltage reflection coefficient $\Gamma = -1$. There is a single input current pulse of *1 mA* which has been given a sloping leading edge so we can see if the pulse by any chance is reversed. This sends equal and opposite currents in *a* and *c* of *1 mA* and creates an input voltage of *50 mV* in agreement with the *50 Ω* line impedance. Because both ends are open-circuit, there are no further external current pulses.

After *50 ns* the pulses reach the far end where we see the incident and reflected waves superimposed to give a pulse of *100 mV* which arrive back at the input after *100 ns* and the current creates a pulse of *100 mV* at that time and with the same polarity because the input also is an open-circuit. This results in a forward wave of *50 mV*, and this continues forever.

Example – 'Tline' Matched Termination

We make $R1 = 50\ \Omega$ in both of the previous schematics, we find the input pulse creates equal and opposite current pulses and a voltage pulse. After *50 ns* these arrive at the far end where we find an in-phase voltage pulse of the same amplitude as the input because there is no reflection. We also find equal and opposite inverted current pulses. There are no more voltage or current pulses after this time because the load adsorbs all the power.

Example – 'Tline' Short-Circuited

Now we have $R1 = 1f\Omega$. The initial current pulses at terminals *a* and *c* again are equal and opposite at *20 mA*. But this time, at the far end, the voltage is zero. And we have an identical reflected current pulse sitting on the incident current pulse which also bounces back and forth along the line for ever. In the case of ('Lossless Gamma v1.asc'), the input is a short-circuit, so there are no more voltage pulses after the initial one, but in the case of a current input, schematic ('Lossless Gamma v2.asc'), the input is an open-circuit, so we see repeated inverted voltage pulses at *100 ms* intervals.

Example – 'Tline' Mismatch Termination
Schematic ('Lossless Gamma V3.asc') terminates the line with a *1 Ω* resistor, and we find a current reflection in agreement with Eq. 10.20.

Example – 'Tline' Effect of Source Impedance
If, with the line short-circuited, we make *R2 = 50 Ω*, the injected voltage is halved to *0.5 V*. The current pulse returns from the far end; in the case of a voltage input, it is completely adsorbed by the matched resistor *R1*. A similar situation arises if we open-circuit the far end.

10.4.4 Multimode Behaviour

A transmission line may be capable of more than one mode, in particular if the characteristics of the conductors are not all identical. We can create as many modes as we wish by placing lines in parallel. Each will be excited independently by the source. However, unless the lines have the same characteristic impedance, it is impossible to match both at the far end so there will always be some reflection. It can often help to make things clearer if we use two plot panes.

An example can be found as ('TransmissionLineInverter.asc') in the **Educational** folder. We now model the inner conductor of a coaxial cable by one *tline* and use another for the outer braid with a different characteristic impedance and delay time.

Example – Two Identical 'Tlines'
This is schematic ('Dual Mode Lossless Line.asc') which has a voltage drive. We see initial equal and opposite currents of *20 mA* at the input terminals of both lines because the input is connected to *T1(a)* and *T2(c)*. We find currents of *40 mA* at the far end of both resulting in a current of *80 mA* in the *1 nΩ* terminating resistor: but remember this is just the peak current at that instant of the reflected wave superimposed on the incident one, and the actual current of *40 mA* is reflected back as *20 mA* on each line.

If we terminate the line with a matched load of *25 Ω*, there is no reflection, and the current is adsorbed by *R1*.

If we leave the line open-circuit, we see no current pulses at the far end, only a voltage pulse of *2 V* which is inverted each time and equal and opposite current pulses of *40 mA* as the start .

Example – Two 'Tlines' with Identical Impedance and Different Delays
Using the same schematic ('Dual Mode Lossless Line.asc'), we increase the delay of *T1* to *75 ns*. With *R1 = 1nΩ* if we plot *Ib(T1)* and *Id(T2)*, we see a repeating pattern of the individual output current pulses. After the first, we find separate pulses from *T1* at 150 ns intervals and from *T2* at 100 ns intevals. This is a result of the choice of delay; if we increase it to *85 ns* we now find a partial overlap of the current pulses and a stepped current in *R1*.

If we open-circuit the load so that the two outputs are joined we find that each drives a current in the other. For example, with $Td=85\ ns$ for $T1$ at $50\ ns$ the current pulse from $T1(b)$ drives an equal opposite current in $T2(d)$, and at $85\ ns$ the pulse from $T2(d)$ arrives and drives an equal opposite current in $T1(b)$ so that after a further $50\ ns$ at $135\ ns$ the reflection occurs at $T1(a)$. This is coincidental with the pulse at $T2(c)$ created by the reflection of the pulse $Id(T2)$ at $50\ ns$ arriving $85\ ns$ later.

Example – Matching Two 'Tlines' with Different Delays
However, if we try to match the lines to a load the situation is different. The schematic is now ('Dual Mode Lossless Line V1.asc'). There is a slight connection difference in that the input is applied to the a terminal of both. It also helps to set a very long delay on $T1$ – something like $1000\ ns$ – so the pulse will not arrive during the transient window of $400\ ns$.

Equal and opposite pairs of current pulses are launched at time $0\ s$. After $50\ ns$ those from $T2$ arrive at the far end and we find $Ib(T1) = 13.33\ mA$, $Id(T1) = -13.33\ mA$, $Ib(T2) = 26.68\ mA$, $Id(T2) = 26.68\ mA$ $Vout = 668\ mV$, $I(R1) = 13.4\ mA$. The $50\ \Omega$ characteristic impedance of line $T1$ is now in parallel with the $50\ \Omega$ load resistor. And as the lines are bidirectional, the original current pulse can drive a current from the end of $T1$ towards its beginning. So the $20\ mA$ pulse divides as (a) 1/3 sent back along line $T1$ (b) 1/3 adsorbed by the load (c) 1/3 creating the reflected wave. This agrees with the measurements above. And the currents add to $+/\ 40\ mA$ which is what we expect from a $20\ mA$ incident wave.

At $150\ ns$ the pulse from $T1$ arrives, and we find the reflected wave of $13.33\ mA$ at $Ia(T2)$ and $-$ of course $-$ $-13.33\ mA$ at $Ic(T2)$. There is no current in $T1$. The difference in current is that adsorbed by the load resistor at $50\ ns$.

This attenuated current wave travels down the line again to arrive after $150\ ns$ where again it is partly adsorbed and partly reflected and we find an output of only $223\ mV$ and a load current of $4.43\ mA$. Thus we find the current rapidly decays. Changing the characteristic impedances does not alter the underlying behaviour.

If we now restore the delay of $75\ ns$ on $T1$, the picture becomes more complicated, but it is not fundamentally changed, and we again see each line passing reverse current pulses along the other.

Example – Two 'Tlines' with Different Impedances
If we use the previous schematic but change $Z0$ of one of the lines, we find that the currents change proportionally to the impedance but otherwise everything is unchanged.

Explorations 2
1. Use the two schematics to explore the effects of the load mismatch with a single line.
2. Try connecting end b to c instead of ground with a single line.
3. Explore the effects of lines in series.
4. Excite the lines with a current instead of a voltage.
5. Try more variations with Dual Mode Lossless Line schematics.

10.4.5   Frequency Response

This now changes from the time domain to the frequency domain. We must no longer imagine a pulse travelling along the line but rather a continuous train of waves. If the line is not matched there will a reflection from the far end as equation 10.21 showed and the impedance is no longer the characteristic impedance and so is not constant for all frequencies.

10.4.5.1   The Simple Explanation

We can make some progress towards understanding the behaviour of the line if we image the frequency of the input signal is such that its wavelength is exactly 1/4 of the wavelength of the signal in the transmission line. This last condition needs to be stated because the velocity in the line is generally 2/3 of the speed of light.

If the line is terminated in a short-circuit, the reflection coefficient is -1, and because the line has no losses, we have an inverted wave of exactly the same amplitude reflected back. When this reaches the start of the line, we therefore have an identical but inverted wave, and so the two will cancel leaving no nett input voltage, hence no current, and thus, although there is the input voltage, the line appears to have infinite input impedance. If the input wavelength is not exactly 1/4 of the wavelength of the line the amplitude of the reflected voltage when it returns the start of the line will not be the same as the input and so there will not be complete cancellation but some residual voltage that drives a current in the line. Then the line will appear to have an input impedance greater than its characteristic impedance, but not infinite.

If we increase the input frequency so that 3/4 wavelengths fit in the line, the reflected voltage will again be in anti-phase and again we expect to find infinite impedance. And this will be true whenever the reflected wave, when it returns to the start of the line, is inverted and of exactly the same amplitude, that is, when $\lambda = (1 + 2n)/4$ where n is an integer and λ is the wavelength.

Now as wavelength is proportional to the reciprocal of the frequency, we should expect to see the wavelengths for infinite impedance to follow a reciprocal relationship, that is, they should get closer together with increasing frequency. However, they are equally spaced in the LTspice *tline* which uses something other than a simple LC ladder of identical values.

But we are still left with the knotty problem of what happens if the reflected wave, when it returns to the input, is in-phase as it will be if the line is a multiple of 1/4 wavelength. Now it will add to the input, and we could imagine, perhaps, twice the current and thus half the input impedance. But in practice we know that the impedance of the line will be zero.

One explanation – though very imprecise – is to consider the very first LC section of the line. At a frequency $\omega = 1/\sqrt{(LC)}$, this will appear as a short-circuit. If we calculate the frequency and compare it to reality, it is wrong; it is too low. But at least it is some insight into what is happening. Similarly, in a vague way, if we halve the frequency, the impedance of the inductors is halved and that of the shunt capacitors doubled. And now the first section is not a short-circuit, but in conjunction with the next section or two, we might still achieve zero impedance. So, although this is a very unsatisfactory explanation, it does at least correctly point to frequencies of zero impedance when the input wavelength is 1/2 the wavelength of the line. And this also, correctly, predicts that if we increase the frequency the impedance finally becomes that of the first inductor. The correct explanation now follows.

10.4.5.2 Impedance with Non-matched Load

If there is a mismatch in the line termination, the impedance is no longer Z_0. Instead we must go back to Eqs. 10.12 and 10.18. If we assume the line is short-circuit, the reflection coefficient is -1, so the forward and reverse voltage amplitudes are identical as v, and we can write Eq. 10.12 as:

$$v(x) = ve^{-\gamma x} + ve^{\gamma x} = v(e^{-\gamma x} + e^{\gamma x}) \Rightarrow v(x) = v2\cos(\gamma x) \qquad (10.24)$$

As the line is lossless, $\gamma = j\beta$ finally we have:

$$v(x) = 2v\cosh(\beta x) \qquad (10.25)$$

From Eq. 10.18, again with a reflected wave equal in amplitude to the incident:

$$i(x) = \frac{v}{Z_0}(e^{-\gamma x} - e^{\gamma x}) = \frac{2v}{Z_0}\sin(\gamma x) \quad v(x) = 2v\sinh(\beta x) \qquad (10.26)$$

and the impedance at any point is:

$$Z(x) = Z_0\tanh(\beta x) \Rightarrow Z(x) = Z_0 j\tan(\beta x) \qquad (10.27)$$

We shall measure the line starting from the far end, and we now are interested in what happens where $x = l$ (the length of the line). We consider the condition when $\beta = n2\pi/\lambda$ where n is an integer and λ is the wavelength and l is exactly 1/4 of the wavelength. Then:

$$Z(l) = Z_0 j\tan(\frac{2\pi n}{\lambda}\frac{\lambda}{4}) \Rightarrow Z(l) = Z_0 j\tan(\frac{n\pi}{2}) \qquad (10.28)$$

and we see that with n the impedance alternates between zero and infinite as:

$$\text{if } \lambda = \frac{1+2n}{4} \Rightarrow Z(l) = 0 \quad \text{if } \lambda = \frac{n}{2} \Rightarrow Z(l) = \infty \qquad (10.29)$$

Moreover, we also see the shape of the impedance graph between these extremes if we use the absolute value. And if the line is open-circuit we find: $Z(l) = Z_0 \coth(\gamma x)$

10.4.5.3   Cut-Off Frequency and Bandwidth

The lumped parameter model is valid when the phase shift in each section is very small. As it is difficult to see a 3 dB point because, as we shall see in the simulations, the output is mainly cusps, an adequate estimate of the maximum frequency is $\omega = \frac{1}{\sqrt{LC}}$ using the values of the first section of the line. It therefore follows that if we double the number of sections, we can halve the values of the inductors and capacitors, still giving the same totals, and thus double the upper frequency limit.

From this we can see that for a good frequency response, we shall need perhaps 100 sections. This imposes a computational overhead, but the components are linear so it is acceptable. The website https://en.wikipedia.org/wiki/Telegrapher%27s_ equations has alternative active circuits to realize the telegrapher's equations.

If we are dealing with a pulse input, the bandwidth requirement will usually be determined by the steepness of the rising and falling edges. Given that the bandwidth $BW = 0.35\, t_r$ where t_r is the rise or fall time, whichever is steeper, and that we need at least *10* sections per wavelength, we have:

$$ n \times \frac{1}{10 \times T_d} > \frac{0.35}{t_r} \tag{10.30}$$

where n is the minimum number of sections.

10.4.5.4   Spatial Distribution of Voltage and Current

From the previous section, we can replace *cosh* by *jcos* and the magnitude of the current and voltage will follow half sine-loops, forming a standing wave pattern when the wavelength of the input voltage in the transmission line is a multiple of 1/4 wavelength, Fig. 10.7, which is drawn for a short-circuit end: for open-circuit, exchange the labels *voltage* and *current*.

Example – 'Tline' Impedance with Matched Load
We change the delay time to $T_d = 5\ ns$ in schematic ('Lossless Gamma v1.asc'), and with an AC input and the directive *.ac dec 1e5 10e5 100 g* and the line terminated by $Rl=50$, we see a correct constant input impedance $Zin(V1)$ of 50 Ω and an output of 1 V, and its frequency response is flat however far we like to go. The phase shift is fairly constant up to some *100 MHz* then falls to a minimum of $-770k°$ at *43 GHz* showing that the line must contain the equivalent of several hundred sections.

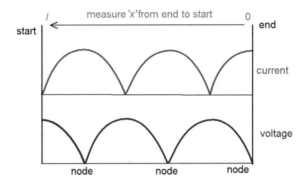

Fig. 10.7 Standing waves

Example – 'Tline' Impedance with Short-Circuit Load

In order to reduce the sharpness of the peaks, we shall terminate the line in a very small impedance of *1 Ω* and reduce the run to *1 g* in schematic ('Lossless Gamma V3.asc'). Then we see a sequence of evenly spaced cusps of *Zin(V1)*. To view these in detail it is better use a linear x-axis. But with a logarithmic scale starting from the left, the impedance rises at *20 dB*/decade implying an inductive behaviour before the first cusp at *50* MHz of *2.5 kΩ* on a linear y-axis. This is because of the load mismatch. The reflection coefficient with *1 Ω* is *−49/50* which will not completely cancel the forward wave to give infinite impedance but leaves *50 × 50 Ω = 2.5 kΩ*. We can check that by substituting other terminating resistances to find the cusps are *5.0 kΩ* with *0.5 Ω* and *25 kΩ* with *0.1 Ω*.

We can similarly explain why the minimum is not zero: that too is due to the load mismatch where the next cusp at *100 MHz* is *1 Ω* and increases to *2 Ω* if the load is increased to *2 Ω*. This pattern continues with the remaining cusps.

It is illuminating to add traces of the currents in the resistors *R1* and *R2* and to cancel the *logarithmic* x-axis and use a *logarithmic y-axis*. We then see a 'U'-shaped trace for *R1* peaking at *1.0 A* on the cusps – which agrees with the *1 V* input and *1 Ω* load resistor – and falling to *20 mA* at the bottom of the 'U' which agrees with the *50 Ω* line and a *1 V* input.

The current circulates backwards and forwards along the line with just a little power being adsorbed by the load resistor each time. At the same time the current in *R2* shows a series of cusps from a minimum of *400 µA* at *5 MHz* when the input impedance is *2.5 kΩ* – which agrees with the input of *1 V* – to a maximum of *1 A* at *10 MHz* with the load resistance of *1 Ω*. We can also see the phase decreasing in steps of *180°* . From Eq. 10.28, the impedance between cusps varies as the tangent, which is what we see.

And if we open-circuit the far end, the impedance falls at *−20 dB/decade* before reaching the sequence of cusps which are now inverted, and the line initially behaves as a capacitor.

Explorations 3

1. Use schematic ('Lossless Gamma V1.asc') with loads from *1 fΩ* upwards.
2. Terminate the line with a reactive load consisting of resistance and inductance or capacitance.

3. Swap the output terminals in schematic ('Lossless gamma V2.asc') and note that it only inverts the pulse.

10.4.6 Discrete Lossless Transmission Line

This can be built from cascaded LC sections, Fig. 10.8. This could well be made into a sub-circuit so that the line can easily be extended. It consists of 10 identical cascaded sections each consisting of a series inductor and a shunt capacitor and will therefore result in 10 maxima of $Z(in)$.

The previous equations of Sect. 10.3 no longer apply because it is not safe to view the line as composed of a very large number of infinitesimally short sections when, in fact, there may be only ten. The result is that the granularity of the line becomes important but can be mitigated by increasing the number of sections. However, the result from Eq. 10.7 that γ is imaginary and so there is no signal loss in the line is still true.

10.4.6.1 Dispersion

This occurs because the waves do not all travel along the line at the same *phase velocity*. We can see this in the schematic ('Group Delay.asc'). At low frequencies there will be almost no phase shift between the input voltage and the output across the capacitor. Hence there is only a very small delay between input and output, and this applies over several decades of frequency and only becomes significant at about *1/10* of the resonant frequency.

The point is that provided we keep to that limit, the line is not dispersive and we should not expect to see significant distortion of the signal because of the different group delays.

Example – Group Delay
Using the schematic ('Group Delay.asc') with the values $C = 50\,pF\ L = 10\,pH$ that are used in the *1000 section* line below, a transient run at *1 GHz* shows no visible phase shift, just a slight reduction in amplitude between input and output. With an ac run, we see the resonance at *7.11 GHz*. And if we look carefully at the first two cycles of the AC run we can see a small ripple voltage superimposed on the signal and estimate its frequency as some *8 GHz* - very close to the resonance. If we set a cursor

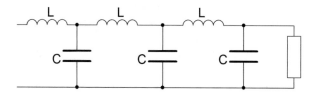

Fig. 10.8 Lossless line

on the output we can also note that the group delay is *53 fs* at *1 GHz* falling to *50 fs* as we go to lower frequencies. Therefore we should expect the line to show an overall group delay of *50 ps* which is about a thousand times less than the signal delay.

10.4.6.2   Impedance

At frequencies up to about *10%* of the resonance, the previous relationship that $Z_0 = \sqrt{(L/C)}$ is still good. But if the input is a pulse with a steep rise the frequency will extend to *1 GHz* . A full exposition, but from a slightly different viewpoint is to be found at *www.sophphx.caltech.edu/Physics.../Experiment_14.pdf*.

Example – Discrete Lossless Line Rise Time
The schematic ('Lossless Delay Line Discrete 10.asc') is a simple ten-stage LC ladder. This will show strong ringing with a steep input pulse with a matched load. If we change the rise and fall times of the pulse to *5 ns*, the ringing is greatly reduced. If we compare it to ('Lossless Delay Line Discrete 20.asc'), the waveform is much cleaner. If we find the frequency of the ringing, it is the resonant frequency of the LC sections.

However, the response in both cases with open-circuit and short-circuit outputs is very poor.

Example – Discrete Lossless Line Frequency Response
We can regard the above line as a ten-stage LC filter, where each section will have an asymptotic slope of *-40 dB/decade* and therefore we expect an asymptotic output slope of *-400 dB/decade*, and this is what we measure with an AC input. It is also noticeable that to see the cusps correctly, we need a small load resistor, perhaps *1 Ω*, and at least *25,000* steps per decade; else the sharp cusps are not all the same amplitude.

A typical set of cusps of maximum impedance is:

MHz	17	49	81	110	136	159	177	190	199
No. 1/4 waves	3	5	7	9	11	13	15	17	19
MHz/cusp	5.6	9.8	11.6	12.2	12.4	12.2	11.8	11.2	10.5

If we assume the first cusp is at 3/4 wavelength rather than 1/4 and given the inaccuracy of measuring the cusps, there is tolerable agreement between the number of 1/4 wavelengths of the input voltage and the position of the cusps. This is sketched in Fig. 10.9 where on a logarithmic scale we would see the upward slope of the impedance of the inductor at *20 dB/decade* from *100 kHz* to *10 MHz* before the first cusp.

Example – Spatial Distribution of the Voltages
We have no access to the internal workings of the LTspice transmission line, but with the discrete line, we can probe the nodes. We use schematic ('Lossless Delay Line Discrete V1.asc') which is the same circuit as ('Lossless Delay Line.asc'). We first make a wide *ac* run to see the cusps.

There is some problem finding the resonant frequency because LTspice drastically reduces the number of measurements, so a linear run over a restricted frequency

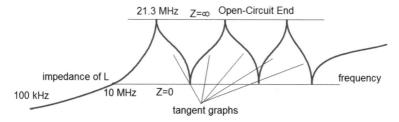

Fig. 10.9 Input Impedance Zin

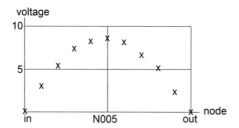

Fig. 10.10 Spatial voltage

range encompassing only the first cusp better. Even so, the resulting frequency of *44.3634 MHz* produces an inferior result than *44.35 MHz* which we use for a sine wave source to create standing waves on the line. As it takes a little time for the amplitudes to settle, we start making measurements after *1.5 µs* and measure the voltages at all the nodes, remembering that for node *V(in)* we must subtract the input voltage to find the contribution from the standing wave – which gives zero. We can plot the graph of Fig. 10.10 where the numbers are typical. These show a good approximation to a half sine wave. We can check by dividing each value by the sine of the angle when we should get the same number each time. This is almost true and is very sensitive to the exciting frequency.

What is perhaps a little surprising is that the voltage at node N0005 is some 100 times larger than the input.

It is interesting to excise each inductor in turn in schematic ('Lossless Delay Line Discrete.asc'). If we remove L2, we are left with the first series resonant circuit and the impedances cancel, when $j\omega L1 = 1/(j\omega C1)$.

Explorations 4
1. Measure the spatial distribution of the voltage on the other delay lines; in particular, try higher harmonics and different line terminations.

1000-Section LC Line Sub-Circuit
This was derived from the 10-stage ladder by using **F6** to defined a rectangle surrounding the ladder and drag a copy clear of the original. This was repeated to make a total of five instances, that is, 500 sections, that is schematic ('Lossless Delay Line Discrete 50.asc'). The net list was saved by **Edit as an Independent Netlist** which was edited to remover everything that was not part of the sub-circuit and the

header and **.end** added. In the first attempt, it was made as a two-pot network with **.subckt LC50 in 0 out 0**, but the two ground pins were not connected together, so the present version has just *3* pins and exists in the 'Schematics(asc)' folder so that it is easily available. The capacitors and inductors are defined as parameters. As the first line comment would automatically be used as the search path for the sub-circuit when the option to automatically create a symbol was chosen, it was edited. There was also a little editing to the positions of the pins and the size of the rectangle.

Example – 1000 Section Lumped Line
This is schematic ('Lossless 1000.asc'). The test values of $L_s = 10\,pH\ C_p = 50\,pF$ are unrealistic but a terminating resistor of $0.45\,\Omega$ does not show reflections. Changing the components to $L_s = 2500pH\ C_p = 1\,pF$ creates a characteristic impedance of $50\,\Omega$, and a delay time 50 *ps/section* = *50 ns* overall and the pulse shape is better preserved than before. Also the 'round trip' time of the pulse from start to end and back is in reasonable agreement, remembering that the time for intermediate nodes is shorter.

A good approximate figure for the delay of a 1 m length of co-axial cable is *5 ns* so we divide both the component values by *10*. We must also reduce the pulse duration **PULSE(0 1 0 1n 1n 2n** and the time to *60 ns*. As before, there is no inversion of the signal at the far end, but, when it returns at about *12 ns*, the input voltage source is a short-circuit, and so there is an inversion before the signal is sent on its way again. We can also note the doubled amplitude at intermediate nodes if the return wave falls on top of the incident wave. If we show node *012* onwards, we can see the two waves partly overlapping to create a castellated wave.

If we replace the high resistance by *1 nΩ*, we find a steady decline in the signal amplitude from node *015* onwards to the end.

Explorations 5
1. Use the various discrete delay lines to explore the effects of line length, delay, input pulse shape and so on – in short, anything that looks interesting.
2. Schematic ('LC 2 Stage.asc') has $Zo = 74.8\,\Omega$, close enough to *75 Ω*. There is some signal delay and considerable distortion of the input pulse and with a matched load almost no reflection. Measure the group delay and the frequency response.
3. Repeat the Example - Spatial Distribution of the Voltages using other discrete lines
4. Repeat 3, but measure the current.
5. It is instructive to alter the length of the line by excising the inductors, and it will be found that the number of cusps depends on the number of sections

10.5   Lossy Transmission Line

This is a more realistic description of practical lines. At RF two factors come into play. The first is the *skin depth* of the conductor, whereby the current cannot penetrate to the core of the conductor and is confined to the outer layer. It is proportional to the square root of the reciprocal of the frequency: the effect is negligible at audio frequencies being *0.66 mm* at *10 kHz* but falls to *0.066 mm* at

1 MHz and is only *0.0066 mm* at *100 MHz* resulting in a rapid increase in the resistance. For a *1 mm* copper wire we find from https://chemandy.com/calculators/round-wire-ac-resistance-calculator.htm that the resistance per metre increases from *0.031 Ω* at *100 k* to *0.825 Ω* at *100 MHz*. The second effect is *dielectric loss* which is significant at high frequencies and is reduced by careful choice of dielectric.

10.5.1 Analysis

This is derived from section 'Impedance with Non-Matched Load' only we now include α. With a short-circuit, at the start of the line where $x = l$, equation 10.27, becomes:

$$Z(l) = Z_0\tanh(\alpha l + j\beta l) = Z_0\left[\frac{\tanh(\alpha l) + \tanh(j\beta l)}{1 + \tanh(\alpha l)\tanh(j\beta l)}\right] \tag{10.31}$$

On the assumption that the losses are small, we can replace *tanh(αl)* by *αl* and we can write *jtan(βl)* for *tanh(jβl)*, and thus we have:

$$Z(l) = Z_0\left[\frac{\alpha l + j\tan(\beta l)}{1 + j\alpha l\tan(\beta l)}\right] \tag{10.32}$$

Then

$$\text{if } \lambda = \frac{1 + 2n}{4} \rightarrow Z(l) = \frac{Z_0}{\alpha l} \quad \text{if } \lambda = \frac{n}{2} \rightarrow Z(l) = Z_0\alpha l \tag{10.33}$$

If the line is open-circuit, the analysis yields:

$$Z(l) = Z_0\left[\frac{1 + j\alpha l\tan(\beta l)}{\alpha l + j\tan(\beta l)}\right] \tag{10.34}$$

and the minimum and maximum impedance conditions are reversed:

$$\text{if } \lambda = \frac{1 + 2n}{4} \rightarrow Z(l) = Z_0\alpha l \quad \text{if } \lambda = \frac{n}{2} \rightarrow Z(l) = \frac{Z_0}{\alpha l} \tag{10.35}$$

The condition for no distortion is that $R/G = L/C$.

10.5.2 The LTspice 'ltline'

This is more comprehensive. It is described by a *.model* card similar to the switch. It would appear that this is 'work in progress' in that some parameters are not implemented, particularly *G*, and we have only *R*,*L* and *C*. Given that *L*/*C* is often more than *1000*, the omission of *G* is not so serious. It seems that some of the switches work. However, unless we are dealing with hundreds of metres of line, or very high frequencies, the shunt conductance can be ignored. But lacking *G* we cannot create situations where there is no distortion.

Example – Z_0 and Delay
We find Z_0 with $R = G = 0$ $C = 100$ *pF* $L = 100$ *nH* and is *19.3 Ω* The delay is calculated as *3.15 ns*. However, with $L = 1$ it is measured as *2.1 ns*, at $L = 2$ it is *5.1 ns*, and at $L = 3$ *8.1 ns* but with $L = 10$ it is correct, and the wave shape is greatly improved. This is schematic ('Lossy Line Z0.asc').

Example – Reflections and Impedance
We can use schematic ('Lossy Transmission Line.asc'). This models a typical co-axial cable using $C = 100$ *pF* $L = 250$ *nH* $Len = 10$ to give *5 ns* per length and a total of *50 ns*. With a termination of *1 fΩ*, the behaviour is the same as the *tline* except for the decaying amplitude.

10.5.3   Discrete Lossy Line

The schematics ('Lossy Delay Line Discrete.asc') is derived from the lossless line with the resistances of the inductor and capacitor added, but the signal is so distorted even without them that it is difficult to see any effect. Modify the 1000-Section sub-circuit: that should be better.

10.6   Summary

In this chapter we started with a more detailed exploration of the URC and then moved on to transmission lines where mainly we used the LTspice lossless *tline*, compared it to a discrete line, and finally briefly looked at lossy lines, both LTspice *ltline* and discrete. The main points of a lossless line are:

- A line can be terminated with a short-circuit when the reflection coefficient is −*1* and an equal and opposite voltage wave is reflected.
- The line can be terminated with a matched load which adsorbs the incident wave, and there is no reflection.
- It can be terminated with an open-circuit, and now the reflection coefficient is *1*, and the reflected wave adds to the incident.
- If the reflected wave is in anti-phase with the incident, the voltages cancel and the line is an open-circuit. This occurs when the line is an odd number of 1/4 wavelengths with a shorted end or an even number with an open end.
- In the general case the amplitude will decrease in time as energy is dissipated by the terminating resistor.

Chapter 11
Inductors and Transformers

11.1 Introduction

This chapter is mainly concerned with the physics of magnetic fields and electric currents which are inextricably intertwined: a current invariably creates a magnetic field and a varying magnetic field always induces an electromotive force which frequently creates a current. The subject can become very complicated, and magnetism as a topic was often avoided or left to physics courses, partly because many of the resulting equations were not easily amenable to manual analysis: all that has changed with SPICE. And although the design of all but simple inductive devices – not just transformers but also chokes – is best left to specialist companies, a realistic analysis is possible, albeit at the cost of quite complex models.

In this chapter we shall first introduce essential magnetic units and the derivation of inductance and then move on to the more vexed problems of practical inductors and transformers and their representations in SPICE.

11.2 Magnetism

In 1820, Hans Oersted found that a magnetic field is produced when a current flows in a wire. And in 1831, during various experiments, Michael Faraday found that a wire placed in a changing magnetic field generated an emf: whole industries have been founded on these two discoveries. We shall use both to explain the phenomenon of inductance. But let it be noted that this is an area where we can find ourselves immersed in abstruse mathematics which, as simple engineers, we shall avoid.

© Springer Nature Switzerland AG 2020
C. May, *Passive Circuit Analysis with LTspice*®,
https://doi.org/10.1007/978-3-030-38304-6_11

11.2.1 Magnetic Effects of an Electric Current

Let us start with Fig. 11.1. This shows the circular lines of force surrounding a conductor carrying a current. The force is directly proportional to the current so we may write:

$$F = kI \qquad (11.1)$$

where k is a constant. Ampere found that the force per unit length of two parallel current-carrying conductors in air was:

$$F = \frac{\mu_0}{2\pi} \frac{I_1 I_2}{d} \qquad (11.2)$$

where I_1, I_2 are the currents in the wires, d is their separation and μ_0 is the rather unfortunately named *permeability of free space* or *vacuum permeability*. It is, however, simply an arbitrary constant and does not imply some mystical property of empty space.

11.2.1.1 The Ampere

The besetting problem before the start of the twentieth century was to incorporate magnetism and electricity into the mass-length-time systems of units. The unit of electric charge, the coulomb, was one attractive possibility, but it is difficult to collect isolated electric charges in any quantity, so, instead, single-charged ions were used, and the ampere was defined as that electric current that would deposit 0.001118 g of silver per second from a solution of silver nitrate. But as silver consists almost equally of two common isotopes, Ag107 and Ag109, differing in mass by about 2%, any imbalance in the ratio would invalidate the result, let alone the difficulties of ensuring the solution was always saturated and that the current was constant, and in drying and weighing the silver.

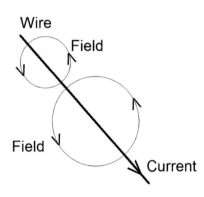

Fig. 11.1 Magnetic field around wire

So from 1946 the ampere has been defined as that current which, flowing in two long parallel conductors of negligible circular cross-section, spaced 1 m apart in a vacuum, generates a force of 2×10^{-7} newtons per metre between them. This has the merit of tying the ampere to a force and thus to the mass-length-time triad, but there are attendant measurement difficulties with such a small force. And as this is the definition of a standard, the results must be accurate and precise to at least seven places. So generally the conductors are closer together and the current more than 1 amp, and then it is important to ensure that the conductors are not distorted but truly straight and parallel, in particular, that the force does not move them apart by even one micron, and corrections must be made for their finite cross-section.

This definition also fixes the value of μ_0 since, from Eq. 11.2 and with 1 A for each current and 1 m for d, we can write:

$$2 \times 10^{-7} = \frac{\mu_0}{2\pi} \tag{11.3}$$

so:

$$\mu_0 = 4\pi \times 10^{-7} NA^{-2} \tag{11.4}$$

But had the ampere been defined differently as, say, 10^{-6} N, μ_0 would then have been $2\pi \times 10^{-6}$. N A^{-2} which is why we say it is an arbitrary constant, not a fundamental property of nature.

11.2.2 Inductance of a Solenoid

As we have just seen, there is a magnetic field associated with every conductor carrying an electric current. If, however, the current is not constant, but changes, the magnetic field also will change. And as the conductor is immersed in its own magnetic field, an EMF will be induced in the conductor in such a direction as to oppose the change in current. This property is so defined that an inductance of 1 henry exists if a current changing at the rate of 1 A/s induces an EMF of 1 V. in the conductor. In general, the induced EMF is given by:

$$v = -L\frac{di}{dt} \tag{11.5}$$

where L is the inductance in henries and the negative sign indicates that the voltage opposes the current. However, this definition disguises the underlying relationship which is that it is the changing magnetic flux, not the current itself, that is responsible for the EMF. But if we assume a linear relationship between current and flux Φ we can write:

$$v = -Lk\frac{d\Phi}{dt} \tag{11.6}$$

where k is some constant.

Inductance is associated with every conductor carrying an electric current. In many cases it is a small effect and can be ignored; in others it is a property we wish to exploit as in a solenoid which is one of a few forms of inductance that is amenable to a simplified analysis. It consists of a long non-magnetic cylinder (although a rectangular cross-section could equally well be used) closely wound with turns of wire - perhaps in several layers - connected to a source of electricity. We can imagine a section through the longitudinal plane of the centre, Fig. 11.2, where the current entering the plane is shown by '+' and current emerging from it by a dot.

We must be careful to make a clear distinction between several effects when a current passes through the solenoid. The first is that the current will create a *magnetomotive force F*. This is determined by the current I and the number of turns on the solenoid N:

$$F = NI \tag{11.7}$$

and is measured in *ampere turns (At)*. This force is spread over the length of the solenoid and will result in a *magnetic field H* or *magnetizing force* which we can think of as the gradient of F, analogous to electric field strength:

$$H = \frac{F}{l} = \frac{NI}{l} \tag{11.8}$$

where l is the length of the solenoid and H is measured in *ampere turns per metre*.

Now let us return to our solenoid: following the direction of the magnetic field shown in the figure, the fields due to the individual turns of the solenoid will reinforce each other inside and outside the solenoid as shown in Fig. 11.3. On the other hand, between the turns, the magnetic fields will oppose each other and cancel as we see in Fig. 11.3.

To calculate the inductance, we now make two assumptions: the first is that the magnetizing force inside the solenoid is uniform. This is not too bad if the solenoid is long and thin. The second is that there is no magnetizing force outside the solenoid.

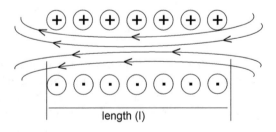

Fig. 11.2 Magnetic field inside a solenoid

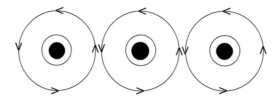

Fig. 11.3 Magnetic field between turns

But note that this does not imply that there is no magnetic field outside the solenoid (which there certainly is, as can easily be demonstrated by a compass needle or iron filings) but rather that the magnetizing force is due only to the current in the closed loops of the solenoid wire and thus is inside rather than outside the solenoid: so this is a reasonable assumption. For those interested in such things, look up the Biot-Savart law or Ampere's circuital law.

The magnetic field will induce a magnetic flux inside the solenoid where the *flux density B* is measured in tesla (T) and is independent of the area of the solenoid:

$$B = \mu_0 H = \mu_0 \frac{NI}{l} \tag{11.9}$$

Thus the flux per turn φ is the flux density multiplied by the area:

$$\phi = \mu_0 HA = \mu_0 \frac{IA}{l} \tag{11.10}$$

where A is the cross-section area of the solenoid. And thus the total flux Φ due to all the turns N is:

$$\Phi = \mu_0 \frac{NIA}{l} \tag{11.11}$$

in the case of air (strictly, a vacuum, but the difference can be ignored) and is measured in Webers (Wb).

If the current changes, the changing flux will link every turn of the solenoid, so from Eq. 11.11, the induced EMF will be the rate of change of the flux multiplied by the number of turns:

$$v = -\left(\frac{\mu_0 N^2 A}{l} \frac{dI}{dt} \right) \tag{11.12}$$

and hence from Eq. 11.5, the inductance L is:

$$L = \frac{\mu_0 N^2 A}{l} \tag{11.13}$$

And substituting Φ in Eq. 11.13:

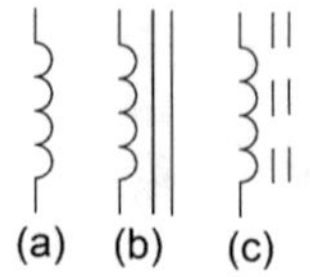

(a) (b) (c)

Fig. 11.4 Solenoid symbol

$$L = \frac{N\Phi}{I} \tag{11.14}$$

Also, if we differentiate Eq. 11.11, we have:

$$\frac{d\Phi}{dt} = \frac{\mu_0 NA}{l} \frac{di}{dt} \tag{11.15}$$

and hence for the voltage v:

$$v = -N\frac{d\Phi}{dt} \tag{11.16}$$

If the solenoid has a ferromagnetic core, perhaps a ferrite or soft iron laminations, we multiply μ_0 by the relative permeability μ_r for the material in Eq. 11.11, which is often of the order of a thousand.

11.2.2.1 Circuit Symbol

Drawn by hand, it is often a series of loops similar to the LTspice symbol. But the preferred IEC symbol for circuit diagrams is a series of touching semi-circles, Fig. 11.4. The nature of the core is shown by adjacent lines, or their absence. Referring to the figure, an air-cored inductor, one with no magnetic core, is shown in '(a)', one with a laminated iron core by '(b)' and one with a dust-iron or ferrite core by '(c)'.

An alternative symbol, mainly European, is a solid rectangle.

11.3 Inductors

So far we have assumed the inductors are ideal, that is, although they may have some resistance (and, in practice, they always will due to the winding wire), the magnetic properties are constant and unaffected by the current. Now we shall look at real inductors.

In many non-critical cases, we can use an off-the-shelf inductor with either an iron core for low frequencies or a ferrite core or just air for high frequencies where its inductance and series resistance are given in the data sheet. Otherwise, if we are dealing with some inductor that we happen to have to hand, we can measure its properties. For this, we need to delve a little more deeply into magnetics.

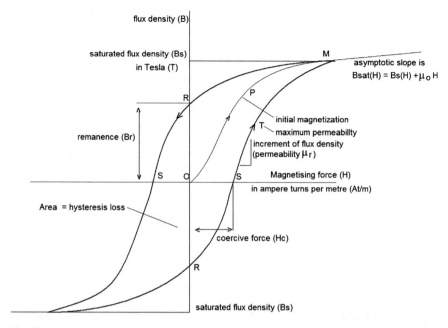

Fig. 11.5 Hysteresis loop

11.3.1   The B-H Curve

If the inductor has a core containing ferromagnetic material, such as iron, or alloys of iron, nickel, cobalt or similar metals, or a ferrite, then if it is magnetized and demagnetized, the magnetization will trace out a B-H curve shown in Fig. 11.5.

11.3.1.1   Magnetic Domains

The simple explanation for the hysteresis loop is that the core material contains myriad small magnetized domains, initially randomly oriented. But this begs the question: 'Why are there domains?'

It is all about the free energy. Suppose we start with a specimen completely magnetized in one direction, Fig. 11.6a. There is now a strong demagnetizing field outside the specimen created by the free magnetic poles at each end and containing a lot of energy. We can show this if we try to bring another specimen alongside it magnetized in the same direction with north to north and south to south, it will require considerable force. However, if its magnetism is in opposition with north to south and south to north, the new specimen will freely move to join the other rather than needing to be forced close to it. And we shall then find that the external magnetic field is less because the opposite free poles at the ends tend to cancel: the free energy has been reduced.

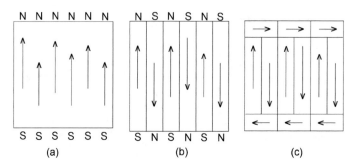

Fig. 11.6 Domains

So taking our original specimen, if we notionally divide it into sections whose direction of magnetization alternate, we can reduce the free energy Fig. 11.6b, and we can reduce it even further if we place closing sections across the ends, Fig. 11.6c. On this basis, it would seem that the lowest energy is to divide the specimen into infinitely small regions. This does not happen. To understand why we need to go to the atomic level.

The Pauli exclusion principle states that only two electrons can occupy an energy level in an atom. Regarding them as billiard balls, they have spin, and they will spin in opposite directions. And because the electrons have electric charge, a moving charge is a current and a current automatically has an associated magnetic field. Thus, if there are two electrons, their magnetic fields will cancel. However, in the case of iron and certain other *ferromagnetic* metals such as nickel and cobalt, there are unpaired outer electrons and the atom will have a nett magnetization.

Now from what we have just said, we would expect adjacent atoms to align with their magnetic fields in opposition. This does not happen because we need to bring in the exchange interaction between the atoms and, in the case of these metals, it favours alignment of their magnetism in the same direction along one of the crystal axes.

We now have the conundrum that, on this basis, we should be back to the first figure with all the atoms aligning their magnetism in the same direction – which also does not happen. Instead, as the number of aligned atoms increases, they create an external magnetic field raising the free energy of the region to the point that it overcomes the exchange energy. This creates a domain. And because the magnetization of all the atoms within a domain is aligned, its magnetism is saturated and cannot increase. But the domains themselves are oriented so that their vector sum is zero – that is, there is no external magnetic field, even though each domain is strongly magnetized.

The surrounding atoms now align in different directions creating new domains, each separated by a *domain wall* perhaps a hundred atoms wide where the direction of magnetization incrementally turns to the direction of magnetization of the new domain. It requires energy to create this wall. Thus the microscopic structure of the ferromagnetic material is that of a large number of domains randomly orientated giving the specimen no overall magnetization. The domains are not necessary co-terminal with the material grains: one grain may contain more than one domain, and, conversely, a single domain may encompass more than one grain.

The Wikipedia entry 'Magnetic_Domain' is extremely illuminating as is their entry 'Domain Wall (magnetism)' with its animation of 'unpinning' and illustration of the domain wall transition from one domain to another.

The final state is a balance between a number of forces, particularly the magnetostatic force of the domains and the energy of the domain walls. It is important to note that at high temperatures, usually more than 200 °C, the thermal agitation breaks down the alignment and the material loses its magnetism. This is the *Curie* point. A useful, but rather technical, article can be found at https://www.phase-trans. msm.cam.ac.uk/2003/Vicky.Yardley/Chapter03.pdf. Also https://www.tf.uni-kiel. de/matwis/amat/elmat_en/kap_4/backbone/r4_3_1.html has a complete physical description of ferromagnetism.

Hence, starting at point *O* on the Fig. 11.5, there is no bulk magnetization of the core. However, if we impose an increasing external magnetizing force *H*, first those domains whose magnetization is most closely aligned to the applied field will grow as their domain walls flex so creating an external magnetic field shown along the path *O-P*. The growth of the grains is limited by *pinning sites* created by foreign atoms, voids, or crystal defects. As the external magnetizing force continues to increase, the domain walls can jump past the pinning sites, shown graphically in the Wikipedia article, and the magnetization increases not at a steady rate, but in small steps which can be made audible with a simple apparatus – the Barkhausen effect. Then the direction of magnetization of non-aligned domains will try to turn: first those whose direction is closest to the applied force and along 'easy' crystallographic axes. Finally there are fewer remaining domains and they are difficult to turn so the flux density increases more slowly until saturation occurs at *Bs* where all the domains are aligned and in fact merge into a single domain. If the magnetizing force is increased beyond that, the flux increases slowly due only to the magnetizing force in air $\mu_0 H$.

If the magnetizing force is slowly removed, the magnetic flux does not follow the same path, but some domains are effectively locked in the aligned position by, for example, carbon atoms, and remain aligned and when the force has reduced to zero, there is some *remanence* at *R*. In fact, to demagnetize the specimen, the magnetizing force must be reversed and increased to point *S,* the *coercive force*. The curve is usually symmetrical so the specimen can be magnetized in the reverse direction with the same coercive force and saturated flux density.

The slope of the B-H curve is the permeability. For ferromagnetic materials, this is not constant but starts at a high value, so that a small increment of the magnetizing force creates a large increment in the flux density and the permeability is substantially constant, implying a linear slope to point *T* where there is a slight increase and thereafter it falls steeply towards *M*.

What is worrying is that if we increase the magnetizing force beyond *Bs*. The relative permeability of the core falls to a very low value (for the physicists, we are now dealing with the *paramagnetism* of the unpaired spins) and so the flux density rises very slowly. Remembering that the inductance is determined by the rate of change of flux, and hence the rate of change of flux density, the inductance becomes very small, and a very large current can result. Hence the power dissipated by the resistance of the winding will increase and can result in overheating and destruction of the inductor unless rapidly checked.

Magnetostriction
The movement of the domain walls can create a change in the physical dimensions of the magnet. As the material is taken through a magnetic cycle, this effect also exhibits hysteresis. It is included in the Jiles-Atherton model but not in SPICE. The effect is generally small in the material used for inductors and transformers but can result in an audible signal from the transformer.

11.3.2   The Magnetic Circuit

This is the analogue of the electrical circuit where flux replaces current. And just as the current is the same throughout the electrical circuit, so the total flux is constant. Even in the case of the simple solenoid, the concentrated flux inside it is equal to the spread-out flux of the return path through the air, so the flux 'leaving' at the north pole is equal to the flux 'returning' at the south. Certainly the flux density is by no means constant – but that is another story.

   Many inductors are wound on magnetic cores which have a closed path, Fig. 11.7, then to a good approximation, all the flux is contained in the magnetic circuit. This is the assumption made by the hysteretic model to be described later.

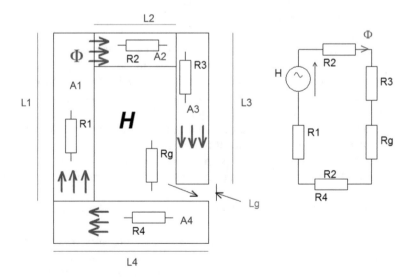

Fig. 11.7 Magnetic circuit

For a uniform cross-section A and including the relative permeability of the core, Eq. 11.11 becomes:

$$\Phi = \frac{NI\mu_r\mu_0 A}{l} \tag{11.17}$$

where we have changed the order of the numerator to show that as the magnetizing force is NI. This is the analogue of voltage and we have a 'magnetic Ohm's Law':

$$I = \frac{V}{R} \quad \Phi = \frac{NI}{R_{mag}} \tag{11.18}$$

where:

$$R_{mag} = \frac{l}{\mu_0\mu_r A} \tag{11.19}$$

is the *reluctance* and is the analogue of electrical resistance, and also $\mu_0\mu_r$ corresponds to the resistivity. Remembering that the magnetizing force is $H = NI$, we can write Eq. 11.18 as:

$$\Phi = \frac{H}{R_{mag}} \tag{11.20}$$

If the magnetic circuit consists of elements with different dimensions and magnetic properties, we can treat each as a 'resistance' using Eq. 11.19. If we take the simple magnetic circuit shown in the Fig. 11.17 we can divide it into sections, not forgetting the all-important air gap lg, and represent each as a reluctance:

$$R_{TOT} = R_1 + R_2 + R_3 + R_{lg} \tag{11.21}$$

The analogy breaks down slightly with H because the magnetizing force is distributed over the whole magnetic path, whereas we show it as a lumped voltage generator. But the calculation is not affected that:

$$NI = \Phi R_{TOT} \tag{11.22}$$

So knowing the magnetic properties and dimensions of every part, we can compute the flux from the number of turns and the current. More complex magnets, such as Fig. 11.8, can be handled in the same way, but generally there is no need because most cores have a constant cross-section. For example, the centre leg in this figure would be twice the area of the outer legs and the closing 'I' piece, Fig. 11.9.

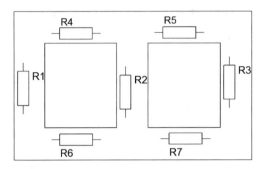

Fig. 11.8 Parallel magnetic paths

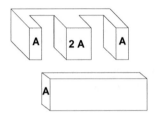

Fig. 11.9 Core section

11.3.2.1   The Effect of the Air Gap

If we suppose the areas are all equal in the figure and the relative permeability of the core is $\mu_r = 2000$, then an air gap of just *0.1 mm* has the same reluctance as *200 mm* of the core. Thus even a very small air gap greatly reduces the effective permeability:

$$\mu_{eff} = \frac{\mu_{init}}{1 + \mu_{init}\left(\frac{g_{len}}{MP_{len}}\right)} \tag{11.23}$$

where μ_{eff} is the effective permeability, μ_{init} is the initial permeability of the core and g_{len}/MP_{len} is the ratio of the gap length to the total magnetic path length as quoted by https://www.ieee.li/pdf/.../fundamentals_magnetics_desi. ...

Example – Effect of Air Gap
If then, we have a magnetic path length of *100 mm* and a gap of *1 mm* and $\mu_{init} = 2000$ then:

$$\mu_{eff} = \frac{2000}{1 + 2000\left(\frac{1}{100}\right)} = \frac{2000}{21} = 95.2$$

or roughly *1/20* of its value without a gap.

We may then wonder why, having carefully chosen a materiel with a high permeability, we now choose to dramatically reduce it. There are several reasons for this. One is that the air gap gives a more linear permeability, Fig. 11.10, so less

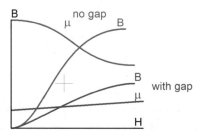

Fig. 11.10 Permeability with air gap

distortion. Secondly, the air gap reduces the risk of saturation. Thirdly, as permeability is not tightly controlled, but air gaps in ferrite cores can be, the effective permeability has better tolerance. On the other hand, there must now be many more turns of wire for the same inductance.

Manufacturers quote the effective dimensions of their ferrite cores but often save us all the bother of doing the sums, for example, by giving the inductance in millihenries per 1000 turns:

https://www.mag-inc/getattachment/Products/Ferrite-Cores/Learn-More-about-Ferrite-Cores/Magnetics-Ferrite-Catalog-2017.pdf?lang=en-US nc.com

Also of importance – and shown in graphs in this reference – is the temperature limit or Curie point at which the permeability rapidly falls to zero. For a more comprehensive treatment, see http://www.nptel.ac.in/courses/Webcourse-contents/IIT%20Kharagpur/Basic%20Electrical%20Technology/pdf/L-21%28TB%29%2928ET%29%20%28%28EE%29NPTEL%29.pdf.

11.3.3   Inductor Losses

Two critical aspects of inductor design are to ensure that the core does not saturate and that it does not overheat. The first is realized by keeping the excursions around the B-H loop well below Bs, the second by careful material selection and design. The losses are the following.

11.3.3.1   Hysteresis Loss

This is the area under the B-H loop and represents energy expended as heat in taking the specimen through the magnetic cycle as the domains turn and the domain walls move. This is minimized by selecting a materiel with a very narrow hysteresis loop. Iron-silicon alloys and iron-nickel alloys (especially *mu-metal*) are used at DC and low frequencies and soft ferrites at high frequencies. The loss is directly proportional to the frequency and can be calculated for sinusoidal excitation using Steinmetz's equation – see the Wikipedia entry. It has been improved, but the paper at www. apec-conf.org/Portals/0/.../2016/1124.pdf seems to have moved. However, in many cases, this loss is the smallest and is often disregarded.

11.3.3.2 Copper Loss

This is entirely due to the resistive losses of the wire. The length of the wire is set by the number of turns to achieve the desired inductance and cannot be changed. For low frequencies, thicker wire will reduce the resistance but at the cost of a larger and heavier core to accommodate the increased cross-section of the winding. There is usually a compromise, whereby the centre of the core is allowed to get hot, but not so hot as to degrade the insulation, typically 45° C.

For high frequencies, over 100 kHz, the *skin effect* comes into play, and the current is forced to the surface of the conductor, so it is better to use multiple strands of wire in parallel, *Litz wire*. In every case it is essential to be careful not to damage the insulation; else a 'shorted turn' may result. Although the induced voltage in this one turn is small, it has negligible resistance compared to the whole winding, and excessive current will flow which will quickly cause the inductor to overheat. A more detailed explanation can be found at Amphone.net.

11.3.3.3 Eddy Current Loses

Alternating currents in the inductor will induce EMFs in the core if it is magnetic. If the core were solid, it would represent a single shorted turn, so even if the resistivity were high, because of the large cross-section, a large current would flow causing considerable heating. The solution at low frequency is to make the core of thin laminations, typically less than 1 mm thick, insulated by varnish, and shaped as 'E' and 'I' which can be alternated to make a stack with a window for the winding. Cores of powdered material in an insulting matrix are used for high-frequency inductors where the small particle size reduces eddy currents. These cores are available in a variety of shapes and sizes. The loss is proportional to the square of the frequency.

11.3.3.4 Dielectric Loss

There is distributed capacitance between the turns with the insulation of the wire as the dielectric, usually polyurethane. This becomes important at high frequencies because this capacitance with the inductance has a resonant frequency.

11.3.4 Choice of Magnetic Material

From this, we can deduce the properties for different types of magnet. For permanent magnets, those used in loudspeakers and magnetrons, we require a 'hard' magnetic material, one with a high coercive force and a high remanence and, ideally, a high permeability. For chokes and transformer cores which are taken through magnetic cycles, we require a small coercive force to make the hysteresis loop narrow and hence the core losses small, also a high saturated flux density and a high permeability are desirable to keep the size small. For recording media, magnetic tapes or computer hard disc drives, a high remanence is needed and a high permeability and also the ability to make very small magnetizable particles since their size limits the data density.

There are just a few materials in common use for constructing inductors and transformers. Iron-cored transformers are quite robust and can sustain even prolonged overloads provided the temperature rise is not so great as to damage the insulation. With ferrite cores, however, overloads can cause the core temperature to rise above its Curie point, and then the inductance falls, and the current increases regeneratively, so protective circuits are essential.

11.3.5 Inductor Design

The days of large, heavy, iron-cored inductors working at mains frequency are largely past in light-current electronics, and most catalogue parts and design efforts are focused on high-frequency applications from a few kHz to over 100 MHz.

The design of an inductor is perfectly possible, and a very detailed article can be found at https://www.ieee.li/pdf/viewgraphs/fundamentals_magnetics_design.pdf which sets out the design steps and which we largely follow here. This is comprehensive and includes all the essential design parameters. The concise slide shown at website https://ninova.itu.edu.tr/en/courses/.../ekkaynaklar?... includes a step-by-step design, and there is a design flow chart at https://www.allaboutcircuits.com/technical-articles/basic-inductor-design-constraints/.

On the other hand, some manufactures of ferrite cores such as TDK have written design programmes to ease the problem of core selection within their product range and the calculation of the inductor parameters. These may – or may not – be user-friendly and helpful. But in any event, the first of the above websites is invaluable. There are other websites dealing with specific applications, often in high-frequency voltage conversion circuits.

Whilst SPICE may be used for the design or analysis of an induction, constructing inductors or 'chokes' is only practical using freely available cores usually in the form of 'pot cores' or toroids and only for simple designs where the wire is wound by hand, often on a plastic bobbin. If the inductor is to be made with several layers of wire with the turns laid side-by-side, a coil winding machine is needed. Low-cost

hand-operated machines are available for prototype inductors; but it can be difficult
to keep the turns close wound, and in any case production quantities are best left to
specialist manufacturers if a suitable catalogue part cannot be found.

11.3.5.1   The Inductance

This is obviously the critical design point. Core manufacturers often quote the
inductance per 1000 turns, A_L, typically in mH. And as inductance L is proportional
to the square of the number of turns, the number of turns N is given by:

$$N = \sqrt{\frac{L}{A_L}} \qquad\qquad (11.24)$$

Example – Inductor Turns Calculation
To make an inductance of *5 mH* given that $A_L = 1.2\ mH$ from Eq. 11.24, we require
$N = \sqrt{(5/1.2)} = 2041$ turns.

11.3.5.2   Power Handling

This is associated with the power loss and brings together a number of considerations. First, we must ensure that core temperature does not rise to the Curie point.
As this is usually at least 135 °C, the more immediate limit is the insulation of the
wire, usually polyurethane, also limited to 135 °C. And this is measured at the hottest
point, usually near the centre of the core. Also, the resistance of the winding wire
will increase with temperature, so as a rough guide, a rise of no more than 40 °C
should be aimed at. Also, a good working guide is an efficiency of 95% with the
losses spread evenly between the core, the DC winding loss of the wire and the AC
loss of the wire.

It is also important to ensure that even under the worst conditions, the core does
not go into saturation, which in turn raises the important question of the variation of
the permeability of the core with temperature. To further complicate matters, the core
characteristics change with frequency. All these are interlinked and are difficult to
model.

11.3.5.3   Core Material Selection

This depends upon the power and the frequency. Manufacturers data sheets quote the
core loss as mW/cm^3 of the core volume and the total volume of the core for each of
their products. This allows us a first guess at the correct core. The data sheet also
gives the winding window area.

Iron Laminations or Tape

For low-frequency power applications, that is, below about 10 kHz, iron laminations are used because for inductances more than a few millihenries, a large number of turns are needed and to keep losses low relatively thick wire must be used resulting in a physically large inductor which may exceed the size of freely available ferrite cores and so we use thin insulated soft iron 'E' and 'I' laminations. Although the laminations are standard size, the stack height is not, and these are custom designs.

Insulated flexible iron alloy tape can be used to form a toroid and then covered with insulation, sometimes as a tape, and can be found as stock parts in various sizes from some manufacturers.

Moly Permalloy Powder (MPP)

This is an alloy of some 80% nickel, 16% iron and 4% molybdenum although the exact composition varies slightly between manufacturers. The material is ground to a powder and mixed with a small quantity of ceramic binder which therefore creates a distributed air gap (or, to be precise, a non-magnetic 'ceramic gap'). It is then pressed into a toroid – the only available shape.

It has the lowest hysteresis loss of all materials. The permeability is relatively low because of the air gap and ranges from 15 up to 550. These cores are useful up to about 200 kHz. The inductance is stable against changes of DC flux, temperature and frequency. The tolerance on permeability is relatively low, and the cost is relatively high.

Iron Alloy Powder

These are toroidal cores made from iron or iron alloy powder mixed with a binder and compressed into toroids. They also have a distributed air gap and relatively low permeability but are usable up the megahertz range. They are cheaper than MPP cores.

Ferrites

There are various grades of ferrite. Murata have a very comprehensive online catalogue. The permeability changes dramatically with the magnetizing force and temperature and everything else. They have high permeability, often a few thousand, and can be supplied with precision air gaps. They are also available in a wide variety of shapes, often with matching plastic bobbins for the winding. Pot cores are especially useful because virtually all the flux is contained within the inductor and some have an adjusting screw to accurately set the inductance. The problems with ferrites are:

- Permeability can vary widely with temperature and falls to zero a little above the Curie point, usually less than 300 °C.
- Core loss varies with temperature from an almost constant small loss for T-type ferrite to 10 times bigger for L-type.
- Core loss can increase by a factor of 100 as flux density increases from 0 to about 50% of *Bmax*.
- *Bsat* falls linearly with temperature.
- Permeability changes with flux density as we would expect from the B-H loop. The graph is curved, rising gently, then falling.

- Core loss increases with frequency by a factor of 10 or more.
- Permeability remains almost constant with frequency until near the maximum when there can be a peak, and in every case there is a rapid fall at higher frequencies.

11.3.5.4   The Winding

Single layer toroids can be wound by hand, but for production quantities, an expensive winding machine is needed; else the work is handed over to a specialist company, and even then, they are only made one by one. For winding on bobbins, simple, very inexpensive, hand machines can be found suitable for prototypes and limited production runs.

Fill Factor
This is how much of the window can be filled with conductor. This can never be 100% because we will use round wire rather than square so the best possible is 78%. For a toroid this will be about 50% because space must be left for the shuttle to pass. The insulating varnish also adds a few percent and so does the bobbin or former and any insulation between layers. If voltages are going to exceed a couple of hundred volts, then the winding will not fill the whole of the window width because some creepage distance must be left to prevent voltage breakdown between layers. In all, a fill factor of 50–60% is usual but even lower for high voltage designs.

Winding Loss
The resistance of the windings is $R_w = \frac{\rho N l_t}{A}$ where ρ is the resistivity of the wire, A its cross-section area, N the number of turns and l_t the mean length of each turn – which is not often supplied by the core manufacturer. This is simplified by using wire tables giving the resistance per 100 m or per km for the wire sizes, either American Wire Gauge (AWG) or Standard Wire Gauge (SWG) as well as cross-section area. The only thing we do not know at this point is the mean length of each turn. A usual design rule is to make the winding loss equal to the core loss.

Self-Resonance and Q-Factor
The distributed capacitance turns the inductor into a tuned circuit damped by the winding loss and typically having a self-resonant frequency (SRF) of some megahertz given by $f_{srf} = 1/(2\pi\sqrt{LC})$, so it is not a problem with low-frequency inductors where we use the thickest wire to fill the window and the Q-factor is given by:

$$Q = \frac{\omega L}{R_{wire}} \tag{11.25}$$

Below the SRF the effective inductance L_{eff} is:

$$L_{eff} = \frac{L}{1 - \left(\frac{f}{f_{SRF}}\right)^2} \tag{11.26}$$

and becomes infinite at the SRF.

On the other hand, with toroids used at high frequencies where the working frequency approaches the SRF, it is good practice to keep on the rising 'Q' part of the curve below the SRF. These curves are published by the core manufactures for their core sizes and materials. Above the SRF the inductor behaves like a capacitor. The best design is to minimize the capacitance by using a single layer of wire.

Critique
The designs mentioned above do not directly take into account temperature changes nor tolerances. And it is also tedious to rework the calculations for different DC bias currents and changes of frequency both of which can seriously change the magnetic properties of the core. Coilcraft has example production statistics at Doc869_Statistics_for_Custom_Designs-1.pdf.

11.3.6 LTspice Inductor

If we are going to simulate an inductor, we need to be aware that the properties of ferrites change dramatically, far more so than iron-cored inductors. LTspice has three models which all accept the parasitic components and temperature coefficients, but frequency effects are problematical.

11.3.6.1 Linear Inductor with Parasitic Resistance and Capacitance

This is shown in Fig. 11.11. It is the model we shall use in the following chapter to explore circuits with inductors; capacitors are resistors where most of these are circuits with low levels of currents and sometimes with air-cored inductors. And as an initial investigation, a linear inductor is a good choice because, as with capacitors, any unexpected behaviour of the circuit will not be due to parasitic properties of the inductor.

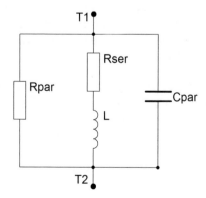

Fig. 11.11 LTspice inductor

On right clicking, the symbol the **Inductor – Ln** dialogue opens. The parameters are:

Parameter	Comment
Inductance (L)	Can be qualified as a Param else by tc1, tc2
Peak current (A)	Not acted on
Series resistance (Ω)	Resistance of winding wire – This has a small default value that generally can be ignored, except in resonant circuits – See Chap. 12 'Tuned Circuits'
Parallel resistance(Ω)	Use for hysteresis and eddy current losses
Capacitance (F)	Between winding and core and ground

The LTspice option **Select Inductor** opens a list of commercial inductors and quotes the inductance and series resistance. However, if we click **Quit and Edit Database** we find more information about the models. To generalize, these have inductance in the range up to about 100 µH; the parallel resistance is a few tens of kilohms; the series resistance is less than 1 Ω; and the parallel capacitance is only a few picofarads, but Coilcraft give no values. These are inductors wound on ferrite cores and mainly used at high frequencies.

If the check box **Show phase dot** is checked, a small open dot will be added to the symbol denoting the start of the winding. Four other parameters can be added:

ic initial conditions, applied if *uic* is set on transient analysis
tc1 linear temperature coefficient of inductance
tc2 quadratic temperature coefficient of inductance
temp temperature of this instance, not global

If the inductance is defined by a parameter, most functions are allowed in the relationship with temperature, so (if you really want to):

$$.param\ L = 1m * (1 + 0.01 * sin\,(temp))$$

is accepted in schematic ('Inductor as parameter.asc'). The same is true for *Cpar*, *Rser* and *Rpar*. But the behaviour of the inductor is linear; it is merely that values are scaled by the temperature: we cannot change the inductance as a function of current to model the changing permeability. This can be important as some ferrites show a large, highly non-linear, change of permeability over their working temperature. Core loss and flux density also change strongly. These effects can be larger than the change of permeability with *H* (and thus with current) which is shown in Fig. 11.12 where the permeability peaks somewhere early in the curve. The shape and the degree of peaking depend upon the specific material.

If we include the relative permeability of the inductor in Eq. 11.13, we have:

$$L = \frac{\mu_r \mu_0 N^2 A}{l} \tag{11.27}$$

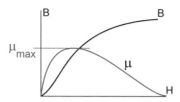

Fig. 11.12 Change of permeabilty with *H*

and for an inductor of given dimensions and number of turns, the inductance is directly proportional to the permeability.

Example – Variable Parallel Capacitance and Series Resistance

A simple test, just to illustrate the point, is schematic ('Inductor Cpar Rser.asc'). We declare them as parameters in the **Inductor – L1** dialogue where we pick appropriate names enclosed in curly brackets:

$$.param\ Rser = 1k * (1 - temp * 0.01)$$

$$.step\ param\ Cpar\ 1p\ 10p\ 2p$$

It is clearer if we activate them one at a time. There is considerable difficulty if we try to plot the voltage first, but none if we add it after plotting the current. Curious!.

Example – Self-Resonance of an Inductor.

If we place an inductor on the schematic then **Select Inductor** we can place a Bourns SDE0403A-100 M on the schematic ('Bourns Inductor.asc'). Bourns state that it is for low amplitude AC use only; transient and DC effects are not modelled. If we click **Quit and Edit the Database** we find the parallel resistance and capacitance which also have been filled in on the schematic: in short, selecting a component automatically fills in the LTspice fields.

   We measure a resonant frequency f_r of *38.6233 MHz*. The model has $L = 10\ \mu H$ $C_p = 1.698\ pF$, so we calculate the resonant frequency as:

$$f_r = \frac{1}{2\pi\sqrt{LC}} = \frac{1}{2\pi\sqrt{(10 \times 10^{-6} \times 1.6984 \times 10^{-12})}} = \frac{10^9}{2\pi 4.1211} = 38.62\ MHz$$

It is interesting the reduce the series resistance to zero and then $f_r = 38.2727\ MHz$ – a difference of *350 kHz*. Likewise with $R_{par} = 10\ M\Omega$, the frequency does not change but there is a sharp notch in the current.

   Well below resonance we see the graph is a straight line with a downward slope of $-20\ dB/decade$ and a phase $-90°$ confirming that the response is inductive. But above *100 MHz*, it slopes upwards at the same rate and with a phase of *+90°*, so now it is capacitative. The effect on the inductor current of series resistance up to *100 Ω* is difficult to see, and even *1 kΩ* produces a very small effect, showing the high impedance of the inductor at resonance. This can be tried by uncommenting the . *step* command and commenting the *.param.*

The measured minimum of −*84.59 dB* is directly attributable to the parallel resistance and is *20 × log₁₀ (1/16970)*.

11.3.6.2   LTspice Inductor Loss

An ideal inductor does not dissipate power: the voltage and current are in quadrature. Given a sinusoidal excitation, their product is also sinusoidal with equal positive and negative peaks as energy is stored in the inductor and returned and the average is very small. If the inductor is given series or parallel resistance, its dissipation will be reported correctly.

Example – Inductor Power Loss
In schematic ('Inductor Power Loss.asc'), we have a 100 Ω resistor in series with a 100 mH inductor excited by a 1 V sinusoid at 100 Hz. The peak inductor voltage is *540 mV* and the peak current *6.5 mA*, yet LTspice returns an average power of *44.6 nW* compared to *3.5843 mW* for the resistor. However, it is important to make the measurement only after any turn-on transients have subsided, in this case by starting to record after *0.01 s*.

If we place an improbably small *100 Ω* parallel resistor in the inductor, the peak inductor voltage falls to *391 mV*, whence we find the resistor power is *764 μW* – the same as the measured value.

With a ramp input *PULSE(0 1 0 2.5m 2.5m 0 5m)*, we find just *80 nW* for the inductor dissipation with no resistor and *181 μW* with it restored, in close agreement with the inductor voltage of *134 mV_RMS*.

11.3.6.3   Behavioural Model of the Flux

This is the companion to the capacitor charge model, and we must be equally careful in using it. It allows us to create any arbitrary flux relationship which can be useful for ferrites. For inductors we use the reserved parameter *x* which is the current through the inductor and is analogous to the factor *x* used in a capacitor which is the voltage across it. If we use a linearly rising current, we may find the inductance at any current from Lenz's law:

$$L = -v\frac{dt}{di} \qquad (11.28)$$

Constant Flux
A simple statement:

$$Flux \; = \; < value >$$

is not a constant inductor but constant flux, and as the voltage depends upon the rate of change of the flux, the voltage across the inductor is zero.

Example – Constant Flux
Run schematic ('Constant Flux.asc'). We see an RMS current of *707 mA* but no voltage across the inductor, although there is a very small input voltage $V1 = 707 \, pV$ which agrees with the voltage across *R1* which is *1 nΩ*.

Flux Linearly Proportional to Current
What is needed for an ideal inductor is flux that is linearly proportional to the current. We modify the previous schematic by adding *x*: ***Flux = 100 m ∗ x***, and using Lenz's law, we should find a constant voltage for a linearly rising current.

Example – Constant Inductance
In ('Constant L.asc') we see a constant voltage of *−100 mV* across the inductor. As the red arrow in the 'current clamp meter' over the inductor is downwards, and the voltage opposes the current, the negative sign is correct. We have a linearly rising current from the PWL current source at the rate of *1 A/min*, so using $v = -L\frac{di}{dt} = 100\,mH \times \frac{1}{1} = -100\,mV$ which is correct. And we can repeat with other inductance values. If we change to a sinusoid excitation, we find a sinusoidal voltage across the inductor.

   If we replace the current source by a voltage source and increase the series resistance *R1* from *1 nΩ* to *1 kΩ* an AC analysis, ('Constant L measure.asc'), shows a 3 dB point of *1.59 kHz* – which is we should expect for a 100 mH inductor.

Flux Proportional to Tanh (Current)
To simulate an inductor whose inductance varies with current, we can try the 'Help' file example, ('Help File Inductor.asc '), using:

$$Flux = 1\,m \, * \, tanh \, (5 * x)$$

This is a nice idea since tanh is an 'S'-shaped curve starting horizontally and then smoothly blending into an almost linear rise until it turns into an asymptotic value. This is a fair approximation to a general B-H graph. The initial voltage is directly proportional to the coefficient of *x* being only *2.5 mV* for 2.5∗x but *10 mV* for 10∗x. At the same time, the initial voltage is directly proportional to the multiplier again being *50 mV* for *10 m∗tanh(5∗x)*, whereas it is only *5 mV* for *1 m∗tanh(5∗x)*.

   What also changes is the rise time of the curve, becoming reciprocally shorter with increasing coefficient of *x*. However, this only works with a positive current. We can try an *IF* statement such as ***Flux = if (x > 0, −10 m ∗ tanh (5 ∗ x) − 0.1, 10 m ∗ tanh (5 ∗ x))*** to handle negative magnetization, but there is a singularity at the origin causing a huge voltage spike. Nevertheless, with some care, we can hope to at least approximately fit the B-H curve of an inductor used only with positive flux, and this does not include hysteresis:

Example – Flux = 1m*tanh(5*x)
The schematic ('Help File Inductor.asc') shows that with *tanh(2*x)* from about *0.2 A* to *0.7 A*, the graph is substantially linear. This is judged by eye and open to debate. Above and below these limits, the trace is highly non-linear, the increase of inductance with current is non-linear and tends to a constant final value. As the current rises at 1 A/s, the y-axis is the inductance. This is a fair approximation to a real inductor. If we run ('Help File Inductor V2.asc'). starting with a negative current, we find the spike at $x = 0$.

The schematic ('Help File L Test.asc') has a sinusoid current applied to the inductor, and increasing the peak current to only *200 mA* in schematic ('Help File L Linearity Test.asc') causes visible distortion.

Arbitrary Flux Relationship
These must be handled with extreme care to avoid very high spikes at change-over points if we use the *if(x,y,x)* statement such as ***Flux = 100 m * if(abs(x), 0.01 * x, x)*** with a pulse voltage source. If we include a parasitic capacitance and resistance, this produces uncontrolled oscillations at the change point, without them it is a huge voltage spike. We must, however, be careful to use only odd powers so that if the current reverses, so does the flux: this will not happen with an even power.

Example – Flux Using Polynomial
An exponential or polynomial flux is worth considering, in particular a high coefficient such as:

$$Flux = 100\ m * x * (1 - abs(0.01 * x * *5))$$

in schematic ('Polynomial Flux.asc'). This shows an almost constant inductance of 100 mH measured by the inductor voltage up to about 0.8 A after which it falls quite steeply to 0 at 1.7 A. And care must be taken not to exceed that current; else the flux reverses. Schematic ('Polynomial Flux Test.asc') shows the distortion of the voltage waveform due to the reduced permeability at high currents. However, this also shows that it works with a negative input.

Example – Flux using 'if(x,y,z)'
We see the problem in schematic ('PWL Flux.asc'). Using the statement **Flux = 100 m * if(abs(x), 0.01 * x, x)**, there are huge voltage and current spikes at *0.5 A* where there is a discontinuous jump in the inductance. We can try adding parallel capacitance and a series resistance to damp the change.

Example – Exponential Flux
Schematic ('Exp Flux.asc') shows permeability falling from a negative value. If we reverse the direction of the current, the graph shows a graph with a rapid initial fall then tending to asymptotically zero, not a very useful example.

Explorations 2
1. Run ('Inductor as parameters.asc'). Try different temperature coefficients and functions. Note that the inductor is always linear, no matter how large the current.

2. Run ('Constant L.asc') and ('Constant L measure.asc') with different values and note that the inductance is constant.
3. Run ('Help File Inductor.asc') and note carefully the curvature of the voltage graph.
4. Run ('Help file L test.asc') and note the current level when distortion appears. Try changing the multiplier to *tanh(2∗x)*, etc. and changing the DC bias and compare with the observed curvature in 3.
5. Explore ('Polynomial flux test.asc') and note that only the node *l* voltage distorts and the other voltages and currents are sinusoidal. Try running with a current input and different series resistors.

The conclusion from this is that simple, linear inductance can be modelled, but it is not easy to associate the current *x* with non-linearities, and attempts at a piecewise inductor fail.

11.3.6.4 Hysteretic Core Model

This models the inductance directly from the hysteresis loop. It shows the distorted current and voltage waveforms, but it requires some knowledge of magnetics. The model does not include core losses. Its main use is to simulate custom inductors where the design involves the magnetic properties of the core rather than off-the-shelf stock items.

The model describes the flux as a function of *H* for the upper branch with positive current as:

$$Bup(H) = B_s \frac{H + H_c}{H + H_c + H_c \left(\frac{B_s}{B_r} - 1 \right)} + \mu_0 H \tag{11.29}$$

where H_c is the coercive force, B_s the saturated flux density, B_r the remanence and *H* the magnetizing force. For the lower branch with decreasing current, it is:

$$Bdn(H) = B_s \frac{H + H_c}{H - H_c + H_c \left(\frac{B_s}{B_r} - 1 \right)} + \mu_0 H \tag{11.30}$$

where the only difference is the sign of *Hc*. This is sketched in Fig. 11.5 where *Bs* is the intersection of the asymptotic slope with the y-axis. In use, the model will always start from zero along the path *OPM* where the magnetization is:

$$B(H) = \frac{Bup(H) + Bdn(H)}{2} \tag{11.31}$$

As well as these magnetic parameters, the model uses physical parameters:

- The length of the magnetic path in metres which can be difficult to measure but is often taken as the mean path around the circuit. Manufacturers usually quote the effective path length for their products which can include paths in parallel for 'E-I' core pairs.
- The length of any air gap in metres and is assumed to be small so that fringing fields are neglected.
- The cross-section area of the path in m². This assumes it is constant which is often not true. Manufacturers of ferrite cores usually quote an average area. This is especially useful for pot cores.
- The number of turns.

So this implies a knowledge of the magnetic parameters.

Example – Hysteretic Core Inductance
Open ('Gapped Inductor.asc'). This is the example inductor from LTspice Help. The horizontal axis is a current rising linearly to 1 A after 1 s, and therefore the voltage at node 001 by definition is the inductance in henries. We ignore the negative sign which is simply because the inductor voltage opposes the rising current. The inductance starts at 47.7 mH and then increases slowly to a measured *47.83 mH* at *42.84 mA* after which the permeability decreases. This is what is shown in Fig. 11.12, but the effect is greatly reduced in a gapped core.

Example – Hysteretic Core Model
The example in the 'Help' file is schematic ('Hysteretic Model.asc'). This shows a smooth S-shaped curve with a fairly linear region from *100 mA* to *350 mA*.

Schematic ('Hysteretic Model Core Loss.asc') shows that the product of current and voltage is very small, some *888 nJ*, but increases greatly with the turns *N*.

Example – Effect of Core Gap
We can see the change of inductance as we reduce the gap length schematic ('Hysteretic Model Test.asc'). The trace is an inverted '*V*' which becomes clearer as we reduce the gap length. With a wide gap, we saw a smooth curve, so perhaps the model is faulty. If we remove the gap and reduce the current ramp to *10 mA*, we have a permeability that rises to a cusp at *316 μA*. This is the same phenomenon of increasing permeability but at a different current. The position of the cusp and the flatness of the graph depend upon the magnetic properties.

Example – Signal Response
The schematic ('Hysteretic Model Test.asc') shows considerable distortion with a sinusoidal input of *300 mA* which greatly reduces at *200 mA* where we only have *1.04%* distortion. An AC run finds the 3 dB point at *3.5 kHz* corresponding to an inductance of *0.455 H*.

Explorations 3
1. Open ('Gapped Inductor.asc'). This is the example inductor from LTspice Help. The horizontal axis is a current rising linearly to 1 A after 1 s, and therefore the voltage at node 001 by definition is the inductance in henries, whence it can be seen that it starts at 47.7 mH, decreases slowly to some 45.8 mH at a current of

160 mA and then falls rapidly to 8.5 mH at 260 mA, finally falling to 1.7 mH at 1 A.

2. Expand the y-axis to run from −47 mV to −48 mV and the x-axis to run from 0 s to 0.1 s and you will find that the inductance (and therefore the permeability) increases slightly to 47.83 mH at 40 ms. This is point T in Fig. 11.5 or μmax in Fig. 11.12.

3. Change the ramp current to run from −1 A to +1 A by **PWL(0–1 2 2)**, and the same inductance is seen with the core magnetized in either direction.

4. Extend the time to **PWL(0 0 2 2)** and start the time to record at 1 s by **.tran 0 2 1 1 m**, and the graph is a straight line falling from 1.561 mH at 1.0 A to 1.546 mH at 2.0 A, a linear reduction of −0.015 mH. Does this agree with the asymptotic slope of $Bs + \mu oH$ of Fig. 11.5?

5. Open ('Gapped Inductor LR.asc') plot the inductor current and voltage. The waveforms are obviously distorted. An FFT will show that odd harmonics have been created because the core starts to saturate twice per cycle. Reduce the voltage to 200 V, and there is no distortion. Hence confirm that this is in agreement with the currents found with the first of these explorations.

6. Increase the voltage to 1 kV and note the dramatic drop in the inductor voltage. Does the current at that point agree with the first exploration?

11.3.7 Other Models

Alternatives have been proposed by individuals and commercial concerns and seek to improve the modelling of the B-H curve. They do not include temperature effects. At high frequencies, the skin effect must be included which increases the effective resistance of the winding. Core losses depend on frequency and so can the inductance.

11.3.7.1 Jiles-Atherton Model

One popular model, developed some time ago and since extended, is the Jiles-Atherton model. This delves into the domain theory of magnetism and has been extended to include anisotropic material. This involves such parameters as 'domain wall pinning' and 'domain wall flexing' which are not usually supplied either by transformer manufacturers or makers of ferrite materials and are difficult to measure.

The equations are taken from https://opus4.kobv.de/.../Biondic_Comparison_Ferroma... which also introduces other models:

$$M_{an} = Ms\left(\coth\frac{H + \alpha M}{a} - \frac{a}{H + \alpha M}\right) \qquad (11.32)$$

$$\frac{dM}{dH} = \frac{1}{1+c} \frac{M_{an} - M}{k.\,sgn\left(\frac{dH}{dt}\right) - \alpha(M_{an} - M)} + \frac{c}{1-c} \frac{dM_{an}}{dH} \tag{11.33}$$

where M_s is the saturated magnetization, M_{an} the anhysteretic magnetization which counteracts the magnetization, c the magnetic reversibility, k the average energy to break pinning sites, a the domain wall density and α the inter-domain coupling describing the shape of the anhysteretic curve.

11.3.7.2   Newport Components Model

In a 1993 report http://www.intusoft.com/articles/inductor.pdf, Martin O'Hara of Newport Components created a sub-circuit with the same topography as LTspice except for a current generator *B1* in parallel with the inductor, ('Newport Subcircuit. asc'), which enables a non-linear inductor to be modelled. The passive circuit models a linear inductor, including parasitics, the same as LTspice.

The B-H loop shows that at high currents the permeability and hence the inductance fall, eventually to a very small value when saturation is reached. If we apply a linearly increasing current drive, ('Newport subckt test.asc'), this implies that the current in the inductor must increase at more than a linear rate. In this sub-circuit, this is handled by a current generator *B1* in parallel with the inductor. Its current has a cubic relationship to the current in the inductor and a small coefficient so that it only makes a significant contribution at high currents when saturation begins.

There is no obvious restriction on the *B1* function, except that if the current exceeds some *3.16* times the rated current, we find a negative inductance and, as is said in the article, it is a good idea to add an IF statement to prevent this. Otherwise the function is best found by using a curve-fitting programme for a polynomial rather than cubic or even a piecewise linear current.

There is an alternative and that is to use the LTspice flux model instead of the current generator, schematic ('Newport flux model.asc') where the cubic term is applied directly to the flux. This has the advantage that we can simply use the LTspice inductor model and not build a sub-circuit.

Example – Newport Model Using a Current Source
If we use a sufficiently large sinusoidal input, 'Newport subckt. asc', we should expect the voltage peaks of the inductor to become rounder as the inductance is reduced and the current peaks to be sharpened. This also is what we see if we use a voltage source and make $R3 = 1\,k$ in ('Newport subckt test.asc').

Example – Newport Model Using Flux
This is schematic ('Newport flux model.asc') which behaves in exactly the same way.

Explorations 4
1. Explore the LTspice model, creating inductors covering a range of values.
2. Test the Newport sub-circuit by an AC analysis. Measure the Q and the self-resonant frequency. Try different functions for the current source. The circuit

represent a 1 mH inductor rated at 4 A DC with a $Q = 49$ and a self-resonant frequency of $fr = 800\ kHz$. The series resistance $R3$ must be at least 1 MΩ.

3. Excite the Newport model with 10 kV at 50 Hz and note the flattening of $V(l+)$. Remove $B1$ and there is no flattening. Step $V(1)$ from 3 kHz and use a FFT to note the current at which the distortion of $V(l+)$ increases. This is a measure of the onset of saturation, but it is not sudden. There is some third harmonic, but at high currents, others are introduced.

4. Excite the Newport sub-circuit with a fast current pulse (rise time 0.1 ms) and observe the damped oscillations at 800 kHz.

5. Repeat with the ('Newport Flux Model.asc').

11.3.7.3 The Coilcraft Advanced Model

We should now be used to finding components modelled by unlikely combinations of parts. This model is a good example. Their simple model can be found in the LTspice inductor database and uses the LTspice inductor. This advanced model does not (https://www.coilcraft.com/modelsltpice.cfm). It uses three voltage-dependent current sources $G1$, $G2$ and $G3$ to model the inductor, the skin effect and the core loss, respectively. These depend on frequency and so are modelled using s parameters and the Laplace command.

Taking them in turn, the inductor $G1$ is described by:

$$LAPLACE = +1/(S * 1e - 6 * (\{L1\_K3\} - (\{L1\_K4\} * LOG(\{L1\_K5\} * (S/(2 * pi))))))$$

This is a little easier to read if we drop the $L1\_$ prefix and just keep the k s:

$$Laplace = \frac{1}{s \times 10^{-6}\left(k3 - k4 \log\left(k5\right)\frac{s}{2\pi}\right)}$$

where $k3 = 3.3$. The 10^{-6} turns this into the correct low frequency inductance of $3.3\ \mu H$. The second factor $k5$ is the change of inductance with frequency.

The skin effect $G2$ models a resistor:

$$LAPLACE = +1/\{L1\_K1\}/(\text{-}S * S/4/PI^\wedge 2)^\wedge 0.25$$

which can be written as:

$$\frac{1}{k1\left(-sx\frac{s}{4\pi^2}\right)^{0.25}}$$

to show that the loss increases with the square root of the frequency. We might also note that $-s*s = -j\omega*j\omega$ and will always be positive.

The core loss formula $G3$ is identical to the skin effect, also modelling a resistor, except for the change of factor to **L1_K2**:

$$LAPLACE = + 1/\{L1\_K2\}/(\text{-}S * S/4/PI^{\wedge}2)^{\wedge}0.25$$

The remaining components are the parasitic capacitance in series with a resistor to dampen the resonance peak and the very high resistance $R3$ to remove the warning that the node between $G1$ and $G2$ is floating.

Example – Coilcraft Model

We need a series resistor to prevent uncontrolled oscillations, schematic ('CoilCraft L AC.asc'). With it, we can measure the self-resonance frequency of *63.9 MHz* and the peak of *70.488 dB*. LTspice of course reports a difficulty finding the DC values of the Laplace statements, but continues afterwards.

The calculated self-resonant frequency is $f_r = \dfrac{1}{2\pi\sqrt{3.3\mu \times 2.26p}} = 58.2\,MHz.$

If we just keep the constant term in the Laplace for $G1$, we find *58.56 MHz*.

We can view the core loss by removing $G1$ and $G2$. With an AC run with a voltage input, we find the current falls from *1.26 mA* at *1 MHz* to *40 μA* at *1 GHz*. This is a straight line plot with a logarithmic y-axis.

As Coilcraft say, we can find the inductance by plotting:

$$Im(V(n001)/I(R2))/(2 * pi * frequency)$$

which is the imaginary part of $Z = j\omega L = \frac{v}{i}$. We find the inductances has an almost constant value below some *30 MHz* and the phase is slightly more than $90°$. Above resonance the phase changes due to the capacitor. It is interesting to view the 'Real' part and the whole impedance.

A transient run fails because the Laplacian is singular at DC.

11.3.7.4   The Murata Models

Murata have a conventional model for small-signal analysis. The components are fixed, Fig. 11.13. However, they also add voltage sources to it, the 'Dynamic Model', where the inductance changes with current:

https://www.murata.com/~/media/webrenewal/tool/library/pspice/note_dynamic-model_mi101e.ashx?la=en

These are encrypted models that can be downloaded for LTspice. The inductor files are bundled with the MLCC models. If we open the *Inductor* folder, we find *.asy* and *.mod* files in pairs. Depending on the version of Windows, the models may be listed as 'Movie Clips'. Rather than splitting them up, the easiest way is to create a new sub-folder as . . .*lib/sym/Murata_inductors* and copy the whole lot there and

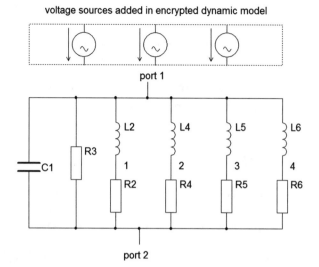

Fig. 11.13 Murata inductor Dynamic Model

then another sub-folder . . .*lib/sub/Murata_inductors* and again copy them all there. Of course, this bloats both files, and it is tidier to split them off by extension. We now need to add these folders to the search paths by **Tools → Control Panel → Sym and Lib Search Paths** and then right click in each large edit box and select **Browse** from the pop-up menu and navigate to the two folders. Note that we must have both – LTspice will not look for a symbol in the library search path. Otherwise we can just *. inc* the model files as needed.

Brief data for the models is contained in the header of the conventional model; else we must trawl through the website to find the data sheet.

Example – Murata Small-Signal Inductor Model

This is schematic ('Murata Small Signal L.asc'). The inductance is distributed as a series of RL elements. Their time constants are widely spread, *10e-12, 32e-11,34e-9,88e-8*, thus shaping the flanks of the self-resonance curve. The inductance is quoted as *0.4 nH* and its working range from *50 MHz* to *20 GHz*. Extending the analysis shows a resonance at *33 GHz*.

Example – Murata Dynamic Model

The schematic is ('Murata Advanced L.asc') where we can step the DC voltage to show the shift of the resonance frequency in response to the changing current and permeability. We should note that setting a constant DC and stepping the temperature has no effect, schematic ('Murata Advanced Permeaility.asc').

The inductor LQM21PN2R2MGH is described as $L = 2.2\ \mu H$, $I_{sat} = 0.45\ A$ $R_{DC} = 0.156\ \Omega$ $SRF > 40\ MHz$.

Example – Murata Dynamic Model Waveform

The schematic ('Murata Advanced Waveform.asc') consists of a *10 MHz* sinusoidal source in series with a *100 Ω* resistor and the inductor. An input of *100 V* results in a peak inductor current of some *1.6 A,* an average dissipation of *4.2474 W* and distortion of *11.3%.*

Reducing the input to a more reasonable *50 V,* we find a peak current of *316 mA,* a distortion of only *1%* and a reactive power of *348 mW.* The DC resistance contributes only 7.8 *mW.*

If we reduce the frequency *10 kHz* and the amplitude to *35 V* to give the same current and ignore the horrible voltage waveform, the loss is *9.4 mW.*

11.3.7.5   The Peretz Model

This is described at http://wwwee.ee.bgu.ac.il/~pemic/publications/jour001.pdf using a Kool Mu iron powder core as an example to set up a model either as a list of values or as a fitting function involving the dimensions of the core such as:

$$(\mathbf{pwr(n,2)} * \mathbf{4} * \mathbf{3.14} * \mathbf{pwr(10,\text{-}7)} * \mathbf{area} * \mathbf{pwr(10,\text{-}4)}/\mathbf{length}) *$$

$$(\mathbf{sqrt((((pwr(125,2))\text{-}56.18u * pwr(125,3)} * \mathbf{v(sw)} + \mathbf{104.3p * pwr(125,4)} * \mathbf{pwr(v(sw),2))}$$

$$/((\mathbf{1} + \mathbf{67.42u} * \mathbf{125} * \mathbf{v(sw)} + \mathbf{62.1n} * \mathbf{pwr(125,2)} * \mathbf{pwr(v(sw),2)))))$$

The voltage-dependant voltage source *E1* asserts an output voltage which is the input voltage divided by the function. This output voltage in turn generates a current in *G* proportional to the output current, Fig. 11.14. In essence it is an extension of the following section. The netlist contains two forms, 'A' as table of values, 'B' as a function. This and other files in Orcad format can be downloaded from http://www.ee.bgu.ac.il/~pel/download.htm.

11.3.7.6   Modelling a Ferrite Core

Useful as the above models are, they are still incomplete, as a glance at the ferrite characteristics mentioned above will have shown. A good starting point for creating a linear model and without going into thermal effects is:

http://youspice.com/spice-modeling-of-magnetic-core-from-datasheet/3/
www

To create a comprehensive model including variations of permeability and core loss with temperature, frequency and DC bias current is not simple. We can try

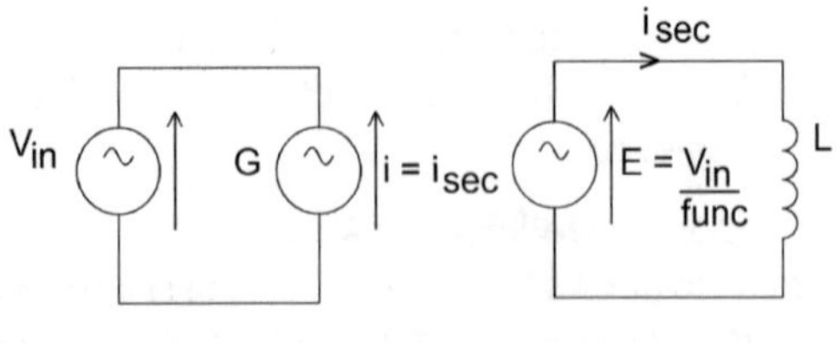

Fig. 11.14 Peretz model

curve-fitting. There are several websites for this allowing a choice of different relationships such as polynomial, exponential and trigonometrical. But we must remember that outside limits of the data points, the function may veer off wildly. A piece-wise linear approach is possible, and as tolerances can be as relaxed as 25%, there is no need to be too precise. And it has the advantage that beyond the last data point, the function uses the final value.

We shall only attempt to model some of the graphs here using the Magnetic Inc. type R ferrite where we find the data at https://www.mag-inc.com/getattachment/ Products/Ferrite-Cores/Learn-More-about-Ferrite-Cores/Magnetics-Ferrite-Catalog-2017.pdf?lang=en-US nc.com.

Permeability vs Temperature
The LTspice model does not implement permeability directly. Instead we can scale the inductance. The permeability at 25 °C is 2300, so if we divide the permeability at any temperature by this, we find the relative inductance as a fraction of its nominal value at 25 °C.

This is best handled by a table: We tabulate the numbers and then the fraction.

Temperature °C	Permeability	Fraction
−50	1200	0.52
50	1700	1.17
120	4500	1.96
150	4000	1.74
175	4500	1.96
210	5800	2.52
230	0	0

This is entered as a table in the schematic:

$$.param\ L = 10\,m * table(temp,\ -50, 0.52, 50, 1.17, 120, 1.96, 150, 1.74, 175,$$
$$1.96, 210, 2.52, 230, 0.01)$$

This is ('R Ferrite mu with temp.asc'). The *230 °C* is given a small value to avoid infinite current. We expect the current to fall as permeability increases which is what we find except with a run to *200 °C*, there are only *five* traces since two are identical.

Core Loss vs Temperature
The graph shows an almost linear decrease from 150 mW/cc at 0 °C to 50 mW/cc at 90 °C followed by a linear rise to 75 mW/cc at 120 °C. We can model this by the Rpar resistor using the IF statement:

$$.param\ R = if(temp - 90, 50 + (temp - 90) * 0.833, 150 - temp * 1.11)$$

and is ('R Ferrite core loss.asc'). This value must be multiplied by the volume of the ferrite core.

Permeability vs Flux Density
Unfortunately, the graph changes shape with temperature. We can try a quadratic but it is not going to be easy and is not attempted here.

Core Loss vs Flux Density
The graph shows a strong increase with frequency as well as flux density. Here, we shall model the flux density aspect. If we take a flux density of 100 mT, we find that core loss increases from 10 mW/cc at 50 mT to 1000 mW/cm^3 at 300 mT. As this is a straight line on log-log paper, we find the slope in logs as (log1000 – log10)/(log 300 – log 50) = 2.56, and the graph is of the form x = ay$^{2.56}$. Substituting in 1000 = a 300$^{2.56}$, we find a = 0.00046.

Permeability vs Frequency
This cannot be handled by the linear model.

Flux Density vs Temperature
The graph shows a drop of B$_{sat}$ from 470 mT at 25 °C to 320 mT at 130 °C, a rate of 1.43 mT/C. So we need a relationship of *470–1.43e−6(temp-25)* in T, but the linear model does not include saturation.

11.4   Mutual Inductance

This is a powerful and convenient way of modelling a transformer, in particular loose-coupled transformers used in communications circuits. Mutual inductance exists whenever the flux from one inductor cuts another. The fractional flux linkage can range from very small, as in air-cored RF transformers (*loose coupled*), to very large in iron-cored power transformers (*tight coupled*). The result is that current in one winding induces a voltage in the other (or others) which therefore can supply current to a load. In other words, power supplied to one winding can be transferred to another.

11.4.1   Theory

We shall divide this into the magnetic description and the circuit properties.

11.4.1.1   Magnetic Description

An interesting exposition, from which the following is adapted, can be found at:
    https://www.youtube.com/watch?v=hoTInTKij0o
    We start with voltage induced in a coil:

$$v = -N\frac{d\Phi}{dt} \tag{11.34}$$

where N is the number of turns. We can extend this to the case where the flux is produced in one coil and some of it links a second coil. Using the subscript 1 for the first coil and 2 for the second, we write:

$$v_2 = -N_2\frac{d\Phi_{12}}{dt} \tag{11.35}$$

where Φ_{12} is the portion of the flux in coil 1 that links with coil 2. We also saw that:

$$v_2 = -L\frac{di}{dt} \tag{11.36}$$

We define the mutual inductance M as the fraction of the inductances linked together, so from Eq. 11.36, we have:

$$v_2 = -M\frac{di}{dt} \tag{11.37}$$

Equating Eqs. 11.35 and 11.37:

$$M_{12}\frac{di}{dt} = \frac{N_2 d\phi_{12}}{dt_1} \Rightarrow M_{12}di = N_2 d\phi_{12} \tag{11.38}$$

where M_{12} is the fraction of the primary flux linking the secondary winding.

We now define the dimensionless *coupling coefficient* K as the ratio of the flux in the linked winding to the flux in the other coil creating the flux. This can never be greater than one and is the same whichever inductor is energized:

$$K = \frac{\phi_{12}}{\phi_2} = \frac{\phi_{21}}{\phi_1} \tag{11.39}$$

If K is greater than 0.5, the coils are said to be *close coupled* or *tight coupled*; this applies to power transformers. Otherwise, they are *loose coupled* which applies to tuned circuits.

We can expand and recast Eq. 11.38 as:

$$M_{12} = \left(\frac{N_2 d\phi_2}{di_1}\right)\left(\frac{d\phi_{12}}{d\phi_2}\right) \tag{11.40}$$

which is:

$$M_{12} = L_2 K \tag{11.41}$$

and similarly:

$$M_{21} = L_1 K \tag{11.42}$$

but the primary flux linking the secondary must be the same as the secondary flux linking the primary, so $M_{12} = M_{21} = M$ and thus:

$$M^2 = L_i L_2 K^2 \tag{11.43}$$

which is usually expressed as:

$$M = K \sqrt{L_1 L_2} \tag{11.44}$$

The T-Model
If M is less than unity, not all the primary flux links the secondary. This means that some of the primary inductance is effectively all on its own and not associated with the secondary. Likewise not all the secondary inductance is linked.

We can incorporate this in a 'T'-model, Fig. 11.15. We replace the mutual inductance by an ideal one with $K = 1$ with an infinite number of turns so that it draws no current due to its inductance but generates a secondary voltage. This is a convenient fiction because now we can add series inductances L_{p_leak} L_{s_leak} in the primary and secondary representing the unlinked portions, not to be confused with the total primary and secondary inductances L_p L_s. This model is useful at one extreme for power transformers where K is about 0.95, and we can separate the leakage inductance, and at the other for inter-stage coupling transformers in communications circuits with M often less than 0.5, and we are interested in the response with capacitors in parallel with primary and secondary.

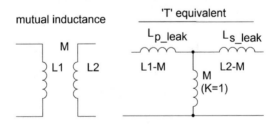

Fig. 11.15 Mutual inductance T Model

11.4.1.2   Circuit Theory

We relabel the windings as primary and secondary, although there is nothing to stop us reversing the connections: this is just for identification. In the time domain, we rewrite Eq. 11.37 for the voltage induced in the primary as:

$$v_p = M \frac{di_s}{dt} \qquad (11.46)$$

where i_s is the secondary current. The impedance of an inductor increases with frequency and is $j\omega L$. Also, unlike a capacitor, the voltage leads the current by 90°. The mnemonic 'CIVIL' is helpful here – C (capacitor), I (current), V (voltage), I (current) and L (inductor) – the current leads the voltage in a capacitor, and the voltage leads the current in an inductor:

$$v_p = j\omega M i_s \qquad (11.47)$$

If we model the primary just by the inductor alone, we will find only the small current predicted by a series LR circuit. But there will be a voltage induced in the secondary, and it can supply a current, and the power must come from the primary. In other words, the primary current should increase in proportion to the load current. The physically correct way of handling this is to reduce the inductance of the primary as we shall see in Sect. 11.5. This is difficult to handle analytically, so we adopt the alternative approach of introducing an artificial series voltage to increase the primary voltage and thus increase the current in the primary. This voltage does not exist and is not shown in the simulation. It can also be very large: suppose the primary inductance of 5 H is excited by the 240 V 50 Hz mains and power taken from the secondary winding is 400 W. That implies a primary current of 1.67 A, and using Eq. 11.47, an additional 2.6 kV must be applied to the primary winding. If this were a real voltage, it would present a significant danger that the insulation would break down.

Nevertheless, we add the voltage from Eq. 11.47 to write the equation for the primary, Fig. 11.16:

$$v_p = i_p \left(R_p + j\omega L_p \right) - j\omega M i_s \qquad (11.48)$$

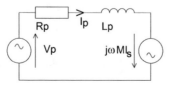

Fig. 11.16 Primary Mutual inductance

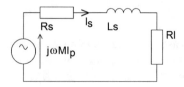

Fig. 11.17 Secondary Mutual inductance

The induced voltage in the secondary is $j\omega Mi_p$, hence Fig. 11.17:

$$j\omega Mi_p = i_s(R_s + R_L + j\omega L_s) \tag{11.49}$$

A special case is if the secondary is open-circuit so that $I_S = 0$, and then from Eq. 11.48 we have:

$$V_p = I_p R_p - j\omega L_p \tag{11.50}$$

or:

$$i_p = \frac{v_p}{R_p + j\omega L_p} \tag{11.51}$$

We can write the secondary voltage, Vs, *as*:

$$v_s = j\omega M i_p \tag{11.52}$$

then substitute for Ip from Eq. 11.51:

$$v_s = \frac{j\omega M v_p}{R_p + j\omega L_p} \tag{11.53}$$

Then if $Rs = 0$ and after replacing M from Eq. 11.44:

$$\frac{v_s}{v_p} = K\sqrt{\frac{L_s}{L_p}} \tag{11.54}$$

And from Eq. 11.24 as $L \propto n^2$ where n is the number of turns:

$$\frac{v_s}{v_p} = K\left(\frac{n_s}{n_p}\right)^2 \tag{11.55}$$

This is an important result because it relates voltage to inductance.

Example – Mutual Inductance Output Voltage
Open ('Mutual Inductance pt1.asc') where the inductance $L2$ is open-circuit, and so there can be no current in $L2$. In effect we simply have a series circuit of $R1$ and $L1$.

We apply a pulse excitation to $L1$, whilst we step the mutual inductance. The result is that the inductor $L1$ current and voltage are unchanged because no power is taken by $L2$. However, the voltage induced in $L2$ increases with the coupling.

Changing the circuit to ('Mutual Inductance pt2.asc') where the times are long enough to find that the rise times of current and voltage agree with an inductor $L1$ of 1 mH with a mutual inductance of one.

If we change the excitation to a 50 Hz sinewave, we can see the current and voltage displaced by 90° in agreement with theory. Note also that the ratio of the output and input voltages is:

$$\frac{v_{out}}{v_{in}} = \sqrt{\frac{L_2}{L_1}}$$

which can be tested with different inductances.

Example – Mutual Inductance Currents
Now let us try the radical move of short-circuiting $L2$, ('Mutual Inductance pt3.asc'). If mutual inductance is zero, we have the previous results. But if the coupling is one, the voltages are zero, but the currents are square waves, and we find the current in $L2$ is 0.577 A, whilst that in $L1$ is 1 A. Changing the coupling to 0.5 and the current in $L1$ remains at 1 A, but the current in $L2$ is only 0.288 A, and we no longer have square waves because now half of $L1$ is not linked to $L2$ so the primary is again a series LR circuit where we anticipate the next chapter to find the rise time t_r is around 0.9 ms. In theory it is:

$$t_r = 2.2\frac{L}{R} = 2.2\frac{500\mu}{10} = 1.1 \; ms$$

which is close enough agreement.

Example – Mutual Inductance Transformer
A transformer is operated from the 240 V, 50 Hz mains, and can deliver 3 A at 24 V_{rms} to a load resistor connected across the secondary. The primary inductance is 20 H and the coupling coefficient is unity. The secondary inductance is found from Eq. 11.54 where $0.1 = \sqrt{(L_S/20)}$ and $L_S = 0.2$ H. From Eq. 11.34, the mutual inductance is 2 H, and from Eq. 11.50, the RMS primary voltage is 2 π50∗3 A∗2 H = 1885 V. This is totally fictitious and meaningless in the sense that it does not represent a voltage in the 'real' transformer, so there is no need to worry about insulation breaking down.

A better approach is to use a Norton model and place a current source in parallel with the inductance, and then there is no excess voltage. This is what we shall do in later sections.

11.4.1.3 Modelling Mutual Inductance in LTspice

This is extremely simple: just place the two inductors on the circuit (and they need not be in close proximity, although that helps for clarity) and add a statement:

$$K < LI >< L2 > \ldots < K \text{ value} >$$

for example:

$$K \ L1 \ L2 \ L3 \ 0.9$$

but do not prefix it with a dot – this is not a SPICE directive. A mutual inductance cannot have a pure voltage source connected across it else there are two voltages in parallel. Add a resistance to the voltage source or an external series resistance in the circuit.

Also LTspice only supports mutual coupling between simple linear inductors. It cannot be used if the inductance changes with current, even less can it model saturation. It is only useful for small-signal AC analysis.

A point to note is that we can have **K L1 L2 1** for two windings with perfect coupling between them. Physically, complete flux linkage is impossible, but supposing it were, that would mean that all the flux was contained in the magnetic path linking the two inductors. It might conceivably be possible to add another winding that did not intercept the whole magnetic flux but that is not allowed by LTspice. Thus **K1 L1 L2 1** and **K2 L1 L3 0.1** for two close-coupled windings and a third loose coupled are not possible. This can be resolved by changing the first to **K1 L1 L2 0.99** or less leaving some leakage flux for coil **L3**.

But it is now time to test some of our theoretical predictions about mutual inductors. We should note that we need a short time step with a transient analysis else there can be an apparent voltage drop, schematic ('Magnetizing Inductance 2. asc'). Also, if the resistance is very small, it takes a very long time to reach steady state. Finally, although the primary current with an open circuit secondary is correct, if we add a load, the difference between the currents falls to only a few percent of the open-cicuit value.

Explorations 4
1. Open schematic ('Mutual Inductance pt1.asc'). Note the primary and secondary voltages and confirm their ratio agrees with Eq. 11.54. Note the primary current and check that it agrees with Eq. 11.48 with effectively an open-circuit secondary. Note also the phase difference between primary current and voltage.
2. Complete the circuit from t with a smaller resistor and check that the currents and voltages agree with theory for different load resistors.
3. Note the effect of increasing the series resistance of the voltage source.
4. Open Schematic ('Mutual Inductance Step K.asc') and check that the secondary voltage agrees with Eq. 11.54.

5. The schematic ('Mutual Inductance expanded.asc') has an ideal transformer *L1 L2* with perfect coupling and both windings in series with uncoupled inductors. This is the 'T' model for transformers. The circuit on the left is the equivalent with $K = 0.8$. Test both with different coupling coefficients and load resistors.

11.5 Power Transformers

These are the embodiment of the abstract mutual inductance. They are close-coupled transformers and were never included in SPICE and must be constructed as sub-circuits. The circuit symbol is derived from an inductor, Fig. 11.18: there is no significance about the size of the windings, but a dot may be added to indicate the winding direction or polarity of the windings, if this is significant, and solid or broken lines to show the construction of the core.

The essential relationships between flux, mutual inductance and voltages and currents have been covered in the previous section, Mutual Inductance. Here we shall extend the theory somewhat.

11.5.1 Magnetics

The essential conundrum which we posed before is that the primary and secondary windings of a transformer possess considerable inductance, perhaps 10 H and 0.1 H respectively, and some small resistance. But if we draw a current from the secondary, by the law of conservation of energy, that power can only come from the primary so the primary current must increase to supply the load current.

There are only two mechanisms that can do that. One is to suppose that the primary inductance remains unchanged, and therefore the voltage across the primary winding must increase. This is the strategy we used above.

The alternative – correct – explanation is to look carefully at the magnetization of the transformer core. A time-varying voltage (it does not have to be a sine-wave) is applied to the primary winding which will cause a current to flow and create a time-

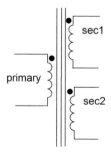

Fig. 11.18 Transformer symbol

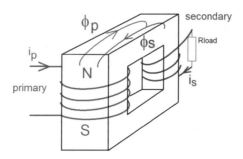

Fig. 11.19 Power Transformer

varying magnetic flux Φ_P proportional to the current and the number of turns on the primary winding. The polarity of the magnetic field will depend upon the direction of the current and the sense of the windings.

Let us suppose that at the instance at which Fig. 11.19 is drawn, the primary current is increasing. Then the top of the primary is north, and the direction of the flux Φ_P is from the top of the primary clockwise around the magnetic path. And because it is changing, it will induce a voltage in the secondary winding.

Now suppose we connect some load resistor *Rload* across the secondary winding and that the secondary is also wound clockwise (i.e. in the same sense as the primary). The induced EMF will make the top of the secondary winding positive too, so current will flow out of the top of the winding and through the load as shown. This current will also create a flux Φ_S but in the opposite sense to Φ_P. And if the secondary is wound anti-clockwise, the induced voltage will be reversed and hence the secondary current, but the flux Φ will still be anti-clockwise. Thus the secondary current causes the total flux to decrease, and, because the inductance of a coil depends on the flux linkage, see Eq. 11.14, the effective inductance of the primary winding falls, and its current increases. We may equivalently argue that the secondary current draws power from the transformer, and this is supplied by the magnetic field which therefore becomes smaller and with it the inductances. Therefore the primary current increases to restore the original flux and with it the inductance and the power drawn by the load. We must now quantify this.

11.5.1.1   Load Current Referred to the Primary

We suppose that all the primary flux links the secondary. This is not too bad an approximation for power transformers. Then for the secondary, according to Lenz's law and ignoring the negative sign, the induced EMF will depend upon the rate of change of flux from the primary and the number of turns linked by the flux:

$$V_S = n_S \frac{d\Phi_P}{dt} \tag{11.56}$$

and conversely we can find the rate of change of primary flux:

$$\frac{d\Phi_P}{dt} = \frac{V_P}{n_P} \tag{11.57}$$

and from Eqs. 11.56 and 11.57, the ratio $V_s/n_s = V_p/n_p$ is the *volts per turn* and is constant for a given transformer. Thus:

$$V_S = V_P \frac{n_S}{n_P} \tag{11.58}$$

where n_s/n_p is the *turns ratio*. Equation 11.58 is often used in transformer calculations but is only valid if our assumption holds good and all the primary flux links the secondary – i.e. there is no flux leakage. Then if we take a load current from the secondary, this will create a changing flux:

$$\frac{d\Phi_S}{dt} = \frac{V_S}{n_S} \tag{11.59}$$

we can substitute from the volts per turn

$$\frac{d\Phi_S}{dt} = \frac{V_P}{n_P} \tag{11.60}$$

so the changing primary flux is exactly balanced by the opposing changing secondary flux; in other words the decreased magnetization of the core due to the secondary current is counterbalanced by the increased magnetization caused by the increased primary current. This is not to say that there is no magnetization of the core – there is – but we shall ignore that for the moment.

If we add a load resistor R_{load} to the secondary, the secondary current Is will be given by:

$$I_S = \frac{V_P}{R_{load}} \frac{n_S}{n_P} \tag{11.61}$$

and the power dissipated by the load is simply:

$$P_L = I_S V_S = \frac{1}{R_{load}} \left(V_P \frac{n_S}{n_P} \right)^2 \tag{11.62}$$

This power can only come from the primary so the load current referred to the primary I_{RP} is given by:

$$I_S V_S = I_{RP} V_P \tag{11.63}$$

whence:

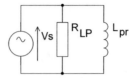

Fig. 11.20 Load Referred to Primary

$$I_{RP} = \frac{V_P}{R_{load}} \left(\frac{n_S}{n_P}\right)^2 \qquad (11.64)$$

so we can model the secondary current by a primary resistor R_{LP} of:

$$R_{LP} = R_{load} \left(\frac{n_S}{n_P}\right)^2 \qquad (11.65)$$

connected across the primary where its value is found using the turns ratio. Thus we have the very simple model of Fig. 11.20 where V_s is the primary supply voltage. If the load includes capacitance and inductance, these also can be referred back to the primary. This model is often used for manual calculations of supply side conditions, but it has the great draw-back that the user must calculate R_{LP} every time the load changes.

Example – Load Current Referred to the Primary
The schematic ('Load Referred to Primary.asc') has a *110.3* V_{RMS} input from a *156* V_{pk} supply to an ideal transformer formed from two inductors with $L_1 = 10H$ $L_2 = 1H$ $K = 1$, so there is total flux linkage between them. The secondary load resistor $R_1 = 10\,\Omega$.

From Eq. 11.54, we expect the secondary voltage to be $V_{sec} = V_{pr}\sqrt{\frac{L_s}{L_p}} = 110.3 \times 3.162 = 34.88\,V_{RMS}$ which is what we find. We also find the secondary current is *3.4878* A_{RMS} and the primary current *1.104* A_{RMS}. The lower circuit has V_2 with the same supply voltage but terminated by a resistor $R_2 = 100\,\Omega$ being the secondary resistance referred to the primary using Eq. 11.65 and this too draws *1.104* A_{RMS}. The small series resistors of the sources are an LTspice requirement to avoid two voltage sources directly in parallel.

11.5.1.2 Magnetizing Current

However, even in the absence of a load current, there will be a primary current because the primary inductance is finite. This current is in addition to any load current and depends only on the primary inductance and the supply voltage and therefore is constant.

If there is no load connected to the secondary, the primary inductance Lp, usually several henries, will draw a small RMS magnetizing current $Imag(RMS)$ although this is not strictly a loss and is given by:

$$I_{mag(RMS)} = \frac{V_{p(RMS)}}{\omega L_P} \qquad (11.66)$$

where $V_{p(RMS)}$ is the primary voltage and L_p the primary inductance. We try to keep the magnetizing current small, no more than 5% of the maximum load current for small transformers and often a lot less for high powers. This is for three reasons: the first is that whilst the magnetizing inductance does not dissipate power, it does draw reactive current from the supply and in power applications that can affect the size and ratings of cable; the second is that the magnetic field created by the magnetizing current creates hysteresis and eddy current losses in the core, albeit usually very small; and the third reason is that this magnetization is not offset by the field generated by the load current, and it represents a real magnetization of the core. This becomes a serious problem if the supply frequency falls for then the impedance of the primary inductance is less and the peak flux can even lead to saturation of the core – with potentially disastrous consequences. This is not usually a problem using the mains, but is a serious consideration for switch-mode power supplies.

Example – Magnetizing Current Schematic
If we take a supply of *110 V 60 Hz* – which is *156 V_{pk}* – a transformer with a *10 H* primary we expect a magnetizing current of $I_{mag} = \frac{110\ V}{2\ \pi\ 60\ Hz\ \ 10\ H} = 29.2\ mA_{RMS}$. The schematic ('Magnetizing Inductance.asc') reports *28.3 mA_{RMS}* with perfect coupling between input and output and negligible secondary current. This is independent of the coupling coefficient and, by extension, of the secondary current. The difference in current is due to Rser of the voltage supply which is 1 Ω.

Example – Magnetizing Current
A small transformer with a 10 H primary winding used on the 50 Hz mains has an impedance of 3.14 kΩ, and with a supply of 240 V_{RMS}, it will draw a magnetizing current of 76 mA.

11.5.2   Special Transformers

There are two of particular interest. The theory is unchanged.

11.5.2.1   Autotransformer

This consists of a single winding with a tapping point, so it provides no electrical isolation between input and output, Fig. 11.21. It has two particular uses. One is in impedance matching; the other is as a substantial power transformer, often weighing a few kilogrammes and with an output rated at several amperes, with an almost continuously variable tapping point where the output voltage is less than or equal to the input voltage. These are sold under the trade name 'Variac'.

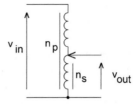

Fig. 11.21 Autotransformer

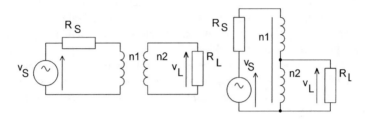

Fig. 11.22 Transformer matching

About the best illustration is at https://www.elcoteam.com/privati_en/metrel-hsg0202-trasformatore-variac-0-260vac-8a-2-08kva.html which shows that it consists of a thick soft magnetic cylindrical core wound with heavy gauge insulated copper wire. The insulation is removed from the wire at one end of the cylinder so that the rotatable wiper can pick off anything from zero to the maximum of the input voltage - or often a little more. Generally the unit is enclosed in a protective earthed case to prevent accidental expose to the live conductors. It is used to test circuits against varying supply voltages.

The transformer equation applies so:

$$v_{out} = v_{in} \frac{n_s}{n_p} \tag{11.67}$$

Example – Impedance Matching
We can use a fixed ratio autotransformer, Fig. 11.22, as an alternative to a double wound transformer to try to transfer the maximum source power to the load where *R1* and *R4* are the source resistances. The left-hand schematic ('Transformer Matching.asc') shows that perfect matching is possible with a double wound transformer, but an autotransformer is problematic and is better matching a low resistance load.

11.5.2.2 Current Transformer

This is used to measure current, often in situations where the current or voltage is very large. In its basic form, the primary consists of a single turn surrounding a conductor carrying AC. The secondary consists of several hundred turns. The governing equation is:

$$\frac{n_p}{n_s} = \frac{i_s}{i_p} \qquad (11.68)$$

From this we see that that the secondary current is much reduced compared to the primary. It is also a step-up voltage transformer with the secondary voltage much higher than the primary and potentially very dangerous if the secondary is open-circuit.

11.5.2.3 Audio Transformers

An audio amplifier is usually designed with a certain load impedance in mind being that of the loudspeaker unit which could range from 4 Ω to 16 Ω. However, a public address system may need several loudspeakers, perhaps some distance from the amplifier. This raises two problems. The first is that the impedance of the speakers in parallel will be very low and not a good match to the output of the amplifier; the second is that the low impedance of the loudspeakers means there will be a large current and hence the need for heavy connecting cables between the amplifier and the loudspeakers. Both of these problems can be solved with matching transformers.

Impedance Matching
If the amplifier has an optimum load impedance Z_{load} and we connect n loudspeakers each with an impedance Z_{LS} in parallel across the output of the amplifier, the load impedance will be Z_{LS}/n and will almost certainly not be optimum, Fig. 11.23. However, we can match them to Z_{load} by using transformers T_{LS} so that their input impedance is:

$$Z_{in} = n\, Z_{load} \qquad (11.69)$$

and this will create the optimum load impedance for the amplifier.

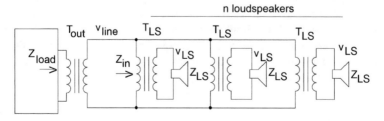

Fig. 11.23 PA System

The transformers must also match the loudspeakers to Z_{in}. Using Eq. 11.65:

$$Z_{in} = Z_{LS} \left(\frac{n_s}{n_p} \right)^2 \tag{11.70}$$

and:

$$v_{line} = v_{LS} \left(\frac{n_p}{n_s} \right) \tag{11.71}$$

It may be possible to connect directly to the amplifier without transformer T_{out}. We shall test this in the example.

Example – Transformer Impedance Matching.
Using RMS values, we want to connect ten speakers in parallel each of impedance 4 Ω and power 100 W to an amplifier whose optimum load is 8 Ω. From this we find that the secondary voltage of the transformers T_{LS} must be *20 V* and the current *5 A*.'

From Eq. 11.69 $Z_{in} = 80 \; \Omega$ and from Eq. 11.70, the turns ratio is $\frac{n_s}{n_p} = \sqrt{\frac{4}{80}} = 0.224$, and the line voltage from Eq. 11.71 is $v_{line} = \frac{20}{0.316} = 89\,V$. This voltage is acceptable so we only need transformer T_{out} for safety reasons to electrically isolate the speakers, and it will have a turns ratio of unity. The output current is $\frac{1000W}{89V} = 11.2\,A$ which we can compare to the *50 A* if we connect them directly.

11.5.3 Models for Manual Analysis

The complexity of the model depends upon our area of interest. If we are concerned with the transformer itself, then we need a comprehensive model including possible saturation and temperature effects which are difficult to handle analytically but possible with SPICE. Light current engineers are mainly concerned with the output side where it is assumed that the transformer is linear and working normally. The input side is of passing interest, mainly in regard to electrical noise created by the output and reflected back to the input. But when the power exceeds a hundred watts or so, the input side matters especially because of power factor considerations.

A transformer has losses just like a solenoid, and these must be included in any comprehensive model. These limit the power handling capacity of the transformer which is given as the product of (volts) × (amps) or *VA* rather than watts. For small transformers, up to some 100 VA, the losses may be as much as 5%, but above that they are generally around 2%.

11.5.3.1 Secondary Circuit Model for Manual Analysis

Often 'light current' engineers just use a Thevenin model for the secondary circuit and neglect the primary. If we connect a load to a secondary of a mains transformer, we find that in going from no-load to full-load, the output voltage falls by a few percent due to all the losses discussed above.

Regulation Model
We can lump all the losses by a resistance in series with the open-circuit secondary voltage. And as electronic engineers often are content to take transformers 'as they come' from the manufacturer rather than going into magnetic design, this is adequate for some purposes.

What is quoted in short-form catalogues is the open-circuit output voltage and the regulation, that is, the fall in output voltage in going from no-load to full-load – usually given as a percentage of the no-load. For a small transformer less than 100 VA, it could be as much as 10%, but 5% is the more usual figure for mains transformers used in electronic equipment. Thus we arrive at the simple Thevenin model for the secondary, Fig. 11.24, where we lump all the losses, both primary and secondary, in R_{reg}. This can be measured by taking the open-circuit output voltage and the full-load output voltage and current, Fig. 11.25.

Copper Loss and Inductance Model
The previous model did not explicitly include leakage inductance but lumped it with the copper loss. This model separates them. The leakage inductance is the second main reason for the fall of output voltage, although in a well-designed transformer, this may only be a few percent. As not all the primary flux links with the secondary and vice versa, we model the difference as a series inductance in the secondary circuit. The distinction is important for two reasons: the first is that the inductive part does not dissipate energy so if we want an accurate figure for the losses, the simple resistor yields too high a figure; the second reason is often more important and is that the leakage inductance helps prevents very rapid changes of load current.

The model is an extension of the previous where we find the resistance of both primary and secondary windings. The resistance of the wire used to wind the primary R_P causes the voltage applied to the ideal transformer V_P to be less than the input voltage V_{in} and hence affects the regulation. The resistance of the secondary causes a volt drop when current is taken, but it is not always so easy to measure it because often it is only a fraction of an ohm.

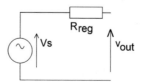

Fig. 11.24 Simple Transformer Secondary Model

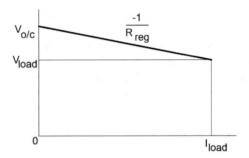

Fig. 11.25 Transformer regulation

We can easily measure the primary resistance. We refer this to the secondary by inserting another resistance R_{PR} in series with R_S which consumes the same power R_P. Since $P = V^2/R$ from Eq. 11.65, the required resistance is:

$$R_{PR} = R_S \frac{n_s^2}{n_p^2} \tag{11.72}$$

and hence we know the equivalent resistance to insert in the secondary circuit. We lump the reflected primary resistance in with R_S as $R_{S(EQ)}$. This will be less than R_{reg} so we now insert an inductance $L_{S(EQ)}$ in series with this to give the correct regulation. In effect, we include both L_S and the effect of L_P in the secondary. This creates a potential divider with the load so that:

$$1 + r = \frac{R_L + R_{S(EQ)} + j\omega L_{S(EQ)}}{R_L} \tag{11.73}$$

where r is the regulation expressed as a fraction. The modulus of this is:

$$1 + r = \frac{\sqrt{\left(R_L + R_{S(EQ)}\right)^2 + \omega^2 L^2_{S(EQ)}}}{R_L} \tag{11.74}$$

from which $L_{S(EQ)}$ can be found. The point to watch is that it is not a simple case of finding the phasor sum of the secondary resistance and inductance.

Example – Secondary Circuit Model of a Transformer.
A transformer has a rated output of 2 A at 24 V and a regulation of 4%. The no-load voltage is $1.04 \times 24 = 25$ V. This is the required voltage source. The voltage falls by 1.0 V for a change in current of 2 A, so the secondary impedance must be 0.5 Ω, and this is R_{rreg}.

Example – Secondary Model Including Inductance
The transformer of the previous example has a DC resistance of the secondary winding of 0.1 Ω and of the primary of 20 Ω. As before we find the secondary impedance is 0.5 Ω when the load resistor is 12 Ω. The secondary resistance is $0.1 + 20 * (24/240)^2 = 0.3$ Ω, and we substitute in Eq. 11.74 to find:

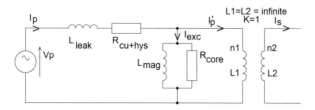

Fig. 11.26 Transformer primary model

$$1.04 = \frac{\sqrt{(12 + 0.3)^2 + 98696\,L^2}}{12}$$

so $155.75 = 151.29 + 98696\,L^2$ and finally $L_{S(EQ)} = 6.72\ mH$. This is schematic ('Secondary Model with L.asc') where the input voltage is $24 \times \sqrt{2}\ z\ 1.04 = 35.3\ V$. There is a slight phase shift between the load and input voltages.

11.5.3.2 Primary Model

The primary flux leakage can be modelled by an inductor in series with the primary Fig. 11.26. But this inductance does not transfer power to the secondary. For that we need an ideal transformer of infinite inductance so that it draws no magnetizing current and with $K = 1$. This transfers power from primary to secondary and vice versa taking account of the turns ratio $n1{:}n2$. We therefore model the magnetizing current by a separate inductor across the primary L_{mag}.

We must also include in the primary side hysteresis losses as the core material is taken through its magnetization cycle each period of the supply and the resistance of the primary winding, the *copper loss*. Both of these depend upon the primary current and can be included in a series resistor $R_{cu\ +\ hys}$.

Additionally, there is the eddy current loss due to induced emfs in the core creating currents. This depends only on the supply voltage and the frequency and can be modelled by a resistor across the primary *Rcore* in parallel with *Lmag*. These two draw the *excitation current Iexc*.

Primary and Secondary Unloaded Model
In the secondary, we again have the copper loss of the winding wire *Rs*, often just a fraction of an ohm if the output is low voltage, in series with the secondary leakage inductance *Lls*. This is Fig. 11.27.

11.5.3.3 Load Referred to the Primary

An excellent discussion in great detail can be found at http://www.vias.org/matsch_capmag/matsch_caps_magnetics_chap5_07.html. This is mainly of interest for

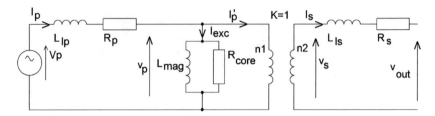

Fig. 11.27 Transformer Equivalent Circuit

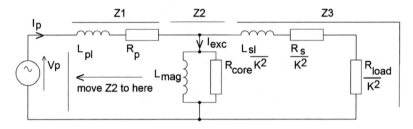

Fig. 11.28 Transformer primary model loaded

power engineers and goes into detail about the various voltage and current components and their phase angles. The following model is largely derived from there.

The secondary circuit plus the load resistance can be referred to the primary through the factor K, Fig. 11.28. From Eq. 11.72, the load resistance Rs referred to the primary Rs' is:

$$R'_S = R_S \frac{V_p^2}{V_s^2} \text{ or } R'_S = R_S = \frac{N_1^2}{N_2^2} \text{ or } R'_S = \frac{R_S}{K^2} \qquad (11.75)$$

And we can similarly transpose Ls. Thus we have the model of Fig. 11.28. However, in many cases, the load is not a simple resistor, and this strategy is of limited application.

To find the input impedance we have:

$$Z_{in} = Z_1 + \frac{Z_2 Z_3}{Z_2 + Z_3} \qquad (11.76)$$

This is difficult to handle. One simplification is to ignore the excitation Z2 as this is usually less than 10% of the full-load input, and then we have:

$$Z'_{in} = j\omega \left(L_{pl} + \frac{L_{sl}}{K^2} \right) + \left(R_p + \frac{R_s}{K^2} + \frac{R_{load}}{K^2} \right) \qquad (11.77)$$

This does not give the correct phase angle between the supply voltage and current. But it is good enough for most purposes.

However, if we do need a more accurate picture, as the volt drop across $Z1$ is small, there is only a small error if we move $Z2$ across the input in Fig. 11.28. Then we have:

$$Z_{\text{in}} = \frac{Z_1(Z_2 + Z_3)}{Z_1 + Z_2 + Z_3} \tag{11.78}$$

and if we replace $j\omega$ by s we can expand this as:

$$Z_{\text{in}} = \frac{\frac{sL_{mag}R_{core}}{sL_{mag}+R_{core}}\left(sL_{pl} + \frac{sL_{sl}}{K^2} + R_p + \frac{R_s}{K^2} + \frac{R_{load}}{K^2}\right)}{\frac{sL_{mag}R_{core}}{sL_{mag}+R_{core}} + sL_{pl} + \frac{sL_{sl}}{K^2} + R_p + \frac{R_s}{K^2} + \frac{R_{load}}{K^2}} \tag{11.79}$$

and after multiplying:

$$Z_{\text{in}} = \frac{\left(sL_{mag}R_{core}\right)\left(sL_{pl} + \frac{sL_{sl}}{K^2} + R_p + \frac{R_s}{K^2} + \frac{R_{load}}{K^2}\right)}{sL_{mag}R_{core} + \left(sL_{mag} + R_{core}\right)\left(sL_{pl} + \frac{sL_{sl}}{K^2} + R_p + \frac{R_s}{K^2} + \frac{R_{load}}{K^2}\right)} \tag{11.80}$$

In many cases the output voltage is less than $1/10$ of the input, so K^2 is of the order of 0.01 and $1/K^2 = 100$. If we keep only the terms in K^2 and ignore leakage inductances and Rs as being much smaller than the load resistor:

$$Z_{\text{in}} = \frac{\left(sL_{mag}R_{core}\right)\frac{1}{K^2}R_{load}}{sL_{mag}R_{core} + \left(sL_{mag} + R_{core}\right)\frac{1}{K^2}R_{load}} \tag{11.81}$$

which reduces to:

$$Z_{\text{in}} = \frac{\left(sL_{mag}R_{core}\right)\frac{R_{load}}{K^2}}{\left(sL_{mag} + R_{core}\right) + \frac{R_{load}}{K^2}} \tag{11.82}$$

which is just the parallel sum of the excitation $Z2$ and the transferred load resistor. But remember that very often the output of a power transformer is connected to a rectifier circuit, so the above analysis must be used with care.

Returning to Fig. 11.28, if you have few sheets of paper to spare and half an hour or so with nothing better to do, you could try the exact analysis.

11.5.3.4 Measuring Parameters

Most catalogues list only the secondary voltage(s) and VA rating. To complete the model, we make open-circuit and short-circuit tests.

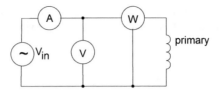

Fig. 11.29 Open circuit test

We start with the open-circuit test, Fig. 11.29, where we either use an ammeter, a voltmeter and a wattmeter as shown to measure the resistive and inductive components or a two-channel oscilloscope with a small series resistor to measure the current and the second channel to measure the voltage, and using Lissajous figures we may find the phase angle and again separate the resistive and inductive parts of the input. Some care needs to be taken with earthing! In effect, we are measuring $Z1 + Z2$ in Fig. 11.28. We can use an ohmmeter to find Rp then $Rcore = Rc - Rp$.

We connect a Variac to the primary to supply a low voltage AC with the secondary short-circuited through an ammeter. Then we increase the primary voltage until the rated secondary current is reached. We may discount $Z2$ as its relative impedance is much larger, so now we are measuring $Z1 + Z3$ in series, and we again resolve the readings into a resistive part Rl and an inductive part Ll. As we know Rp, we can find Rs by subtraction $Rs = (Rl - Rp)K^2$. The leakage inductances we ascribe by the turns ratio.

Example – Transformer Measurements
The peak no-load input current for a transformer was 128 mA lagging the supply by $58.5°$. The resolved peak components of the current are 66.88 mA in phase and 109.1 mA lagging. Hence Rc is $340 V/66.88 mA = 5.08 k\Omega$, and the inductive impedance is $340 V/109.1 mA = 3.12 k\Omega$, whence the inductance is $9.93 H$. Note that the inductance is the sum of the leakage and magnetizing inductances. However, the magnetizing inductance is by far the larger, so we may neglect the error in just assigning it to $Lmag$.

11.5.4 LTspice Models

Mutual inductance lends itself to a linear model, but not if we want to include the change of permeability with current shown in Fig. 11.12.

11.5.4.1 Linear Model

A simple mutual inductance will serve with a coupling constant of K which means it has a leakage inductance of $Lleak = L(1 - K^2)$, schematic ('Mains Transformer.asc'). The core loss can be handled by a resistor across the primary.

11.5.4.2   Non-linear Models

The 'examples//Educational' folder contains ('NonLinear Transformer.asc'). We can compare this with the Fig. 11.27. The ideal transformer consists of the four voltage-controlled current sources where $G1$ and $G2$ drive currents through the 1 MΩ resistor Rx proportional to the input and output voltages, respectively, whilst $G3$ and $G4$ drive currents proportional to the voltage across Rx through the source and load. This means the transformer is reversible, that is, the excitation $V1$ and the load resistor $Rload$ could be interchanged. The inductor $L1$ represents the excitation arm of Fig. 11.27 and models the leakage inductance and the core losses. Leakage inductances and copper loss could be included in the schematic as extra series components.

We can also use the comprehensive model developed above. We can give the transformer the correct primary and secondary inductances and so remove L_{mag}. We may also make everything non-linear and functions of temperature and current.

Example – Non-linear Transformer

With a 10 V input from $V1$, the inductor $L1$ is already well into saturation. Reduce the voltage to 3 V and all is well. The transposed load current is in parallel with the inductor $L1$ and correctly flows through the low impedance path of series resistor $R1$, and the inductor current remains constant at about 0.2 A for loads from 1 Ω to 1 MΩ although the current through $R1$ ranges from 3 A to 0 A and accords with the transformed load current.

The inductor $L1$ will saturate correctly due to increased current from the source $V1$. This starts at about 7.5 V. What could be argued is that the current waveform is more heavily distorted and should be used instead of the voltage across $L1$.

Explorations 5

1. If we have $K = 1$ in schematic ('Mains Transformer.asc') and a large load, we find the secondary voltage is 24 Vpp and in phase with the very small current. (Note the polarity of $L2$ – we need to draw $-I(L2)$.)
2. Try different values for K and load resistor, and the load voltage and current will remain in phase as they should do, but we cannot probe leakage inductances.
3. Measure the primary current and confirm that it agrees with the transposed load.
4. Reduce the input voltage $V1$ of ('NonLinear Transformer.asc') to 3 V and measure the (a) the voltage across $L1$, (b) the current through $L1$, (c) the current through $R1$ and (d) the load current for different values of $R1$ and $Rload$ and confirm that the inductor current is almost constant and that the current through $R1$ agrees with the load current.
5. Alter the values of $G2$ and $G4$ to model a step-down transformer where the secondary voltage is less than the primary.

A second non-linear model, schematic ('NonLinear Transformer V2.asc'), is left for the reader to explore. It uses a current-dependent voltage source $H1$ to transfer the primary current in $L1$ measured by the dummy voltage source $V2$ as a voltage in the secondary by setting its **Value2** field to 10 to make the transformer ratio.

Meanwhile the secondary current is measured by the dummy voltage source *V3* and transformed as a current in parallel with the inductor *L1*. In LTspice we can dispense with the dummy voltage sources and use the current in resistors R_1 R_2 to control the voltage sources. Essentially this is the previous schematic except that it only is supposed to work primary-to-secondary, and the secondary voltage is controlled by the primary current, not the voltage. It still needs a lot of development!

11.6 Summary

This chapter has covered the magnetic effects of a current. The key points are:

- The inductance of a solenoid is $L = \frac{\mu_r \mu_0 N^2 A}{l}$.
- LTspice accepts arbitrary functions for an inductor and temperature coefficient.
- LTspice also defines an inductor by the flux, e.g. ***Flux = 100 m ∗ x ∗ (1 − abs (0.01 ∗ x ∗ ∗ 5))***, where *x* is the current, or by the magnetic properties and dimensions of the core and the number of turns.
- Sub-circuit models can also be used for more accurate modelling of parameter variations with current and temperature. Many are provided by inductor manufacturers such as CoilCraft and Murata.
- Mutual inductance exists between two coils and is given by $M = K\sqrt{L_1 L_2}$ and is specified in LTspice by
  K < LI > < L2 > ... < K value> on the schematic without a leading dot.
- Power transformer have a *K* value close to unity.
- They have a constant *volts per turn* so that $V_S = V_P \frac{n_S}{n_P}$.
- The *copper loss* is due to the resistance of the winding wire and depends on the load current, whilst the *core loss* depends on the primary voltage and is due to hysteresis and eddy currents.
- There is also the constant *magnetizing current* drawn by the primary inductance.
- A linear model of a transformer has an ideal mutual inductance to transfer the primary voltage to the secondary at a suitable turns ratio together with the magnetizing inductance (almost equal to the primary inductance) and a resistor representing the core loss in parallel with the primary inductance. Leakage inductances and resistors representing the copper losses are added in series with the primary and secondary.

Chapter 12
LR and LCR Circuits

12.1 Introduction

In some ways, inductors are the inverse of capacitors. They, too, are frequency-sensitive components. The physics of how they and transformers work were covered in the previous chapter. Here we shall just assume linear devices and not bother about saturation, winding loss and all the other practical complications. And this is not so reckless as it might seem; many circuits use inductors with air cores or low currents and hence low flux densities and little power dissipation. And, having once proved that the circuit works, we can always introduce the parasitic inductor components, or even a sub-circuit, if need be.

So, assuming we are dealing with, if not exactly ideal inductors, at least ones that are well-behaved, we can calculate several basic properties and their time and frequency responses. From there it is interesting to explore bridge methods to measure an inductance. The remainder of the chapter is something of a rag-bag of circuits, mainly simple filters of some sort or octher, and delay lines and loudspeakers.

12.2 Inductors

In practice, these are not as widely used as CR circuits at low frequencies since inductors are more difficult to make and cannot be made as small as some capacitors, nor does an inductor give DC isolation. They are, however, widely used at high frequencies. In the main they consist of a number of turns of wire on a core which is either filled with air or a ferromagnetic medium. The main application of single inductors is in frequency-selective circuits, and we should note that inductance is often a parasitic effect associated with other devices as we saw with capacitors.

© Springer Nature Switzerland AG 2020
C. May, *Passive Circuit Analysis with LTspice*®,
https://doi.org/10.1007/978-3-030-38304-6_12

12.2.1   Energy Stored in an Inductor

This can be quite considerable and quite destructive. It is stored in the magnetic field but we will calculate it using the current which is directly proportional to the inductance.

Whilst we know that the current in a circuit is not due to the wholesale flow of electrons, nevertheless, it is convenient to consider the work done in moving a small quantity if charge q again the opposing voltage v of the inductor. This opposing voltage in itself implies that the current is increasing. So representing the extra charge by $i.dt$, the work done on it is:

$$i.dt\,v = i.dt\,L\frac{di}{dt} \tag{12.1}$$

we integrate to find the work done in establishing the current:

$$W = L\int_0^I i\,di = \frac{1}{2}LI^2 \tag{12.2}$$

where I is the final current.

12.2.2   Inductors in Series and Parallel

We assume that there is no interaction between the inductors. If we place inductors in series, Fig. 12.1, the same current flows through each, so they will all experience the same rate of change of current. Therefore, using lower case for time-varying parameters, we have for the voltage v developed across the inductors:

$$v = -L_1\frac{di}{dt} - L_2\frac{di}{dt} - L_3\frac{di}{dt} + \ldots \tag{12.3}$$

which reduces to:

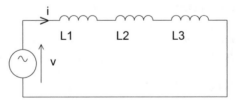

Fig. 12.1 Series Inductors

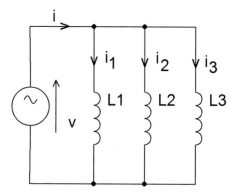

Fig. 12.2 Parallel Inductors

$$v = -(L_1 + L_2 + L_3 + \ldots)\frac{di}{dt} \tag{12.4}$$

and so the inductances add, just like resistors.

If the inductors are in parallel, Fig. 12.2, the current will divide between them as:

$$i = i_1 + i_2 + i_3 + \ldots \tag{12.5}$$

and they will all have the same voltage:

$$v = -L_1\frac{di_1}{dt} = -L_2\frac{di_2}{dt} = -L_3\frac{di_3}{dt} + \ldots \tag{12.6}$$

but for the equivalent total inductance L:

$$v = -L\frac{di}{dt} \tag{12.7}$$

or:

$$v = -L\frac{d}{dt}(i_1 + i_2 + i_3 + \ldots) \tag{12.8}$$

and on expanding:

$$v = -L\left(\frac{di_1}{dt} + \frac{di_2}{dt} + \frac{di_3}{dt} + \ldots\right) \tag{12.9}$$

then substituting from Eq. 12.6:

$$v = L\left(\frac{v}{L_1} + \frac{v}{L_2} + \frac{v}{L_3} + \ldots\right) \tag{12.10}$$

and dividing by v and rearranging:

$$\frac{1}{L} = \frac{1}{L_1} + \frac{1}{L_2} + \frac{1}{L_3} + \cdots \quad (12.11)$$

which is the same form as resistors in parallel, and if we have only two inductors:

$$L_{eq} = \frac{L_1 L_2}{L_1 + L_2} \quad (12.12)$$

Example – Inductors in Series and Parallel
Given inductors of *1,2* and *3 H*, these will add to *6 H* in series and *0.545 H* in parallel.

Example – Inductors in Parallel
Run schematic 'Parallel Inductors.asc' and confirm that the currents and voltages are correct. Use **Tools -> Control Panel -> Hacks** to check '*Always default inductors to Rser=0*'.

12.2.3   Time Response of an Inductor

We take the simplest circuit of a voltage source, a resistor, and an inductor in series with a switch, Fig. 12.3. The resistor is inevitable and may only be the resistance of the wire used to wind the inductor. Thus with the switch at *A* we have:

$$V = iR - L\frac{di}{dt} \quad (12.13)$$

and we have the current and its differential, and just as for a capacitor, we seek a solution that will be dimensionally correct, and once again we have either an exponential or a sinusoid. One possibility is that $di/dt=0$ and then the current is a steady *V/R* and it looks like it could be difficult to fit that with a sinusoidal solution so we turn to an exponential solution and we have two possibilities. The first is $i = Ae^{-bt}$ where *A* and *b* are constants. Let us test it. At time zero this has a maximum *A*, and at infinite time it has decayed to zero. But we have just seen that there is a substantial final current, so that will not do. The other possibility is $i = A(1 - e^{-bt})$. At time zero

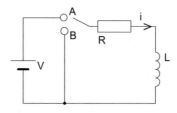

Fig. 12.3 Inductor Time Response

this is zero and rises to a final value A at infinite time. This looks more promising, so we will try it. Equation 12.13 becomes:

$$v = A(1 - e^{-bt})R + bALe^{-bt} \tag{12.14}$$

and at infinite time this is:

$$v = AR \tag{12.15}$$

so $A = I_f$, the final current. Then at time zero we have:

$$v = I_f bL \tag{12.16}$$

whence $b = R/L$ and finally we have:

$$i = I_f\left(1 - e^{\frac{Rt}{L}}\right) \tag{12.17}$$

where L/R is the time constant.

If we now switch the voltage to zero at B we have:

$$iR + L\frac{di}{dt} = 0 \tag{12.18}$$

and as a result the current will reduce to zero, so the simple negative exponential is a likely solution. However, as the current in an inductor cannot change instantaneously, the initial current is the previous final current:

$$I_f e^{-bt}R + I_f bLe^{-bt} = 0 \tag{12.19}$$

or:

$$I_f e^{-bt}(R + bL) = 0 \tag{12.20}$$

and $b = R/L$ as before. From Eq. 12.19 we see that the voltage also is a decaying exponential:

$$v = RI_f e^{\frac{-Rt}{L}} \tag{12.21}$$

so at the instant the voltage is removed, the back EMF is $v = I_f R$, the same as the original supply voltage. The big danger, though, is if the voltage is removed by opening a switch leaving no path for the stored energy and the resistance is infinite. In this case the energy will flow in a spark or an arc. This can be intentional as in the ignition coil of a car or potentially destructive when some means of restricting the voltage must be found.

We may define rise and fall times exactly as for capacitors, namely, the time from 10% to 90% of the final value. And we may use the sag in the current in a similar fashion to the sag in voltage on a capacitor.

Example – Inductor Turn-off

The inductor has a resistance of *0.1 Ω*, and the voltage across the inductor is *12 V*, so its asymptotic current is *120 mA*. At the instant that the switch opens in schematic 'Inductor Turn-off.asc', this current diverts to the *2 kΩ* resistor in parallel with the inductor, and therefore the instantaneous voltage is *0.12 A × 2 kΩ = 240 V*.

The energy stored in the inductor is *1/2 × 0.1 H × (0.12 A)² = 720 μJ*. The *.meas* directive reports *718 μJ*. The fall time using the cursor is *110 μs* in agreement with *2.2 × (time constant) = (2.2 × 0.1 H)/2 kΩ*.

If we put a *1 μF* capacitor in parallel with the inductor, the total turn-off time is about *10 ms* because of the damped oscillations and the peak voltage is reduced to *34 V*.

Explorations 1

1. Open the circuit 'Pulse Excitation LR.asc'. Measure the rise and fall times and the time constant.
2. Change *Ton(s)* to 1 ms and the current waveform is almost triangular whilst the inductor voltage shows a sag.
3. With *L = 0.05 H* the inductor current is almost a square-wave limited by the resistor, whilst the voltage is sharp positive and negative exponential spikes.
4. Use a symmetrical −10 V to +10 V excitation.
5. Reduce *Ton(s)* to 0.2 m with a 0–10 V pulse.
6. The schematic 'Inductor Turn-off.asc' uses a switch to turn off the current. Explore the magnitude of the back EMF and time constant for different values of *R2* and inductor.

12.2.4   *Frequency Response of an Inductor*

This follows the same methodology as for a capacitor. For the circuit in Fig. 12.4 we have:

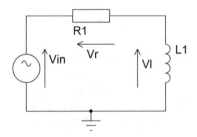

Fig. 12.4 LR Series Circuit

$$v_{in} = v_r + v_l \qquad (12.22)$$

Suppose the excitation is of the form $V_{pk}\sin\omega t$. The voltage across the resistor must be in phase with the excitation, so we seek an oscillatory solution of the form $i = I_{pk} \sin \omega t$. Thus we can write for Eq. 12.22:

$$V_{pk} \sin (\omega t) = I_{pk} \sin (\omega t)R - L\frac{d(I_{pk} \sin (\omega t))}{dt} \qquad (12.23)$$

and performing the differentiation and Eq. 12.23 becomes:

$$V_{pk}\sin(\omega t) = I_{pk}\sin(\omega t)R + \omega I_{pk}\cos(\omega t)L \qquad (12.24)$$

and writing *jsin* for *cos* we have:

$$V_{pk}\sin(\omega t) = I_{pk}\sin(\omega t)R + j\omega I_{pk}\sin(\omega t)L \qquad (12.25)$$

and after cancelling *sinωt*:

$$V_{pk} = I_{pk}(R + j\omega L) \qquad (12.26)$$

12.2.4.1 Reactance and Impedance of an Inductor

From Eq. 12.26 we see that the inductor presents a reactance of ωL and an impedance of $j\omega L$. Again, this has the dimensions of resistance and again, like a capacitor, this is not a power dissipation by the inductance itself; but unlike a capacitor, both increase with frequency.

Thus we may use exactly the same techniques as before to determine the 3 dB point, phase angle and the like. The inductor's impedance is drawn along the positive j-axis (note it is *j*, not −*j*), so we have from Fig. 12.5 that the circuit's impedance is:

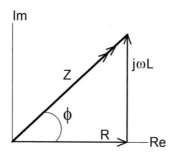

Fig. 12.5 LR Impedance

$$Z = \sqrt{R^2 + (\omega L)^2}$$ (12.27)

from which we may calculate the total current drawn by the circuit. The inductor forms the lower arm of a complex potential divider in Fig. 12.4 so:

$$v_L = v_{in} \frac{j\omega L}{R + j\omega L}$$ (12.28)

or:

$$v_L = v_{in} \frac{1}{1 - \frac{jR}{\omega L}}$$ (12.29)

and this has a 3 dB point at $\omega = \frac{R}{L}$ (the reciprocal of the time-constant). And Eq. 12.28 clearly predicts that the output voltage will tend to zero as the frequency gets lower and to v_{in} at high frequency. The phase angle φ is found from:

$$\phi = \tan^{-1}\left(\frac{\omega L}{R}\right)$$ (12.30)

in Fig. 12.5.

'.net' Command for a Single Port
The difference is that we only reference the input source as *.net V(in)* and we can draw traces of *Y(in)* and *Z(in)*. There is no need to add Rout.

Example – Inductor Reactance and Impedance
We use the schematic 'Series Inductor.asc' to measure the input impedance. Taking a frequency of 1 kHz using eq. 12.27, we find $Z = \sqrt{10^8 + \left(2\pi 10^2\right)^2} = 11.81 \; k\Omega$, and the phase angle from eq. 12.30 is: $\phi = \tan^{-1}\left(\frac{2000\pi}{10^4}\right) = 32.14^0$ both measured and calculated values agree.

Example – Potential Divider with Inductance
If we have a resistor in parallel with the inductor, Fig. 12.6, we can take the inductor as an external component and convert the two resistors and voltage source into their Thevenin form. If we put some simple number $R1 = R2 = 1 \; k\Omega$ and $L1 = 100 \; mH$, we have an equivalent of *5 V* in series with *500 Ω* and the inductor. schematic 'LR Potential Divider.asc'.

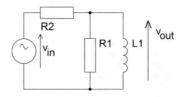

Fig. 12.6 LR Potential Divider

Explorations 2
1. Plot the frequency response showing phase and magnitude in 'Series Inductor. asc', and compare this with measurements from Transient analyses spread over the same frequency range.
2. Just to add to the joys of life, try adding the series resistance of the inductor.
3. Try different combinations of R,L and exciting frequency and amplitude, and compare your calculated voltages, currents and phase angles with the results of transient and frequency analyses.
4. Measure the impedance and reactance of an inductor at various frequencies. Do this by an AC analysis, and note that the AC frequency range will override any setting of *V1*. Change the vertical axis to *Logarithmic*.
5. Build and test the above example and try different values.

12.3   Settling Time

The inductor voltage does not reach its final value immediately, but only after a few cycles.

12.3.1   Pulse Train Input

During the ON time of the pulse in figure 'Inductance Pulse Train' up to time a, the inductor current increases and hence the voltage across it decreases to y. Then as the pulse is suddenly removed, the voltage across the inductor falls by vin so that it undershoots to a negative value z. This negative voltage drives a reverse current through R_L with the voltage increasing towards zero at a slower rate because the

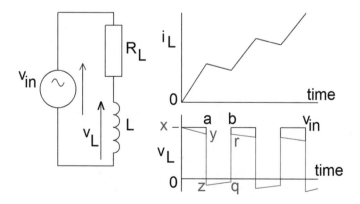

Fig. 12.7 Inductance Pulse Train

inductor voltage is less, reaching q at time b when the next pulse arrives. The inductor voltage again suddenly increases by vin but, because it started from a negative value, it only reaches r. Thus we see a steady decrease in the inductor voltage over time until the average voltage is zero.

Example – Inductor Pulse Settling Time

With the initial setting of schematic 'Inductor Pulse Settling Time.asc', we see the initial voltages and currents. The alternative settings show an increasing settling time with duty cycle to about 2 s.

12.3.2   Sine Wave Train Input

We shall content ourselves with a qualitative explanation using figure 'Inductor Sine Train' from schematic 'Inductor Sine Settling Time.asc'. The inductor and input voltages and the inductor current all start at zero, time a. And here is the problem – we know that the current and voltage are at 90^0 (remember CIVIL), so during the first quarter cycle the peak inductor voltage at b leads the peak input voltage which occurs at *1.25 ms*, and we therefore find that the peak of the inductor current is at time c when the inductor voltage is zero. The difference between the two voltages is the voltage across the series resistor and therefore corresponds to the inductor current. Hence at time d when the difference is zero, the inductor current is instantaneously zero. This occurs slightly before the negative peak of $v(s)$. After that point, the inductor voltage is greater than the input so it drives a reverse current through the resistor. However, the distance between the voltages is smaller than during the initial period from a to d and therefore the magnitude of the peak of the reverse current when the inductor voltage is zero at e is f and is less than c.

The next point of interest is *5.00 ms* when the input voltage is zero. When we first started with zero input voltage at time a, the inductor current was zero. Here it is

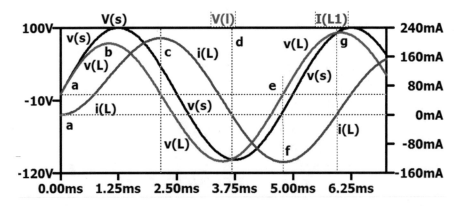

Fig. 12.8 Inductor Sine Train

negative: when the current again becomes positive after time g, the positive peak will be less than c, and over the next few cycles the current waveform will move downwards and settle at an average of zero.

Example – Inductor Sine Settling Time

If we run schematic 'Inductor Sine Settling Time.asc', we reproduce the figure. Adding the trace **V(s)-V(l)** and we see the difference in voltages is sinusoidal (as we should expect from adding sinusoids). Taking the longer transient time and we see the average current settling to zero.

12.4 Bridges to Measure Inductance

This is something of theme and variations and is well worth working through as a case study of how to improve a basic design. In every case it is necessary to avoid other inductors in close proximity else they would create mutual inductance. Unlike the resistance bridge, the balance here depends not only on getting the magnitudes of the two voltages across the inductor equal but also their phases.

With all of these bridges, we need two variable controls, one to balance the inductance of the inductor, the other its resistance. One of the problems is that of finding a calibrated variable capacitor – not easy and expensive. So it is better if we can balance the bridge just with a fixed one. We should also note that the right-hand side of the bridges consists of a resistor and the inductor in series, so in order to drop approximately half the excitation voltage across the resistor, we must switch either the frequency in decades or the resistor.

A great advantage of these bridges is that we easily can amplify the out-of-balance AC signal since there is no DC offset and possible drift to bother us. An oscilloscope or a digital voltmeter will enable us to see 1 mV, and we can precede it with an amplifier for even greater sensitivity. In fact, noise is likely to be the limiting factor or thermal drift of the values of the components.

12.4.1 The Maxwell Bridge

This was created by James Clerk Maxwell in 1873, Fig. 12.9, before the days of radio and coils of much less than 1 mH, but when telegraphy was well-established and the problem of voice distortion over long lines was acute and solved by inserting iron-cored inductors known as *loading coils* (which we shall explore later as Delay Lines).

In preparation for the following bridge variations, let us denote the four elements as general impedances $Z1$, $Z2$, $Z3$, $Z4$ and at balance:

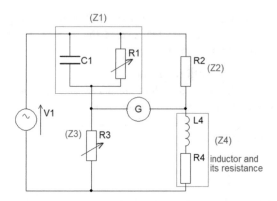

Fig. 12.9 Maxwell Bridge

$$\frac{Z_1}{Z_3} = \frac{Z_2}{Z_4} \tag{12.31}$$

or:

$$Z_1 Z_4 = Z_2 Z_3 \tag{12.32}$$

for $Z1$ we have:

$$\frac{R_1 \frac{1}{j\omega C_1}}{R_1 + \frac{1}{j\omega C_1}} = \frac{R_1}{1 + j\omega C_1 R_1} \tag{12.33}$$

and substituting in equation 12.32:

$$\left(\frac{R_1}{1 + j\omega C_1 R_1}\right)(j\omega L_4 + R_4) = R_2 R_3 \tag{12.34}$$

$$j\omega R_1 L_4 + R_1 R_4 = R_2 R_3 + j\omega C_1 R_1 R_2 R_3 \tag{12.35}$$

Equating real and imaginary parts at balance:

$$L_4 = C_1 R_2 R_3 \quad R_4 = \frac{R_2 R_3}{R_1} \tag{12.36}$$

and the balance does not depend upon the exciting frequency or the amplitude, but it is a good idea to use a frequency close to the working frequency of the inductor. Ferrite or iron-cored inductors may need a standing DC to model the working flux.

Example – Maxwell Bridge Calculation
A bridge is balanced with $C_1 = 100\ nF$, $R_1 = 100\ \Omega$, $R_2 = 53\Omega$, $R_3 = 27\ \Omega$. Calculate the value of the inductor and its resistance.

From Eq. 12.36 the inductance is $L = 100\,n \times 53 \times 27 = 143\,\mu H$ $R = (53 \times 27)/100 = 14.3\,\Omega$.

Maxwell Bridge Critique

The balance conditions interact, so it is best to balance for the inductance and then for the resistance. The balance conditions are independent of frequency, so we do not need an accurately calibrated oscillator. An important measure of an inductor is its *Q-factor* defined as:

$$Q = \frac{\omega L}{R} \tag{12.37}$$

In this case from Eq. 12.36 it is:

$$Q = \frac{\omega L_4}{R_4} = \omega C_1 R_2 R_3 \frac{R_1}{R_2 R_3} = \omega C_1 R_1 \tag{12.38}$$

This does use the frequency of the oscillator but is not usually critical. Inductors used in telecommunications can have a Q-factor of over 100. Inserting that in Eq. 12.38 with our 100 nF capacitor and a frequency of 1000 r/s then $R_1 = 100/(1000 \times 100\,n) = 1\,M\Omega$ which is not unreasonable, but if the frequency is lower, or the capacitor is smaller, there is a problem finding an accurate resistor of perhaps 10 MΩ.

Example – LTspice Inductor Default Series Resistance

If we start with an improbable inductor of *1 H 1 Ω* in schematic 'Maxwell Bridge Default Rser'. If we use a DC excitation and find the operating point by *.op* and make *R3* a parameter then:

.step param R3 9.99 10.11 0.01

and plotting *V(c)-V(l)* will result in a straight line graph sloping upwards with resistance on the horizontal axis and voltage on the vertical. We find a balance at $R3 = 10.005\,\Omega$. Then from Eq. 12.36 the resistance of the inductor is *2.001 Ω*. But as we have not added any resistance to the inductor itself, only added a series resistor $R4 = 2\,\Omega$ to represent the resistance of the inductor, it follows that the default resistance added by LTspice is *1 mH*. Using **Tools→Control Panel→Hacks→Always default inductors to Rser=0** and the balance now is exactly at $R3 = 10\,\Omega$.

Example – Maxwell Bridge

Using schematic 'Maxwell Bridge.asc', we can find the resistance of the inductor as before (in contradiction to the previous critique). That fixes the values of *R2* and *R3*, so with a sinusoidal excitation we vary *C1* to find the balance:

.step param C1 15m 30m 0.5m

A transient run will show the waveform and we can take a guess at the correct value. Better is to add:

$$.meas\ \textbf{Vbal}\ \textbf{RMS}\ \textbf{V}(c)\text{-}\textbf{V}(I)$$

to measure the error voltage. The listing in the *Spice Error Log* file will show this is exactly step *21*, and if we right click and plot the stepped measured data, we find the result is exactly zero at $C1 = 25\ mF$. It would be extremely difficult to find an accurate capacitor of such a large value, so, in practice, it would be necessary to scale up the product *R2R3* by at least 10^4. We also find the Q-factor is $Q = 2\pi fC1R1$ which is an improbable *3142*.

12.4.2   The Hay's Bridge

This replaces the parallel resistor-capacitor combination with a series one, Fig. 12.10. There is therefore no DC path, so we cannot balance the inductors resistance with a DC supply. We can again invoke Eq. 12.32, and this time we have:

$$\left(R_1 + \frac{1}{j\omega C_1}\right)(R_4 + j\omega L_4) = R_2 R_3 \qquad (12.39)$$

expanding:

$$R_1 R_4 + j\omega R_1 L_4 - j\frac{R_4}{\omega C_1} + \frac{L_4}{C_1} = R_2 R_3 \qquad (12.40)$$

equating real and imaginary parts:

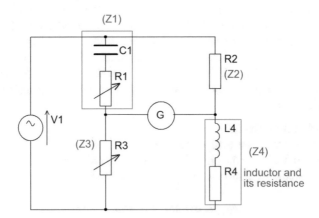

Fig. 12.10 Hay's Bridge

$$\omega R_1 L_4 = \frac{R_4}{\omega C_1} \quad \text{then} \quad L_4 = \frac{R_4}{\omega^2 R_1 C_1} \quad \text{and} \quad R_4 = L_4 \omega^2 R_1 C_1 \qquad (12.41)$$

but R_4 is the resistance of the inductor so we cannot find the inductance from this equation.

However,

$$R_1 R_4 + \frac{L_4}{C_1} = R_2 R_3 \quad \text{then} \quad R_4 = \frac{1}{R_1}\left(R_2 R_3 - \frac{L_4}{C_1}\right) \qquad (12.42)$$

and we can equate Eqs. 12.41 and 12.42 and collect terms to solve for L_4:

$$L_4\left(\omega^2 R_1 C_1 + \frac{1}{R_1 C_1}\right) = \frac{R_2 R_3}{R_1} \qquad (12.43)$$

multiplying by $R_1 C_1$

$$L_4\left(\omega^2 R_1^2 C_1^2 + 1\right) = C_1 R_2 R_3 \qquad (12.44)$$

Finally,

$$L_4 = \frac{C_1 R_2 R_3}{\left(\omega^2 R_1^2 C_1^2 + 1\right)} \qquad (12.45)$$

And substituting in Eq. 12.41:

$$R_4 = \frac{\omega^2 C_1^2 R_1 R_2 R_3}{\left(\omega^2 R_1^2 C_1^2 + 1\right)} \qquad (12.46)$$

Then,

$$Q = \omega \frac{L_4}{R_4} = \frac{\omega C_1 R_2 R_3}{\omega^2 C_1^2 R_1 R_2 R_3} = \frac{1}{\omega C_1 R_1} \qquad (12.47)$$

Hay Bridge Critique

The frequency of the AC excitation must be known, and the expressions for the inductance and its resistance are cumbersome. However, even with $Q = 100$ and supposing $\omega = 1000\ rad/s$, the capacitor and resistor can be quite small. Also, we can note that in Eq. 12.45 $\omega^2 R_1^2 C_1^2 = \frac{1}{Q^2}$ so if Q is large, we can approximate Eq. 12.45; by ignoring this factor, then $L_4 = C_1 R_2 R_3$ and similarly $R_4 = \omega^2 C_1^2 R_1 R_2 R_3$.

Example – The Hay Bridge

Because of the number of variables, we shall handle this in reverse by starting off with an inductor of *10 mH* having $Q = 200$ in schematic 'Hay's Bridge.asc' so that we can use the approximate expressions and a *1 kHz* source. We shall also set $C_1 = 10\ nF$ on the basis that it is easier to find a close tolerance fixed capacitor and variable resistor than the other way round. Then using Eq. 12.47 $R_4 = \frac{2\pi \times 1kHz \times 0.02H}{200} = 0.628\Omega$ and $R_1 = \frac{1}{2\pi 1kHz 10nF 200} = 79.58\Omega$. Then using Eq. 12.45 $R_2R_3 = 2M\Omega$

It is interesting to note that it is the product of these two resistors that matters, and we can assign them as $R_2 = 1\ k\Omega\ R_1 = 2\ k\Omega$ or the other way round.

We can test the sensitivity by stepping C_1, R_1 and R_2 where it will be found that changes of 0.1% are easily detected. In fact, stepping by 1 mΩ is detectable, and with $R_1 = 1000.026\ \Omega$ the output is about *1.6 µV*. As before, we can plot the stepped data to easily see the null point.

Also of note is the great difference in the currents of the two arms of the bridge, about *10 mA*.

12.4.3   The Owen Bridge

This requires a calibrated variable capacitor as well as a fixed one, Fig. 12.11. And again we cannot find the resistance of the inductor by using DC excitation. The balance condition is:

$$(R_4 + j\omega L_4)\left(\frac{1}{j\omega C_1}\right) = R_2\left(R_3 + \frac{1}{j\omega C_3}\right) \tag{12.48}$$

Expanding and separating real and imaginary parts:

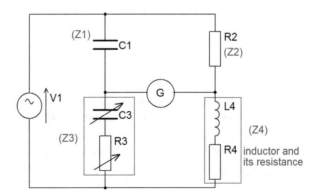

Fig. 12.11 Owen Bridge

$$-j\frac{R_4}{\omega C_1} = -j\frac{R_2}{\omega C_3} \quad R_4 = \frac{R_2 C_1}{C_3} \tag{12.49}$$

$$\frac{L_4}{C_1} = R_2 R_3 \quad L_4 = C_1 R_2 R_3 \tag{12.50}$$

then:

$$Q = \omega\frac{L_4}{R_4} = \frac{\omega C_1 C_3 R_2 R_3}{C_1 R_2} = \omega C_3 R_3 \tag{12.51}$$

Owen Bridge Critique

The balance does not depend upon frequency, but calibrated variable capacitors are expensive and less than 1 nF, so if $Q = 100$, we require a very large R_3. An interesting, albeit ancient, article on how to calibrate a capacitor can be found at *nvlpubs.nist.gov/nistpubs/jres/. . ./jresv64Cn1p75_A1b.p.*

The accuracy depends on that of the variable capacitor, usually about 1% although it proved difficult to find any on the Internet. However, this bridge can be used over a wide range of inductance.

Example – Owen Bridge Schematic

Equation 12.49 is a good starting point. Calibrated variable capacitors are limited in size, so let us take *100 pF* as typical. Suppose the inductor has a resistance of *0.628 Ω* as before, then from Eq. 12.49, the product $C_1 R_2 = 0.628 \times 10^{-10}$. Insert that in Eq. 12.50 assuming the same *20 mH* as before and $R_3 = 3.18 \times 10^{10}$ Ω. And this is totally impractical. Not only is R_3 excessively large, but almost the entire excitation voltage is dropped across C_1 and R_1, schematic 'Owen Bridge.asc'.

If we start by making $R_2 = 125$ Ω so that it equals the modulus of the inductor's impedance we find a balance correctly with $C_1 = 100$ pF, $C_3 = 19.9$ nF and R_3 $1.6 = MΩ$. With these capacitor values, C_1 must be the variable.

12.4.4 Anderson Bridge

This adds complexity to the Maxwell Bridge, Fig. 12.12. At balance there is no current in the galvanometer and the currents in the inductor and R_2 are identical:

$$i_3 = i_1 + i_{C1} \tag{12.52}$$

Also $v_a = v_b$ therefore:

$$i_2 Z_4 = (i_{C1} + i_1)R_3 + i_{C1}R_5 \tag{12.53}$$

And equating the voltages across C_1 and R_2:

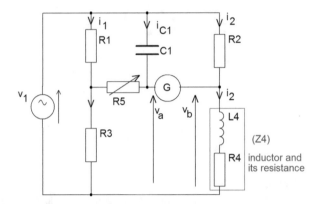

Fig. 12.12 Anderson Bridge

$$\frac{i_{C1}}{j\omega C_1} = i_2 R_2 \quad then \quad i_{C1} = i_2 R_2 j\omega C_1 \tag{12.54}$$

These are the two balance conditions. Substituting for i_{C1} in Eq. 12.53 from Eq. 12.54:

$$i_2 Z_4 = i_2 R_2 j\omega C_1 (R_3 + R_5) + i_1 R_3 \tag{12.55}$$

or:

$$i_1 = i_2 \frac{1}{R_3} (Z_4 - R_2 j\omega C_1 (R_3 + R_5)) \tag{12.56}$$

Equating the voltage across the series branch $C_1 R_5$ with that across R_1:

$$i_{C1} \left(R_5 + \frac{1}{j\omega C_1} \right) = i_1 R_1 \tag{12.57}$$

Substituting for i_{C1} in Eq. 12.57 and dividing by R_1:

$$i_1 = i_2 \frac{1}{R_1} (R_2 R_5 j\omega C_1 + R_2) \tag{12.58}$$

Equating 12.56 and 12.58:

$$R_1 (R_4 + j\omega L_4 - j\omega C_1 R_2 (R_3 + R_4)) = R_3 (R_2 R_5 j\omega C_1 + R_2) \tag{12.59}$$

Equating real parts:

$$R_1 R_4 = R_2 R_3 \quad R_4 = R_2 \frac{R_3}{R_1} \tag{12.60}$$

which means that we can find the resistance with DC excitation and ignore the reactive components.

Equating Imaginary parts:

$$R_1 L_4 - C_1 R_1 R_2 (R_3 + R_4) = C_1 R_2 R_3 R_5 \tag{12.61}$$

$$L_4 = \frac{C_1 R_2}{R_1} (R_3 R_5 + R_1 (R_3 + R_5)) \tag{12.62}$$

Equation 12.62 shows us that we have C_1 and R_5 at our disposal to find a balance. It is better to use a fixed capacitor – preferably less than $1\ \mu F$ because at lower values it is easier to find close tolerance stable components with a low temperature coefficient.

Example – Anderson Bridge Schematic
Open 'Anderson Bridge.asc'. The first thing we see is that with an inductor of $1\ H$, $2\ \Omega$ and an excitation of $100\ Hz$, the impedance of the inductance is $j628\ \Omega$. If we are to drop about half the supply voltage across the inductor, we need $R2 = 600\ \Omega$. But from Eq. 12.60 that means that $R_3/R_1 = 1/300$ and there will be almost no voltage at the other side at node 3. Therefore we need to reduce the frequency of the excitation to just a few hertz. This is a limitation.

So we shall take a smaller inductor of $20\ mH$ $5\ \Omega$ and an excitation of $100\ Hz$. The bridge balances with the values in the schematic, and this agrees with theory.

Critique
This has been 'theme and variations'. We may well ask why there are so many different bridges. The Maxwell Bridge, of course, was the first, so what, then, are the advantages of the later developments? Some of the pertinent questions are:

- How easy is it to arrive at a balance? That is, how far do the variables interact?
- What is the range of inductance that can be measured? As an example, the Maxwell bridge was invented during the days of telegraphy and therefore was designed to cope with the existing iron-cored inductors of millihenries.
- What range of Q-factors can be measured? As an example the Hay's Bridge can cope with high Q-factors without extreme component values.
- How accurate is the measurement? Remember that calibrated variable capacitors are expensive and only accurate to about 1%.

Explorations 3
1. Explore and compare all the inductor bridges using the above criteria.

12.5   Power Supply Filters

These are an important application of inductors. Most power supplies use rectifier diodes to convert AC to a lumpy DC which must be smoothed to an almost constant DC for use in electronic circuits such as radios and television sets; else there will be an audible hum from the loudspeaker at the mains frequency or, more usually, at twice than frequency. We can approach these circuits in either the time or the frequency domains.

12.5.1   Capacitor Input

The schematic 'FW Capacitor Input Filter.asc' models the output of a typical power supply working off the AC mains. The voltage source *B1* converts the AC voltage from *V1* into its absolute value of positive half sine loops. The switch *S1* is needed rather than a diode so that we can make it with no forward resistance and no volt drop, so it will conduct whenever *Vabs* is greater than *Vout*. Thus this rather odd circuit acts like an ideal full-wave rectifier, but that is not our concern at present.

12.5.1.1   Ripple Voltage

All we need to say now is that to a good approximation, the capacitor discharges linearly into the load resistor for the full half cycle implying a constant load current. This is from the peak of the input voltage *c* to 15 ms but in fact, it only discharges to *a* where the rising input voltage equals the capacitor voltage. This is not strictly true, the diode will not stop conducting at *c* but at *b* where the capacitor voltage is equal to

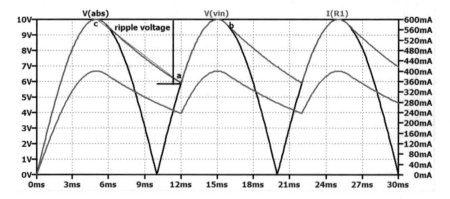

Fig. 12.13 FW Ripple

the falling input voltage. But as the capacitor has a tolerance of perhaps 20%, it is adequate.

In this case, the charge lost by the capacitor each period is due to the load current (which we assume to be constant) so

$$\frac{I_L t_{period}}{2} = CV_r \quad \text{or} \quad V_r = \frac{I_L t_{period}}{2C} \tag{12.63}$$

where V_r is the *ripple voltage*. The factor 2 arises because the capacitor recharges every half cycle.

Example – Full-Wave Rectifier Capacitor Input
Given a supply of *10 V_{pk} 50 Hz* and a load resistor of *100 Ω*, we can calculate the magnitude of the ripple voltage if the capacitor is *C1 = 500 μF*.

The load current is *10 V/100 Ω = 0.1 A*. The period is *1/50 = 20 ms*. Therefore from Eq. 12.63 V_r = *0.1 A × 20 ms/(2 × 500 μF) = 2 V*. By measurement on the schematic, it is 1.5 V. But if we double the capacitor size, both theory and measurement using schematic 'FW Capacitor Input Filter.asc' agree at a ripple of *1 V* and this is a more realistic figure of 10% ripple.

Example – Full-Wave Rectifier Capacitor Loss
If we plot the power we find *205.6 mW*. This is wrong – the perfect capacitor has no losses. If we add a *1 Ω* **Equiv. Series Resistance** to the capacitor, we find *288 mW*. A better estimate of the actual loss is the RMS current of *(321 mA)^2 × 1 Ω*. This causes the capacitor to get warm or even hot – and is potentially dangerous. If we increase the capacitor to *1 mF*, the ripple is approximately halved but the average capacitor current doubles and so does the RMS current and the capacitor loss.

Of course, this is a highly artificial circuit with no diode loss and no resistance of the voltage source, whereas in reality it would be the secondary of a transformer with at least the winding resistance. In the pre-SPICE days, we had to use graphs to estimate the capacitor loss.

12.5.2   Capacitor Input Filter

If we now follow the capacitor with a series inductor and a shunt capacitor, we can reduce the ripple further, Fig. 12.14, where we ignore the load resistor:

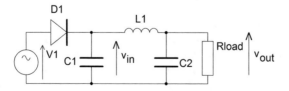

Fig. 12.14 Capacitor Input Filter

$$v_{out} = v_{in} \frac{\frac{1}{j\omega C_2}}{j\omega L_1 + \frac{1}{j\omega C_2}} \quad \text{or} \quad \frac{v_{out}}{v_{in}} = \frac{1}{1 - \omega^2 L_1 C_2} \qquad (12.64)$$

This appears to show that if $\omega = \frac{1}{L_1 C_2}$, the output will be infinite. However, this fails to take account of C_1. If the load is open-circuit, there will be no current, and C_1 will charge to a steady voltage, so there will be no AC applied to the inductor filter. On the other hand, if there is a load, this will damp any possible oscillations and again the situation is perfectly safe.

In normal operation the idea is that the $L1C2$ combination with the load resistor will reduce the ripple voltage. Including this resistor we have:

$$\frac{v_{out}}{v_{in}} = \frac{\frac{R_{load}}{1+j\omega C_2 R_{load}}}{\frac{R_{load}}{1+j\omega C_2 R_{load}} + j\omega L_1} = \frac{1}{1 + j\omega \frac{L_1}{R_{load}} - \omega^2 L_1 C_2} \qquad (12.65)$$

We will see in a later chapter that this is a damped oscillation which will reach steady state after a few seconds. This initial period is a little difficult to handle analytically and is generally overlooked. It is only a potential issue with light or disconnected loads when the voltage overshoot can be 50%.

For the fundamental the attenuations is:

$$\frac{v_{out}}{v_{in}} = \frac{1}{\sqrt{\left(1 - \omega^2 L_1 C_1\right)^2 + \left(\omega \frac{L_1}{R_{load}}\right)^2}} \qquad (12.66)$$

What also is difficult to handle analytically is the initial diode current turn-on surge. In this ideal model, it has a peak of *1.6 A*, and if we make $C1 = 5\ mF$, we find it is *16 A*. This is unrealistic; in fact there would be resistance associated with the voltage source and the diode modelled by the ideal switch, but still it is important to note these things.

Example – Capacitor Input Filter Overshoot

We have added an LC filter in schematic 'FW CLC Input Filter.asc'. And we have an overshoot of the load voltage by nearly 100%. So it is good policy to be generous with capacitor voltage ratings. Changing the $L1C2$ product alters the number of oscillations, but the peak remains substantially the same. Reduce the load resistor to *100 Ω*, and there is no surge. We should also note the intial large current pulse 1.6 A even though the steady-state current is but a few mA - and of course this is unrealistic because it ignores the resistance of the diode.

Example – Capacitor Input Filter

We wish to reduce the ripple by a factor of 10 in 'FW CLC Filter.asc' compared to the ripple on capacitor C_1 with a fundamental frequency of *100 Hz* and a *100 Ω* load resistor and keeping the *1 H* inductor.

As $\omega = 2\ \pi f$ we have $\omega = 628\ rad/s$ and the second term in the denominator of Eq. 12.66 is $(628/100)^2 = 39.4$, and therefore the first term must equal *60.6*. Taking the square root $1 - \omega^2 C1 = 7.78$ and thus $C_2 = 17.1\ \mu F$. By measurement it is *1.5 V/ 0.135 V = 11*.

However, if we increase the load resistor to a few kilohms, there can be a 50% overshoot in the output voltage which is not resolved by increasing the size of the capacitor.

Example – Capacitor Input Filter Oscillation
Schematic 'FW CLC Input Osc.asc' is essentially the previous schematic but with an added LC series circuit on the left. This has no capacitor C_1 nor a load resistor and the excitation frequency has been changed to *15.91 Hz* – the resonant frequency of the LC combination. We can see the voltage building up to more than *3.5 kV* with $R_2 = 1e10\ \Omega$. And even with $R_2 = 1k\Omega$, it is still *100 $V_{p\text{-}p}$*. And we see an initial overshoot in the right-hand circuit with a large load resistor.

12.5.3   Inductor Input Filter

This omits *C1*, schematic 'FW Choke Input Filter.asc'. It is widely used with switch-mode power supplies with a rectangular wave input because the output voltage is proportional to the duty cycle. If the *L1C1* product is too small, there will be a voltage overshoot which can be somewhat larger than with a capacitor input.

There is also the potential danger illustrated by the previous example that if the input frequency changes, there could be an excessive output voltage.

Example – Choke Input Filter
The output voltage of schematic 'FW Choke Input filter.asc' shows damped oscillations. In reality, the input frequency would be a few tens of kilohertz, but with some extreme values, it hangs, and it may be necessary to add a rise time to the pulse.

Explorations 4
1. Open and run schematic 'FW Capacitor Input Filter.asc' and note the RMS current in the capacitor. This can be a limiting parameter in power supply design. Add an ESR and find the capacitor loss.
2. Check that the ripple voltage approximately agrees with Eq. 12.63, and note how the RMS capacitor current increases with the reduction in ripple voltage.
3. Run schematic 'FW Capacitor Input Filter.asc' with different inductor and capacitor combinations. In particular, note the voltage overshoot and the 'diode' surge current at turn-on. Increase *Ron* of the switch to a few ohms and note the effect on the surge current.
4. Open and run schematic 'FW Capacitor Input Filter.asc', and find the best distribution of capacitance value between the two capacitors – is a large *C1* and small *C2* better than vice versa or both equal?
5. Explore the relationship between duty cycle and output voltage with schematic 'FW Choke Input Filter.asc'. This is best done by changing *Ton* to a parameter

and using *.step param Ton 1m 9m 1m.* Try running it with a *40 kHz* input. Note the damped oscillations and overshoot.

6. Remove the *abs* from the voltage source to create a half-wave rectifier and repeat the measurements.

12.6   LR Filters

These are the complement of the RC filters developed previously in 'Second Order RC Filters' where now we replace capacitors by inductors. It is generally cheaper to use a capacitor; also they are often smaller and in some cases offer DC isolation. On the other hand, at high frequencies surface mount small inductors are an attractive possibility. The analysis is very similar to that for RC circuits. We may choose to turn the equations into standard form or just solve them as they stand as quadratic equations.

12.6.1   Cascaded LR Filters

We can use the general analysis developed for RC filters which resulted in the unwieldy equation which we repeat here:

$$\frac{v_{\text{out}}}{v_{\text{in}}} = \frac{Z_2 Z_4}{Z_1 Z_3 + Z_2 Z_3 + Z_1 Z_4 + Z_2 Z_4 + Z_1 Z_2} \qquad (12.67)$$

12.6.1.1   Cascaded LR Low-Pass Filter

Using the circuit of Fig. 12.15, we substitute in Eq. 12.67 neglecting the source resistance R_1:

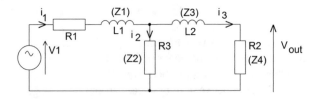

Fig. 12.15 Cascaded LR Low Pass Filter

$$H(s) = \frac{R_3 R_2}{sL_1 sL_2 + R_3 sL_2 + sL_1 R_2 + R_2 R_3 + sL_1 R_3} \tag{12.68}$$

Collecting terms:

$$H(s) = \frac{R_2 R_3}{s^2 L_1 L_2 + s(L_2 R_3 + L_1 R_2 + L_1 R_3) + R_2 R_3} \tag{12.69}$$

And this is a second-order system because s^2 is the highest power of s. We turn Eq. 12.69 into standard form by dividing throughout $L_1 L_2$:

$$H(s) = \frac{\frac{R_2 R_3}{L_1 L}}{s^2 + s \frac{(L_2 R_3 + L_1 R_2 + L_1 R_3)}{L_1 L_2} + \frac{R_2 R_3}{L_1 L_2}} \tag{12.70}$$

We write the denominator as:

$$H(s) = s^2 + s\Sigma + \Pi \tag{12.71}$$

where the product and the sum are:

$$\Pi = \frac{R_2 R_3}{L_1 L_2} \quad \Sigma = \frac{(L_2 R_3 + L_1 R_2 + L_1 R_3)}{L_1 L_2} \tag{12.72}$$

We use Eq. 12.71 to find the poles:

$$p_1, p_2 = \frac{-1}{2} \left[\Sigma \pm \sqrt{(\Sigma^2 - 4\Pi)} \right] \tag{12.73}$$

Then we can write the transfer function as:

$$H(s) = \frac{\frac{R_2 R_3}{L_1 L_2}}{(s + p_1)(s + p_2)} \tag{12.74}$$

and by expanding the denominator of Eq. 12.74, we have $s^2 + s(p_1 + p_2) + p_1 p_2$, and by comparison with Eq. 12.71 we have:

$$\Sigma = p_1 + p_2 \quad \Pi = p_1 p_2 \tag{12.75}$$

Filter Design

As the filter has two breakpoints and four components, it would seem that we have a free choice of two. This is not necessarily so. One approach is to make the impedance of the second stage much greater than the first or vice versa so that we can treat each separately as we did for RC filters. We can rewrite the Σ term in Eq. 12.72 as $\Sigma = \frac{R_3}{L_1} + \frac{R_2}{L_2} + \frac{R_3}{L_2}$. The first two terms are the breakpoints we want, so if we can make the last term small, Σ will closely approximate to the sum of the two poles. If we want coincidental breakpoints, then $\frac{R_3}{L_1} = \frac{R_2}{L_2}$, and if we take $R_2 = aR_3$ then $L_2 = aL_1$ and $R_3/L_2 = R_3/aL_1$ and $\Sigma = (2+1/a)p$. The practical limit to a is about 100 then $\Sigma = 2.01p$. Using that in Eq. 12.73 and with $\Pi = p^2$, we find the actual poles differ by about 4%. If we stretch a to a larger value, the error becomes less, but we can find ourselves with rather large inductors or very small resistors. This can be adapted to situation where the breakpoints are not identical.

An alternative is to note that from Eqs. 12.72 and 12.75:

$$\frac{\Sigma}{\Pi} = \frac{p_1 + p_2}{p_1 p_2} = \frac{L_2 R_3 + L_1 R_2 + L_1 R_3}{R_2 R_2} \quad = \frac{L_2}{R_2} + \frac{L_1}{R_3} + \frac{L_1}{R_2}$$

which is the inverse of the previous and leaves us four parameters to choose. But it is not a free choice because the numerator of Σ in Eq. 12.72 is often a quadratic. For example if we have $R_3 = \frac{L_1 L_2}{R_2}$ and substitute for Σ, we must ensure that its roots are real.

Example – Cascaded Low-Pass Filter

Taking the values of schematic 'Cascaded LR Low Pass.asc', we have $R_2 = R_3 = 50\,\Omega$ $L_1 = L_2 = 0.1$ H. The output trace seems to show breakpoints quite close together and a high frequency slope of $-40\ dB/decade$ – which is what we should expect from a second-order network. Using the cursor we find an upper 3 dB, of $30\ Hz$.

Then from Eq. 12.72 $\Pi = 2.5e5$ $\Sigma = 1500$. Substituting in Eq. 12.73, we have:

$$p1, p2 = \frac{1}{2}\left(-1500 \pm \sqrt{1500^2 - 1e6}\right) = \frac{1}{2}(1500 \pm 1118)$$

$$p_1 = 208.3Hz \quad p_2 = 30.4Hz$$

showing the second pole much higher than the first. The phase trace of $V(r)$ shows an inflexion at something more than $100\ Hz$ confirming the calculated results. The low frequency voltage gain is $A_v = \frac{\Pi}{p_1 p_2} = \frac{\Pi}{\Pi} = 1 = 0dB$.

Example – Cascaded LR Low-Pass Filter Design

We decide on two identical poles at $20\ Hz = 125.66\ rad/s$. If we take $a = 100$ and $R_3 = 5\,\Omega$, then $R_2 = 500\,\Omega$ and we find $s = 125.66 = R_3/L_2$ and $L_2 = 39.8\ mH$. Then $L_1 = 100L_2 = 3.98$ H. This is schematic 'Cascaded LR Low Pass Filter Design.asc' where the measured 6 dB point is at $20\ Hz$, and we see a smooth phase transition from $-1°$ to $178°$ indicating that the poles at least are close together.

Z(in) shows an impedance in the pass band of $5\ \Omega$ above $1\ Hz$ as we should expect, but if we extend the low frequency we find, it tends to about $833\ m\Omega$. Do not forget the default terminating $1\ \Omega$ resistor kindly inserted by LTspice. We can override this by $Rout=1e100$. In the stop band we find the impedance rising at $20\ dB/decade$ because the two inductors are effectively in series. We should also note the strange phase shift at about $100\ mHz$ again due to the terminating $1\ \Omega$ resistor. Make this very large .**net V(out) V1 Rout=1e100** and the problem goes away.

If, however, we use Z(11), there is no terminating resistor, and at low frequency we find *4.95 Ω* which is exactly the parallel sum of R_2 and R_3, and the phase shift behaves itself properly.

12.6.1.2   Cascaded High-Pass Filter

For this we interchange inductors and resistors and Eq. 12.67 becomes:

$$\frac{sL_1 sL_2}{R_1 R_2 + sL_1 R_2 + R_1 sL_2 + sL_1 sL_2 + R_1 sL_1} \tag{12.76}$$

After collecting terms and dividing:

$$\frac{s^2}{s^2 + s\frac{(L_1 R_1 + R_2 L_2 + L_1 R_1)}{L_1 L_2} + \frac{R_1 R_2}{L_1 L_2}} \tag{12.77}$$

and we should note the similarity of Eqs. 12.77 and 12.69. Rather than converting to standard form, we can just solve the quadratic equation.

Example – Cascaded High-Pass Filter
Using the values from schematic 'Cascaded LR High-Pass.asc', we have $L_1 = L_2 = 0.1\ H\ R_2 = R_3 = 50\ \Omega$. Therefore for the quadratic equation, we have $a = 1\ b = \frac{150}{0.01} = 1500\ c = \frac{2500}{0.01} = 2.5e5$

$$p_1, p_2 = \frac{1}{2}\left(-1500 \pm \sqrt{1500^2 - 1e6}\right) \quad = -750 \pm 559$$

$$p_1 = 30.4Hz \quad p_2 = 208.4Hz.$$

and we have the same breakpoints as before only in reverse order.

12.6.1.3   Cascaded Band-Pass Filter

This can be built either as RLLR or LRRL just like the filters with capacitors, Fig. 12.16. If the breakpoints are widely separated, we can calculate each separately

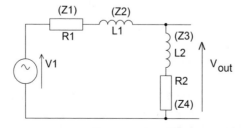

Fig. 12.16 Cascaded LR Band Pass Filter

with an error of just a few per cent. Otherwise we can analyse it as a complex potential divider using:

$$Z_1 = sL_1 + R_1 \quad Z_2 = \frac{sL_2R_2}{sL_2 + R_2} \tag{12.78}$$

then:

$$H(s) = \frac{\frac{sL_sR_2}{sL_2+R_2}}{sL_1 + R_1 + \frac{sL_2R_2}{sL_2+R_2}} = \frac{sL_2R_2}{(sL_1 + R_1)(sL_2 + R_2) + sL_2R_2} \tag{12.79}$$

Expanding and collecting terms:

$$H(s) = \frac{sL_2R_2}{s^2L_1L_2 + s(L_1R_2 + L_2R_1 + L_2R_2) + R_1R_2} \tag{12.80}$$

Then the denominator can be written as:

$$s^2 + s\frac{L_1R_2 + L_2R_1 + L_2R_2}{L_1L_2} + \frac{R_1R_2}{L_1L_2} \tag{12.81}$$

and we can again use the product and sum as we did before along the lines of equation 12.75.

Example – Cascaded Band-Pass Filter

Using the values of schematic 'Cascaded LR Band Pass.asc' and taking the breakpoints in turn, we have $f_l = \frac{50\Omega}{2\pi \times 1H} = 7.96Hz$ $f_h = \frac{50\Omega}{2\pi \times 1mH} = 7.96kHz$, and these are in fair agreement with the measured breakpoints. Otherwise we have:

$$\Pi = \frac{R_1R_2}{L_1L_2} = \frac{50\Omega \times 50\Omega}{1H \times 1mH} = 2.5e6$$
$$\Sigma = \frac{L_1R_2 + L_2R_1 + L_2R_2}{L_1L_2} = \frac{50 + 0.01}{0.001} = 5.001e4$$

and using .meas directives we find:

$$f_l = 7.934Hz \quad f_h = 7.981kHz$$

12.6.2   Bridged-T Filters

We may replace the capacitors with inductors as Sulzer did in the paper referenced in the chapter on 'Second Order RC Filters'. The equations and formulae quoted here are copied from that chapter. The essential equation is:

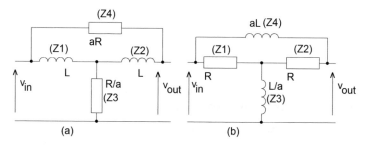

Fig. 12.17 Bridged-T with Inductors

$$\frac{v_{out}}{v_{in}} = \frac{Z_1 Z_2 + Z_1 Z_3 + Z_2 Z_3 + Z_3 Z_4}{Z_1 Z_2 + Z_1 Z_3 + Z_2 Z_3 + Z_3 Z_4 + Z_1 Z_4} \qquad (12.82)$$

There are two variations, Fig. 12.17.

12.6.2.1   With Series Inductors

Taking Fig. 12.17a, we substitute in Eq. 12.82, and this reduces to:

$$H(s) = \frac{s^2 L^2 + \frac{2sLR}{a} + R^2}{s^2 L^2 + sLR\left(\frac{2}{a} + a\right) + R^2} \qquad (12.83)$$

We can turn this into standard form by dividing throughout by L^2:

$$H(s) = \frac{s^2 + \frac{s2R}{aL} + \frac{R^2}{L^2}}{s^2 + \frac{sR}{L}\left(\frac{2}{a} + a\right) + \frac{R^2}{L^2}} \qquad (12.84)$$

It is interesting to compare this with that chapter. And also following that chapter, we must test the numerator of Eq. 12.84 to see if the roots are negative. We do this by examining the $(b^2 - 4ac)$ term which is $\frac{4R^2}{a^2 L^2} - \frac{4R^2}{L^2}$, and as for any useful notch $a \gg 1$, there are no real roots.

Notch Frequency

We require the minimum transfer function of Eq. 12.84. This occurs when we substitute $\omega_0 = R/L$, then $s = iR/L$, and the transfer function is:

$$H(s_{min}) = \frac{\frac{-R^2}{L^2} - j2\frac{R^2}{aL^2} + \frac{R^2}{L^2}}{\frac{-R^2}{L^2} + \frac{jR^2}{L^2}\left(\frac{2}{a} + a\right) + \frac{R^2}{L^2}} = \frac{\frac{2}{a}}{\frac{2}{a} + a} = \frac{2}{2 + a^2} \qquad (12.85)$$

Notch Depth

We have this directly from Eq. 12.85, and if a is large, this approximates to:

$$H(s_{min}) = \frac{2}{a^2} \qquad (12.86)$$

Compare this to the version using capacitors.

Example – Bridged-T with Series Inductors

The schematic is 'Bridged-T with Inductors.asc'. The resistors are $R_1 = 20\ k\Omega$ and $R_2 = 100\ \Omega$. As $R_1/R_2 = a^2$ we find $a = 14.14$, so we expect a notch depth of approximately $1/14.14^2 = 1/200 = -46\ dB$ which agrees with the measurement. The circuit has $R = 1414\ \Omega$ and $L = 0.01\ H$. Using $s = jR/L$ we calculate the notch frequency as $22.5\ kHz$, whereas it is measured as $22.288\ kHz$.

12.6.2.2   Bridged-T with Series Resistors

This is Fig. 12.17b. Substituting in Eq. 12.83 we have:

$$H(s) = \frac{R^2 + \frac{sRL}{a} + \frac{sRL}{a} + s^2L^2}{R^2 + \frac{sRL}{a} + \frac{sRL}{a} + s^2L^2 + saLR} \qquad (12.87)$$

Collecting terms and converting into standard term by dividing by L^2:

$$H(s) = \frac{s^2 + s\left(2\frac{R}{aL}\right) + \frac{R^2}{L^2}}{s^2 + \frac{sR}{L}\left(\frac{2}{a} + a\right) + \frac{R^2}{L^2}} \qquad (12.88)$$

This is exactly the same as Eq. 12.85 and we again substitute $s = iR/L$ to find the same equations.

So which form is better? The first uses inductors of 10 mH which are not too difficult to make and quite small, but how closely do they need to be matched? And is it possible to compensate for a mismatch by altering a resistor? Using series resistors we have a large difference in inductor values. The larger inductor could be a pot core whose inductance can be changed somewhat by a screw-actuated tuning plug.

Example – Bridged-T with Series Resistors

To keep the same performance, with the same resistance and inductance, we require $R_1 = R_2 = 1414\ \Omega$ and $L_1 = 0.01 \times 14.14 = 0.1414\ H$ $L_2 = 0.01/14.14 = 707.2\ \mu H$. This is schematic 'Bridged T with Inductor V2.asc'. Using the *.meas* statement we find substantially the same results as before.

Explorations 5

1. The deciding factor between the two forms of the Bridged-T could turn on the effect of mismatches. Explore both.

2. To really complicate things, add temperature coefficients. Is it possible to choose resistors whose temperature coefficients balance the inductor's?
3. Add a .*net* statement and compare the impedances of the two versions.

12.7 Impedance Matching

In much small signal design this is not an issue, and there can be significant mismatches between stages. However, when the signal is very small, typically where the input is from an antenna, it is important to gather as much of the signal power as possible. Similarly, when we are dealing with substantial power, it can be important to create a good match; if nothing else, to prevent power being wasted as heat.

The maximum power transfer will occur when the load impedance is the complex conjugate of the source impedance. We have seen that we can use two capacitors for this, 'Tapped Capacitor'. And we have seen that we also can use transformers; however, these tend to be expensive and at low frequencies can be heavy. But they do have the great advantage that they can provide matching over a wide range of frequencies. The point is that if we simply connect a high impedance source to a low impedance load, much of the power will be wasted.

Example – Impedance Mismatch

We can demonstrate this very easily with schematic 'Load Mismatch.asc' where we have a *1 kΩ* source impedance and a *50 Ω* load. If we hold down **Alt** and move the cursor over the components so that it turns into a thermometer and then left click, we will have traces of the power in the component. These show that the power in the source resistor is *453.61 μW*, whilst that in the load is only *22.68 μW*, and power from the voltage source is *476.68 μW*. This shows that only *4.76%* of the source power is transferred to the load.

12.7.1 LC Matching

If the source and load are both resistive, we can modify the load by adding reactive components so that the total impedance seen by the source is the same as the source impedance.

This can be achieved at one frequency using a parallel combination of an inductor and a capacitor. In effect, we create a tuned circuit, and a fuller discussion will be found in the chapter 'LCR Tuned Circuits'. This will result in a large circulating current in the combination, but if the capacitor and inductor have no resistance, there will not be an energy loss. This circuit can only match a high impedance source to a low impedance load.

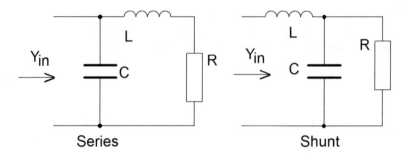

Fig. 12.18 Impedance Matching

The following discussion is taken from *158.132.149.224/.../FM_saveAs.php?...=/impedancemat...* by Michael Tse, which also includes other matching methods.

Referring to Fig. 12.18 where we use the series form, it is more productive to write the input admittance rather than the input impedance. We then have:

$$Y_{in} = j\omega C + \frac{1}{R + j\omega L} \tag{12.89}$$

We multiply top and bottom of the second term by its complex conjugate:

$$Y_{in} = j\omega C + \frac{R - j\omega L}{R^2 + (\omega L)^2} \tag{12.90}$$

We can now separate the real and imaginary parts of the second term:

$$Y_{in} = j\omega C + \frac{R}{R^2 + (\omega L)^2} - \frac{j\omega L}{R_2 + (\omega L)^2} \tag{12.91}$$

Then collecting real and imaginary terms:

$$Y_{in} = \frac{R}{R^2 + (\omega L)^2} + j\omega \left[C - \frac{L}{R^2 + (\omega L)^2} \right] \tag{12.92}$$

From this we see that if:

$$C = \frac{L}{R^2 - (\omega L)^2} \tag{12.93}$$

the input admittance is real. If we rearrange Eq. 12.93, we can find ω:

$$R^2 C - (\omega L)^2 C = L \quad \rightarrow \quad (\omega L)^2 = \frac{1}{C}\left(R^2 C - L\right) \tag{12.94}$$

and we shall denote this by ω_0

$$\omega_0 = \sqrt{\left[\frac{1}{LC} - \frac{R^2}{L^2}\right]} \tag{12.95}$$

Returning to Eq. 12.92, when this condition applies, the input impedance rather than admittance is:

$$Z_{in} = \frac{R^2 + (\omega_0 L)^2}{R} \tag{12.96}$$

and this can be written as:

$$Z_{in} = R\left(1 + \left(\frac{\omega_0 L}{R}\right)^2\right) \quad Z_{in} = R(1 + Q^2) \tag{12.97}$$

where we have used the conventional designation:

$$Q = \frac{\omega_0 L}{R} \tag{12.98}$$

And if Z_{in} equals the source impedance, we have the best match.

12.7.1.1   Design Method

Given the load resistance R and the source resistance equal to R_{in}, we start with Eq. 12.97 to find Q. We then use Eq. 12.98 with the operating frequency to find L and finally Eq. 12.92 to find C.

Example – LC Load Matching
We design for a load resistance of $Rload = 50\ \Omega$, a source resistance of $Rsource = 1\ k\Omega$ and a frequency of $1\ MHz$. Following the design method, we find $L = 34.7\ \mu H$ $C = 693\ pF$. The results of schematic 'LC Matching.asc' are that the load power is $100.7\ \mu W$ and the source power $282.7\ \mu W$ which is 35.6% – a considerable improvement over direct connection of source and load. And it is not too sensitive to the value of the capacitor, or, to put it another way, to the frequency.

Explorations 6
1. Make a frequency sweep with schematic 'LC Matching.asc'.
2. Explore other configurations of L and C and compare them in terms of how well they match and sensitivity to frequency or component changes.

3. Some variations can be found at http://analog.intgckts.com/impedance-matching/
   l-matching/ with equations but no theory (which essentially follow the same
   method as above) including upward matching.
4. Explore the use of T and Π sections as described by Tse.

12.8   Crystals

A crystal is made from a thin slice of piezoelectric material with contact made to both
faces and sealed in a vacuum to prevent damping from the air. If a voltage is applied
to it, it can oscillate by alternately exchanging the strain energy of the slightly
deformed crystal for electrical energy.

12.8.1   Equivalent Circuit

The equivalent circuit consists of a parallel arm representing the physical capacitance of the leads and the electrodes as Cp and a series arm modelling the crystal
itself. This consists of a resistor Rs representing the mechanical damping loss as the
crystal flexes, an inductor Ls representing the mass of the crystal, and a capacitor Cs
for the resilience or 'spring' of the crystal slice. The series arm is purely an electrical
analogue of the physical crystal. Thus we arrive at the equivalent circuit,
Fig. 12.19 and we can use the LTspic capacitor to model it.

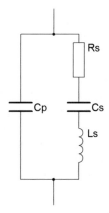

Fig. 12.19 Crystal Rquivalent Circuit

Ignoring the resistor, the series-parallel combination has an impedance of:

$$Z = \frac{\left(j\omega L_s + \frac{1}{j\omega C_s}\right) \frac{1}{j\omega C_p}}{j\omega L_s + \frac{1}{j\omega C_s} + \frac{1}{j\omega C_p}} \tag{12.99}$$

and after multiplying by $j\omega C_p$:

$$Z = \frac{j\left(\omega L_s - \frac{1}{\omega C_s}\right)}{-\omega^2 L_s C_p + \frac{C_p}{C_s} + 1} \tag{12.100}$$

This has a zero at:

$$\omega_s = \frac{1}{\sqrt{L_s C_s}} \tag{12.101}$$

which represents one resonant frequency and a pole at:

$$\omega_p^2 L_s C_p = \frac{C_p + C_s}{C_s} \tag{12.102}$$

or

$$\omega_p^2 = \frac{C_p + C_s}{L_s C_p C_s} \tag{12.103}$$

which we can write as:

$$\omega_p = \frac{1}{\sqrt{L_s C_t}} \tag{12.104}$$

where C_t is the sum of the capacitors in parallel and is the second, higher, resonant frequency. So if C_s is much smaller that C_p (which is nearly always true), the two resonances will be close together.

We can also write Eq. 12.102 as:

$$\omega_r = \frac{1}{\sqrt{L_s C_s}} \sqrt{1 + \frac{C_s}{C_p}} \tag{12.105}$$

or:

$$\omega_p = \omega_p \sqrt{\left(1 - \frac{C_s}{C_p}\right)} \tag{12.106}$$

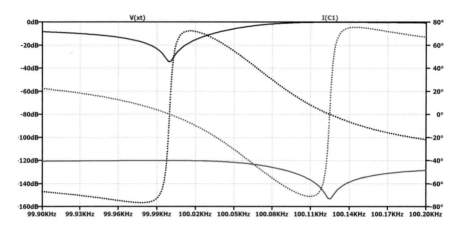

Fig. 12.20 Crystal Resonance

which shows, with Eq. 12.101, that the separation of the resonances depends upon the ratio Cs/Cp. Outside of the resonance range, the crystal behaves as a capacitor C_p, Fig. 12.20 which is derived from a screen shot.

12.8.2 Parameter Extraction and LTspice Model

We can measure Cp easily enough, then we find the two resonant frequencies. Because the ratio of C_s/C_p is of the order of 1/1000, we may safely approximate Eq. 12.106 to $\omega_p = \omega_2 \left(1 - \frac{C_s}{2C_p} \right)$ from which we can find Cs. And from Eq. 12.101 we find Ls. That only leaves Rs which comes from the measurement of the Q-factor.

When creating a SPICE model, it should be noted that the values for Cs and Ls are often given to five figures or more since the resonant frequency is specified to the same accuracy. Take, for example, a crystal for use in clocks and watches: there are 86400 s in a day so an error of only 1 digit in the crystal frequency of 32768 Hz is not acceptable.

LTspice has a symbol in the 'Misc' page which uses a capacitor as the model where we do not use *Rpar* and *RLshunt*. This should simulate faster than a discrete component model, but it is not possible to add temperature coefficients to the inductor. A library of some popular crystals can be found at http://www.gunthard-kraus.de/Spice_Model_CD/Vendor%20List/Spice-Models-collection/xtal.lib from which the two examples below are taken.

Explorations 7
1. Open 'Watch Xtal.asc'. Plot $V(xt)/I(C1)$ to see the resonances. They can be separated as *Vxt* and *I(C1)*. As the crystal is designed for series resonance, the correct frequency is at *Vxt*. Expand the trace and measure the Q-factor. Measure the difference between the resonance frequencies and check it agrees with theory.

2. Open '100 kHz Xtal.asc' and repeat. Expand the sweep range and note the shape of the curves away from resonance.
3. Open '100 kHz Xtal pulled.asc' and note that the frequency can be pulled slightly by a series capacitor. If we measure the resonant frequency and the capacitor, we can see if the effect is linear. Try other values and note that reasonable values of parallel capacitor have no effect on the series resonance. This is good to know because it means that variable circuit stray capacitance can be ignored.

12.9 Assorted Circuits

These are a few circuits involving inductors and capacitors which do not seem to fit in other chapters.

12.9.1 Electromagnetic Interference (EMI) Filter

The suppression of electrical noise transmitted down mains cables is a matter of considerable importance. Electric drills and other small domestic appliances with an electric motor invariably contain a mains filter to reduce the high frequency noise to a permissible level. And as almost every computer uses a switch-mode power supply, the problem is universal. A useful article on filter design is *https://pdfs.semanticscholar.org/. . ./b08b9bb991dd8511*.

For IT equipment standards EN55022 and CISPR22 are appropriate and EN55032 for multi-media, radio, TV sets and the like. As the standards seem to keep changing, a good rule is to aim at less than 70 μV average.

12.9.1.1 Common Mode and Differential Mode

Common mode interference travels in the same direction on both line and neutral conductors. It can be reduced by a series inductor in each line and a capacitor to ground from each line, capacitors C_1 and C_2 in Fig. 12.21. Differential interference is attenuated by a capacitor between line and neutral, C_3.

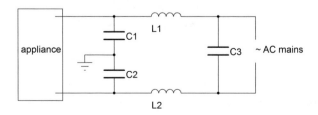

Fig. 12.21 Mains Suppressor

12.9.1.2   Capacitor Types

These are specially designed for use on the AC mains with the attendant voltage spikes. The capacitors C_1 and C_2 are type 'Y' which go open-circuit on a fault. This is a safety feature that there is no danger of electrocution nor of a fire – only the interference will not be attenuated. More information can be found at: https://allaboutcircuits.com/technical-articles/safety-capacitor-class-x-and-class-y/

On the other hand, capacitor C_3 is type 'X' which will go short-circuit on an over-voltage fault in order to blow the mains fuse or open the trip on the supply line.

Example – EMI Filter

We often find the filter directly built into the mains input of the equipment. We can also find mains sockets incorporating filters. There are several component values in use and schematic ('EMI Mains Filter.asc') is typical. The voltage source represents the noise and R_{load} approximates to the prescribed line conditioning.

We find a sharp resonance in the current from V1 at *31 kHz* due to C_2 L_1 and C_3 L_2. The 3 dB point is at *1.1 kHz*. This is very sensitive to changes in the load resistor.

12.9.2   Power Factor Correction

The idea is to make the power factor as close to unity as possible. Let us suppose, for example, that a certain item consumes 480 W operating from the 240 V 50 Hz mains, Fig. 12.22. If the power factor is unity, it will draw a line current of 2A. But if the power factor is 0.7 (corresponding to a phase angle of 45°), the RMS line current is 2.86 A.

If the waveforms are not substantially sinusoidal, power factor correction is difficult. Often we are dealing with small domestic appliances, usually motors which are inductive, and hence a capacitor connected close to the motors terminals can be used. SPICE helps us see the original and improved waveforms.

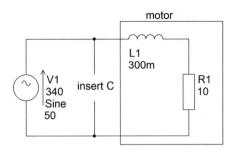

Fig. 12.22 PowerFactor Correction

12.9.2.1 Real and Apparent Power

If the voltage and current are in-phase, the power is simply:

$$P = I_{rms} V_{rms} \tag{12.107}$$

whereas if there is a phase difference between the current and voltage, the instantaneous power is:

$$p = V_{pk} \sin \omega t . I_{pk} \sin (\omega t + \varphi) \tag{12.108}$$

which can be written:

$$p = \frac{1}{2} V_{pk} I_{pk} (\cos \varphi - \cos (2\omega t - \varphi)) \tag{12.109}$$

The average power is found by integrating Eq. 12.109 and dividing by the period where $cos(2\omega t - \varphi)$ will average to zero and we are left with $\frac{1}{2} V_{pk} I_{pk} \cos \varphi$ or:

$$P = V_{rms} I_{rms} \cos \varphi \tag{12.110}$$

The term $V_{rms} I_{rms}$ is the *apparent power* and $cos\varphi$ the *power factor*. We can correct the power factor by a parallel capacitor where $j\omega L = \frac{1}{j\omega C}$ or:

$$C = \frac{1}{(2\pi f)^2 L} \tag{12.111}$$

Example – Power Factor Correction
The schematic 'Power Factor.asc' implements the figure with $C1 = 33.77\ \mu F$. We need to delay the start of recording the data to about *500 ms* to avoid the turn-on surge. A point that arose when building the circuit is that $V1$ needed to be rotated by *180°* so that its voltage and current were substantially in phase and not anti-phase. The same may be needed for the capacitor. The load power is *65.6 W* with or without the capacitor. Without the capacitor the RMS source current is *2.56 A*.

It appears from the trace that the phase angle is nearly zero with $C = 33.25\ \mu F$ rather than the calculated value. Then the RMS current from the source is *0.265 A* giving a load power of *63.7 W*. We can note the power in all the components with and without the capacitor by moving the cursor over the components in turn so that it turns into a current clamp meter with a red arrow indicating the direction in which the current is measured. Then holding down **Alt** and the cursor turns into a thermometer and left clicking on the component will plot its power. We find *63.81 W* drawn from the supply, *63.513 W* dissipated by the resistor, *0.292 mW* dissipated by the inductor and *1.099 mW* by the capacitor. We also see large and almost identical currents in the capacitor and the inductor.

The correction is not particularly sensitive to changes in the capacitor value, which can be explored by stepping its value.

12.9.3   LCR Notch Filter

This is just a simple filter and is nothing more than a parallel tuned circuit in series with a resistor Fig. 12.23 and will be discussed at length in the chapter 'Tuned Circuits'. We mention it here because it can be used as an 'L' half-section to form more interesting filters in chapter 'Filters'. The analysis is:

$$Z = \frac{j\omega L \frac{1}{j\omega C}}{j\omega L + \frac{1}{j\omega C}} \quad Z = \frac{j\omega L}{1 - \omega^2 LC} \tag{12.112}$$

$$\frac{v_{out}}{v_{in}} = \frac{R}{R + Z} = \frac{R}{R + \frac{j\omega L}{1 - \omega^2 LC}} = \frac{(1 - \omega^2 LC)R}{(1 - \omega^2 LC)R + j\omega L} \tag{12.113}$$

We note that if $\omega^2 LC = 1$ the output is zero and we denote that frequency by ω_r. Then we can write Eq. 12.113 as:

$$\frac{v_{out}}{v_{in}} = \frac{1 - \frac{\omega^2}{\omega_r^2}}{1 - \frac{\omega^2}{\omega_r^2} + \frac{j\omega L}{R}} \tag{12.114}$$

And the effect of R is broaden the flanks of the notch.

Example – LCR Notch Filter
We set $L= 10mH$ $C= 10nF$ $R= 10\ k\Omega$ and we find $f_r = \frac{1}{2\pi\sqrt{10^{-10}}} = 15.9kHz$ which is what we read from schematic 'LCR L Notch Filter.asc'. Reducing R broadens the notch.

Explorations 8
1. Open 'EMI Mains Filter.asc' and note that attenuation starts above mains frequencies. There is a second breakpoint at about 100 kHz causing a steeper attenuation reaching 66 dB at 1 MHz

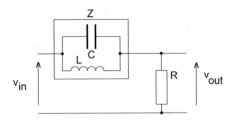

Fig. 12.23 Notch Filter

2. 'Motor Power Factor Correction.asc' uses the values from the Fig. 12.22. The
   capacitor is effectively zero, and we find that the supply and load currents are
   identical but with a phase angle of about 90° with respect to the supply voltage
   (and notice that it takes a few cycles for the circuit to settle down to an RMS load
   current of some 2.88 A,) but this is measured during the start-up so it does not
   agree exactly with the 2.86 A.
3. Now calculate a suitable value of capacitor so that the load impedance is purely
   resistive and test it. You should find that the circuit is quite sensitive to its value,
   for example, there is a considerable phase difference even with 30 µF compared
   to the ideal 33.4 µF.
4. Plot a graph of real and apparent power as a function of capacitance.

12.9.4   Interrupted Continuous Wave Transmission

Before the days of transistors, even before the days of thermionic valves (tubes),
some ingenious scientists and engineers found a way to create and transmit radio
waves: they used a *spark transmitter*. The properties of tuned circuits were well
known, but the problem was how to maintain the oscillations. The solution was to
use a large battery *B1* in Fig. 12.24 that charged a Leyden Jar *C1* which was
connected to a spark gap *G* whose other terminal was connected to another Leyden
Jar *C2* in parallel with a coil *L1*.

Dry air is a good insulator, but if the electric field exceeds 3 kV/mm, it will break
down and an electric arc will form: this has a low resistance. So, by carefully choosing
the spark gap width, when the Leyden Jar *C1* was almost fully charged, the air broke
down, a spark (or arc) formed, and the jar discharged into the jar *C2* and the coil
causing them to resonate. This was a damped resonance because some of the energy
was transmitted by the aerial. However, before the oscillations had died away, Jar *C1*
had recharged, and another arc formed and the process repeated continually.

A second version of the circuit, Fig. 12.25, had a telegraph key *K* in series with
the battery, but this time with an induction coil with many more turns on the
secondary than the primary. When the key was opened, the energy stored in the
primary created a large back EMF in the secondary, a spark formed, and the energy

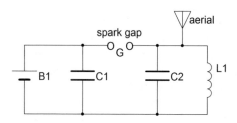

Fig. 12.24 Spark Gap Transmitter Concept

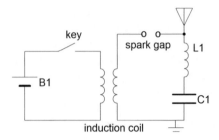

Fig. 12.25   Spark Transmitter

was transferred to the resonant circuit as before. But this time the radio wave only lasted a short time, it was not automatically maintained. So, presumably, the distinction between a short 'dot' in Morse code and a long 'dash' had to be made not by the length of the transmission, but by the time between the next burst.

Example – Spark Transmitter
The schematic 'Spark Transmitter.asc' uses a switch to model the spark gap and a resistor R_2 for the aerial. It implements Fig. 12.24. This creates a continuous train of pulses with measured period *320 kHz* which agrees with the resonant frequency of C_2 and L_1. It could be rebuilt with a telegraph key as in Fig. 12.25.

12.10   ESD Simulation IEC 61000-4-2

Electrostatic electricity is not generally harmful to humans – with the usual caveat about kissing if you both are in a dry atmosphere and have become charged through non-conducting floor coverings and clothing – but lethal for many electronic components.

The IEC test simulates a charged human body by a capacitor of 150 pF charged to 2 kV, 4 kV, 6 kV or 8 kV, depending on the test regime, in series with a 330 Ω resistor. From this we can deduce that the energy is very small indeed: using $Q = CV$ and $E = \frac{1}{2} CV^2$, the capacitor holds 300 nC of charge and 600 uJ of energy. The hazard is that the high voltage can cause localized breakdown of insulators and a spark discharge, which, because the energy is concentrated in a very small area, leads to very high temperatures, thus destroying the structure of the insulator and often rendering it conductive.

A simple RC circuit cannot replicate the required current pulse rise time of 0.8 ns $\pm 25\%$, and a more sophisticated circuit is needed. This schematic 'ESD Generator. asc' is copied directly from https://www.youspice.com/simple-spice-esd-generator-circuit-based-on-iec61000-4-2-standard/2/.

The circuit consists of two series LCR circuits in parallel. We shall see in the next chapter that these form a damped resonant circuit. For the present it is sufficient to note that the right-hand circuit consists of a small capacitor in series with a small inductor from which we may deduce that the capacitor will discharge quickly. It is this circuit that creates the initial fast pulse of 15 A. The left-hand circuit consists of the required 150 pF in series with 330 Ω and is responsible for the slower, broader pulse.

The specifications state that the test can be through a resistor of no more than 2.1 Ω – where the circuit has 2 Ω – or through an air discharge. The website has a sub-circuit and assembly for the simulator and a choice of output voltage.

Example – ESD Simulation

The above website and *https://www.silabs.com/documents/public/application-notes/AN895.pdf* both have graphs of the test current waveform. The measurements on the schematic confirm the required peak of 15 A at 1.32 ns, but as the pulse is delayed by 0.1 ns, this is just in the specification limits. The currents after 30 ns and 60 ns are within the broad limits of ±30%.

Probing the capacitor voltages and their sums helps to explain the output waveforms.

12.11 Loudspeakers and Crossovers

A loudspeaker is an electromechanical system that can be modelled by passive components of inductors, capacitors and resistors. That is why we introduce it here. But that is only the starting point. To analyse or design a complete acoustic system, we need to move on to the acoustic properties of the loudspeaker enclosure and from thence to the acoustic properties of the room, lecture hall, theatre or whatever. This involves such parameters as the acoustic reflection coefficients of walls and ceiling, use of drapes or curtains, the spacing of loudspeaker enclosures and suchlike. An example website is https://www.slideshare.net/abhishek201165/acoustic-design-process. But here we concern ourselves only with the loudspeakers and crossovers.

The acoustic output of a driver is described by the Sound Pressure Level (SPL) measured dBs at 1 m in front of the driver. The reference is an AC of 2.83 V at 1 kHz corresponding to 1 W in an 8 Ω load.

12.11.1 Loudness

This is the subjective experience of hearing sounds as adjudged by a human ear. An important point to make here is that the human ear is not equally sensitive to all frequencies. It is most sensitive in the region of *1 kHz* which, in musical terms, is two octaves above middle C. The Wikipedia article is informative *https://en.wikipedia.org/wiki/Loudness*. We may generalize vaguely to say that below a few 100 kHz, we need an SPL some 40 dB greater to achieve the same loudness. This can be seen by inspecting a LP record with a magnifying glass where the bass note excursions are far greater than the high. Therefore typically we find in a hi-fi system a large, powerful woofer and a smaller tweeter.

12.11.2   Driver Construction

The conventional unit consists of a substantial rigid frame or chassis with a very strong circular permanent magnet with a central cylindrical pole, Fig. 12.26. The magnetic is very important because the force acting on the voice coil depends linearly on the current and the field strength. From this we also see that the magnetic field must be uniform; else there will be distortion when the voice coil moves.

The voice coil of the driver is firmly attached to the cone, and this freely moves vertically in the annular gap in the magnet. To keep it centred, so that it does not touch the pole, the cone has a flexible suspension or 'spider', partly augmented by a flexible foam surround at its outer edge. And clearly the displacement of the voice coil must be limited so that all of it remains within the magnetic field. The cone must be rigid else if it flexes that too will introduce distortion. And we want to avoid mechanical resonances if at all possible.

That is the basic structure. Modern driver designs often use a die-cast aluminium chassis, and as drivers are typically less than 10% efficient, some incorporate special heat-dissipation techniques, and cones can be made from aluminium for improved rigidity. It is an area of continual research to create units that more faithfully convert the electrical signal into sound.

12.11.2.1   Multi-driver Systems

If we try to encompass the full acoustic range with one driver the Doppler Effect comes into play: if we output a bass note and treble note together, there will be several oscillations of the treble note during one period of the bass, so as the

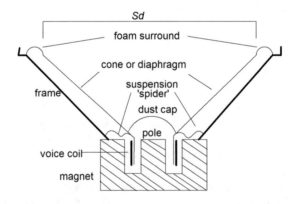

Fig. 12.26 Loudspeaker

loudspeaker cone moves forward on a bass note, it will compress the treble and raise its pitch, and as the cone moves back on the bass note, it will lower the pitch of the treble. This is one reason to divide the frequency spectrum into three parts – bass, middle and treble – with a driver for each so that individually they only handle two or three octaves and the problem is ameliorated. Additionally, to create the increased bass SPL, we need a large cone and therefore a large mass which makes it difficult for it to respond to high frequencies. And as the SPL for the treble is much less we use a smaller, lighter cone that can respond more rapidly. To put the numbers in perspective, *middle C* is 256 Hz, which is near the top of the bass voice range, and *concert A* (the note used by orchestras when tuning up) is 440 Hz.

Incidentally, we also run into problems with the spatial distribution of the sound. If the dimensions of the driver are much less than the wavelength of the sound, it will radiate over a hemisphere as a point source. The velocity of sound in air can be taken as 340 m/s. So given a 100 Hz bass note, the wavelength is 3.4 m. The dimensions of the average bass driver are around 30 cm, so it effectively acts as a point source, and the sound will be heard comfortably over the whole room. But a 10 kHz note has a wavelength of only 34 cm, and the bass driver will create a more-or-less focused beam of sound. Therefore a tweeter with a cone diameter of only a couple of cm is needed or an auxiliary cone in a wide-range driver.

12.11.3 Thiele-Small Driver Parameters

These are the industry standard description of a driver. They can be divided into the two basic electrical parameters and fundamental mechanical parameters that can be measured on the driver, and derived small-signal parameters. The latter then can be converted into an electrical equivalent circuit. The Wikipedia article is very comprehensive and has been followed here.

12.11.3.1 Electrical Parameters

These two can be measured directly on the driver unit.

Le
The inductance of the coil in mH measured at 1 kHz with a small signal input so that the unit is working linearly.

Re
The resistance of the coil in ohms measured at DC

12.11.3.2 Fundamental Mechanical Parameters

Manufacturers do not always give values for everyone, and sometimes the notation is slightly different, but it is generally easy enough to identify which is which.

Bl

The product of the magnetic field strength and the length of wire used to make the voice coil. With the current this gives the force in Newtons as $F = BIL$. Clearly it is very advantageous to have a strong magnetic field to reduce the power needed and hence the heat dissipated by the resistance of the coil. Adding more turns to the coil is an option, but will increase the inductance, and if this means thinner wire the resistive losses will increase.

Sd

The projected area of the cone in cm^2 or m^2. This is important for calculating how much air will be moved and thus the change of air pressure in the enclosure.

Mms, Mmd

The first, Mms, is the mass of the cone and voice coil in gm or kgm. The second is the total moving mass. Often we are only given one in which case we can assume they are the same. This is an inertial load that must be accelerated by the force created by the voice coil. Using Newton's law of motion $F = ma$, here we have $BIL = M_{ms}\frac{dv}{dt}$, and if we compare that to Lenz's Law $v = L\frac{di}{dt}$, then M_{ms} behaves like an inductor.

Cms

The compliance of the suspension is m/N and is the reciprocal of its stiffness. This relates to the restoring force needed to return the displaced voice coil to its neutral position which depends upon the distance it has moved and is the integral of the velocity v or $F = \frac{1}{C_{ms}} \int v dt$, and hence C_{ms} can be modelled by a capacitor.

Rms

The mechanical resistance of the suspension. This is energy lost in flexing the suspension in N s/m. If it is not given, it can be calculated from:

$$R_{ms} = \frac{2\pi F_s M_{md}}{Q_{ms}} \tag{12.115}$$

12.11.3.3   Small-Signal Parameters

These are derived from the previous values.

Fs

The mechanical resonant frequency in Hz. It is given by:

$$F_s = \frac{1}{2\pi\sqrt{C_{ms}M_{ms}}} \tag{12.116}$$

This is roughly the low frequency limit of the unit and is less than 50 Hz for woofers, a few hundred Hz for mid-range units and a few kHz for tweeters. It can be lowered by increasing the mass of the cone or the compliance of the suspension.

Qes

Electrical quality factor given by:

$$Q_{es} = \frac{2\pi F_s M_{ms}}{B_l^2} = \frac{R_e}{B_l^2}\sqrt{\frac{M_{ms}}{C_{ms}}} \quad (12.117)$$

where we substitute from Eq. 12.116 for F_s. This assumes a driving source impedance of zero. This is closely related to the mechanical resonance below in that the increased excursion of the voice coil at resonance results in a larger back emf which reduces the current and increases the impedance.

Qms

The mechanical quality factor of:

$$Q_{ms} = \frac{2\pi F_s M_{ms}}{R_{ms}} \quad \frac{1}{R_{ms}}\sqrt{\frac{M_{ms}}{C_{ms}}} \quad (12.118)$$

where again we substitute for F_s. The main factor is R_{ms}.

Qts

The total quality factor:

$$Q_{ts} = \frac{Q_{es}Q_{ms}}{Q_{ts} + Q_{ms}} \quad (12.119)$$

in effect, putting them in parallel.

Vas

This is the *equivalent compliance volume* and is the volume of air in m^3 of a cylinder of area Sd that has the same compliance as the suspension of the driver.

$$V_{as} = \rho c^2 S_d^2 C_{ms} \quad (12.120)$$

where ρ is the density of air (1.184 kg/m^3) and c the speed of sound in air (340 m/s at 25 °C). Its importance is in the volume of the enclosure.

12.11.3.4 Large-Signal Parameters

There are several but the most often quoted is Xmax.

Xmax

This is the maximum excursion of the voice coil for low distortion. It is usually possible to go further without damage, typically on peak sounds, but distortion will be much greater.

12.11.4   Equivalent Circuits

Many years ago KEF industries published equivalent circuit for their drivers. The
topology was similar but not identical to that derived from the Thiele-Small parameters. That for their tweeter, mid unit and woofer are Figs. 12.27, 12.28 and 12.29,
and it will be seen that the structure is identical for all, and the woofer values are of
the same order as those for the woofer made by 'Volt', the BM165.1.

The Thiele-Small parameters can be reduced to a simple passive circuit. The coil
resistance and inductance are inserted in series as R_1 and L_1. We then need to model
the suspension compliance by an inductor L_2, the mechanical damping of the

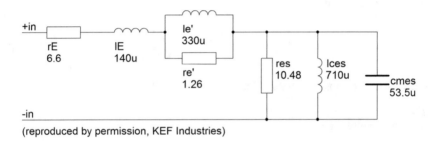

(reproduced by permission, KEF Industries)

Fig. 12.27   Tweeter

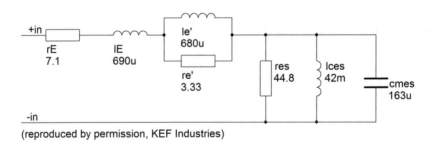

(reproduced by permission, KEF Industries)

Fig. 12.28   Mid Unit

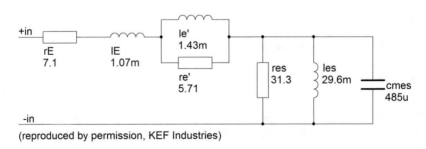

(reproduced by permission, KEF Industries)

Fig. 12.29   Woofer

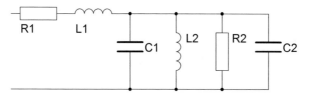

Fig. 12.30 Driver Equivalent Circuit

suspension resistance by a resistor R_2, the moving mass of voice coil and cone by a capacitor C_1 and the air mass by another capacitor C_2. The air mass here is notional and depends on the enclosure and models the sound pressure. We should regard it as a relative rather than an absolute figure. These are all in parallel, Fig. 12.30. A slightly different circuit is show in schematic 'Woofer.asc'.

12.11.4.1 Component Formulae

These are taken from https://circuitdigest.com/electronic-circuits/simulate-speaker-with-equivalent-rlc-circuit. The derivation of these equations is not given but more can be found at http://projectryu.com/wp/2017/07/23/electrical-model-of-loud speaker-parameters/. Note that these are in SI units and ρ is the density of air.

$$C_1 = \frac{M_{md}}{B_l^2} \text{ and use } M_{ms} \text{ if } M_{md} \text{ is not given} \tag{12.121}$$

$$L_2 = C_{ms}B_l^2 \tag{12.122}$$

$$R_2 = \frac{B_l^2}{R_{ms}} \tag{12.123}$$

$$C_2 = 8\rho\frac{A_d^3}{3B_l^2} \quad \text{where} \quad A_s = \sqrt{\frac{S_d}{\pi}} \tag{12.124}$$

We must remember that this is a linear model. However, it is well-suited to an exploration of the behaviour of the driver and enclosure at modest power levels.

Example – Driver BM 165.1
This is manufactured by 'Volt' in the UK. https://voltloudspeakers.co.uk/techspecs/. The Thiele-Small parameters are:

$F_s = 30 \ Hz \ R_s = 5.5 \ \Omega \ L_e = 1.20 \ mH \ Q_{ms} = 4.12 \ Q_{es} = 0.26 \ Q_{ts} = 0.24 \ C_{ms} = 1.23$
$mm/N \ B_l = 9.8 \ N/A \ M_{ms} = 25 \ g \ V_{as} = 40.75 \ l \ S_d = 159.94 \ cm^2 \ V_d = 96.88 \ cm^3$
$X_{me} = \pm 6.3 \ mm.$

Then $C_1 = 0.025 \ kg/9.8^2 = 258 \ \mu F, \ L_2 = 118 \ mH, \ R_{ms} = 1.14 \ Ns/m \ R_2 = 84 \ \Omega$
$A_s = 0.0714 \ C_2 = 12 \ \mu F$

The schematic 'Volt 165-1.asc' shows a similar response to the KEF Woofer. The impedance peaks at *28 Hz*- close to the data sheet *30 Hz*. And if we remove C_2, it is *29 Hz*.

The trace overall is a smooth approximation to the published graph made by measurements on a unit but lacks the slight resonance between 2 kHz and 3 kHz https://voltloudspeakers.co.uk/loudspeakers/bm165-1-6-5/.

12.11.5   Enclosures

How the speaker is mounted greatly affects the performance of the system by changing such factors as the resonant frequency. The physics of this are discussed by http://www.silcom.com/~aludwig/Sysdes/Thiel_small_analysis.htm. In general, the designs assume that the air is moved as a column by the driver without compression and that the wavelength of the sound is larger than the dimensions of the enclosure. We also assume that the enclosure walls are rigid. On that basis, for a sealed enclosure, we can calculate the mass of the air it contains and thus its compliance. Then from Eq. 12.116, this added compliance will reduce F_s.

We can model the air in a sealed enclosure by increasing C_2. A bass-reflex port adds a little more complexity. But the good news is that the intricate designs of some decades ago with multiple ports no longer seem to be needed. If we want to move on to modelling the 'listening room' (if you can afford one), then things get really physical and complicated.

Example – Enclosure Effect
We increase C_2 to *120 µF* and note the reduction in F_s. This, however, does not explain what volume of air corresponds to 120 µF.

12.11.6   Crossovers

These must separate the frequency ranges to suit the drivers and must handle considerable power. One due to KEF is Fig. 12.31. The components must be of high quality and stability to avoid introducing distortion. In particular, inductors must be strictly linear so they are operated well away from saturation on the linear portion of the B-H curve.

This must not only be matched to the driver's impedance but also to its frequency range. So the design of a speaker system must be taken as a totality of drivers, crossover and enclosure.

Explorations 9
1. Build the crossover shown in Fig. 12.31, and note the response of each section and the overall response from bass to treble. Use the simplified schematics

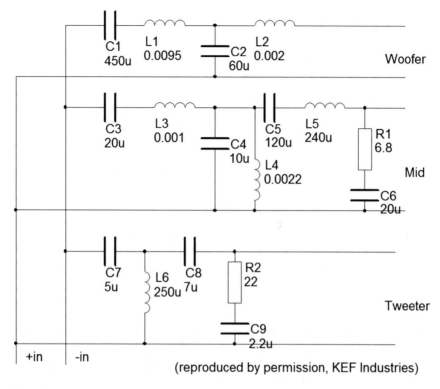

(reproduced by permission, KEF Industries)

Fig. 12.31 Crossover

'Woofer.asc' and 'Woofer Crossover.asc' to explore effects of mismatches on the frequency response.
2. Add the three KEF drivers and review the total response by adding the currents in capacitors C_{mes}
3. Create sub-circuits for other drivers using schematic 'KEF LS Test.asc' as a guide.
4. Calculate the Q-factors from the mechanical parameters for a BM 165.1 and compare with the data sheet.

12.12 Summary

In this chapter we have seen:

- Inductors store energy in their magnetic field $E = \frac{1}{2} LI^2$ and have an impedance of ωL.
- Inductors in series and parallel add like resistors.

- They have a time constant of R/L.
- An LC combination in series or parallel has a resonance frequency of $\omega = 1/\sqrt{LC}$ and a quality factor Q of $Q = R\sqrt{\frac{C}{L}}$.
- Resistance introduces damping and critical damping occurs when $i(t) = Ae^{\frac{-Rt}{2L}}$.
- Bridges to measure inductance use a capacitor as the reference. Only the Hay Bridge uses the excitation frequency in the balance equations, but all use it for the Q-factor.
- Large capacitors are used to smooth the output voltage of rectifier circuits. Their RMS current rating is important.
- The output DC voltage of a choke-input filter depends upon the duty cycle of the input pulses.
- Cascaded LR filters are similar to cascaded RC filters, and an LR notch filter is similar to an RC notch filter.
- An LC circuit can be used to match a high impedance source to a low impedance load.
- A crystal has two resonances – parallel and anti-parallel – and is modelled by a capacitor in LTspice.
- EMI from IT equipment can be reduced to legal limits by two inductors and three capacitors.
- Electrostatic discharge (ESD) may contain little energy but is destructive of semi-conductor devices.
- Loudspeaker drivers are described by their Thiele-Small parameters supplied by the manufacturers, and these model the acoustic performance of the unit.

Chapter 13
LCR Tuned Circuit

13.1 Introduction

We now move on to the frequency and time response of tuned circuits consisting of an inductor, capacitor, and resistor in series or parallel. These are second-order systems. Such systems have wide applications not only in electronics but in mechanical engineering where we can ascribe electrical analogues to the mechanical components in order to model them, Fig. 13.1. Indeed, second-order systems are pervasive in engineering.

It is surprising that so few components can give rise to such complex mathematics which, perforce, involve not just quadratic but differential equations. It may be felt that this is a somewhat esoteric topic not of general interest. In defence, it must be said that these circuits offer a valuable insight into using LTspice to examine aspects of a signal that are not possible in the laboratory, for example, by splitting an equation for a current into its component parts. They are also good practice at comparing simulated results with calculated where it will be found that the accurate LTspice analysis reveals features that are easily overlooked by manual methods.

Of course, it can be argued that, as we shall see, the difference between the response of a circuit with and without an added load resistor is but a few percent, of the same order as component tolerances, and so there is no need for a detailed analysis, and we can just rely on the simulation results. But on the other hand, it is intellectually satisfying to be able to reconcile the two, or, at least, to understand why they differ, and thereby to gain a better understanding of the relative importance of component values.

These circuits also show the very small – almost vanishingly small – difference between the results including the default series resistance of the LTspice inductor and without it.

© Springer Nature Switzerland AG 2020
C. May, *Passive Circuit Analysis with LTspice®*,
https://doi.org/10.1007/978-3-030-38304-6_13

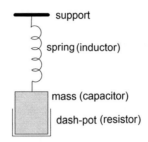

support

spring (inductor)

mass (capacitor)

dash-pot (resistor)

Fig. 13.1 Mechanical Damping

13.2   Series Tuned Circuit

Series circuits of an inductor, capacitor and resistor are used to create frequency selective networks. They are widely found in communication equipment such as radios and television sets. In some cases, the resistor is unwanted and is just the resistance of the wire used to make the inductor. Their frequency response has a peak at the *resonant frequency* falling away on either side. This is the most important feature of the circuit. There is also a large phase change between the input and output signals as the input frequency is swept through the resonance. All of this is standard textbook theory which we will now review. Also of note is that there is a small difference between the resonance measured as the largest current through the circuit and the maximum voltage across the different elements, and this is dramatically different to the voltage measured across the inductor and capacitor in series. Then interesting things happen when we excite the circuit with a pulse, which is not always mentioned in books. It is possible to go into great detail on this, http://rftoolbox.dtu.dk/book/Ch2.pdf – we shall not.

13.2.1   Frequency Response

The circuit is shown in Fig. 13.2 and consists of an inductor and capacitor in series with a resistor, which, as we have said, could just be the resistance of the inductor and can also include the resistance of the signal source. We will first find the impedance of the circuit.

13.2.1.1   Impedance

Going round the circuit using KVL where $R = R_S + R_L$ we have:

$$v_{in} = i\left(R + j\omega L + \frac{1}{j\omega C}\right)$$ (13.1)

from which we see the impedance Z is:

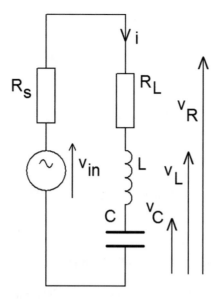

Fig. 13.2 Series Tuned Circuit

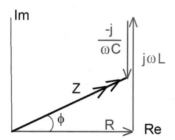

Fig. 13.3 Series Tuned Circuit HF Phasor Diag

$$Z = R + j\omega L + \frac{1}{j\omega C} \tag{13.2}$$

To handle this we need to bring the j term of $1/j\omega C$ to the numerator. We do this by multiplying throughout by j to bring j to the numerator of $1/j\omega C$ and as $j.j = -1$ in the denominator of this term is now negative, not positive. The impedance, then, is:

$$Z = R + j\left(\omega L - \frac{1}{\omega C}\right) \tag{13.3}$$

To show this graphically we use Fig. 13.3. This represents the case when the inductive impedance is greater than the capacitative. First we draw the resistor R along the real axis, and then, at the tip, we draw the inductive impedance in the positive imaginary direction, and from the tip of that we draw the capacitative

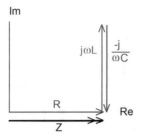

Fig. 13.4 Series Tuned Circuit Resonance Phasor Diag

impedance in the negative imaginary direction, and the circuit impedance is the closing phasor Z at a phase angle Φ.

13.2.1.2   Resonant Frequency for Maximum Current

Following on from Eq. 13.3 the impedance has a minimum when:

$$\omega L = \frac{1}{\omega C} \tag{13.4}$$

which we show in Fig. 13.4 where the two components in the imaginary direction exactly cancel out so $Z = R$. Conventionally this frequency is denoted by ω_n, the natural resonant frequency of the LC combination. This needs a word of explanation: we shall later see that LCR combinations can resonate at a slightly different frequency because of the resistance, and we shall reserve ω_0 for those cases. Certainly the difference is small, in practice usually just 1% or 2%, and textbooks may or may not make a distinction or use different notation. But returning to Eq. 13.4, we have:

$$\omega_n = \frac{1}{\sqrt{LC}} \tag{13.5}$$

The impedance then is a minimum and is merely R, and hence it is called an *acceptor circuit*. At this point the phase shift between current and voltage in the network is zero because the j terms cancel. However, remember CIVIL (C = Capacitor, I = current, V = voltage, I = current, L = inductor) – the current leads the voltage across the capacitor by 90°, and the current lags the voltage across the inductor by 90°. These relationships between current and voltage are always true; only at resonance do these two impedances cancel because their magnitudes are identical, and hence, for the circuit as a whole, there is no phase shift at the resonant frequency. From Eqs. 13.1 and 13.2, the current is:

$$i = \frac{v_{in}}{Z} \tag{13.6}$$

and at resonance this is:

$$i = \frac{v_{in}}{R} \tag{13.7}$$

which is the largest circuit current, and if the resistor is small, this will be very large showing that this circuit is best suited for supplying a current rather than a voltage; otherwise it is very inefficient. If we take the voltage across the capacitor at resonance this is:

$$v_C = \frac{v_{in}}{R} \frac{1}{j\omega_n C} \Rightarrow \frac{v_C}{v_{in}} = \frac{-j}{R}\sqrt{\frac{L}{C}} \tag{13.8}$$

after substituting for ω_n, and similarly we find for the inductor:

$$\frac{v_L}{v_{in}} = \frac{j\omega_n L}{R} = \frac{j}{R}\sqrt{\frac{L}{C}} \tag{13.9}$$

confirming that the voltages are equal and opposite and can be much greater than the input voltage. This is viewing the circuit as a whole, and from the above, we see that the resonant frequency is independent of the resistor which only serves to reduce the height of the resonance and broaden the curve.

Example – Series Resonance Maximum Current (Effect of Inductor Default Rser)
If we create a circuit with $C1 = 10\,nF$, $L1 = 25\,\mu H$ and a variable resistor and using the default series resistance of the inductor, we observe the effect of changing the resistor value, schematic ('Series Resonance 1.asc')., where the resistor is stepped in decades with 5 values per decade to give a better graph when we plot the resonant frequency. We can extend beyond $10\,\Omega$, but then the resonant peak becomes difficult to see and finally vanishes.

.step dec param R 1 10 5

The *.meas* statements first measure the maximum of the capacitor voltage for each value of resistance:

.meas maxI MAX I(C1)

and then we find the resonant frequency:

.meas fRes FIND freq WHEN I(C1) = maxI

Then **View→SPICE Error Log** enables us to see the result, and the resonant frequency is constant at *318305 Hz* including the case where the resistance is *1 Ω*.

There are very small phase angles starting at *0.0862°* and reducing to *0.00863°* at
10 Ω. However, we must beware that LTspice assigns current directions depending
on the orientation of the component. The current, unfortunately, is only given in dBs.
But we can change the y-axis on the plot to read current, and then we find the ratio
I(C1)/1 A almost agrees. For example, we read *99.444 mA* for a resistor of *10 Ω*, and
this is *−20.0048 dB* compared to *20.0009 dB* from the **Spice Error Log**.

If we remove the default series resistance of the inductor, the simulation results
are in exact agreement; for example, the current with *1 Ω* is *0.99999887 A* which
LTspice resolves as *−1.#INFdB*. Using a hand calculator the result is
10.4 × 10^{−6} dB, and with *10 Ω* it is *100.036 mA*, and we find exactly *−20 dB*.
Certainly, the difference is very small, but as LTspice can return very accurate
answers, it is worth bearing this in mind.

If we right click on the **Spice Error Log** data, the pop-up menu offers us the
option of plotting the stepped data.

13.2.1.3 Resonant Frequency for Maximum Voltage

We may alternatively view the circuit as a potential divider and take the maximum
output voltage across either the capacitor or the inductor instead of the maximum
current. We shall take the capacitor:

$$v_C = v_{in} \frac{\frac{-j}{\omega C}}{j\omega L - \frac{j}{\omega C} + R} \tag{13.10}$$

and if *R = 0*, this reduces to $\frac{1}{\omega^2 LC - 1}$ and the solution is again Eq. 13.5. We multiply
Eq. 13.10 throughout by *ωC/−j* and rearrange:

$$\frac{v_C}{v_{in}} = \frac{1}{1 - \omega^2 LC + j\omega CR} \tag{13.11}$$

This shows that the real and imaginary parts, being orthogonal, cannot cancel, so at
the maximum capacitor voltage, the phase angle will not be *90°*. Instead we need to
find the minimum impedance, Fig. 13.5. The two real factors subtract and *ωCR* is
drawn in the imaginary direction so that the impedance is the closing phasor *Z*. It is
sufficient to find its magnitude squared as:

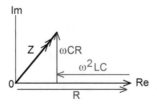

Fig. 13.5 Series Resonance Maximum Impedance

$$Z^2 = \left(1 - \omega^2 LC\right)^2 + (\omega CR)^2 \tag{13.12}$$

then expanding:

$$Z^2 = 1 + \omega^4 L^2 C^2 - 2\omega^2 LC + \omega^2 C^2 R^2 \tag{13.13}$$

and we differentiate wrt ω to find the turning point:

$$0 = 4\omega^3 L^2 C^2 - 4\omega LC + 2\omega C^2 R^2 \tag{13.14}$$

Cancelling and rearranging we see that the resonant frequency falls with increasing resistance:

$$\omega_0 = \sqrt{\frac{2L - CR^2}{2L^2 C}} \Rightarrow \omega_0^2 = \omega_n^2 (1 - \frac{R^2 C}{2L}) \tag{13.15}$$

Example – Series Resonance Maximum Voltage
Using schematic ('Series Resonance 1.asc'), we change the measurements to find the maximum voltage across the capacitor and thence the corresponding frequency:

$$.meas\ maxV\ MAX\ V(c)$$

$$.meas\ fRes\ FIND\ freq\ WHEN\ V(c) = maxV$$

This is saved as schematic ('Series Resonance 2.asc') and is without the default series resistance of the inductor. From Eq. 13.15, we expect the resonant frequency to fall with increasing resistance, which is what we find. Substituting $R=10\ \Omega$ in Eq. 13.15 yields:

$$\omega_0 = \sqrt{\frac{50.10^{-6} - 1.10^{-6}}{2.(25.10^{-6})^2 \times 10^{-8}}} = \sqrt{\frac{49.10^{-6}}{1250.10^{-20}}} = 1979899\ rad/s$$

$$= 315111 Hz$$

and this compares well with the reported *315109 Hz*.

13.2.1.4 Resonance Frequency for Minimum Voltage

If we take the voltage *V(lc)* across the resistor and capacitor in schematic ('Series Resonance 3.asc') we have:

$$\frac{v_{LC}}{v_{in}} = \frac{j\left(\omega L - \frac{1}{\omega C}\right)}{j\left(\omega L - \frac{1}{\omega C}\right) + R} \tag{13.16}$$

And the resonance is at $\omega_n = \frac{1}{\sqrt{LC}}$ as before and the voltage is zero. This is also the sum of the voltages from Eqs. 13.8 and 13.9.

Example – Series Resonance Voltage
We use the simulation ('Series Resonance 4.asc'). This has three identical circuits; only the order of the components is changed so that, in turn, the capacitor, inductor and resistor can be probed directly without having to subtract, for example, $V(c)$ from $V(lc)$.

If we first probe $V(c)$, $V(l)$ and $V(r)$, the capacitor and inductor voltages are left-to-right mirror images and show no signs of resonance. There is, however, a very flat peak to the resistor voltage showing that there is, in fact, a slight resonance effect.

Using the default decibel y-axis the trace shows $0\ dB$ for the capacitor voltage at low frequencies and it is very difficult to measure, even zooming in. However, we can use the *.meas.* statement. Rather than writing a statement for each frequency, we can handle it concisely by declaring a dummy variable x and stepping that:

.step dec param x 1 1e8 1

to give 1 step per decade from 1 Hz to 100 MHz. Then

meas AC Vc FIND V(c) WHEN freq = x

will measure the capacitor voltage at decades of frequency and show that it is decreasing at $-40\ dB/decade$ relative to the $0\ dB$ at the origin. This is because LTspice refers voltage gains to an AC input of $1\ V$, so when the voltage across the capacitor is also $1\ V$, the result is $0\ dB$. If we change to a logarithmic y-axis we see voltages rather than dBs.

If we reduce the series resistor, we find a voltage peak for all three components. We also find that the currents in the capacitor and inductor are identical (not surprising, really, with a series circuit). Making the series resistance 1 nΩ, we find an unrealistic current of nearly 1 kA.

Example – Series Resonance Frequency for Voltage
With the same schematic we find the peak voltage *Vpk* as the maximum of the voltage across the capacitor *V(c)* by two *.meas* statements. The first is to find the peak voltage:

.meas AC VpkC MAX V(c)

over the whole AC run, then we measure the frequency at the peak voltage, *fresC*. We might note that LTspice does not insist on any order in which the measurements are made:

$$.meas\ fresC\ FIND\ freq\ WHEN\ V(c) = VpkC$$

'.meas' MAX Problem

LTspice will often reduce the number of measurement points by a factor of ten or more so if the resonance is sharp, it will be missed. For example, in schematic ('Series Resonance 4.asc') the points are reduced from 100000 to 8077. The solution is to decrease the frequency span to start at 10 kHz, not 1 Hz, and perhaps reduce the maximum frequency from 100 MHz to 10 MHz and the reduction in the number of data points is much less.

If $R = 10\ \Omega$ in ('Series Resonance 3.asc'), the simulation reports a resonant frequency measured across $C1$ of $315103\ Hz$ and a voltage gain of $14.022\ dB$. Using Eq. 13.15 we have:

$$f = \frac{\omega}{2\pi} = \frac{1}{2\pi}\sqrt{\frac{50.10^{-6} - 10^{-8}10^2}{12.510^{-20}}} = 315110 Hz$$

If we substitute $\omega=1.98e6\ rad/s$ in Eq. 13.11 and remembering that the terms are orthogonal we have:

$$\frac{v_c}{v_{in}} = \frac{1}{\sqrt{(1 - 0.98)^2 + (0.19799)^2}} = \frac{1}{0.19899} = 5.023 = 14.023 dB$$

and the phase is $\phi = \tan^{-1}\frac{0.19799}{0.02} = 84.23°$ compared to $84.22°$. The differences are insignificant. Other resistance values also agree.

Example – Series Resonance Phase Relationships

If we choose Cartesian for the y-axis and reduce the AC range from 50 kHz to 1 MHz, we can see the resonant frequency is measured as $318306\ Hz$ and is not quite that same as the previous example. This is the point where the inductor and capacitor voltages are exactly equal at $0.5\ V$. Setting a cursor on this point shows that the capacitor and inductor voltage phases are at $-90°$ and $90°$, respectively, and the current in $R1$ is at $0°$

Example – Series Resonance Voltage Magnification

Substituting for ω_n and using the values from the simulation except that we reduce R to $1\ \Omega$ in ('Series Resonance 3.asc') $L = 25\ \mu H$, $C = 10\ nF$, the magnitude of the voltage across the capacitor is:

$$\frac{v_C}{v_{in}} = \frac{1}{R}\sqrt{\frac{L}{C}} = \sqrt{\frac{25^{-6}}{10^{-8}}} = 50 = 34 dB$$

and is 50 times larger than the input. The measurement statement agrees, giving $33.97\ dB$ at $-89.1°$. If we make $R = 10\ \Omega$, the voltage magnification is $5 = 14\ dB$, and the measurement statement gives a phase of $-84.2°$. In both cases the phase is with respect to the input voltage, and although the phase relationship between the inductor and capacitor voltages is exact, the resistor disturbs the overall phase relationship as we saw in Eq. 13.11.

We may also note that if we make $L = 2.5\ \mu H$, $C = 100\ nF$, we have the same resonant frequency but a much flatter peak. This is what we found with schematic ('Series Resonance 1.asc'). The explanation is to be found in Eq. 13.28.

This is a good point to explore these very important features of the circuit

Explorations 1
1. Open the schematic ('Series Resonance 4.asc'). This consists of three identical circuits differing only in the order of the components so that it is easier to probe the voltages across the capacitor, inductor and resistor. Run the circuit and note the resonant frequency is the same for all.
2. Note the resonant voltages and phase shifts for the capacitors and inductors. These are 90° and accord with theory.
3. Measure the inductor voltage and show that it, too, changes at the same rate as the capacitor voltage
4. Measure the voltage and current for *R3* and show that they are in phase at resonance.
5. Explore other combinations of L, C and R. Try stepping the values and compare the results with theory.

13.2.1.5   3 dB Points

We are also interested in the points where the output across the capacitor has fallen by 3 dB or $1/\sqrt2$ from its maximum. This is an arbitrary figure just is it was for CR circuits and implies that the impedance has increased by $\sqrt2$ of its resonance value. Therefore the modulus of the real and imaginary parts must be equal, Fig. 13.6, giving a result of $\sqrt{2R}$. There are two possibilities; one is that the inductive impedance is greater, the left-hand diagram, and the other that the capacitative impedance predominates, the right-hand diagram. And this also shows the phase differences of 45°. Thus we have:

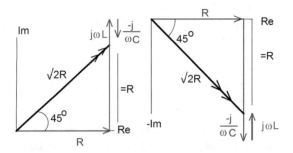

Fig. 13.6   Series Tuned Circuit 3 dB Phasor Diagrams

$$R = \omega L - \frac{1}{\omega C} \tag{13.17}$$

which we will turn into a quadratic equation in standard form by multiplying by ωC:

$$\omega C R - \omega^2 L C + 1 = 0 \tag{13.18}$$

then dividing by LC, and replacing $1/LC$ from Eq. 13.5, we have:

$$\omega^2 - \frac{R}{L}\omega - \omega_n^2 = 0 \tag{13.19}$$

As Eq. 13.19. is a quadratic in ω, we can now write the upper and lower 3 dB points as a product of two terms, the first being the resonant frequency minus the lower 3 dB frequency and the second being the resonant frequency plus the upper 3 dB frequency:

$$(\omega_0 - \omega_l)(\omega_0 + \omega_h) \tag{13.20}$$

which we expand as:

$$\omega_0^2 - \omega_0(\omega_h - \omega_l) - \omega_l\omega_h = 0 \tag{13.21}$$

where ω_h and ω_l are the higher and lower 3 dB frequencies. By comparison with Eq. 13.19:

$$\omega_h - \omega_l = \frac{R}{L} \tag{13.22}$$

showing that for a narrow peak, we need a low resistance. A large inductor also helps, and in the following section we shall see that it also improves the Q-factor. We may also note that:

$$\omega_h\omega_l = \omega_0^2 \tag{13.23}$$

or:

$$\omega_0 = \sqrt{\omega_h\omega_l} \tag{13.24}$$

and ω_n is the geometric mean.

For the voltage across the inductor and capacitor in series, $V(lc)$, we cannot find a drop of -3 dB because the theoretical minimum is $-\infty$ dB. But we can find the points where the voltage is 3 dB down from the off-resonance limit of 0 dB, the *base BW* in Fig. 13.7. However, let us be clear that this is by no means the same measure. The *base BW* is much wider than the *3 dB bandwidth*

To find this, we start with Eq. 13.16; we require the magnitude of the denominator to be $\sqrt{2}$ that of the numerator. This will occur if:

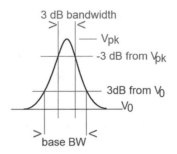

Fig. 13.7 Bandwidth Definitions

$$j\left(\omega L - \frac{1}{\omega C}\right) = R \qquad (13.25)$$

because the two sides of the equation are orthogonal and this is identical to Eq. 13.17 and hence this resonance is much sharper.

Example – 3 dB Points for Voltage
If we make $R = 10\ \Omega$ in schematic ('Series Resonance 3.asc'), the measured bandwidth is *63662 Hz*.

The directive:

.meas $V(lc)3dB$ TRIG $V(lc) = 1/2$ FALL = 1 TARG $V(lc) = 1/2$ RISE = 1

returns the two frequencies when $V(lc)3dB$ is 1/2 of $V(lc)$. This is not $1/\sqrt{2}$ because we are not measuring the magnitude of the voltage. As the voltage decreases, it finds the first **FALL** and then the first **RISE**.

By calculation from Eq. 13.22, it is the same. As we have found the resonant frequency to be *318306 Hz (Example – Series Resonance Phase Relationships)*, the frequencies are *286475 Hz* and *350137 Hz*.

13.2.1.6   Q-Factor

Considering the voltage across the capacitor, a measure of the sharpness of the resonance is the *Q-factor* defined as:

$$Q = \frac{\omega_n}{\omega_h - \omega_l} \qquad (13.26)$$

which is a pure number. Substituting from Eq. 13.22 we have:

$$Q = \omega_n \frac{L}{R} \qquad (13.27)$$

This is a useful form because it tells us that for a sharp resonance, we require a small resistance. We also can substitute for ω_n from Eq. 13.5 and now:

$$Q = \frac{1}{R}\sqrt{\frac{L}{C}} \qquad (13.28)$$

and for a high Q-factor we want a large inductor with little resistance.

13.2.1.7 Voltage Cusp

If we look at the voltage across the inductor and capacitor in series $V(lc)$ we find a sharp peak. Its existence is predicted by Eq. 13.16 where at resonance the voltage across them is 0. However, that does not explain the shape. For that we need to explore the voltage across the inductor alone as $V(lc) - V(c)$ in schematic ('Series Resonance 3.asc') and the voltage across the capacitor $V(c)$ when it will be found that they are identical at resonance (as we have just seen) but that at frequencies very slightly off resonance they no longer cancel because the graphs are highly non-linear and quickly diverge. This is shown in Fig. 13.8 where the existence of a slight peak of the capacitor and inductor voltages at X depends on the size of R and is not seen if it is *100 Ω*. If the resistance is very low, 1 Ω, there is a quite sharp voltage cusp for each but not as sharp as for the series combination.

Example – Series Resonance Cusp Voltage
If we probe the voltage across the inductor and capacitor in series, $V(lc)$, in schematic ('Series Resonance 3.asc') we find a very sharp resonance, far sharper than the capacitor, of about *37 dB*. If we zoom in to try and resolve it using the full AC scan range, we find in the **SPICE Error Log** that the *Number of points per decade* has been reduced from *1e5* to just a few thousand. Changing from *dB* to a *logarithmic* y-axis and moving the cursor over the trace shows about *14 mV*. However, with the shorter AC range, we have nearly fifty thousand steps, and we find the voltage is only *2.9 µV* or so.

Explorations 2
1. Probe the voltage across the resistor and capacitor in ('Series Resonance 3.asc') with $R1=10Ω$. The voltage $V(lc)$ is a sharp cusp. Explore by plotting the outputs

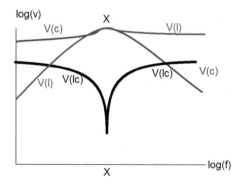

Fig. 13.8 Voltage Cusp

$V(lc)$-$V(c)$ and $V(c)$ as explained above. You may find it better to use a linear y-axis to measure the voltages. Note also the resistor current peak.
2. Repeat with other resistance values and also for current.

13.2.1.8   Power Dissipation

We may also find the energy lost per cycle, that is, the power dissipation. This is of passing interest because in most cases it is very small. The average power is found from the RMS current squared and the resistor; and the RMS current is found from the RMS voltage divided by the impedance. That is:

$$P_{av} = I_{RMS}^2 R = \frac{V_{RMS}^2}{Z^2} R \qquad (13.29)$$

we have the impedance Z in Eq. 13.3, so Eq. 13.29 becomes:

$$P_{av} = V_{RMS}^2 \frac{R}{R^2 + \left(\omega L - \frac{1}{\omega C}\right)^2} \qquad (13.30)$$

Focusing on the terms in ω, we expand the bracket and extract the factor L^2:

$$L^2 \left(\omega^2 + \frac{1}{(\omega LC)^2} - \frac{2}{LC}\right) = L^2 \left(\omega^2 + \frac{\omega_n^4}{\omega^2} - 2\omega_n^2\right) \qquad (13.31)$$

if we now extract the factor ω^2 we have:

$$\left(\frac{L}{\omega}\right)^2 \left(\omega^4 + \omega_n^4 - 2\omega_n^2 \omega^2\right) = \left(\frac{L}{\omega}\right)^2 \left(\omega^2 - \omega_n^2\right)^2 \qquad (13.32)$$

and now we can substitute in Eq. 13.30

$$P_{av} = V_{RMS}^2 \frac{R}{R^2 + \left(\frac{L}{\omega}\right)^2 \left(\omega^2 - \omega_n^2\right)^2} \qquad (13.33)$$

This tells us that at resonance, this is just $P_{av} = \frac{V_{RMS}^2}{R}$. But we can deduce that because the inductor and capacitor do not dissipate power. Off resonance the power is less, schematic ('Series Resonance Power.asc') because the impedance is higher and the current less.

An alternative approach, https://www.kullabs.com/classes/subjects/units/lessons/notes/note-detail/2036, is to use the power factor, that is:

$$P_{av} = V_{RMS} I_{RMS} \cos(\theta) \qquad (13.34)$$

but in any case, the power is usually very small.

13.2.1.9   Off-Resonance Response

SPICE can help us to understand a little better what is happening. At low frequencies the impedance of the inductor will be very small, whilst the impedance of the capacitor will be very large, Fig. 13.9. If we take the voltage across the capacitor, v_c in Fig. 13.2, the capacitor approximates to an open-circuit so the voltage across it is the input voltage with a gain of very slightly less than 0 dB, Fig. 13.10, and the phase will tend to 0°.

At high frequencies, the impedance of the inductor will predominate for the circuit as a whole, but if we look at the capacitor voltage, it will be attenuated by 20 dB due to the impedance of the inductor, and at the same time, the impedance of the capacitor has fallen by 20 dB, so the overall effect is a slope of −40 dB/decade and the phase will tend to 180°. This can be seen in the AC analysis of the circuit, schematic ('Series Resonance 3.asc'), and further tested by a transient analysis. The voltage across the inductor V(l2) is a left-to-right mirror image of the capacitor voltage.

SPICE cannot derive the mathematics of the previous section, but it can enable us to see the shape of the response curves. Mathematicians are able to predict the shape of the graph from the above equations, but for lesser mortals LTspice is a life-saver. And the graphs can show not only the amplitude response but also the phase shift.

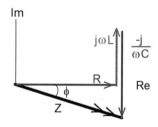

Fig. 13.9 Series Tuned Circuit LF Phasor Diag

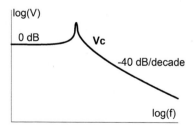

Fig. 13.10 Series Tuned Circuit Capacitor Voltage

Example – Series Tuned Circuit Design

Suppose we want a series turned circuit with a resonance of 100 kHz and a bandwidth of 3 kHz.

This specification does not define a unique set of values, only that the Q-factor is 33. Substituting in Eq. 13.27 we have $33 = 628318*L/R$ so $R = 19040*L$. If we take a standard value $C = 500$ pF, we find $L = 5.066\ mH$ (which would have to made specially) and hence $R = 96.45\ \Omega$ (not a standard value). These are not unreasonable figures. This is schematic ('Series Resonance Example.asc') where the measurements of the resistor current confirm a resonant frequency of $100000r\ Hz$ (the r indicates this is the real part) and a bandwidth of $3030\ Hz$ seen in the **SPICE Error Log**. If we plot the voltage between the top of the inductor and ground, we find the very sharp notch predicted above. If we plot $V(c)$ and $V(l)$-$V(c)$, we can see the two voltages cancelling exactly at the resonant frequency.

An interesting sidelight is that running a transient analysis the circuit takes a few cycles to reach steady state. This is not usually important but it does not come out of the above analysis and takes a bit of work to calculate.

13.2.1.10   Inductor with Parallel Resistor

It was said above that the resistance of the signal source can be added directly to the series resistor. The difficulty is if there is a resistance in parallel with the inductor shown as Rp in Fig. 13.11. It is more efficient to subsume it in the LTspice inductor but we shall retain it as a separate component. We now have:

$$Z = R + \frac{1}{j\omega C} + \frac{j\omega L R_p}{R_P + j\omega L} \qquad (13.35)$$

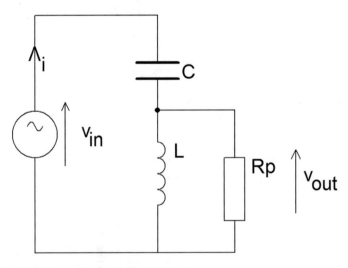

Fig. 13.11 Series Tuned Circuit with Parallel Resistance

where the third term is the parallel sum of C and Rp. If Rp is very large, this reduces to Eq. 13.2. However, in the general case, its effect is to greatly lower or even eliminate the resonance. In the extreme that $Rp = 0$, the inductor has no effect and the circuit is just a resistor in series with a capacitor.

We have seen that the series resistor R limits the current at resonance but does not change the resonant frequency, so we can focus on the second two terms ignoring the series resistor. This is schematic ('Series Res Loaded L.asc').

It is quite surprising just how tricky it can be to analyse this circuit of just three components. The easier approach is to regard it as a potential divider, so we have:

$$\frac{v_{out}}{v_{in}} = \frac{\frac{j\omega L R_P}{R_P + j\omega L}}{\frac{-j}{\omega C} + \frac{j\omega L R_P}{R_P + j\omega L}} \tag{13.36}$$

We can cancel j and multiply throughout by $R_P + j\omega L$:

$$H(j\omega) = \frac{\omega L R_P}{\frac{-(R_P + j\omega L)}{\omega C} + \omega L R_P} \tag{13.37}$$

we now multiply by ωC:

$$H(j\omega) = \frac{\omega^2 C L R_P}{-j\omega L - R_P + \omega^2 L C R_P} \tag{13.38}$$

As a quick check, if R_P is very large, then we can discount the j term to find $1/(\omega^2 C L - 1)$ which is correct for the unloaded case.

We can solve the denominator using the same strategy as Sect. 13.2.1.3 'Resonance Frequency for Maximum Voltage' by finding the impedance of the denominator, then differentiating it to find the turning point. First we turn this into standard form by dividing by LCR_P:

$$H(j\omega) = \frac{\omega_n^2}{\omega^2 - \frac{j\omega}{CR_P} - \frac{1}{LC}} \tag{13.39}$$

After squaring the denominator and differentiating we have:

$$\omega^4 + \frac{1}{(LC)^2} - \frac{2\omega^2}{LC} + \frac{\omega^2}{(CR_P)^2} \Rightarrow 4\omega^3 - \frac{4\omega}{LC} + \frac{2\omega}{(CR_P)^2} \tag{13.40}$$

Divide by 4ω:

$$\omega^2 - \frac{1}{LC} + \frac{1}{2(CR_P)^2} \tag{13.41}$$

$$\omega = \sqrt{\omega_n(1 - 2\zeta^2)} \tag{13.42}$$

Example - Series Resonance Loaded Frequency Response

Simulation ('Series Res Loaded L.asc') shows that to preserve a high Q-factor, the external load must be at least 1 MΩ. The schematic places a resistor in parallel with the inductor. The measurements show that the resonant frequency decreases from *60700 Hz* with a resistor of *400 Ω* to *50646 Hz* with a resistor of *2 kΩ* and *50329 Hz* if the resistor is removed. Note that the resonant frequency occurs at a phase angle of *64.5°* with *400 Ω* and is still only *85.45°* with *2 kΩ* showing that the resonance is no longer when the impedance is real. As a check on Eq. 13.42, substituting the circuit values with $R_P = 400\ \Omega$ and the natural resonant frequency of *50330 Hz* or *316232 rad/s*, we find – *60700 Hz* identical to the simulation – and a phase angle of *64.8°*. That is close enough. And the bandwidth increases with the resistance.

Schematic ('Series Resonance Example C.asc') has the resistor across the capacitor, and the analysis is similar.

Explorations 3

1. Run ('Series Resonance Example.asc'), and check that measurements agree with theory. Can the values be changed to keep the same performance but use standard values?
2. Make an AC analysis from 1 kHz to 10 MHz using ('Series Resonance 3.asc'). Note the voltages for $V(c)$ and $V(l2)$. At low frequencies the impedance of the inductor is very low so the attenuation is almost constant and is due to the resistor and the phase shift is zero. The converse is true for the capacitor. And in both case the asymptotic slope is 40 dB/decade.
3. Measure the Q-factor and explore it with different resistor values. Check against calculations.
4. Measure the 3 dB point for the inductor as the point where the attenuation has increased by 3 dB from its low or high frequency value and see if it is the same as the 3 dB point for the capacitor.
5. Redesign the circuit to have the same resonant frequency but with an inductor ten times larger. How does that affect the Q-factor?
6. Devise a way of measuring the real and apparent power. A good starting point is to use a transient analysis.
7. Test ('Design Example fr=100k.asc') and find the bandwidth. Check with calculation.
8. Probe the phase of the response, and note that below resonance the circuit is inductive and above it is capacitive. Confirm this with transient runs.
9. Run ('Series Res Loaded L.asc') and plot the stepped data. Check against calculations.

13.3   Series Tuned Circuit Time Response

So far we have explored tuned circuits in the frequency domain. This is a very popular and very fruitful approach, but we must remember that an AC analysis gives us no information about the behaviour of the circuit to anything other than a sine wave input. In real life, we often find sudden transitions in the signal – spikes or square waves – and it is important to understand how the circuit will respond to those.

We can qualitatively understand what will happen if we suppose the signal source and the tuned circuit have no resistance and we apply an input voltage pulse. The capacitor will charge up – instantaneously, in theory – and after the pulse has gone it will discharge through the inductor, converting electrical energy into magnetic energy. Then the magnetic field will collapse creating a voltage that will recharge the capacitor, and this will continue forever with the energy oscillating between the capacitor and the inductor at the resonant frequency. However, if there is resistance in the circuit, then energy will be dissipated as heat, and the oscillations will eventually cease. We will now quantify this as we turn from the frequency domain to the time domain.

13.3.1 The Differential Equation

The previous equations are not applicable here: we need to find the current as a function of time. If we take the series tuned circuit with no parallel load resistor Rp, no existing current or charge, and apply a step voltage to the circuit, we have:

$$V_{\text{in}} = iR + L\frac{di}{dt} + \frac{1}{C}\int \frac{di}{dt} \tag{13.43}$$

where the current i is a function of time. It is awkward trying to solve an equation with both integrations and differentiations, and therefore we differentiate it to remove the integration. As V_{in} is constant, when we do this, it becomes zero, and then re-ordering terms and dividing by L gives:

$$\frac{d^2i}{dt^2} + \frac{R}{L}\frac{di}{dt} + \frac{1}{LC} = 0 \tag{13.44}$$

which is a second-order differential equation. A thorough, clear, exposition of these with examples can be found at http://epsassets.manchester.ac.uk/medialand/maths/helm/19_3.pdf where the first 38 pages or so are sufficient for our purposes. We can, however, use our previous strategy for solving RC and LR circuits, and that is that any solution must be dimensionally correct which – as we saw previously – generally is an exponential of the form $i(t) = Ae^{st}$ so that $\frac{di}{dt} = sAe^{st}$ and $\frac{d^2i}{dt^2} = s^2Ae^{st}$, and then Eq. 13.44 is:

$$s^2Ae^{st} + \frac{sR}{L}Ae^{st} + \frac{Ae^{st}}{LC} = 0 \tag{13.45}$$

We can check this dimensionally; $s^2 = T^{-2}$, $R/L = T^{-1}$ (see Eq. 13.15) and $1/LC = T^{-2}$, and the exponents are identical and applied to every term. Then if we divide 13.45 by Ae^{st}, we have the *characteristic equation*:

$$s^2 + s\frac{R}{L} + \frac{1}{LC} = 0 \tag{13.46}$$

And this is what we shall solve. As an aside, this is a form we used before for the bi-quadratic transfer function only now we have just the numerator. We can substitute from Eq. 13.27 $Q = \omega_0 \frac{L}{R}$ and generalize:

$$s^2 + s\frac{\omega_0}{Q} + \frac{1}{LC} = 0 \tag{13.47}$$

This is a useful form when the two roots coincide. But to return to Eq. 13.46. We use the standard formula to find the roots of a quadratic equation. The coefficient a of s^2 is 1. Therefore if we divide Eq. 13.46 by 2, we simplify the standard formula since $2a = 1$ and $4ac = 4 \times \frac{1}{2} \times (1/2LC) = 1/LC$, and we have Eq. 13.48 which has two roots:

$$s = \frac{-R}{2L} \pm \sqrt{\left(\frac{R}{2L}\right)^2 - \frac{1}{LC}} \tag{13.48}$$

These are the positions of the poles. And from Eq. 13.48 we can write the roots as:

$$r_1, r_2 = \frac{-R}{2L} \pm \sqrt{\left(\frac{R}{2L}\right)^2 - \frac{1}{LC}} \tag{13.49}$$

and then the exponential terms are $Ae^{r_1 t}, Ae^{r_2 t}$ so if we write Eq. 13.46 as a product:

$$(s + r_1)(s + r_2) = 0 \tag{13.50}$$

finally we have:

$$i(t) = (s + A_1 e^{r_1 t})(s + A_2 e^{r_2 t}) \tag{13.51}$$

which we shall use in the following sections.

13.3.1.1   Series Circuit Decay Factor and Damping Ratio

These are useful adjuncts to the Q-factor. The decay factor is defined as $\alpha = b/2a$ in the usual notation for a quadratic equation. As b in Eq. 13.46 is R/L, and $a = 1$ we have:

$$\alpha = \frac{R}{2L} \tag{13.52}$$

in nepers per second (Np/s), which expresses how rapidly the energy is dissipated by the resistor and appears as a factor in Eq. 13.48 which we can write as:

$$s = -\alpha \pm \sqrt{\left(\alpha^2 - \omega_n^2\right)} \tag{13.53}$$

We also may write Eq. 13.22 as:

$$\omega_h - \omega_l = 2\alpha \tag{13.54}$$

or from Eq. 13.26:

$$\omega_h - \omega_l = \frac{\omega_n}{Q} \quad so \quad \alpha = \frac{\omega_n}{2Q} \tag{13.55}$$

as an alternative expression. These show that the separation of the 3 dB points increases linearly with the damping ratio or, in other words, that the Q-factor lessens.

For the damping ratio, we return to Eq. 13.43. The cause of energy loss (i.e. damping) is the i^2R power dissipated by the resistor R. This is the *actual damping*.

The damping ratio is usually give the symbol ζ (Greek zeta) and is defined as:

$$\zeta = \frac{actual\ damping}{critical\ damping} \tag{13.56}$$

Anticipating the following sections, critical damping occurs when the roots coincide; then from Eq. 13.74, it is $R = 2\sqrt{\frac{L}{C}}$ and we can write Eq. 13.56 as:

$$\zeta = \frac{R}{2\sqrt{\frac{L}{C}}} = \frac{R}{2}\sqrt{\frac{C}{L}} = \frac{\alpha}{\omega_n} = \frac{1}{2Q} \tag{13.57}$$

Now $\zeta\omega_n = \frac{R}{2}\sqrt{\frac{C}{L}} \cdot \frac{1}{\sqrt{LC}} = \frac{R}{2L}$, so we can write Eq. 13.49 as:

$$r_1, r_2 = \omega_0\left(-\zeta \pm \sqrt{\zeta^2 - 1}\right) \tag{13.58}$$

and

$$r_2 - r_1 = \omega_n\left(-\zeta - \sqrt{\zeta^2 - 1}\right) - \omega_n\left(-\zeta + \sqrt{\zeta^2 - 1}\right)$$

$$= 2\omega_n\sqrt{\zeta^2 - 1} \tag{13.59}$$

which shows that with $\zeta=1$ the roots are co-incidental and as ζ increases the roots become farther apart.

Example – Damping Ratio

Schematic ('Series Res Damping Ratio.asc') has an underdamped series circuit with a series resistor whose value is stepped. We measure the maximum current and find in the first set of measurements that the maximum current decreases with increasing resistance and the phase shifts very slightly.

The second set of measurements is the difference in frequency between the 3 dB points, and this can be compared to Eq. 13.55 or 13.59. If we step the resistor by *1 Ω* to give a smoother trace, it is interesting to plot the stepped response. We see that the maximum current falls very steeply with resistance from hundreds of amperes at *1 nΩ* to *1 A* at *1 Ω* and thereafter preserves that same ratio of current to resistance.

13.3.2   Damping Conditions

We can define four with corresponding equivalent circuit conditions where we need test only one, because if that is true, so are the others. For example, if $D > 0$ then also $\zeta > 0$ and $\alpha > \omega_n$. Excluding the first they all include the value of the discriminant:

- $\zeta = 0, \alpha = 0$ undamped there is no resistance and the frequency is the natural frequency ω_n of the circuit: $\omega_0 = \omega_n$
- $\zeta > 1, D > 0, \alpha > \omega_n$ overdamped where there are two real roots
- $\zeta = 1, D = 0, \alpha = \omega_n$ critically damped where there are two coincidental real roots
- $0 < \zeta < 1, D < 1, \alpha < \omega_n$ underdamped where there are two complex conjugate roots

The currents corresponding to the last three cases are sketched in Fig. 13.12 showing a large initial current surge into the capacitor which, in the underdamped case, can result in a current reversal. The capacitor voltage is shown in Fig. 13.13 where we see that in the underdamped case, there could be decaying oscillations. A brief derivation of the results for these three cases can be found at http://info.ee.

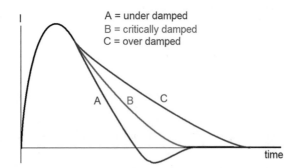

Fig. 13.12 Damping Current

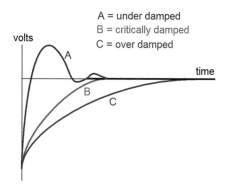

Fig. 13.13 Damping Voltage

surrey.ac.uk/Teaching/Courses/ee1.cct/circuit-theory/section7/index.html. We shall now explore them.

13.3.2.1 Overdamped Current (D >0)

There are two different real roots. For this to be true, we require the discriminant in Eq. 13.48 to be positive or $R^2 > 4\frac{L}{C}$. If we insert the equality where $R^2 = 4\frac{L}{C}$ in the discriminant of Eq. 13.48, we have $\sqrt{\left(\frac{4L}{C}\frac{1}{4L^2}\right)} - \frac{1}{LC} = \frac{1}{LC} - \frac{1}{LC} = 0$ so with the inequality the discriminant will be positive.

To solve Eq. 13.51 at time $t = 0$, we apply a pulse input, not a sine wave, with the capacitor discharged. The initial current is zero and thus $A_2 = -A_1$, and the solution for the current is the sum, not the product, because they are exponentials, and including the minus sign, we have:

$$i(t) = A(e^{-r_1 t} - e^{-r_2 t}) \tag{13.60}$$

Using $v = L\frac{di}{dt}$ and noting that at time $t = 0$ $v = V_1$, we differentiate Eq. 13.60 and rearrange it to find the initial rate of change of current:

$$\frac{di(0)}{dt} = -A(r_2 - r_1) = \frac{V_1}{L} \tag{13.61}$$

whence:

$$A = \frac{-V_1}{L(r_2 - r_1)} \tag{13.62}$$

substituting in Eq. 13.60:

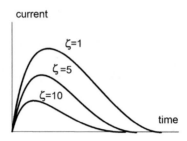

Fig. 13.14 Series Damping Zeta

$$i(t) = \frac{-V_1}{L(r_2 - r_1)}\left(e^{-r_1 t} - e^{-r_2 t}\right) \tag{13.63}$$

and the current is the difference between the exponentials. We should note here that it is also possible to express the current in terms of the damping ratio. Substituting for $(r_2 - r_1)$ in Eq. 13.62 using Eq. 13.59, we have:

$$A = \frac{-V_1}{2L\omega_n\sqrt{\zeta^2 - 1}} \tag{13.64}$$

In every case the root with the higher negative coefficient of e will decay faster, Fig. 13.14, which is not to scale. The height of the peak decreases with damping since the increased resistance reduces the peak current.

Example – Series Resonance Overdamped

Let us try an example choosing easy numbers to work with rather than starting with a circuit having a defined resonance. This, of course, is not the way circuits are designed, but that is of no importance here. We chose $R/2L = 3$, $1/LC = 5$, so from the discriminant we have $\sqrt{(9-5)} = 2$ which is half the difference between the roots. Possible components are $L = 1$, $C = 0.2$, $R = 3$, and the two roots are $r_2 = 5$, $r_1 = 1$. Then with an input voltage of 1:

$$i = \frac{1}{4}\left(e^{-t} - e^{-5t}\right)$$

The two currents are shown in Fig. 13.15 where the e^{-5t} term is drawn in the negative current axis to emphasise that it is subtracted. This also can be found as schematic ('Series Resonance Overdamped Example.asc'). Here we can plot the response of the actual circuit and that of its two roots by separating the roots as two sources. This is possible by two auxiliary circuits: the left-hand one uses current generators and vanishingly small resistors; the other uses voltage generators and 1 Ω resistors to generate the two currents. Some of the currents are negative because of the way the components were placed. It will be found that the difference in currents I $(B1)-I(B2)$ exactly matches the current $I(V1)$.

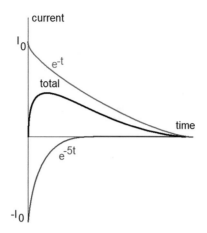

Fig. 13.15 Overdamped Series Tuned Circuit

There are several interesting points to note:

- It is important to set the rise and fall times of the pulse to a small value, not zero else the default values give a slow rise and fall.
- The separate exponential terms combine exactly to form the current in the circuit.
- The e^{-5t} term decays the fastest, Fig. 13.15, and is responsible for the initial rising current.
- The time constant of the circuit approximates to that of the longer time constant, so we can use that to estimate the rise and fall times.
- The maximum current of 13 mA is at time 410 ms corresponding to 0.8 V dropped across the resistor. The capacitor voltage is then 0.2 V and because at this point $\frac{di}{dt} = 0$ the inductor voltage is 0 V.

Taking the limiting value for the resistor $R = 2\sqrt{\frac{L}{C}}$ when we substitute in Eq. 13.28 we have $Q < 0.5$. This is worth remembering. And so is that $\zeta > 1$.

Explorations 4

1. Run simulation ('Series Resonance Overdamped.asc'), and note the capacitor voltage as a function of resistance. Explore the voltages and currents.
2. Run simulation ('Series Resonance Overdamped Example Voltages.asc'), and compare the synthesized values with those of the circuit on the left.
3. Explore with different input voltage levels (remember to change the coefficients from ¼).
4. Calculate component values to give a large separation of the poles, and check that the time constant approximates to the one with the longer decay.
5. Change to an AC analysis and find the resonant frequency and α and ζ. Hence find the largest resistor for overdamping.

13.3.2.2   Critically Damped Current (D = 0)

The website http://www2.ensc.sfu.ca/~glennc/e220/e220l15b.pdf goes into great detail, including the case where the initial current is not zero. In our case it is zero so:

$$\left(\frac{R}{2L}\right)^2 = \frac{1}{LC} \tag{13.65}$$

or

$$\alpha^2 = \omega_n^2 \tag{13.66}$$

Now there are two real roots which coincide:

$$r_1 = r_2 = \frac{-R}{2L} = \alpha \tag{13.67}$$

then Eq. 13.51 suggests the solution is just $i(t) = A_0 e^{-\alpha t}$. Not so. We need a second constant and we must add the term $A_1 t e^{-\alpha t}$. Most websites simply quote this rather than going into the rather long proof. We shall do the same. Therefore the solution is:

$$i(t) = (A_0 + A_1 t)e^{-\alpha t} \tag{13.68}$$

where at time zero the current is zero and as $e^\circ = 1$ we have $A_0 = 0$. Also there is no initial charge on the capacitor so:

$$i(t) = A_1 t e^{-\alpha t} \tag{13.69}$$

the initial rate of change of current on re-arranging $V_1 = L\frac{di}{dt}$ is:

$$\frac{di}{dt} = \frac{-V_1}{L} \tag{13.70}$$

so differentiating Eq, 13.69 at time $t=0$ and comparing it to Eq, 13.70 we find $A_1=V_1/L$ and the current at any time is:

$$i(t) = \frac{V_1}{L} t e^{-\alpha t} \tag{13.71}$$

Notice now that the coefficient of the exponent includes time, unlike the overdamped case. We can differentiate Eq. 13.71 to find the turning point which is the peak current:

$$\frac{di}{dt} = \frac{V_1}{L}e^{-\alpha t}(1 - \alpha t) \quad \text{then } I_{max} \text{ is at } \quad t = \frac{1}{\alpha} \tag{13.72}$$

Also there is just one time constant:

$$\tau = \frac{1}{\alpha} = \frac{L}{R} \qquad (13.73)$$

so we can calculate the rise and fall times accurately.
From Eq. 13.65

$$R = 2\sqrt{\left(\frac{L}{C}\right)} \qquad (13.74)$$

and substituting in Eq. 13.28 we find $Q = 0.5$. This is the condition for the fastest rise to the final state without overshooting. It is not particularly helpful to find the resonant frequency. It is of interest, however, to note that $\zeta = 1$ by definition.

Example – Series Resonance Critical Damping

For example, if we have a series circuit of schematic ('Series Resonance Critical Damping.asc') with $L = 25\ \mu H$ $C = 10\ nF$ we find from Eq. 13.73 $R = 2\sqrt{(25\ \mu/10\ n)} = 2\sqrt{2500} = 100\ \Omega$. And from Eqs. 13.67, we find that $\alpha = R/(2L) = 100/50\ \mu = 2.10^6$. If we insert these numbers in Eq. 13.71 and apply a *10 V* input step, we find $i(t) = 4e5te^{-2e6t}$.

We can find the turning point when the current is maximum when we have:

$$\frac{V_1}{L}e^{-\alpha t}(1 - \alpha t) = 4.10^5 e^{2.10^6 t}(1 - 2.10^6 6t)$$

and the negative or positive turning point is at *0.5 μs*.

And the current at that point is $4e5 \times 0.5e - 6e^{-2e6\ \times\ 0.5e\ -\ 6} = 0.2e^{-1} = 0.073576A$.

Example – '.meas' FIND...WHEN

We used this previously with schematic ('Series Resonance 3.asc') where we saw that we handle this in two stages. First we use a **MAX** measurement to find the peak current where simulation gives *0.0735694 A*. Then we use **FIND..WHEN** measurement to find the time at which the current is the peak, and this is the turning point. And although the FIND can be an expression, the *WHEN* must be a simple variable.

With a step time of 2 *ps*, we find *6.50027 μs* − 6 *μs* = *0.50027 μs*. We subtract 6 *μs* because it measures from the positive pulse. Of course, we can also change the values to measure on the initial negative pulse.

The peak current reduces to *0.0735127 A* if we use a step time of *0.2 ns*, but more significantly, because the turning point is not so well resolved, the time is *0.51049 μs*. And with a step time of *20 ps*, the turning point is *0.49116 μs*.

Also from Eq. 13.72 the initial slope at $t = 0$ is $V_1/L = -4e8\ A/s$.

We can test this with the schematic ('Series Resonance Critical Damping.asc').

Example – '.meas' DERIV

This returns the time derivative. If we want the slope of the current graph, we can put the result in 'graph' as:

$$.meas\ slope\ DERIV\ I(V1)\ AT\ 0$$

and notice it is simply *AT 0*, putting *AT time=0* is an error.

Explorations 5
1. Run ('Series Resonance Critical Damping.asc') and note that the time for the capacitor to reach 9 V is shortest when the circuit is underdamped with a resistor of 80 Ω but it takes longer to settle to the final value of 10 V than the critically damped case.
2. Change the **.meas** statement to measure the negative pulse instead of the positive.
3. In the above circuit, the left-hand side, the RCL circuit, makes no assumption about the damping – it only changes the resistance; but the right-hand circuit with a current generator uses the critical damping equation, and therefore there is generally a difference between the two plots because it will be the wrong equation. However, in the specific case of critical damping with 100 Ω, the two coincide.
4. Calculate the rise time from the component values, and check with the simulation by adding a '.meas' statement.

13.3.2.3   Under Damped Current (D < 1)

If we run simulation ('Series Resonance Example Pulse IP.asc') and look at the current, we find very serious ringing because the circuit is underdamped. We might have predicted this from the Q-factor of 33. We can mitigate it by adding rise and fall times to the pulse of 10 μs which is tending to a triangular wave input and is fairly close to a sinusoid and so – qualitatively – we can see why there is no overshoot or ringing with a sinusoidal input.

The mathematics are a little more difficult and are set out very clearly at http://info. ee.surrey.ac.uk/Teaching/Courses/ee1.cct/circuit-theory/section7/dampedcases.htm.

The discriminant D now is negative so from Eq. 13.49 the roots are

$$r_1, r_2 = \frac{-R}{2L} \pm \sqrt{-D} \qquad (13.75)$$

which we can write as:

$$r_1, r_2 = \frac{-R}{2L} \pm j\sqrt{D} \qquad (13.76)$$

For convenience we write this as: $r_1 = -\alpha + jD_r$ $r_2 = -\alpha - jD_r$ where $D_r = \sqrt{D}$. The solution then is:

$$i(t) = A_1 e^{-(\alpha - jDr)t} + A_2 e^{-(\alpha + jDr)t} \tag{13.77}$$

or:

$$i(t) = e^{-\alpha t} \left(A_1 e^{jDrt} + A_2 e^{-jDrt} \right) \tag{13.78}$$

To solve this we use the Euler Relationship:

$$e^{jx} = \cos(x) + jsin(x) \tag{13.79}$$

and:

$$e^{-jx} = \cos(x) - jsin(x) \tag{13.80}$$

then each of the terms in Eq. 13.78 is of the form:

$$Ae^{-j\alpha t} = A(\cos(\alpha t) \pm j\sin(\alpha t)) \tag{13.81}$$

and Eq. 13.78 becomes:

$$i(t) = e^{-\alpha t}[(A_1(\cos(D_r t) + jsin(D_r t))) + A_2(\cos(D_r t) - jsin(D_r t))] \tag{13.82}$$

We can collect terms:

$$i(t) = e^{-\alpha t}[(A_1 + A_2)\cos(D_r t) + j(A_1 - A_2)\sin(D_r t)] \tag{13.83}$$

and solve it at time t = 0 where the current is zero and therefore the *cos* term must be zero leaving us with:

$$i(t) = Ae^{-\alpha t}\sin(D_r t) \tag{13.84}$$

where we have merged the two constants in one. Differentiating Eq. 13.84 by parts we have:

$$\frac{di}{dt} = -Ae^{-\alpha t}(-\alpha \sin(D_r t) + D_r \cos(D_r t)) \tag{13.85}$$

And at time zero $sin = 0$ $cos = 1$ $e^\circ = 1$ leaving $\frac{di}{dt} = AD_r$ and $V = L\frac{di}{dt}$, so finally we substitute for A and:

$$i(t) = \frac{V}{LD_r} e^{-\alpha t}\sin(D_r t) \tag{13.86}$$

The current, then, consists of three components. The first is the time zero maximum amplitude of V/LD_r; the second is the decaying exponential that shapes the envelope of the third component which is the sinusoidal waveform whose frequency is D_r rad/s. And we can now see why α is called the 'decay factor'.

In the special case of $R = 0$, the decay factor also is zero, and the oscillations will not die away. Recalling that $D_r = \sqrt{\left(\frac{R}{2L}\right)^2 - \frac{1}{LC}}$ in this case also $D_r = \sqrt{1/(LC)}$ and the circuit will oscillate at the natural undamped, frequency, usually designated ω_n. But in the general case, D_r is a forced resonant frequency ω_0 where:

$$\omega_0 = \sqrt{\alpha^2 - \omega_n^2} \qquad (13.87)$$

showing it is slightly less than the natural frequency. The difference is usually just a small fraction of a percent and can be ignored safely; see the following example.

We can make the importance of damping more explicit by substituting in Eq. 13.86:

$$i(t) = \frac{V}{L\sqrt{\alpha^2 - \omega_n^2}} e^{-\alpha t} \sin\left(\sqrt{(\alpha^2 - \omega_n^2)}t\right) \qquad (13.88)$$

or using Eq. 13.57:

$$i(t) = \frac{V}{L\omega_n\sqrt{\zeta^2 - 1}} e^{-\zeta\omega_n t} \sin\left(\omega_n\left(\sqrt{\zeta^2 - 1}\right)\right) \qquad (13.89)$$

And remembering that $Q = \omega_n \frac{L}{R}$ we can understand why a high Q-factor entails ringing or overshoot. This is no problem if we have a sinusoidal input in a radio or television set and we can use circuits with high Q-factors.

Example – Series Resonance Pulse Input
The schematic ('Series Resonance Pulse.asc') realizes Eq. 13.86 as:

$$I = 10/25e - 6 * time * exp\left(-R * time/(50e - 6)\right)$$

and if we set $R1 = 100$ the current from source $B1$ exactly fits the circuit current.

Example – Series Resonance Under damped
Using the numbers from ('Series Resonance UnderDamped.asc') of $L = 0.1\ mH$ $C = 10\ nF$ we find $\omega_n = \sqrt{1/LC} = 10^6 = 159.155\ kHz$. In this simulation, $R = 10\ \Omega$

$$D_r = \sqrt{\left(\frac{10}{2.10^{-4}}\right)^2 - 10^6} = \sqrt{2.5 \times 10^9 - 10^6} = 49990$$

compared to 50000 in the undamped case. The difference is so small that it cannot be resolved with an AC analysis by substituting $1\ n\Omega$ for the resistor even with $1e11$ points per decade (which are reduced to less than $1e6$ by LTspice) but it can be seen by plotting the difference between the two for the first 80 μs.

From Eq. 13.28 the Q-factor is $1/10\sqrt{(10^{-4}\ 10^{-8})} = 10$, and we can check by measuring the bandwidth and using Eq. 13.16 giving $Q = 159\ kHz/(167.2\ kHz-151.3\ kHz) = 10$ which is the same.

13.3.3 Voltages

So far we have found the circuit current. The voltages are found from it.

13.3.3.1 Overdamped

We can consider the three elements in turn.

Resistor Voltage
The voltage across the resistor is in phase with the current and is Eq. 13.63 for the current multiplied by R:

$$v_R(t) = -R\frac{V_1}{L(r_2 - r_1)}(e^{-r_1 t} - e^{-r_2 t}) \tag{13.90}$$

If we differentiate this we find:

$$\frac{dv_r}{dt} = -R\frac{V_1}{L(r_2 - r_1)}(-r_1 e^{-r_1 t} + r_2 e^{-r_2 t}) \tag{13.91}$$

and this has a turning point t_R when:

$$r_1 e^{-r_1 t} = r_2 e^{-r_2 t} \tag{13.92}$$

then,

$$t_R = \frac{\ln(r_1) - \ln(r_2)}{(r_2 - r_1)} \tag{13.93}$$

And this causes a single overshoot, the magnitude of which can be found by inserting t_R in Eq. 13.90. We also see that at time $t = 0$, the voltage is zero; likewise after infinite time it is zero.

Inductor Voltage
We can use Eq. 13.91 minus the factor R to give di/dt then multiply by L to find $v_L = L\,di/dt$

$$v_L(t) = \frac{V_1}{(r_2 - r_1)}(-r_1 e^{-r_1 t} + r_2 e^{-r_2 t}) \tag{13.94}$$

To find the turning point, we again differentiate:

$$0 = \frac{V_1}{(r_2 - r_1)}(r_1^2 e^{-r_1 t} - r_2^2 e^{-r_2 t}) \tag{13.95}$$

and solve:

$$t_L = 2\frac{\ln(r_1) - \ln(r_2)}{(r_2 - r_1)} \tag{13.96}$$

and this gives an undershoot after twice the time for the resistor. At time $t = 0$, there is no voltage across the resistor and as the capacitor is uncharged, there also is no voltage across it, so the whole voltage appears across the inductor. After infinite time the voltage is zero.

Capacitor Voltage
For the capacitor voltage, we must integrate the current:

$$v_C(t) = \frac{1}{C}\int_0^t \frac{V_1}{L(r_2 - r_1)}(e^{-r_1 t} - e^{-r_2 t}) + c \tag{13.97}$$

where c is the constant of integration:

$$v_C(t) = \frac{1}{C}\frac{V_1}{L(r_2 - r_1)}\left(\frac{-1}{r_1}e^{-r_1 t} + \frac{1}{r_2}e^{-r_2 t}\right) + c \tag{13.98}$$

where at time $t = \infty$ $v_C = V_1$ and therefore $c = V_1$. Or, if we have found the resistor and inductor voltages, we can simply subtract:

$$v_C(t) = V_1 - v_R(t) - v_L(t) \tag{13.99}$$

Example – Series Resonance Overdamped Voltage Overshoot
If we take the schematic ('Series Resonance OverDamped Example.asc'), we found $r_1 = 1$, $r_2 = 5$ then from the above equations:

$$v_R = \frac{6}{4}(e^{-t} - e^{-5t}) = 1.5(e^{-t} - e^{-5t})$$

$$v_L = \frac{1}{4}e^{-t} + \frac{5}{4}e^{-5t}$$

$$v_C = 1.25e^{-t} + 0.25e^{-5t} + 1$$

Using schematic ('Series Resonance Overdamped Voltages.asc'), we can separate the voltage components using voltage sources so that $B1$ creates voltage $V(R-1) = e^{-t}$, and B2 creates voltage $V(R-5) = e^{-5t}$ so that if we create the trace $V(r-1)-V(r-5)$, which is the sum of the two resistor voltages, we find it is identical to the voltage across the resistor $V(s)-V(l)$, and so on for the inductor and capacitor voltages, only remember that LTspice is rather cavalier about converting uppercase to lowercase.

Using Eq. 13.93, the resistor voltage overshoot occurs at $(0-1.609)/4 = 0.4$ s and
from Eq. 13.90 the voltage is $6/4(0.6703 - 0.1353) = 0.8024$ V.

Using Eq. 13.96, the inductor undershoot voltage occurs at 0.8 s and the voltage is
$1/4(-0.447 + 5 \times 0.018) = -0.089$ V.

The capacitor voltage is not a simple exponential rise but starts horizontally and is clearer if we increase the value of the inductor. But there is no overshoot and for this reason it is underdamped.

Explorations – 6
1. Open simulation ('Series Resonance OverDamped Example Voltages.asc'). The voltage sources $B1 \ldots B6$ generate voltages for Eqs. 13.90, 13.91 and 13.95. Confirm that their sums almost exactly add to the actual voltages across the components and trim them so the match is exact.
2. Solve Eq. 13.93 for the measured overshoot and show that it agrees with the simulation. Try other resistors greater than the critical value of 4.47 Ω, and note the increase in overshoot as the resistor is smaller.
3. Try different input waveforms such as triangular.

13.3.3.2   Critically Damped

We shall again deal with the three components in turn. Essentially, this is just the limiting case of the previous. The schematic ('Series Resonance Critical Damping. asc') makes runs with resistors above and below the critical value as well as the critical value, and it shows a slight overshoot of the capacitor voltage if the resistor is too small and the faster rise to steady state when the resistor value accords with the critically damped value of $100 \, \Omega$.

Resistor Voltage
In this case the equation for the current is 13.71, so the voltage across the resistor is:

$$v_R(t) = Rt\frac{V_1}{L}e^{-at} \tag{13.100}$$

and again shows an over-voltage surge. At time $t=0$, it is zero and again zero after infinite time. We can differentiate Eq. 13.100 to find:

$$\frac{dv_r}{dt} = R\frac{V_1}{L}e^{-\alpha t}(1 - \alpha t) \tag{13.101}$$

and the turning point is at time:

$$t = \frac{1}{\alpha} \tag{13.102}$$

from which we find the largest voltage is:

$$v_{Rmax} = R\frac{2L}{R}\frac{V_1}{L}e^{-1} \quad = 2V_1 \times 0.36788 \tag{13.103}$$

Inductor Voltage

For the inductor we must differentiate the current to find:

$$v_L(t) = L\frac{di}{dt} \quad = V_1(e^{-\alpha t} - \alpha t e^{-\alpha t}) \quad = V_1 e^{-\alpha t}(1 - \alpha t) \tag{13.104}$$

And as before this will tend to zero after infinite time as at $t=0$ it is V_1. If we differentiate Eq. 13.104, we find:

$$\frac{dv_L}{dt} = \alpha V_1 e^{-\alpha t}(-1 - 1 + \alpha t) \quad = \alpha V_1 e^{-\alpha t}(\alpha t - 2) \tag{13.105}$$

and this has a turning point at:

$$t = \frac{2}{\alpha} \tag{13.106}$$

and is twice that for the resistor, just as we found for the overdamped case. And at that time, the voltage is:

$$v_{Lmax} = V_1 e^{-\frac{\alpha 2}{\alpha}}\left(1 - \alpha\frac{2}{\alpha}\right) \quad = -V_1 e^{-2} \tag{13.107}$$

Capacitor Voltage

We can either subtract from the resistor and inductor voltages or use Eq. 13.96, and then we must again integrate where in this particular case $\alpha = \omega_n$:

$$v_C(t) = \frac{-V_1}{LC}\int_0^t t e^{-\alpha t}dt \quad = V_1 \alpha^2 \int_0^t t e^{-\alpha t} \tag{13.108}$$

and integrating by parts using:

$$\int u\,dv = uv - \int v\,du \quad \text{where} \quad u = t \quad du = dt \quad dv = e^{-\alpha t} \quad \text{and} \quad v = \frac{-1}{\alpha}e^{-\alpha t}$$

then:

$$uv = \frac{-t}{\alpha}e^{-\alpha t} \qquad \int v\,du = \int \frac{-1}{\alpha}e^{-\alpha t}dt \quad = \frac{1}{\alpha^2}e^{-\alpha t} + c$$

Substituting in Eq. 13.108 and tidying up:

$$v_C(t) = V_1(-t\alpha e^{-\alpha t} + e^{-\alpha t}) + c \tag{13.109}$$

where at time $t = \infty$ the capacitor voltages is V_1 and therefore $c = V_1$, and this is essentially the same as Eq. 13.101. A good explanation of integration by parts can be found at http://tutorial.math.lamar.edu/Classes/CalcII/IntegrationByParts.aspx.

In this case, given V_1 we only need $\alpha = 2.10^6$ to solve the equations.

Example – Critically Damped Voltage
Using schematic ('Series Resonance Critical Damping Example Voltages.asc') with the critical resistor of $100\,\Omega$, we find from Eq. 13.106 that the largest resistor voltage occurs at $0.5\,\mu s$ and from Eq. 13.100 it is $100\frac{0.510^{-6}}{25\times10^{-6}}e^{-1} = 0.7358V$.

The largest inductor voltage occurs at $1.0\,\mu s$ and is $-0.1353\ V$, and these agree with simulation.

13.3.3.3   Underdamped

We expect oscillatory solutions.

Resistor Voltage
The equation for the current is 13.86 and for the resistor we have:

$$v_R(t) = R\frac{V}{LD_r}e^{-\alpha t}\sin(D_r t) \tag{13.110}$$

we can divide this into three parts: the first is the constant group RV/LD_r; the second is the decaying exponential which reduces the amplitude of the sine wave, Fig. 13.16, and depends only on the resistance and inductance through the relationship $\alpha = R/(2L)$; and the last is the constant frequency sine wave whose amplitude is controlled by the exponential.

To a very good approximation (which we shall demonstrate with an example below), $D_r = 1/\sqrt{(LC)}$ and so the frequency of oscillation is ω_n.

It is easy to assume that the maximum will occur after a quarter cycle when $\sin(D_r t) = 1$. This is not quite true: the decaying exponential predominates in the early stages, rapidly reducing the amplitude, so that the maximum occurs slightly before

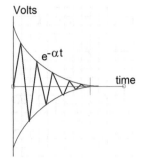

Fig. 13.16 Underdamped Series Tuned Circuit

$90°$. This reduction has steadily attenuated the maximum by the times of the following peaks and they occur more closely on the 1/4 cycle and the time between them increases slightly. Therefore we must differentiate Eq. 13.110 as we did before to get Eq. 13.85:

$$\frac{dv_R}{dt} = R\frac{V}{LD_r}e^{-\alpha t}(-\alpha \sin(D_r t) + D_r \cos(D_r t)) \qquad (13.111)$$

and this has a turning point when:

$$\alpha \sin(D_r t) = D_r \cos(D_r t) \qquad (13.112)$$

or:

$$\tan^{-1}(D_r t) = \frac{D_r}{\alpha} \qquad (13.113)$$

Inductor Voltage
Then for the inductor we have:

$$v_L(t) = \frac{V_1}{D_r}\frac{di}{dt}(e^{-\alpha t}\sin(D_r t)) \qquad (13.114)$$

Then differentiating by parts:

$$v_L(t) = \frac{-V_1}{D_r}e^{-\alpha t}(-\alpha \sin(D_r t) + D_r \cos(D_r t)) \qquad (13.115)$$

The inductor voltage, then, is just the differential of the resistor voltage – as we should expect. The V_1/D_r is simply a multiplier in exactly the same way as RV/LD_r was for the resistor. The second term is the same decay envelope as before. Then we come to the brackets. As D_r is usually an order of magnitude greater than α, the cosine term predominates, which is what we would expect for the voltage across an inductor referred to the current; and as we reduce the damping by making R smaller, D_r tends to ω_n, and the angle tends to $90°$, and α tends to zero – so in the limit we are just left with the cosine and undamped oscillations at the natural frequency of the circuit. However, in the presence of some resistance but less than that for critical damping, there is a small contribution from the sine term, typically less than 1%. This has two effects: first, it adjusts the amplitude; second, it shifts the phase. We can see this from the relationship:

$$\sin(x) + \cos(x) = \sqrt{2}\sin\left(\frac{\pi}{4} + x\right) \tag{13.116}$$

To find the turning point of the inductor voltage, we must differentiate Eq. 13.115:

$$\frac{dv_L}{dt} = \frac{-V_1}{D_r}e^{-\alpha t}\left(\alpha^2\sin(D_r t) - \alpha D_r\cos(D_r t) - \alpha D_r\cos(D_r t) - D_r^2\sin(D_r t)\right) \tag{13.117}$$

collecting terms:

$$\frac{dv_L}{dt} = \frac{-V_1}{D_r}e^{-\alpha t}\left((\alpha^2 - D_r^2)\sin(D_r t) - 2\alpha D_r\cos(D_r t)\right) \tag{13.118}$$

and the turning point is at:

$$\tan^{-1}(D_r t) = \frac{2\alpha D_r}{(\alpha^2 - D_r^2)} \tag{13.119}$$

Capacitor Voltage

For the capacitor we again must integrate by parts:

$$v_c(t) = \frac{V}{LD_r}\int e^{-\alpha t}\sin(D_r t)dt \tag{13.120}$$

This is awkward because we have to integrate by parts, so it is far easier just to subtract the resistor and inductor voltages from the supply as we have done before.

Example – Series Resonance Underdamped Voltages

If we take the schematic ('Series Resonance UnderDamped Example Voltages.asc'), we can calculate both approximate and exact values for some useful parameters. First we calculate:

$\alpha = 30/2.10^{-4} = 1.5.10^5$
$(R/2L)^2 = 2.25.10^{10}$

thence:

$$|D_r| = \sqrt{(2.25.10^{10} - 10^{12})} = 9.88686.10^5 \approx 10^6$$
$$V/(LD_r) = 10/(10^{-4}\ 9.88686.10^5) = 0.1011$$

Now we can find the natural frequency by just using the capacitor and inductor:

$$\omega_n = 1/\sqrt{(10^{-4}.10^{-8})} = 10^6$$

and hence the natural frequency is $f = 159155\ Hz$ and the period $= 6.28318\ \mu s$ and $T/4 = 1.57080$.

And we can calculate the actual frequency $\omega_0 = 9.88686.10^5\ rad/s$ and the exact frequency $= 157354\ Hz$ and the period $= 6.35509\ \mu s$ and $T/4 = 1.58877$ and we can note that there is a very small difference between the natural resonant frequency $\omega_n = 1/\sqrt{(LC)}$ and the exact frequency ω_0 using the exact value of D_r. It is about 1%. Simulation gives the same period of $6.35\ \mu s$ for the resistor current.

We now use Eq. 13.112 to find the angle for maximum voltage is $81.373°$ and therefore the time is $t_{max} = 81.373/369 \times 6.35509 = 1.43648\ \mu s$ and using the accurate value for D_r:

$$v_{R(max)} = 30 \times 1.0114 \times e^{-0.15 \times 1.43648} = 2.4185V$$

We can measure the difference between voltages on the simulation, also we can measure the current, which is 0.0806145 A, and multiply by the resistance to confirm that the voltage is 2.4184 V. If we use the approximate D_r and a quarter period, we find the answer is $30 \times e^{-0.15 \times 1.57080} = 2.37V$. Admitted, this is a difference of only 2% and is probably not significant given the tolerances of most components.

Turning to the inductor, the solution of Eq. 13.119 is $tan^{-1}\ (D_f\ t) = (3.10^5 \times -9.88686.10^5)/(2.25.10^{10} - 97.75.10^{10}) = -29.66/-95.5 = 0.3106 = 17.254°$ and is actually $180 - 17.254 = 162.746°$. However, this is very sensitive to the value of the sin term. We can illustrate this by tabulating a few results where we use the previous values of $2\alpha D_r = -29.66$ and $(\alpha^2 - D_r^2) = -95.50$ ignoring the common factor of 10^{10}:

Angle	$29.66cos(D_r\ t)$	$-95.50sin(D_r\ t)$
163.5	28.44	27.77
163.3	28.41	28.09
163.1	28.38	28.42
162.764	28.328	28.964

where we see that the cosine term changes by less than 0.6% over the range $163.5°$–$163.1°$, whereas the sine term changes by some 2.4%. So it is better to use the table from which we find the peak inductor voltage angle is $163.1°$, and this agrees with the simulation where we can also see that the simulated inductor voltage of $V(L-2) - V(L-1)$ derived from Eq. 13.115 exactly falls on top of the voltage $V(l) - V(c)$.

Explorations 7

1. Open simulation ('Series Resonance Critically Damped Example Voltages.asc') and measure the voltages. The various generators create the voltages across the components – not the currents. Thus the voltage at node $R-1$ is $i*R1$. Note the phase between the currents and voltages.
2. Open simulation ('Series Resonance UnderDamped Example Voltages.asc'), and check that the voltage generators create the correct voltages for the individual components.
3. Make $R1 = 6\Omega$ to give light damping. Measure the time between two or three peak amplitudes and hence show that the decay envelop agrees with theory.
4. Open ('Series Resonance UnderDamped.asc') and make an AC analysis to find the resonant frequency and bandwidth and hence Q.
5. Make a transient analysis and confirm that the currents in $R1$ and $R4$ are almost the same. Note that a short **Maximum Timestep** is essential and 10 ns is a good choice. Note that the AC setting of $V1$ is ignored. It will be seen that the traces do not quite agree. Change the $1e6$ in the sin to $9.9875e5$, and they agree exactly.
6. Measure the period by taking 9 or 10 cycles and confirm that it is 6.29 μs compared to 6.28 μs for the undamped oscillations. Repeat for a resistor $R1 = 50\ \Omega$ and some other values.
7. Change the damping by increasing $R1$, and note that with heavy damping, $\zeta < 2$, there is only an overshoot and no oscillations.
8. Measure the voltage (Vc) and $V(l)-V(c)$, and they are in anti-phase and their sum is $V(l)$.
9. Calculate component values for a natural resonance ten times higher and ten times lower.
10. Explore schematic ('SR Damping.asc').

13.3.3.4 Response with Precharged Capacitor

The input voltage falls to zero at the end of each pulse after the capacitor has fully charged and the current is zero. The capacitor discharges through the series resistance, and the signal source whose resistance has been subsumed in Rs. The circuit parameters are unchanged, only the initial conditions differ from before. With no prior charge on the capacitor, the situation at any time is:

$$V_1 = iR + L\frac{di}{dt} + \frac{1}{C}\int idt \qquad (13.121)$$

If we remove the input voltage when the capacitor voltage is V_1 at any time, we have $V_1 - \frac{1}{C}\int idt$, and the response is the same waveform but inverted and starts from V_1.

Explorations 8
1. Open either of the previous overdamped simulations. If the capacitor is initially charged, there will be a single overshoot before the voltage settles. Uncomment . *ic V(c)=10* to see it and add a short delay to the pulse. Do the analysis inserting the initial condition in the previous equations.
2. Repeat for critical and underdamping.

13.4   Parallel Tuned Circuit

This is similar but now at resonance, the impedance is a maximum, and it is a *rejecter circuit*. It is much more widely used that series resonance. The circuit consists of an inductor and capacitor in parallel and both in series with a resistor, Fig. 13.17, so the circuit forms a potential divider where we take the output voltage across the tuned circuit. We ignore the resistance of the inductor R_L in the following analysis which is very similar to that for the series tuned circuit (Fig. 13.18).

13.4.1   Frequency Response

As before, this assumes linear elements.

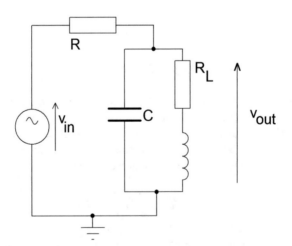

Fig. 13.17 Parallel Tuned Circuit

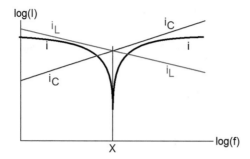

Fig. 13.18 Cusp Currents

13.4.1.1 Impedance

Considering the circuit as a whole, the impedance is that of the resistor in series with the inductor and capacitor in parallel:

$$Z = R + X \tag{13.122}$$

where X is the parallel sum of the inductor and capacitor, that is:

$$X = \frac{\frac{1}{j\omega C} \cdot j\omega L}{\frac{1}{j\omega C} + j\omega L} \tag{13.123}$$

which simplifies to:

$$X = \frac{j\omega L}{1 - \omega^2 LC} \tag{13.124}$$

Then the impedance Z of the circuit is:

$$Z = R + \frac{j\omega L}{1 - \omega^2 LC} \tag{13.125}$$

where we see that there is a constant term due to the resistor and an imaginary term which is the inductor and capacitor in parallel.

There are three interesting points. The first is that there is a frequency at which $\omega^2 LC = 1$, and then the impedance is infinite, and there is no current. The second is that at high and low frequencies, the impedance of either the inductor or the capacitor becomes large, and the circuit is effectively the resistor in series with the capacitor or in series with the inductor. Therefore we expect a 20 dB/decade fall in voltage, and if the impedance of the resistor is much greater than the inductor or capacitor, the current will be virtually constant. The third is that if we focus on the imaginary term, we are able to explain the cusp. In Fig. 13.18, we look just at the region close to the cusp where the inductor and capacitor currents are changing and find the inductor current falling linearly with frequency on logarithmic scales as we should expect. Likewise the capacitor current increases linearly. And their contribution to the total current is their difference. Thus at low frequencies, the inductor current predominates, whilst at high frequencies, the capacitor current is larger. However, at frequency X they cancel. On linear axes, it can be seen that these currents themselves show resonance with a large series resistor, and in any case, the currents are changing rapidly in the region near the cusp so even a slight departure from X means that they no longer cancel.

13.4.1.2   Resonant Frequency

We see from Eq. 13.125 that if $\omega^2 LC = 1$, then the impedance of the tuned circuit is R. Calling this frequency ω_n, the natural frequency of the LC combination, we have:

$$\omega_n = \frac{1}{\sqrt{LC}} \tag{13.126}$$

or in terms of frequency:

$$f_n = \frac{1}{2\pi\sqrt{LC}} \tag{13.127}$$

The circuit is a complex potential divider where the output is:

$$v_{out} = v_{in}\frac{X}{R + X} \tag{13.128}$$

Substituting from Eq. 13.128:

$$v_{out} = v_{in}\left[\frac{\frac{j\omega L}{1-\omega^2 LC}}{R + \frac{j\omega L}{1-\omega^2 LC}}\right] \tag{13.129}$$

which simplifies to:

$$v_{out} = v_{in}\frac{j\omega L}{R(1 - \omega^2 LC) + j\omega L} \tag{13.130}$$

and at resonance:

$$1 = \omega^2 LC \quad \text{or} \quad \omega = \frac{1}{\sqrt{LC}} \tag{13.131}$$

and

$$v_{out} = v_{in} \tag{13.132}$$

and also at this frequency the phase difference between input and output is zero.

Example – Parallel Resonance Frequency

The schematic ('Parallel Resonance.asc') has a very small resistance in series with the voltage source. Given that $L1 = 500\ \mu H$, $C1 = 500\ pF$ we expect a resonant frequency of $f_r = 318\ kHz$ which is independent of the value of the resistor. An AC analysis shows scarcely any change in output voltage. But the resistor current is attenuated by more than -153 dB at the calculated resonant frequency. The sharpness of the notch depends somewhat on the time step of the analysis.

Explorations 9

1. Open ('Parallel Resonance Ve.asc') and measure the output voltage for various values of R using .step **dec param R 1 100k 1** to step the resistor from $1\ \Omega$ to $100\ k\Omega$ in decades with one point per decade. The resonance only becomes sharp for large values, and the output is 0 dB, that is, it equals the input.
2. Open ('Parallel Resonance.asc') and probe the capacitor and inductor currents and the current in $R1$. The picture is clearer using a linear y-axis. Confirm that the currents agree with theory.
3. The previous two explorations indicate that if we use the circuit with a low source resistance, the current in $R1$ shows a resonance peak but the voltage v_{out} is more like a band-pass filter.
4. Change the AC range to *.ac dec 1e8 1k 1e8* and probe $I(R1)\ I(L1)\ IC1)$, and note the currents above and below resonance. Make $R1 - 10\ \Omega$ and repeat.

13.4.1.3 3 dB Points

We can again take the 3 dB points as the measure of the bandwidth. Starting with Eq. 13.130 after dividing and moving the j to the denominator of the fraction we have:

$$\frac{v_{out}}{v_{in}} = \frac{1}{1 - \frac{jR}{\omega L}\left(1 - \omega^2 LC\right)} \tag{13.133}$$

At the 3 dB points, the modulus of the imaginary part equals the real, just as for the series case, and in this instance the imaginary term equals one:

$$\frac{R}{\omega L}\left(1 - \omega^2 LC\right) = 1 \tag{13.134}$$

Multiply by ω/RC:

$$\frac{1}{LC} + \omega^2 - \frac{\omega}{RC} = 0 \tag{13.135}$$

and after a little manipulation and substituting for $1/LC$:

$$\omega^2 + \frac{\omega}{RC} - \omega_n^2 = 0 \tag{13.136}$$

We factorize Eq. 13.136 just as we did for series resonance:

$$(\omega_n - \omega_l)(\omega_n + \omega_h) \tag{13.137}$$

which we expand as:

$$\omega_n^2 - \omega_n(\omega_h - \omega_l) - \omega_l\omega_h = 0 \tag{13.138}$$

where ω_h and ω_l are the higher and lower 3 dB frequencies then by comparison with Eq. 13.136:

$$\omega_h - \omega_l = \frac{1}{RC} \tag{13.139}$$

and:

$$\omega_h\omega_l = \omega_n^2 \tag{13.140}$$

or:

$$\omega_n = \sqrt{\omega_h\omega_l} \tag{13.141}$$

which are the same as series resonance.

13.4.1.4   Q-Factor

If we define it in terms of voltage, this follows directly remembering that the term in ω in Eq. 13.136 can be written as $\frac{\omega_0}{Q}$ and then:

$$\frac{1}{RC} = \frac{1}{\sqrt{LC}Q} \quad so \quad Q = R\sqrt{\frac{C}{L}} \quad also \quad Q = \frac{\omega_n}{\omega_h - \omega_l} \quad or \quad Q$$
$$= \frac{\sqrt{\omega_h\omega_L}}{\omega_h - \omega_l} \tag{13.142}$$

showing that for a high Q-factor we need a large series resistor (which will also severely attenuate the signal if the load across the tuned circuit is not also very large) or – better – a large capacitor and a small inductor. This is the inverse of the series case. A good alternative is to use the current instead where we have seen that this is a very sharp notch falling to zero at resonance.

Using '.meas' and 'WHEN' to Find the Bandwidth and Q-Factor

Remembering first to turn off compression, we then use the directive:

$$.meas\ AC\ resV\ MAX\ V(out)\ FROM\ 0\ TO\ 1000k$$

to find the voltage at the resonant frequency as *resV*. then:

$$.meas\ AC\ resVfr\ WHEN\ V(out) = resV$$

and here we should note that there is no **Measured Quantity** which has the effect of returning the time that the condition *V(out)=resV* is met as the value *resVfr* – the frequency. This is worth noting as a way to find the abscissa for a measurement. Now we can measure the bandwidth using the interval from the first rise of *resV/2* to its first fall.

And note that we can use either the magnitude or real part of the voltage and divide by √2, or remembering that the real and imaginary parts are equal at the 3dB points, we can use the whole value and divide by 2. Note also that we must supply a number for the rise and fall; if it is left empty the directive will fail.

$$.meas\ AC\ BW3dB\ TRIG\ V(out) = resV/2\ RISE = 1\ TARG\ V(out)$$
$$= resV/2\ FALL = 1$$

We should also note that the LTspice 'Help' manual says that, due to rounding errors, exact *WHEN* statements can fail in which case we can subtract a small value as *resV/2-0.0001*. Finally we find the Q-factor as:

$$.meas\ AC\ Qfactor\ PARAM\ resVfr/BW3dB$$

where we use a **PARAM** to calculate the Q-factor which is reported in dBs.

Example – Parallel Resonance Q-Factor

Still using schematic ('Parallel Resonance.asc'), then Eq. 13.142 is $Q = R/10^3$. For $Q = 10$ we need $R1 = 10\ k\Omega$. If we measure the current, we find that off-resonance it is *80 dB* because of the attenuation produced by the resistor, but the notch has the same depth. The *.meas* statement has been used to find the resonant frequency and the 3 dB points and thus the Q-factor of 20.0001 dB.

13.4.1.5 Off-Resonance Response

This has been covered in the discussion in section 'Off-Resonance Response' for the series circuit.

Explorations 10

1. Open ('Parallel Resonance.asc') and measure the resonant frequency, Q-factor and impedance. Note the effect of changing *R1* and that at small values, just few a

ohms, in terms of the voltages, the circuit is almost a band pass filter albeit with a very broad pass band.

2. Probe the current in *R1* and note that there is a very sharp resonance with a resistor up to at least 100 Ω. Open ('Parallel Resonance Damped BW.asc') and measure the bandwidth in current.

3. One of the nice things, we can do in SPICE is to build impractical circuits. Insert $C = L = 1$ and $R = 100$. The circuit should have $Q = 10$. Try swapping to $C = 2$, $L = 0.5$ and see if $Q = 200$.

13.5   Parallel Resonance Damping

Practically, this can occur in two ways. One is by a resistor in parallel with the tuned circuit; the other is the resistance of the inductor. Of course we could have a resistor in series with the capacitor – or all three resistors – but these two will suffice. For this analysis, we turn the circuit into its Norton form, Fig. 13.19.

13.5.1   Including a Parallel Resistor

A resistor may be placed in parallel to increase the bandwidth and reduce the Q-factor and thereby increase the damping and reduce or eliminate any overshoot or ringing. And here we must be clear about where we measure the overshoot. The input current using a current source to excite the circuit will be constant. Generally we should expect some damped oscillations, but as the inductor current cannot change quickly, under certain circumstances, its current may not overshoot, whereas the resistor and capacitor currents do. Therefore we shall use the inductor current as a measure of the damping.

We are going to analyse the situation with no initial charge on the capacitor. An alternative situation occurs when we remove the input leaving the capacitor charged

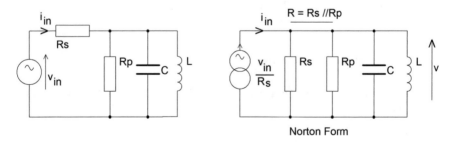

Norton Form

Fig. 13.19 Loaded Parallel Resonance

and a current through the inductor. A very clear explanation can be found at https://www.ius.edu.ba/sites/default/files/u772/ch8_rlc.pdf, and identical circuits and derivations can be found at http://www.ee.nthu.edu.tw/~sdyang/Courses/Circuits/Ch08_Std.pdf.

Example – Parallel Resonance Damping
The schematic ('Parallel Resonance Damped.asc') illustrates the approach we shall use in the following sections. First we see that the output voltage is a pulse, and here we can use .*meas* directive to get accurate figures. These are stored in a text file with the same name as the schematic and are:

maxop: MAX(v(out))=1.57481 FROM 0 TO 0.002
maxopt: v(out)=maxop AT 9.32422e-005r
minop: MIN(v(out))=0.000161313 FROM 0.001 TO 0.002

In the following sections, we shall correlate these measurements with calculations. Here we can just note the trailing *r* of the *maxopt* which means that this is the real part of a complex variable. We can also note that although the current pulse ended at *1 ms* the current has not yet decayed completely at *2 ms*: this is difficult to see on the trace panel unless we zoom in.

   Also of note is splitting the inductor current into parts as $I(L1) = 0.01-I(R3)-I(R4)$. In the following sections, we shall see that very often the current or voltage associated with a component consists of two components – usually exponentials – and a good test of the validity of the analysis is if this trace exactly fits that of the component itself, in this case the inductor *L1*.

13.5.1.1 Damping Characteristic Equation

If we try to visualize the situation when we apply a step input I_{in} at time $t = 0$ with no initial charge on the capacitor or current in the circuit, we find that at that very first instance, the capacitor is effectively a short-circuit, and the voltage across the circuit is zero, and so all the current will flow into the capacitor. Then finally, when the voltage is constant, the inductor will be a short-circuit, and the voltage will have fallen to zero again with all the constant input current passing through the inductor. So we can predict that the voltage will rise to some value and then decreases to zero, and the current will divide between all three components in a time-varying manner, and we need a solution for the current in each component at any time. Taking the capacitor first, at any time *t*, we have:

$$v_C = \frac{1}{C} \int \frac{di}{dt} \quad thus \quad i_C = C \frac{dv}{dt} \tag{13.143}$$

and for the inductor:

$$v_L = L\frac{di}{dt} \quad thus \quad i_L = \frac{1}{L}\int \frac{dv}{dt}dt \tag{13.144}$$

Then using KCL on the circuit we have:

$$I_{in} = C\frac{dv}{dt} + \frac{1}{L}\int_0^t \frac{dv}{dt}dt + \frac{v}{R} \tag{13.145}$$

and now we have the current in terms of the voltage whereas we want the individual currents. A solution used by the above websites is to take the inductor current as the reference then uses the voltage across the inductor (which, of course, is the voltage across all the components) to find the currents in the other two. Thus, equating the voltages in Eqs. 13.143 and 13.144, we have for the capacitor:

$$L\frac{di_L}{dt} = \frac{1}{C}\int \frac{di_C}{dt} \tag{13.146}$$

and if we differentiate Eq. 13.146:

$$L\frac{d^2i_L}{dt^2} = \frac{1}{C}i_C \quad and\ so \quad i_C = LC\frac{d^2i_L}{dt^2} \tag{13.147}$$

and similarly:

$$L\frac{di_L}{dt} = i_R R \quad and\ so \quad i_R = \frac{L}{R}\frac{di_L}{dt} \tag{13.148}$$

Thus Eq. 13.146 becomes:

$$I_{in} = LC\frac{d^2i_L}{dt^2} + i_L + \frac{L}{R}\frac{di_L}{dt} \tag{13.149}$$

We might well ask why the inductor current was chosen and not the resistor current. The answer is found by repeating Eqs. 13.147 and 13.148 using the other two in turn, and it will be found that they involve an integration. For example, starting with the resistor, we have:

$$i_R R = \frac{1}{C}\int \frac{di_C}{dt} \quad and\ so \quad i_c = CR\int i_R dt \tag{13.150}$$

This is not to say that it cannot be done – just that it is more convenient to have the equations in the form of 13.149, because by dividing by LC and rearranging, we have a homogeneous equation:

$$\frac{I_{in}}{LC} = \frac{d^2 i_L}{dt^2} + \frac{1}{RC}\frac{di_L}{dt} + \frac{1}{LC} \tag{13.151}$$

Then the characteristic equation is:

$$s^2 + \frac{s}{RC} + \frac{1}{LC} = 0 \tag{13.152}$$

13.5.1.2 Decay Factor and Critical Damping

Equation 13.152 has roots:

$$r_1, r_2 = \frac{-1}{2RC} \pm \sqrt{\left(\frac{1}{2RC}\right)^2 - \frac{1}{LC}} \;\; = -\alpha \pm \sqrt{\alpha^2 - \omega_n^2} \tag{13.153}$$

and therefore:

$$\alpha = \frac{1}{2RC} \quad \omega_n = \frac{1}{\sqrt{LC}} \tag{13.154}$$

Returning to Eq. 13.56 (the expression for critical damping) and Eq. 13.57, in this case we have:

$$\zeta = \frac{\alpha}{\omega_n} \;\; = \frac{1}{2R_p C}\sqrt{LC} = \frac{1}{2R_p}\sqrt{\frac{L}{C}} \;\; = \frac{1}{2Q} \tag{13.155}$$

So there is no change in the natural resonant frequency nor the Q-factor compared to the undamped case, as we expected. However, α is not the same as series resonance.

13.5.1.3 Overdamped Currents ($\alpha > \omega_n$)

We again look for a solution for the current in the inductor of the form:

$$i_L(t) = A_1 e^{-r_1 t} + A_2 e^{-r_2 t} \tag{13.156}$$

and as before, we need two conditions in order to find the constants. The situation now is a little different.

At time infinity, we know that $i_L = I_{in}$. So we need to modify Eq. 13.156:

$$i_L(t) = I_{in} + A_1 e^{-r_1 t} + A_2 e^{-r_2 t} \tag{13.157}$$

then at time $t = 0$:

$$I_{in} + A_1 + A_2 = 0 \tag{13.158}$$

after differentiating Eq. 13.157 and setting $t=0$ at which time $\frac{di_L}{dt} = 0$:

$$r_1 A_1 + r_2 A_2 = 0 \tag{13.159}$$

then from Eq. 13.159:

$$A_2 = -A_1 \frac{r_1}{r_2} \tag{13.160}$$

and from Eq. 13.158:

$$A_1 = -I_{in} \frac{r_2}{r_1 - r_2} \quad A_2 = I_{in} \frac{r_1}{r_1 - r_2} \tag{13.161}$$

and finally:

$$i_L(t) = -I_{in} \frac{r_2}{r_1 - r_2} e^{-r_1 t} + I_{in} \frac{r_1}{r_1 - r_2} e^{-r_2 t} \tag{13.162}$$

The voltage across all the components is the same at any time:

$$v_L = v_R = v_C \quad \text{therefore} \quad L\frac{di_L}{dt} = i_R R = \frac{1}{C} \int i_C dt \tag{13.163}$$

So for the resistor, we first differentiate Eq. 13.162 and divide by the resistance:

$$i_R(t) = I_{in} \frac{L}{R} \frac{r_1 r_2}{r_1 - r_2} (e^{-r_1 t} - e^{-r_2 t}) \tag{13.164}$$

And for the capacitor, the easiest is to subtract the previous two currents from I_{in}. Otherwise we have:

$$i_R R = \frac{1}{C} \int i_C dt \quad \text{then} \quad i_C = RC \frac{di_R}{dt} \tag{13.165}$$

$$i_C = I_{in} LC \frac{r_1 r_2}{r_1 - r_2} (-r_1 e^{-r_1 t} + r_2 e^{-r_2 t}) \tag{13.166}$$

Example – Parallel Resonance Overdamped
We now have sufficient information to find the two constants using schematic ('Parallel Res OverD.asc'). The component values are $C = 0.2~\mu F$, $L = 50~mH$, $R = 200~\Omega$ $I_{in} = 10~mA$. We find:

$$\alpha = 2.5.10^3$$

$$\omega_n = 1.10^4$$

$$r_1 = 2.10^4 ~ r_2 = 0.5.10^4$$

$$r_2/(r_1 - r_2) = 0.3333$$

$$r_1/(r_1 - r_2) = 1.3333$$

$$r_1 r_2/(r_1 - r_2) = 0.666$$

$$A_1 = -0.0033A ~ A_2 = 0.0133A$$

$$RI_{in}/(r_1 - r_2) = (200 * 0.01)/1.5e4 = 1.333.10^{-4}$$

We find the components of the currents as:

$$iR = I(R1) - I(R5)$$

$$iL = 0.01 - I(R3) - I(R4)$$

$$iC = I(R6) - I(R7)$$

where:

$I(R1) =$ second term in Eq. 13.164
$I(R5) =$ first term in Eq. 13.164
$I(R3) =$ first term in Eq. 13.162
$I(R4) =$ second term in Eq. 13.162
$I(R6) =$ first term in Eq. 13.166
$I(R7) =$ second term in Eq. 13.166

   These are the values used in the schematic. Note that the capacitor current has a slight undershoot of *1 mA* which can be calculated in the previous way as can the peak of the resistor current by finding the point when $\frac{di}{dt} = 0$. The traces of the sums of the partial currents fall exactly over the component current traces.

13.5.1.4   Critically Damped ($\alpha = \omega_0$)

The schematic is ('Parallel Res Example Crit D.asc'). The equation now is of the form of 13.157 with the initial I_{in} because at time $t = \infty$ $i_L = I_{in}$. And with two identical roots equal to the natural resonance, just as for the series resonance case, we also need a second term as in Eq. 13.68:

$$i_L(t) = I_{\text{in}} + A_0 e^{-\alpha t} + A_1 t e^{-\alpha t} \qquad (13.167)$$

and at time $t = 0$ $i_L = 0$ we again have $I_{\text{in}} = -A_0$:

$$i_L(t) = I_{\text{in}} - I_{\text{in}} e^{-\alpha t} + A_1 t e^{-\alpha t} \qquad (13.168)$$

differentiating again and at time $t = 0$ then $t A_1 e^{-\alpha t} = 0$ and $\frac{di_l}{dt} = 0$:

$$0 = \alpha I_{\text{in}} e^{-\alpha t} + A_1 e^{-\alpha t} \qquad (13.169)$$

whence:

$$A_1 = -\alpha I_{\text{in}}$$

So finally:

$$i_L(t) = I_{\text{in}} - \alpha I_{\text{in}} t e^{-\alpha t} - I_{\text{in}} e^{-\alpha t} \qquad (13.170)$$

Equation 13.163 shows that we must differentiate Eq. 13.170 for the resistor current, leaving only:

$$i_R(t) = \frac{L}{R} I_{\text{in}} \alpha^2 t e^{-\alpha t} \qquad (13.171)$$

because the two terms in $\alpha I_{\text{in}} e^{-\alpha t}$ cancel.

Finally, for the capacitor, we can again subtract the previous two currents from the input; else using Eq. 13.163 and differentiating, we have:

$$i_C(t) = LC I_{\text{in}} \alpha^2 e^{-\alpha t} - LC I_{\text{in}} \alpha^3 t e^{-\alpha t} \qquad (13.172)$$

Critical damping occurs when $\zeta = 1$ and using Eq. 13.155 this:

$$1 = \frac{1}{2R} \sqrt{\frac{L}{C}} \quad \text{or} \quad R = \frac{1}{2} \sqrt{\frac{L}{C}} \qquad (13.173)$$

Example – Parallel Resonance Critically Damped

The components for ('Parallel Res Example Crit D.asc') are $C1 = 0.2\ \mu F$ $L1 = 50\ mH$ $I_{in} = 10\ mA$. And we calculate the resistor from Eq. 13.173 to be $R = 250\ \Omega$. To test this in the schematic, we step the resistor current, and this shows both under- and overdamped conditions. It is difficult to see exactly which trace does not overshoot. One possibility is to measure the maximum current – if it is greater than the steady state, then clearly there has been an overshoot. This we do. As any overshoot will occur before 1 ms, then the directive:

.meas TRAN maxI MAX I(L1) FROM 0 TO 1m

will find the maximum in that interval. The simulation shows that at *step 3 (R = 250 Ω)*, the inductor current is *0.00999497 A*, and the previous step is smaller (underdamped), and the following is greater (an over-shoot). This is not conclusive but it does indicate that 250 Ω is close to the mark. We could repeat with smaller resistance steps.

We convert the *.step* to a **comment,** then setting $R = 250\ Ω$, we calculate the parameters:

$$\alpha = 1.10^4\ \alpha^2 = 1.10^8\ \alpha^3 = 10^{12}$$

$$iR = I(R5)$$

$$iL = 0.01 - I(R3) - I(R4)$$

$$iC = I(R6) - I(R7)$$

where:

$I(R3) =$ Eq. 13.170 first exponential
$I(R4) =$ Eq. 13.170 second exponential
$I(R5) =$ Eq. 13.171
$I(R6) =$ Eq. 13.172 first term
$I(R7) =$ Eq. 13.172 second term

The analysis follows the same general methodology as series resonance, and the schematic has the factors of the equations set up as current generators where it will be seen that the components fall exactly over the currents of the circuit itself.

13.5.1.5   Underdamped

This is not quite the same as Sect. 13.3.2.3 because now the final current is I_{in}, not zero, so we cannot discard the *sin* term. But from the Euler relationship, we still have:

$$i_L(t) = I_{in} + A_0 e^{-at} \cos\left(D_r t\right) + A_1 e^{-at} \sin\left(D_r t\right) \tag{13.174}$$

And at time $t = 0$ the current is zero so $A_0 = -I_{in}$:

$$i_L(t) = I_{in} - I_{in} e^{-at} \cos\left(D_r t\right) + A_1 e^{-at} \sin\left(D_r t\right) \tag{13.175}$$

We now differentiate Eq. 13.175:

$$\frac{di_L}{dt} = I_{in}e^{-at}D_r \sin(D_r t) + I_{in}\alpha e^{-at} \cos(D_r t) - \alpha A_1 e^{-at} \sin(D_r t)$$
$$+ A_1 D_r e^{-at} \cos(D_r t) \tag{13.176}$$

and at time $t = 0$ this is zero so:

$$A_1 = -I_{in}\frac{\alpha}{D_r}$$

and finally:

$$i_L(t) = I_{in} - I_{in}e^{-at} \cos(D_r t) - I_{in}e^{-at}\frac{\alpha}{D_r} \sin(D_r t) \tag{13.177}$$

For the resistor using Eqs. 13.163 and 13.176, the current is:

$$i_R(t) = \frac{L}{R}I_{in}e^{-at}\left(D_r \sin(D_r t) + \alpha \cos(D_r t) + \frac{\alpha^2}{D_r} \sin(D_r t) - \alpha \cos(D_r t)\right) \tag{13.178}$$

and finally:

$$i_R(t) = \frac{L}{R}I_{in}e^{-at}\left(\left(D_r + \frac{\alpha^2}{D_r}\right) \sin(D_r t)\right) \tag{13.179}$$

and once again we can find the capacitor current by subtraction.

Example – Parallel Resonance Underdamped
For the schematic ('Parallel Res Example UnderD.asc') with the values $C = 0.5\ \mu F$, $L = 20\ mH$, $R = 500\ \Omega$ the parameters are:

$\alpha = 1/(2RL) = 2.000.10^3$

$\omega_n = 1/\sqrt{(LC)} = 1/\sqrt{(20.10^{-3} \times 0.5.10^{-9})} = 10^4$

$D_r = \sqrt{(\omega_n{}^2 - \alpha^2)} = 9.798.10^3$

$\alpha/D_r = 0.2041$

$(D_r + \alpha^2/D_r) = 9.798.10^3 + 4.10^6/9.789.10^3 = 4.08.10^3$

$D_r{}^2 + \alpha^2 = 9.600.10^6 + 4.000.10^6 = 1.000.10^7$

The resistor current in Eq. 13.179 is realized by the current generator B3. From Eq. 13.177 the inductor current is given by $0.01 - B1\text{-}B5$ where B1 is the first exponential term in Eq. 13.177 and B5 is the second. The capacitor current is $I(B1)\text{-}I(B5)$.

Explorations 11

1. Open schematic ('Parallel Res Example OverD.asc') and measure the resonant frequency and 3 dB points with different degrees of overdamping by stepping *R2*.
2. Try alternative inductor and capacitor values also changing the ratio of capacitance/inductance. Calculate new damping resistor values and test. Recalculate the or generators *B1…B6*.
3. Open and run ('Parallel Res Example CritD.asc') and confirm that the appropriate current generators add to the currents in the components.
4. Open and run ('Parallel Res Example UnderD.asc'). Change the inductor and capacitor values and recalculate a few resistor values for different degrees of underdamping. Measure the decay constant by fitting and exponential to the envelop.

13.5.2   Including the Resistance of the Inductor

This is a practical situation where the resistance is usually the resistance of the wire used to make the inductor, Fig. 13.20, but could include a physical resistor as well. If you really want resistors in series with both the inductor and the capacitor, http://hyperphysics.phy-astr.gsu.edu/hbase/electric/parres.html quotes the result which is partly worked through at http://hyperphysics.phy-astr.gsu.edu/hbase/electric/rlcpar.html#c1.

We may excite the circuit either with a voltage source or a current source. Taking a voltage source, the response is shown in Fig. 13.21. Below resonance the current is limited by the series combination of the resistor and inductor. If the resistance is small the current is essentially that of the rising impedance of the inductor and shows the typical −20 dB/decade slope. If the resistance is large, the curve flattens and the current is determined by the resistance. At frequencies above resonance the graphs merge into one with a slope of 20 dB/decade due to the falling impedance of the capacitor.

With a current excitation, we find a similar shaped graphs only there is an increased voltage at resonance.

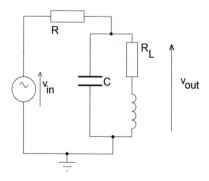

Fig. 13.20 Parallel Resonance with Inductor Resistance

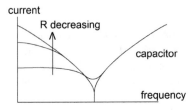

Fig. 13.21 Parallel Resonance Current

13.5.2.1   Self-Resonance of an Inductor

Practical inductors have resistance, of course, but they also have capacitance, usually very small, just a few picofarads, so this analysis applies to them. In general the self-resonance frequency is tens of megahertz or higher and is the highest frequency at which it is still inductive. Graphs of some commercial inductors can be found at https://www.sagami-elec.co.jp/file/tech/coil_doc_06e.pdf.

13.5.2.2   Resonant Frequency

We can define this is two ways, there being a difference of a few percent between them. One is to take the frequency at which the impedance is real so there is no phase difference. This is what we shall do now. The other is to take the maximum impedance, which is what LTspice does, and is the more productive approach.

Impedance Is Real
The impedance of the tuned circuit is:

$$Z = \frac{\frac{-j}{\omega C}(j\omega L + R_L)}{j\left(\omega L - \frac{1}{\omega C}\right) + R_L} = \frac{\frac{1}{C}\left(L - j\frac{R_L}{\omega}\right)}{j\left(\omega L - \frac{1}{\omega C}\right) + R_L} \qquad (13.180)$$

We convert the numerator to a real number by multiplying throughout by its complex conjugate $L + j\frac{R_L}{\omega}$

$$Z = \frac{\frac{1}{C}\left(L^2 + \left(\frac{R_L}{\omega}\right)^2\right)}{\left(L + j\frac{R_L}{\omega}\right)\left(j\left(\frac{-1}{\omega C} + \omega L + R_L\right)\right)} \qquad (13.181)$$

then after multiplying out and collecting terms we arrive at:

$$Z = \frac{\frac{1}{C}\left(L^2 + \left(\frac{R_L}{\omega}\right)^2\right)}{\frac{R_L}{\omega^2 C} + \omega L^2 + L R_L + j\left(\frac{-L}{\omega C} + \omega L^2 + \frac{R_L^2}{\omega}\right)} \qquad (13.182)$$

and since the terms in the numerator are all real and positive, we turn to the imaginary terms in the denominator where their sum can be zero: this is another definition of resonance, that the impedance is real:

$$\frac{-L}{C} + \omega^2 L^2 + R_L^2 = 0 \qquad (13.183)$$

This yields:

$$\omega_0^2 = \frac{L - CR_L^2}{L^2 C} \qquad (13.184)$$

which can be written as:

$$\omega_0 = \sqrt{\frac{1}{LC} - \left(\frac{R_L}{L}\right)^2} = \sqrt{\omega_n^2 - \left(\frac{R_L}{L}\right)^2} \qquad (13.185)$$

making plain that the resonant frequency decreases with increasing resistance, unlike the series tuned circuit. And the change in frequency is generally just 2 or 3%. If $R_L = 0$ this reduces to $\omega_n^2 = \frac{1}{LC}$ the natural resonance.

Example – Parallel Resonance with Inductor Resistance

If we take schematic ('Parallel Res with R inductor.asc'), the natural frequency is $\omega_n = \frac{1}{\sqrt{500pF \times 500\mu H}}$ $\omega_n = 2.10^6$ $f_r = 318310$ and this is exactly what we find if we simulate – provide we set *.ac dec 1e6 200k 400k*. With only 1000 steps per decade, we find a frequency of *318295 Hz*. However, LTspice will usually greatly reduce the steps to less than 1e5. We also find the resonant frequency decreases to *312005 Hz* with $R = 200\ \Omega$ which is a change of *1.972%*. Inserting the values in Eq. 13.184 yields; $\omega^2 = \frac{500.10^{-6} - 500.10^{-12} \times 4.10^4}{2.5.10^{-7} 500.10^{-12}}$ $\omega = 1959592\ f_r = 311879$.This is an error of *126 Hz* or *0.04%*. Reducing the AC range and adding *.OPTIONS numdgt=10* only reduces the simulation result to *312004 Hz*. But as this final trace of *V(l)* is so flat that there are no 3 dB points, we can be satisfied with this result. With *100 Ω* the resonance is at *316714 Hz* compared to the simulated *316722 Hz*.

The resonance is clearer if we change the resistance to *.step param R 1p 20 5* and reduce the AC range to *.ac dec 1e6 310k 330k*. We learn from this is that there is virtually no significant resonance if the resistor is more than *10 Ω*, that is, about 1% of the impedance of the inductor, which is not unreasonable if the resistance is that of the wire. This is useful to know.

We also find the resonance is less heavily reduced with a voltage source instead of current, and there is still an appreciable resonance curve at *200 Ω*. We should note that the 3 dB points are not symmetrical about the resonance frequency.

Impedance Is Maximum

Because the resonant frequency only changes by 2% over the range of load resistors, we substitute ω_n in Eq. 13.180 for $\omega L - \frac{1}{\omega C}$, and this is zero. Then we need the modulus of the numerator and we have:

$$Z_{max} = \frac{1}{R_L} \sqrt{\left(\frac{L}{C}\right)^2 + \left(\frac{R_L}{\omega_n C}\right)^2} \quad Z_{max} = \frac{1}{R_L} \sqrt{\left(\frac{L}{C}\right)^2 + R_L^2 \frac{L}{C}} \qquad (13.186)$$

and in many cases, the second term contributes less than 1% as we shall see in the following example so a good approximation is:

$$Z_{max} = \frac{L}{R_L C} \qquad (13.187)$$

Example – Maximum Impedance

Using schematic ('Parallel Res with R Inductor V Source.asc') and substituting in Eq. 13.186 using the natural resonance frequency and $R_2 = 200\ \Omega$, $Z_{max} = \frac{1}{200} \times \sqrt{10^{12} + (200)^2 \times 10^6} = 74.1497 dB$ which agrees with the measured value of 73.995 dB. The phase angle is $\tan^{-1} \frac{R_L}{\omega L} = \tan_1\left(\frac{200}{10^3}\right) = 11.31°$ compared to 11.09°. For a resistor of 100 Ω, we find 80.0432 dB which agrees with the simulation, but the angle is 5.711° compared to 5.683°. The errors are due to taking the natural frequency.

If we use Eq. 13.187 with all the stepped load resistors, the maximum impedance agrees with the simulation to three significant figures – that is, better than 0.1%.

13.5.2.3 Q-Factor and 3 dB Points

The 3 dB points are also the half-power points. This, however, is difficult with a voltage source; even with a small inductor series resistance, the product of the current and the voltage across the tuned circuit is shaped like a letter 'M' with a strong dip at resonance. This dip reduces and finally vanishes as the resistance of the inductor increases. This is seen clearly in the schematic ('Parallel Res with R Inductor V Source.asc'). On the other hand, the voltage across the tuned circuit and its impedance are well-behaved and show nice resonance curves, so perhaps we could look instead for the fall in voltage.

We can write the impedance of the tuned circuit as:

$$Z = \frac{\frac{1}{j\omega C}(j\omega L + R_L)}{\frac{1}{j\omega C} + j\omega L + R_L} = \frac{j\omega L + R_L}{1 - \omega^2 CL + j\omega CR_L} \qquad (13.188)$$

Equation 13.188 can be solved at any frequency to find the impedance of the tuned circuit. If the resistance of the inductor is small, it can be approximated by:

$$Z \approx \frac{j\omega L}{1 - \omega C(\omega L + R_L)} \approx \frac{j\omega L}{1 - \omega^2 CL} \qquad (13.189)$$

Then the voltage across the inductor is:

$$v_L = \frac{Z}{Z + R_S} = \frac{\frac{j\omega L + R_L}{1 - \omega^2 CL + j\omega CR_L}}{\frac{j\omega L + R_L}{1 - \omega^2 CL + j\omega CR_L} + R_S} \qquad (13.190)$$

This expands to:

$$v_L = \frac{j\omega L + R_L}{j\omega(L + CR_L) + R_L + R_S - \omega^2 CL} \qquad (13.191)$$

So having once found the voltage at resonance, we can try to solve this. But, frankly, it is far easier just to simulate it.

13.5.2.4   Damping

The schematic ('Parallel Res with R Inductor.asc') shows that even a small resistance effectively destroys the resonance peak with a current source, and there is considerable reduction of the peak with a voltage source up to *200 Ω*. On the other hand, we need a series resistance of at least *100 Ω* to reduce the ringing. The mathematics for this get very messy so it is probably better just to simulate it.

Explorations 12
1. Insert a resistor of something around 10 Ω in series with an inductor to model its resistance. How does that affect the resonant frequency and the Q-factor?
2. Open ('Parallel Resonance with R Inductor.asc') and explore different values for the resistance of the inductor. Explore also different ratios for the capacitor and inductor keeping the same resonant frequency, e.g *C = 50 pF, L = 5 mH*.

13.6 Summary

Resonant circuits, both series and parallel are important. If we ignore parasitic components, the frequency analysis is straightforward, but it takes more work to find the time response. The key points are:

- A series resonant circuit is an acceptor circuit where the current is maximum at resonance when $\omega = \frac{1}{\sqrt{LC}}$ and the voltage across the circuit is a minimum, but the current is a maximum, and the voltage across the inductor or the capacitor is maximum.
- We can characterize the response by the 3 dB points where the current or the voltage across the inductor or capacitor has fallen by $1/\sqrt{2}$.
- We can define a Q-factor to describe the sharpness of the resonance as $Q = \omega_0 \frac{L}{R}$.
- We define a damping ratio $\zeta = \frac{actual\ damping}{critical\ damping}$ and a decay factor $\alpha = \frac{R}{2L}$ to describe the time response with the discriminant D as:

  $\zeta = 0, \alpha = 0$ undamped where there is no resistance and the frequency is ω_n

  $\zeta > 0, D > 0, \alpha > \omega_n$ overdamped where there are two real roots

  $\zeta = 1, D = 0, \alpha = \omega_n$ critically damped where there are two coincidental real roots

  $0 < \zeta < 1, D < 1, \alpha < \omega_n$ underdamped where there are two complex conjugate roots

- Similar equations apply to a parallel resonant circuit only now it is a rejector circuit where the current is a minimum at resonance and the voltage is a maximum.
- Including the resistance of the inductor and a possible resistor in parallel with the tuned circuit increases the complexity of the analysis, and it can be quicker just to simulate it.

Chapter 14
The Fourier Series and Fourier Transform

14.1 Introduction

We encountered the Fourier series in passing in Chap. 5. Then it was just to illustrate the importance of sine waves as a fundamental waveform from which more complex ones such as a square wave could be constructed by adding them with different frequencies and amplitudes. Now we shall develop the formalism that, starting with an analytic function, we can replicate it by a series of waveforms – the Fourier series. We can well ask – why bother when we can just use an FFT? A good question! Is there a good answer?

We then move on to the more useful Fourier transform which acts in a similar fashion where we have a waveform but no equation to describe it. Indeed, the waveform may be one, such as a sound recording, that is not amenable to an analytic description. The problem, though, is that to handle the resulting equations is computationally expensive and even with the latest CPUs it takes too long. This is where the fast Fourier transform comes to our aid: it still computes the same results but in a far more efficient way, reducing what would have taken years by direct evaluation of the Fourier transform to a matter of seconds.

14.2 The Fourier Series

Historically, the problem arose with the solution of the equation for a vibrating string where several eminent mathematicians had sought to describe its behaviour by a series of harmonics. The work was developed by a French mathematician and physicist Jean-Baptise Fourier who was studying heat transfer. An excellent history can be found in the introduction of the book https://archive.org/details/introductiontoth00carsrich/page/n4 by H S Carslaw, which is freely available as a download. And starting on page 196 and for the next 100 pages, he gives a detailed

© Springer Nature Switzerland AG 2020
C. May, *Passive Circuit Analysis with LTspice*[®],
https://doi.org/10.1007/978-3-030-38304-6_14

and rigorous development of the Fourier series. There are many websites that have examples: https://www.math24.net/fourier-series-definition-typical-examples/ and http://lpsa.swarthmore.edu/Fourier/Series/WhyFS.html also has an excellent introduction to the Fourier series. For more examples, see also http://tutorial.math.lamar.edu/Classes/DE/FourierCosineSeries.aspx

The process can become quite complicated, but there are websites that will do the work for us where we input the function and it returns the harmonics. One use of the series is in the large-signal response of circuits which can no longer be regarded as linear. Granted that the LTspice transient analysis and a *.four* directive will show us the waveform and its harmonic composition, provided we know the frequency of the excitation, nevertheless, it can be interesting to compute it ourselves. Also it is intellectually satisfying to find the series and overlay the result on the waveform itself. The series can be found in two forms: the trigonometrical form is easier to get to grips with, but the exponential form is more compact. Both yield the same results.

It will be seen that much of this part of the chapter is an exercise integration by parts with a sprinkling of trigonometrical identities. There are several websites which offer a Fourier series calculator for piecewise functions, but these (understandably) will not aid us when we come, for example, the half sine-wave loops from rectifier circuits.

14.2.1   The Concept

We start with a repetitive waveform for which we have some analytic description: this may be an equation, or a piece-wise description, or a combination of both. All that matters is that we can describe it mathematically and that it repeats. In engineering, we often deal with events in time. The fanciful waveform in Fig. 14.1 repeats after time T_p. We will decompose the waveform into a series of sinusoids where the fundamental is drawn; that is, it has exactly the same period as the waveform.

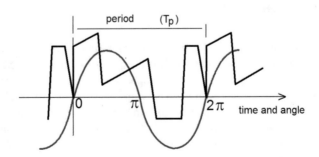

Fig. 14.1 Period of a Waveform

But first we must ask what types of waveforms are needed to describe the original waveform. We know that sine waves are one, but do we need others? If we do not include a phase angle then it is clear that sine waves alone are not enough. But as we recall that a cosine wave is just a sine wave shifted by 90°, then adding sines and cosine of the same frequency with the appropriate amplitudes will enable us to construct a sinusoidal wave with any phase. Thus we see that we can use sine and cosine waves but without the complication of phase angles.

Example: Sine Wave Phase Angle
LTspice does not have a cosine wave. But we can use schematic ('Sin(A) + Cos(A). asc') with a sine wave at 90° to see the effect. We can add the two waveforms to create a wave of the same frequency at any phase and amplitude. We can also use ('Sine Phase.asc') to show that if we add a cosine of different amplitudes the composite wave is a sine with a phase shift up to 90°. We need to adjust the amplitudes to maintain the same amplitude of the resultant.

14.2.2   Harmonics and DC

We must next explore the frequency relationship between the sinusoids. Since the waveform is repetitive, if we add the components of the derived Fourier series, these too must result in a repetitive waveform. Therefore, each wave in the series must contain an exact integral number of cycles of the individual waves comprising it. If this were not so, some waves would contain more or less of a particular frequency and the resultant waves would not all be identical. This means that the waves must be harmonics of the fundamental, just as we found for the square wave. We may not need every harmonic, but certainly we will not find ourselves using a wave with a frequency of, say, 2.4 times the fundamental. We can satisfy ourselves on this point by a little LTspice in the example below.

The longest wavelength, or fundamental, must have exactly the same period as the original waveform, which runs from 0 to 2π and which we have shown in Fig. 14.1 and is more clear in Fig. 14.2 where the fundamental has a frequency of f_0. The upper figure shows that a frequency of half that, $f_0/2$, will not do because it

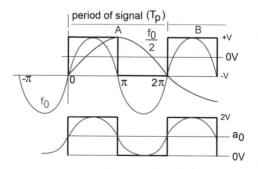

Fig. 14.2 Fundamental and DC

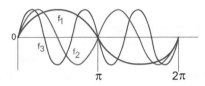

Fig. 14.3 Harmonics

will be positive during one period (A) and negative during the next (B). The lower figure shows that we may need a DC offset a_0 but often $a_0/2$ is used. Higher harmonics (Fig. 14.3) showing the second and third will always start and end at zero, and we can smoothly add more cycles, which also will start and end at zero. The fundamental, so to speak, sets the overall shape of the Fourier series and the other frequencies adjust it to be a better fit. It also follows that we need only analyse one period of the waveform to describe the whole. There is a problem that if the original signal does not start from a time of minus infinity and run to plus infinity, because during the period before the start of the signal and during the period after the termination of the signal the resultant must be zero and this will change the harmonic content. This is an *aperiodic* signal to be discussed later. However, LTspice assumes an infinite series of waves even if we just perform the analysis over one cycle.

Example: Fundamental + 1.4 Fundamental
The schematic ('Fundamental + Not Harmonic.asc') has*1 kHz* and *1.4 kHz* waves. Certainly the pattern of waves repeats, but not at the fundamental frequency, as we can see by comparing it to −*I(2)*, so not every wave in the Fourier series will be an exact replica of the original wave. If we make the frequency an exact multiple of *1 kHz*, it does repeat correctly.

Harmonics and Overtones
This is just an aside to explain the terminology. The fundamental is the first harmonic, the wave at twice the frequency is the second harmonic, that at three times is the third harmonic and so on, as we have used above. However, there is a second terminology used by musicians and some acoustic specialists which is that the wave with twice the frequency is the *first overtone* and that with three times the frequency is the *second overtone* from which the overtone is one less than the harmonic.

Integration by Parts
This is used in virtually every example. The formula is:

$$\int udv = uv + \int vdu \qquad (14.1)$$

L'Hôpital's Rule
We shall need this a few times in later waveforms. It rescues us from situations where the function evaluates to *0/0*. For example, *sin(πx)(πx)* when $x \rightarrow 0$. The

method is to differentiate the numerator and denominator separately and take the limit x → 0. In this case, we have $\pi cos(0)/\pi = 1$.

14.2.3 Trigonometrical Form

From the foregoing we see that we can describe a function $f(x)$ as a DC term plus a sum of cosine and sine terms:

$$f(x) = \frac{a_0}{2} + \Sigma_{n=1}^{\infty} a_n \cos(nx) + \Sigma_{n=1}^{\infty} b_n \sin(nx) \tag{14.2}$$

There are certain technical conditions attached to this, which are discussed in the Wikipedia article, and are the Dirichlet Conditions that (1) it must be absolutley integrable over the period; (2) it must have a finite number of maxima and minima; (3) it must have a finite number of discontinuities. These are satisfied by the examples we shall explore.

It is convenient to replace the summation by integration. This is valid because the summation is to infinity and this gives us a definite integral. More information on integration can be found at: http://tutorial.math.lamar.edu/Classes/CalcI/DefnOfDefiniteIntegral.aspx.

$$f(x) = \frac{a_0}{2} + \int_{n=1}^{\infty} a_n \cos(nx)dx + \int_{n=1}^{\infty} b_n \sin(nx)dx \tag{14.3}$$

Angles, Time, Frequency, and LTspice

Equation 14.2 makes no stipulation about the nature of x. In the examples to be found on the Internet or in textbooks, the x-axis may just be a number, or an angle, or time. But LTspice deals in time, not angles, and real-life engineering signals occur in time. If we normalize the frequency so that $\omega_0 = 1$ we have $T_p = 2\pi$ and the fundamental frequency is: $f_0 = 1/T_p$, Fig. 14.1 and this corresponds to 2π radians. Then at any instance t the fundamental voltage or current is:

$$x(t) = \cos(2\pi t/T_p) \tag{14.4}$$

The simplest is to make the fundamental period 1 s to set the frequency to 1 Hz. Then in LTspice notation, $x(t) = cos(2*pi*time)$ and the harmonics are $cos(n*2*pi*time)$. We also have

$$\omega_0 = 2\pi f_0 \quad \omega_0 = \frac{2\pi}{T_p} \quad T_p = \frac{2\pi}{\omega_0} \tag{14.5}$$

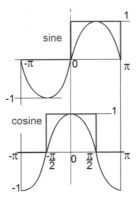

Fig. 14.4 Odd and even functions

By this means, the analysis is quite general and the actual frequency is only a scaling factor, just like amplitude.

Odd and Even Functions

By making n an integer, we find two useful relationships: the first is that the sine of integer multiples of $\pi = 0$. The second is that the cosine of integer multiples of π oscillates between -1 and 1, which we can express concisely as $cos(n\pi) = (-1)^n$ if n is not zero. In summary:

$$\sin(n\pi) = 0; \quad \cos(n\pi) = (-1)^n \text{where} \quad n \neq 0 \tag{14.6}$$

And because $cos(a) = cos(-a)$, Fig. 14.4, this means it is an *even function* and is symmetrical about the y-axis. On the other hand, sine is an *odd function* because *sin (a) $\neq$ sin(-a)* or *sin(a) = -sin(-a)*. Therefore, if we are dealing with an even function the sine terms are zero and we only need the cosine terms in Eq. 14.3, and if it is odd the cosine terms are zero and we only need the sine terms. If we can recognize a function as being odd or even, we can often simplify the mathematics by just evaluating over half the period and doubling the result. And, as is also shown in Fig. 14.4, we can sometimes turn an odd waveform into an even one by shifting the origin to the mid-point. The upper waveform is odd and uses sines; the lower one with the origin shifted uses cosines and is even. Both will yield the same result.

Derivation of the Coefficients

This is rather mathematical and requires some knowledge of calculus. It is in order to skip this and go straight to the end of the section and the Summary Definitions of the Coefficients. On the other hand, the mathematics is not profound, only exhaustive (and exhausting?), and there is some intellectual pleasure in following Carslaw's ingenious procedure as we do here.

There are two things we need to discover. The first is if the function contains a particular frequency, and if so, what is its amplitude. Carslaw achieves both.

The a_n Coefficient

Excluding the DC term $a_0/2$ and assuming unit amplitude, we can expand the terms of Eq. 14.3 as:

$$f(x) = a_1\cos(x) + b_1\sin(x) + a_2\cos(2x) + b_2\sin(2x) + a_3\cos(3x)$$
$$+ b_3\sin(3x)....a_n\cos(nx) + b_n\sin(nx) \tag{14.7}$$

which we integrate from $-\pi$ to π. This equation is not very helpful; we need to separate the individual a_n and b_n terms using some picking function to isolate each so that we can evaluate them one by one. This is done by the clever trick of multiplying each term by $cos(mx)$ where m is an integer and $m \neq n$ so each pair of terms becomes:

$$a_n \cos(mx) \cos(nx) + b_n \cos(mx) \sin(nx) \tag{14.8}$$

We take the $cos(mx)cos(nx)$ term which expands to:

$$a_n\cos(mx)\cos(nx) = \frac{a_n}{2}[\cos(m+n)x + \cos(m-n)x] \tag{14.9}$$

We now integrate from $-\pi$ to π:

$$\int_{x=-\pi}^{\pi} \frac{a_n}{2}[\cos(m+n)x + \cos(m-n)x]dx$$
$$= \frac{a_n}{2}\left[\frac{1}{m+n}\sin(m+n)x + \frac{1}{m-n}\sin(m-n)x\right]_{x=-\pi}^{\pi} = 0 \tag{14.10}$$

where from Eq. 14.6 each *sin* term is zero. Thus when $m \neq n$, every $a_n \cos(mx)cos(nx)$ term is zero and hence every $a_n \cos(nx)$ is also zero. This looks promising: we are left only with $m = n$ which is $cos(mx)^2$ and we use:

$$\cos^2 x = \frac{1}{2}[1 + \cos(2x)] \tag{14.11}$$

and again we integrate:

$$\int_{x=-\pi}^{\pi} \frac{a_n}{2}(1 + \cos2mx)dx = \frac{a_n}{2}\left[x - \frac{1}{2}\sin(2mx)\right]_{x=-\pi}^{\pi} \tag{14.12}$$

From Eq. 14.6, the sine of any integer multiple of π is zero, so $sin(2mx)$ is zero and we are left with:

$$\frac{a_n}{2}\left[(\pi - 0) - ((-\pi) + 0)\right] = \frac{a_n}{2}2\pi = a_n\pi \qquad (14.13)$$

and we have successfully selected the case when $m = n$, but the amplitude is multiplied by π and therefore we need to divide by π:

$$a_n = \frac{1}{\pi}\int_{x=-\pi}^{\pi} f(x)\cos(nx)dx \qquad (14.14)$$

Cos(nx) has a peak amplitude of unity, and thus Eq. 14.14 shows us that multiplying the function by it not only selects the harmonic but correctly returns its amplitude.

Example: cos(m)cos(n)

It is a little difficult to demonstrate Eq. 14.14, but we can go some way towards it with schematic ('cos(m)cos(n).asc'). This has two voltage sources to represent a component of the signal $V(m)$ and the 'picking function' $V(n)$ where we create the cosine waves by a phase shift of $90°$ applied to sine waves.

Using LTspice notation, first we should note that we need to turn time into an angle by $2*pi*time$ then we can integrate a voltage by using the $idt(V)$ function to integrate a voltage (or current) over time. But as that includes time, we only need to multiply by $2*pi$: $V = idt(V(m)) * 2 * pi$. The integral of a cosine voltage m is a sine voltage with an amplitude $1/m$ – and this is what we see. It follows, then, that to integrate the product $V(m)*V(n)$ we must also multiply by $2*pi$. But remember that we must divide by pi in Eq. 14.14 so the required measurement is:

.meas h INTEG 2 * V(m) * V(n) FROM 0 TO 1

where we only multiply by 2. It is important to remember that the multiplication $cos(mx)cos(nx)$ is not zero if $m \neq n$ at any arbitrary point in time – it is only the integral over a complete cycle of the fundamental that it is zero. And as the fundamental is $1\ Hz$ this is exactly what we have done.

The schematic shows the integration of the product of the two waves $V(v12)$ and their integral as time progresses. We see that the integral at any instance generally is not zero but over a period of the fundamental it is. A transient run with $m = n$ is a steadily increasing ramp with some ripple. Also that if we introduce a phase shift Phi degrees for one voltage, especially if it is $90°$ so that one wave is a cosine, the analysis fails.

Example - Cosine Series

Let us suppose we have a function where we already know it is only composed of cosines and is:

$$f(x) = 2\cos(\omega t) + 0.5\cos(3\omega t) + 1.5\cos(6\omega t)$$

We create the waveform in schematic ('Simple Fourier.asc') where we can pick out the three components as a function of time by multiplying the function $f(x)$ by each frequency in turn:

$$f(x)\cos(\omega t) \quad f(x)\cos(3\omega t) \quad f(x)\cos(6\omega t)$$

To find the a components we must integrate over one period, for example, by using the idt function: $V = idt((2 * cos (2 * pi * time) + 0.5 * cos (6 * pi * time) + 1.5 * cos (12 * pi * time)) * cos (2 * pi * time))$ to create the original function above which we multiply in turn by $cos(2 * pi * time)$, then $cos(6 * pi * time)$ and finally by $cos(12 * pi * time)$ and now we find increasing waveforms with some ripple exactly like the previous Example. The individual amplitudes are in the correct ratios, and of the correct magnitude after one period having values of *1 V, 0.25 V* and *0.75 V*, and twice the amplitude after two periods and so on.

Explorations 1
There is a lot of juice that can be sucked out of these examples (and the next set also). Here are some ideas.

1. With schematic ('cos(m)cos(n).asc'), use the cursor (or write a '.meas' statement) to find the maximum of V(im) with different values of m. Note the result of different transient run times. Note the phase relationship between m and its integral.
2. With the same schematic, explore the h results with $m = n$ and with $m \neq n$.
3. Again, with the same schematic, change the integration period from *1* s and explain what happens when $m = n$ and when they are not equal.
4. Yet again, change the amplitude of the signal component V(m) from *3 V*.
5. Using schematic ('Simple Fourier.asc'), measure the harmonics as functions of time and show they agree with the Example.
6. Add statements/directives in that schematic to return the three a_n coefficients.
7. Add a few more harmonics to ('Simple Fourier.asc') and explore.

The b_n Coefficient
We must now check the $b_n \cos(mx)\sin(nx)$ term in Eq. 14.8 to see if that behaves in the same way. We expand this as:

$$b_n\cos(mx)\sin(nx) = \frac{b_n}{2}[\sin(m+n)x + \sin(m-n)x] \qquad (14.15)$$

and integrate:

$$\int_{x=-\pi}^{\pi} \frac{b_n}{2}[\sin(m+n)x + \sin(m-n)x]x\,dx$$

$$= \frac{b_n}{2}\left[\frac{-1}{m+n}\cos(m+n)x - \frac{1}{m-n}\cos(m-n)x\right]_{x=-\pi}^{\pi} = 0 \qquad (14.16)$$

Now $cos(a\pi) = (-1)^a$ and we have:

$$\frac{b_n}{2}\left[\frac{(-1)}{m+n}(-1)^{m+n} + \frac{(-1)}{m-n}(-1)^{m-n} - \frac{(-1)}{m+n}(-1)^{m+n} - \frac{(-1)}{m-n}(-1)^{m-n}\right] = 0$$

$$(14.17)$$

and the integral evaluates to zero, which is what we would like.

If, however, we take $m = n$ then we can use the identity:

$$\sin^2(mx) = \frac{1}{2}[1 - \cos(2mx)]$$

$$(14.18)$$

and after integrating:

$$\int_{x=-\pi}^{\pi} \frac{b_n}{2}[1 - \cos(2mx)dx] = \frac{b_n}{2}\left[x + \frac{1}{2mx}\sin(2mx)\right]_{x=-\pi}^{\pi} = 2\pi$$

$$(14.19)$$

The *sin* terms vanish because $sin(m\pi) = 0$ as before for the a_n terms. So finally we are left only with the one instance when $m = n$ and

$$\int_{x=-\pi}^{\pi} f(x)\cos(nx)dx = b_n\pi$$

$$(14.20)$$

whence we again must divide by π to find the correct value:

$$b_n = \frac{1}{\pi}\int_{x=-\pi}^{+\pi} f(x)\sin(nx)dx$$

$$(14.21)$$

Example sin(m)sin(n)

Schematic 'sin(m)sin(n).asc is the same as 'cos(m)cos(n).asc only using sines. It also correctly returns the result of equal and non-equal frequencies.

Example – Square Wave Harmonics

To see how we can extract the harmonics of a more complex wave form, we use the schematic ('SquareWave Harmonics.asc'). This does not decompose the square wave into its harmonics – that comes later; this only shows how they can be extracted.

The square wave has a period p and a positive voltage V for the first half of the period and zero for the second: **PULSE(0 {V} 0 1f 1f {p/2} {p}).** Equation 14.4 means that for the harmonics, we need to divide their frequencies by the period. Thus where n is the number of the harmonic, the picking function is: **SINE(0 1 {n/p});** else we shall end up with the wrong frequency. For example, with $p = 2$ and $n = 5$ **SINE(0 1 {n/p}** has a period of 0.4 s and a frequency of *5 Hz*. Likewise we divide the harmonic we are seeking by p and we use $V(m)/p$.

The measure statement is exactly that same as Example cos(m)cos(n): *.meas h 2 ∗ INTEG 2 ∗ V(m)/{p} ∗ V(n) FROM 0 TO {p}* to integrate over one period. The *.four {1/p} V(n)* returns *3.8196, 1.273, 0.7637,* and *0.5454* for the first five harmonics with a *6 V* amplitude. The **SPICE Error Log** shows that the amplitude *h* is exactly equal to that derived from the *.four* statement. This, of course, is a circular argument since both are based on the same analysis.

We can uncomment *.step param n 1 7 2* to measure several harmonics at once.

Explorations 2

8. Change the amplitude of the square wave by *.param V =??* and note that the Fourier component and *h* are the same so that Eq. 14.14 correctly returns the amplitude.
9. Change the period and the Fourier component and *h* are the same.
10. Change the picking function to **SINE(0 1 {2/p}** or some other integral multiplier of *p* so that $m \neq n$ and note that the result is very small, far less than 1% when $m = n$. Try turning off compression to see if that reduces it.
11. Although we have explicitly stated that *m* is a multiple of *n*, something like $m = 1.7n$ does not return a zero result.
12. Set a cursor on *V(i)* which integrates continually and it returns twice the harmonic voltage at the second period.
13. Test for higher harmonics and show that they too accord with the Fourier component.

The a_0 Coefficient

As we saw previously, the first term $a_0/2$ handles the case when $n = 0$ and is neither sine nor cosine but a steady, constant, DC and is simply the average. In some texts this is given as a_0 but if we remember it is the average we shall not go wrong and it is:

$$a_0 = \frac{1}{\pi} \int_{x=-\pi}^{\pi} f(x)dx \qquad (14.22)$$

Therefore, we integrate over the whole period from $-\pi$ to $+\pi$ but only divide by π, not 2π so that $a_0/2$ will correctly supply the average. If the amplitude is not unity, we simply multiply by it. The frequencies do not change.

Summary Definitions of Coefficients

We can summarize all of the above as:

- The a_n coefficients if $n > 0$ are picked out by multiplying the function by cos(nx). This will reject every term where $m \neq n$. That is, we are left with: $a_n = \frac{1}{\pi} \int_{x=-\pi}^{\pi} f(x) \cos(nx)dx$.
- The b_n coefficients are likewise selected by multiplying the function by sin(nx) so that: $b_n = \frac{1}{\pi} \int_{x=-\pi}^{+\pi} f(x)\sin(nx)dx$.
- The DC component is found by: $a_0 = \frac{1}{\pi} \int_{x=-\pi}^{\pi} f(x)dx$.
- Any starting and finishing angles (or times) can be used so long as they are one full cycle. It is often more convenient to integrate from 0 to 2π.

- We must also be aware that the Fourier series returns the correct harmonics but says nothing about their relative phases so when we come to simulate the waveform in LTspice we may need to add phase shifts.

14.3   Simulation of Common Waveforms

This is a somewhat artificial definition to select those that can be described by one equation over a full period, such as a ramp, or those that must be handled in two parts such as a triangle. The simulations consist of a voltage or current source generating the original waveform and an arbitrary behavioural source to create the Fourier series. As the result is plotted in time, angles are converted using Eq. 14.4. In some of the following examples, the individual harmonics are given so that the Fourier series can be built term by term. In which case the schematic is 'cleaner' if we use currents because we can add them directly rather than leaving 'floating' outputs of voltage generators.

Often the difficult part is performing the integrations where we may need to integrate by parts several times. If the waveform runs for a full cycle, this will often simplify since sines and cosines of $0°$ or multiples of π are either 0 or ±1. It is only later when we deal with pulse waveforms that it becomes messy. Some examples are carefully explained in detail at: https://www.math24.net/fourier-series-definition-typical-examples/. Several are reproduced here with a slightly different notation.

14.3.1   Constant Waveform f(x) = k

This is a trivial example where we do not know if this is odd or even. We shall take the limits of $-\pi$ and $+\pi$ and evaluate both the a and b factors with a height of k.

$$a_0 = \frac{1}{\pi}\int_{x=-\pi}^{\pi} kdx = \frac{k}{\pi}\int_{x=-\pi}^{\pi} dx \quad a_0 = \frac{k}{\pi}[x]_{x=-\pi}^{\pi} \quad a_0 = 2k \quad \frac{a_0}{2} = k \quad (14.23)$$

so the average is just the value itself, which is what we should expect.

$$a_n = \frac{1}{\pi}\int_{x=-\pi}^{\pi} k\cos(nx)dx \quad b_n = \frac{1}{\pi}\int_{x=-\pi}^{\pi} k\sin(nx)dx \quad (14.24)$$

To integrate by parts we take:

$$u = k \quad du = 0 \quad dv = \cos(nx) \quad v = \frac{1}{n}\sin(nx)$$

then:

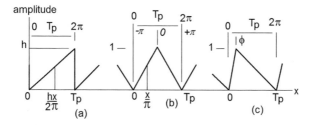

Fig. 14.5 Triangular Waveforms

$$a_n = \frac{k}{n\pi}[\sin(nx)]^{\pi}_{x=-\pi} = 0 \quad b_n = \frac{-k}{n\pi}[\cos(n\pi)]^{\pi}_{x=-\pi} = 0 \qquad (14.25)$$

And we are just left with the average k. But we should take note that this only applies of we are dealing with the full period. If not, see the much later 'Example - Trapezoid Waveform from Ramps'.

14.3.2   The Triangular Wave

Conceptually this is the easiest non-trivial example to handle since the function is defined by two straight lines. They are of the general form $f(x) = d + hx$ where h is the amplitude and d is a constant. Both of the lines are contained within the period T_p.

There are three possible forms, ignoring any DC offset, and they are: a sawtooth with a vertical drop at the end of the period (Fig. 14.5a); a symmetrical wave (Fig. 14.5b); and a scalene triangle with unequal slopes (Fig. 14.5c). The figure shows both the periods of the waveforms T_p and the start and finish angles for the integration.

Sawtooth

We shall handle this rather ponderously to emphasis every step in the process. Although in Fig. 14.5a the waveforms are defined by period, it is easier to use angles so that we integrate from 0 to 2π. We can see that this is not symmetrical about the time axis and is an odd function, so we use sines. The first straight line extends from 0 to 2π; the second is merely a vertical line which we can ignore since it does not extend in time or angle, and as there is no DC offset the amplitude at any point between 0 and 2π is:

$$f(x) = \frac{hx}{2\pi} \quad 0 \le x \le 2\pi \qquad (14.26)$$

First we find a_0:

$$a_0 = \int_{x=0}^{2\pi} f(x)dx \Rightarrow a_0 = \int_{x=0}^{2\pi} \frac{hx}{2\pi} \quad a_0 = h \tag{14.27}$$

for the b_n components:

$$b_n = \frac{1}{\pi} \int_{x=0}^{2\pi} f(x)\sin(nx)dx \quad \Rightarrow b_n = \frac{1}{\pi} \int_{x=0}^{2\pi} \frac{hx}{2\pi}\sin(nx)dx \tag{14.28}$$

If we take the constant $h/2\pi$ out of the integration then the b_n coefficients are:

$$b_n = \frac{1}{\pi} \frac{h}{2\pi} \int_{x=0}^{2\pi} x\sin(nx)dx \tag{14.29}$$

Then from Eq. 14.1 we have: $u = x \quad du = dx \quad dv = \sin(nx) \quad v = \frac{-1}{n}\cos(nx)$. We integrate Eq. 14.29 by parts using Eq. 14.1 and taking the first part under the integral sign we have:

$$uv = \left[\frac{-x}{n}\cos(nx)\right]_{x=0}^{2\pi} = \left[\frac{-2\pi}{n}\cos(2n\pi) - \frac{0}{n}\cos(0)\right] = \frac{-2\pi}{n} \tag{14.30}$$

Then for the second part, we take the $1/n$ outside the integration:

$$\int vdu = \frac{1}{n} \int_{x=0}^{2\pi} -\cos(nx)dx = \frac{-1}{n}\left[\frac{1}{n}\sin(nx)\right]_{x=0}^{2\pi}$$
$$= \frac{-1}{n^2}[\sin(2\pi) - \sin(0)] = 0 \tag{14.31}$$

And we only have the result of Eq. 14.30 to insert in Eq. 14.29:

$$b_n = \frac{1}{\pi} \frac{h}{2\pi} \frac{-2\pi}{n} = \frac{-h}{n\pi} \tag{14.32}$$

Finally, inserting b_n in Eq. 14.26 and changing to summation:

$$f(x) = \frac{h}{2} + \Sigma_{n=1}^{\infty} - \frac{h}{n\pi}\sin(nx) \tag{14.33}$$

It is computationally better to take the constant h/π outside the summation to leave:

$$f(x) = \frac{h}{2} - \frac{h}{\pi}\Sigma_{n=1}^{\infty}\frac{1}{n}\sin(nx) \tag{14.34}$$

and the waveform is:

$$f(x) = \frac{h}{2} - \frac{h}{\pi}\left[\sin(x) + \sin\frac{(2x)}{2} + \sin\frac{(3x)}{3} + \sin\frac{(4x)}{4} + \dots\right] \quad (14.35)$$

A point to note here is that the amplitude h is just a multiplier and does not affect the harmonic content, so we can leave it out of the equations and add it at the end.

Viewing the Amplitude and Phase

The overlay of the sine and cosine series on the waveform itself is a good indication if the decomposition is correct. But we can also use an FFT to measure the individual components.

Right click in the trace panel and from the pop-up menu select **View→FFT**. We can right click on the FFT trace and select **View->Mark Data Points**. It is often better to select a linear **Left Vertical Axis**. For the present we can leave all the settings at their default values. This will be taken up in detail in Sect. 14.8.1

Example: Sawtooth Fourier Series

To convert to a sequence in time, we replace x by *2∗pi∗time*. In schematic ('Sawtooth.asc'), the first term *(2∗pi∗time)* is *1 Hz*. The first 10 terms are:

V = {h}/2-
{h}/pi ∗ (sin (2 ∗ pi ∗ time) + sin (4 ∗ pi ∗ time)/2 + sin (6 ∗ pi ∗ time)/3+
sin (8 ∗ pi ∗ time)/4 + sin (10 ∗ pi ∗ time)/5 + sin (12 ∗ pi ∗ time)/6+
sin (14 ∗ pi ∗ time)/7 + sin (16 ∗ pi ∗ time)/8 + sin (18 ∗ pi ∗ time)/9+
sin (20 ∗ pi ∗ time)/10)

If we build this function term by term, we will find that the steepness of the drop is controlled by the higher harmonics. An FFT run shows the amplitudes and phases agree with the series for **V**.

To idle away an odd few minutes, extend the series to the 40th term – or skip to the next Example.

Parameters and Functions

We have used parameters extensively before to allow the same value applied to several functions to be changed easily, for example in the delay lines where we could change the values of the many capacitors and inductors just by giving two parameters new values. Another important advantage is that a set of parameters such as *2∗pi* used several times in a simulation can be turned into a parameter by *.param tpi = 2∗pi* which is evaluated once and the result inserted in the equation rather than making the computation every time. To be sure, it only makes a small difference to the run time, but it is neater.

However, a parameter may not contain a variable, so *.param w = 2∗pi∗time* will fail. This is because parameter substitution occurs before the simulation and *time* is not constant. This is where we can create a function. The syntax is:

.func < name > (variables) = {< expression >}

The curly brackets seem to be optional with simple expressions. The great advantage is in not having to type out the whole list of terms.

Example – Sawtooth Using '.func'
Taking the previous example, we can declare a parameter

$$\text{.param tpi} = 2 * \text{pi}$$

to calculate the value of 2π once to save doing it every time the function is called. The function to generate the series above is:

$$\textit{.func co}(n) = \{ \sin (n * tpi * time)/n \}$$

Then the expression for the waveform is:

$$V = \{h\}/2-$$
$$\{h\}/pi * (co(1) + co(2) + co(3) + co(4) + co(5) + co(6) + co(7) + co(8) + co(9) +$$
$$co(10) + co(11) + co(12) + co(13) + co(14))$$

This is schematic ('SawtoothA.asc') where we have easily added more harmonics using the function *.func co(n)*.

An Alternative View
We can move the y-axis to half the height and the integration limits to $-\pi$ and π (Fig. 14.6). The term a_0 is unchanged, but the equation for the height is $(x + \pi)h/(2\pi)$, and Eq. 14.30 now becomes:

$$uv = \left[\frac{-(x+\pi)}{n} \cos(nx) \right]_{x=-\pi}^{\pi} = \left[\frac{-2\pi}{n} \cos(n\pi) - \frac{0}{n} \cos(-n\pi) \right] \qquad (14.36)$$

$$b_n = \frac{-2\pi}{n} \cos(n\pi) \quad b_n = \frac{-2\pi}{n}(-1)^n \qquad (14.37)$$

Equation 14.31 now has the limits $-\pi$ and π but as the sine of any multiple of $\pi = 0$ that still does not contribute. So Finally we have:

$$x = \frac{h}{2} + \frac{h}{2\pi^2} \Sigma_{n=1}^{\infty} \frac{-2\pi}{n} (-1)^n \sin (nx) \qquad (14.38)$$

which we can write as:

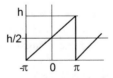

Fig. 14.6 Alternative Sawtooth

$$f(x) = \frac{h}{2} + \frac{h}{\pi} \left[\sin(x) - \sin\frac{(2x)}{2} + \sin\frac{(3x)}{3} - \sin\frac{(4x)}{4} + \dots \right] \tag{14.39}$$

And rather curiously, this also creates a sawtooth.

Example – Odd/Even Sawtooth

This is schematic ('SawtoothC.asc') which embodies Eq. 14.39.

Reversed Sawtooth

To reverse it so that it starts at maximum amplitude and decreases to zero the function becomes:

$$f(x) = h\left(1 - \frac{x}{2\pi}\right) = \frac{h}{2\pi}(2\pi - x) \tag{14.40}$$

and we again take $h/(2\pi)$ out from the integration and Eq. 14.29 becomes:

$$b_n = \frac{1}{\pi} \frac{h}{2\pi} \int_{x=0}^{2\pi} (2\pi - x) \sin(nx)dx \tag{14.41}$$

Equation 14.30 becomes:

$$uv = \left[-(2\pi - x)\cos\frac{(nx)}{n} \right]_{x=0}^{2\pi} = \left[-(0)\cos\frac{(2n\pi)}{n} + 2\pi\frac{\cos(0)}{n} \right] = \frac{2\pi}{n} \tag{14.42}$$

which is the negative of Eq. 14.30. Equation 14.31 still evaluates to zero, so finally we have:

$$b_n = \frac{h}{n\pi} \tag{14.43}$$

which is the negative of Eq. 14.32.

Thus, reversing the sawtooth only involves reversing the sign of b_n.

Example – A Reverse Sawtooth

Thus we see that for the ramp to start at the maximum, the only difference is that we reverse the sign of the coefficients and the sawtooth is flipped left to right. Taking schematic ('Sawtooth Reversed.asc'), we change the function to:

$$V = \{h\}/2 + \{h\}/pi * (co(1) + co(2) + co(3) + co(4) + co(5) + co(6) +$$
$$co(7) + co(8) + co(9) + co(10) + co(11) + co(12) + co(13) + co(14))$$

and it starts with the vertical edge of height h.

Isosceles Triangular

Here we have a choice (Fig. 14.5b). If we leave the origin at the start, it is not symmetrical. But if we move the origin of time to the apex as shown by 0 in italics, it is symmetrical and therefore an even function of unit height so we only need the cosine terms. The function is:

$$f(x) = \frac{x}{\pi} \quad -\pi \leq x \leq 0 \quad f(x) = \left(1 - \frac{x}{\pi}\right) \quad 0 < x \leq 2\pi$$

Rather than integrating in two parts, we can use symmetry and just integrate from 0 to π and double the result. It might be questioned if this is valid since the two functions are different. To answer this we should note that, as we saw with the example above 'A Reverse Sawtooth', the harmonic content of a falling ramp is the same as a rising ramp only the signs are swapped; but as the function from 0 to π has $-x$, the signs will be swapped back again and the two parts will add. The analysis is very similar to the previous only using cosines, not sines. It is neater to integrate the rising waveform and take π out of the integration:

$$a_n = \frac{1}{\pi^2} \int_{x=0}^{\pi} x\cos(nx)dx \tag{14.44}$$

We integrate by parts using:

$$u = x \quad du = dx \quad dv = \cos(nx) \quad v = \frac{1}{n}\sin(nx) \tag{14.45}$$

$$a_n = \frac{1}{n\pi^2}\left(-x\sin(nx) - \int_0^\pi \sin(nx)dx\right) \tag{14.46}$$

$$a_n = \frac{1}{n\pi^2}\left[-x\sin(nx) + \frac{\cos(nx)}{n}\right]_{x=0}^{\pi} \tag{14.47}$$

As $\sin(nx) = 0$, we are left only with the cosine term which evaluates to 1 when $x = 0$:

$$a_n = \frac{1}{n^2\pi^2}[1 - \cos(n\pi)] \quad = \frac{1}{n^2\pi^2}[1 - (-1)^n] \tag{14.48}$$

We can write $[1-(-1)^n]$ as $2(-1)^{n+1}$ and we double a_n to account for the interval 0–π:

$$a_n = \frac{4}{n^2\pi^2}(-1)^{n+1} \; [n \quad odd] \quad a_n = 0 \; [n \quad even] \tag{14.49}$$

We can merge the two parts of Eq. 14.49 by using $(2n + 1)$ for the power of -1 which will always be odd and start the summation from $n = 0$ so that it will select the fundamental.

$$x = \frac{a_0}{2} + \frac{4}{\pi^2} \Sigma_{x=0}^{\infty} \sin \frac{((2n+1)x)}{(2n+1)^2} \tag{14.50}$$

and:

$$a_0 = \frac{1}{\pi} \int_{x=0}^{\pi} f(x) dx \quad = 1 \tag{14.51}$$

Finally we have:

$$f(x) = 0.5 + \frac{4}{\pi^2} \left[\sin(x) + \frac{\sin(3x)}{9} + \frac{\sin(5x)}{25} + \frac{\sin(7x)}{49} + \cdots \right] \tag{14.52}$$

If the amplitude is not *1*, we simply multiply Eq. 14.52 by it

Example: Isosceles Triangle Fourier Series
This is schematic ('Isosceles Triangle.asc') which implements Eq. 14.52. The function for the components is:

$$func\ h(n) = \{ \sin(n * 2 * pi * time)/n * *2 \}$$

and the required voltage is:

$$V = 0.5 + 4/pi * *2 * (h(1)\text{-}h(3) + h(5)\text{-}h(7) + h(9))$$

The series is a remarkably good fit because there are no steep edges; it only fails significantly to accurately follow the peaks for which more terms are needed. This is generally true that sharp discontinuities require high harmonics to follow them accurately.

Scalene Triangular Wave

This is (Fig. 14.5c) where we have a unit height triangle. We mention it now only because it is neither even nor odd; we need both *a* and *b* terms. We shall return to it later.

Explorations 3
1. Schematic ('SawtoothB.asc') has a different phase to ('SawtoothA.asc') because the even terms are subtracted. Do the analysis.
2. Note the effect of adding and removing harmonics on all the sawtooth schematics. Try changing the amplitude and the frequency.
3. Open schematic ('Isosceles Triangle.asc') and explore different amplitudes and frequencies. Observe the phase of the harmonics relative to the fundamental and how the offset of π brings them into line.
4. Check the values by an FFT.

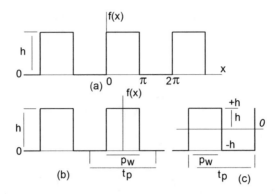

Fig. 14.7 Square wave

14.3.3 Square Waves

This takes us to a different type of waveform. We shall start with a wave which is zero for half the time. We shall define the waveform in two ways (Fig. 14.7) and show that they are equivalent. In the first example '(a)' we define the wave as: $f(x) = h$ for $0 \leq x \leq \pi$ and $f(x) = 0$ for $\pi < x \leq 2\pi$. A very important point to note is that in every definition of a square wave these are step transitions or discontinuities. A sine wave has a finite maximum slope and therefore it is impossible to replicate a square wave no matter how high a range of harmonics we use.

But to return to our definition, to find a_0 we only need consider $x \geq 0$ so:

$$a_0 = \frac{1}{\pi} \int_{x=0}^{\pi} f(x)dx = \frac{1}{\pi} \int_{x=0}^{\pi} h.dx = \frac{h}{\pi} \pi = h \qquad (14.53)$$

And this is odd so we only need the sine terms and we can extract h as a common factor. Also, by symmetry, we can change the integration to 0 to π and double the result:

$$b_n = \frac{h}{\pi} \int_{x=-\pi}^{\pi} \sin(nx)dx = \frac{-2h}{\pi} \left[\frac{\cos(nx)}{n} \right]_{x=0}^{\pi} \qquad (14.54)$$

Then using relationship 14.6 for the cosines:

$$b_n = \frac{-2h}{n\pi} [\pi \cos(n\pi) - \cos(0)] \quad = \frac{2h}{n\pi}(1 - (-1)^n) \qquad (14.55)$$

meaning that if n is odd $b_n = 2h/(n\pi)$, else it is 0 which we realized by $1 - (-1)^n$ which is zero if n is even else it is 1 and the function is:

$$f(x) = \frac{h}{2} + \frac{2h}{\pi} \Sigma_0^\infty (1 - (-1)^n) \sin \frac{(nx)}{n} \qquad (14.56)$$

where we have extracted $2h/\pi$ from inside the summation and the first few terms are:

$$f(x) = \frac{h}{2} + \frac{2h}{\pi} \left[\sin(x) + \frac{1}{3} \sin(3x) + \frac{1}{5} \sin(5x) + \frac{1}{7} \sin(7x) + \ldots \right] \quad (14.57)$$

To turn it into a time sequence, we replace x by $2*\pi*time$:

$$f(2\pi t) = \frac{h}{2} + \frac{2h}{\pi} \left[\sin(2\pi t) + \frac{1}{3} \sin(6\pi t) + \frac{1}{5} \sin(10\pi t) + \frac{1}{7} \sin(14\pi t) + \ldots \right]$$
$$(14.58)$$

and the amplitude of the terms decays slowly as $1/n$, so an acceptable approximation may require a large number of terms.

Example - Square Wave Fourier Series as Sines

This is schematic ('Fourier Squarewave 1.asc'). This has a wave of unit height of width 2 s meaning that its frequency is half that of the triangular waves we explored previously, so we have replaced x by $n\pi$, rather than the $2n\pi$ which we used before, to halve the frequency of the simulation.

V = 1/2 + 2/pi * (sin (pi * time) + 1/3 * sin (3 * pi * time)

+1/5 * sin (5 * pi * time) + 1/7 * sin (7 * pi * time) + 1/9 * sin (9 * pi * time))

which we simplify on the schematic by defining the function:

.func t(n) = { sin (n * tpi * time)/n}
then : *V = 1/2 + 2/pi * (t(1) + t(3) + t(5) + t(7) + t(9) + t(11) + t(13))*

We might also note the result of a *.four* analysis at the correct fundamental where the even harmonics are greatly attenuated and the odd ones agree with the voltage expression and have a phase of $90°$. If we compare this with one at a frequency 10% higher, we see that the even amplitudes are significant and the phase angles wrong. If we change the sign to − *2/pi*(sin(pi*time) +.....)*, the waveform is inverted.

The Gibb's Phenomenon

We should notice the slight overshoots at the beginning of the wave and undershoot at the end. These are not removed by increasing the number of harmonics, only they becomes narrower and closer to the start and finish of the pulse. This occurs whenever there is a discontinuous jump in the waveform and is about 9% of the amplitude. It is well explained in the Wikipedia article. We shall note this in later simulations,

A Different Approach

Instead of starting at zero and solving for distance x, this time we shall solve for time using a pulse of height h. We shall also generalize the analysis by not insisting that the pulse width p_w is half of the periodic time t_p. Moreover, we shall define the origin not at the start of the pulse, but at the mid-point so it is symmetrical about the time axis and the waveform is non-zero only between the times $-p_w/2$ and $+p_w/2$ (Fig. 14.7b). This makes the waveform even so we use the cosine terms. Also we shall use $\frac{2\pi}{t_p} = \omega$ rad/s instead of the frequency. First we have:

$$\frac{a_0}{2} = \frac{ht_p}{p_w} \tag{14.59}$$

Then because of symmetry, we only need integrate over half the pulse width and multiply by 2. We extract the constant h and instead of dividing by π we divide by half the period:

$$a_n = 4\frac{h}{t_p}\int_{t=0}^{p_w/2} \cos\left(\frac{n2\pi}{t_p}t\right)dt \tag{14.60}$$

Then performing the integration:

$$a_n = 4h\frac{t_p}{n2\pi t_p}\left[\sin\left(\frac{n2\pi}{t_p}t\right)\right]_{t=0}^{\frac{p_w}{2}} \tag{14.61}$$

$$a_n = \frac{2h}{n\pi}\sin\left(\frac{n\pi p_w}{t_p}\right) \tag{14.62}$$

Taking the particular case of a square wave so that $p_w = t_p/2$; and for $n > 0$

$$a_n = \frac{2h}{n\pi}\left(\frac{n\pi}{2}\right) \quad = \frac{2h}{n\pi} \tag{14.63}$$

We can take $2h/\pi$ outside the integration and use Eq. 14.6 to pick only odd values to yield:

$$f(x) = \frac{h}{2} + \frac{2h}{\pi}\Sigma_{n=0}^{\infty}(-1(-1)^n)\cos(nx) \tag{14.64}$$

and the first few terms are:

$$f(x) = \frac{h}{2} + \frac{2h}{\pi}\left[\cos(x) - \frac{1}{3}\cos(3x) + \frac{1}{5}\cos(5x) - \frac{1}{7}\cos(7x)\ldots\right] \tag{14.65}$$

where the terms have the same amplitude as Eq. 14.58 because $\sin(nx)/n = \cos(nx)/n$ shifted by $\pi/2$. And because of the behaviour of $\cos(n\pi)$ the terms alternate in sign and the result is always positive.

Example - Square Wave as Symmetrical Fourier Series
We implement Eq. 14.65 by defining a *function:*

$$.\textbf{func } \textbf{t(n)} = \{1/\textbf{n} * \textbf{cos}\,(\textbf{n} * \textbf{tpi} * \textbf{time})\}$$

then:

$$V = 2 + 2 * h/pi * (t(1)\text{-}t(3) + t(5)\text{-}t(7) + t(9)\text{-}t(11) + t(13)\text{-}t(15))$$

where we have a pulse height of $h = 4$, schematic ('Fourier Squarewave as Cos.asc'). And because we have made a symmetrical analysis about the centre of the pulse, that is where the simulation starts, and we need to delay the 'real' square wave by $\pi/2$.

Square Wave as Cosines

Equation 14.55 picks out the values $2h/(n\pi)$ or 0. So, in this particular case, with integer values of π, if we can find an alternative function that returns 1 or -1 for odd integer multiples of π and 0 for even, we can use that. A cosine will work as its value only alternates between 1 and -1 for multiples of π, and as we saw above, we need to subtract alternate terms so that the result is always positive:

$$f(x) = \frac{h}{2} + \frac{2h}{\pi}\left[\cos(\pi t) - \frac{1}{3}\cos(3\pi t) + \frac{1}{5}\cos(5\pi t) - \frac{1}{7}\cos(7\pi t) + \ldots\right] \quad (14.66)$$

and we have Eq. 14.65 again.

Example: Square Wave as Cosine Fourier Series
This is schematic ('Fourier Series as Cos 2.asc') The height has been increased to 10 so the function for the first few terms is:

$$V = 5 + 20/pi * (\cos(pi * time)\text{-}1/3 * \cos(3 * pi * time) + 1/5 * \cos(5 * pi * time)\text{-}$$
$$1/7 * \cos(7 * pi * time) + 1/9 * \cos(9 * pi * time)\text{-}1/11 * \cos(11 * pi * time))$$

Square Wave with No DC

Now the wave travels equally positive and negative as shown in Fig. 14.7c and by inspection the average is zero. The wave is no longer zero outside the pulse time but $-h$ and we can again use symmetry and integrate over p_w and double the result to encompass the period from p_w to t_p. Then Eq. 14.61 is:

$$a_n = \frac{2h}{t_p} \int_{t=t_p}^{P_w} \cos\left(\frac{n2\pi}{t_p} t\right) dt \qquad (14.67)$$

and for a square wave:

$$a_n = \frac{2h}{n\pi}\left[\sin\left(\frac{n\pi t}{t_p}\right)\right]_{t=t_p}^{P_w} = \frac{2h}{n\pi} - 0 \qquad (14.68)$$

which is the same as Eq. 14.63, so adding the two periods we have $4h/(n\pi)$ and then:

$$f(x) = \frac{4h}{\pi}\left[\cos\left(\pi t\right) - \frac{1}{3}\cos\left(3\pi t\right) + \frac{1}{5}\cos\left(5\pi t\right) + \ldots\right] \qquad (14.69)$$

where the difference is the absence of DC and the amplitude is double because it is measured from the centre.

Example: Fourier Series of Square Wave with No DC

This is schematic ('Fourier Series as Cos 3.asc'), where $h = 10$, but this time it is measured from 0 V. The first terms of the series are:

V = 20/pi ∗ (cos (pi ∗ time)-1/3 ∗ cos (3 ∗ pi ∗ time) + 1/5 ∗ cos (5 ∗ pi ∗ time)-
1/7 ∗ cos (7 ∗ pi ∗ time) + 1/9 ∗ cos (9 ∗ pi ∗ time)-1/11 ∗ cos (11 ∗ pi ∗ time))

The 'real' waveform is:

$$\text{PULSE}(5 \ -5 \ 0.5 \ 1f \ 1f \ 1 \ 2)$$

and is delayed by 0.5 s which is $\pi/4$.

Explorations 4

1. Open the schematics mentioned above and explore the effects of different frequencies and amplitudes.
2. Create a function to replace the sequence ....*3 ∗cos(3∗pi∗time)*....
3. Increase the number of terms and notice how the overshoot approaches the dislocation. Estimate the magnitude of the overshoot as a percent of the wave height.
4. Check the results with an FFT.
5. Compare the results to a *.four* measurement.
6. Repeat the analyses using angles instead of time.

Rectangular Waves

Equation 14.68 enables us easily to see the effect of different duty cycles. Writing δ for the duty cycle

$$a_n = \frac{2h}{n\pi}\sin(\delta n\pi) \text{ and the function is :} \tag{14.70}$$

$$f(x) = \frac{h}{2} + \frac{2h}{\pi}\Sigma_0^\infty \frac{\sin(\delta n\pi)}{n}\cos(nx) \tag{14.71}$$

Example: Fourier Series of Rectangular Pulse

The schematic is: ('Fourier Pulse.asc') where Eq. 14.71 is implemented as a series of current sources in parallel. The terms are realized by the function:

$$\textit{func co}(n) = \{(2 * h)/(pi * n) * \sin(n * dc * pi) * \cos(n * tpi * time)\}$$

where we use the now familiar parameter **.param tpi = 2*pi** and **.param dc** is the duty cycle. For the pulse source we have: **PULSE(0 1 {1-dc/2} 1f 1f {dc} 1)** which is *1 V* amplitude where we start the pulse train after *1 s* minus half the width of the pulse by **{1-dc/2}**. We may not enter calculations in the parameters of a voltage source, so trying: **Tdelay (s) = 1-{dc}/2** will fail. Correct is **{1-dc/2}**.

The current in resistor *R1* is the sum of all the harmonics. We can also see the frequency composition of the individual terms by an FFT.

Explorations 5

1. Run schematic 'Fourier Pulse.asc' with different duty cycles. Check the DC current *I(dcv)* against different duty cycles and adjust the current of *I(1)* to match.
2. Compare the FFT harmonics and their amplitudes with those of the individual current sources.
3. In particular, note how reducing the number of harmonics affects the wave shape.
4. Also note that a very short duty cycle of *dc = 0.01* results in a single central pulse (the trace needs to be expanded to see this clearly). Try adding extra harmonics with *R3* and see if there is a relationship between duty cycle and the number of harmonics needed to resolve the pulse.

14.3.4   The Parabola

This shows that sometimes we need to integrate by parts more than once. The equation is: $f(x) = x^2$ and is an even function. Then:

$$a_0 = \frac{1}{\pi}\int_{x=-\pi}^{+\pi} x^2 dx = \frac{2}{\pi}\left[\frac{1}{3}x^3\right]_{x=-\pi}^{\pi} = \frac{2\pi^2}{3} \tag{14.72}$$

and the average is not zero. Because it is an even function, we have changed the integration limits from *0* to π and multiplied the result by *2*.

$$a_n = \frac{2}{\pi}\int_{x=0}^{\pi} x^2\cos(nx)dx \qquad (14.73)$$

Integrating by parts with

$$u = x^2 \quad du = 2xdx \quad dv = \cos(nx) \quad v = \frac{1}{n}\sin(nx) \qquad (14.74)$$

$$a_n = \frac{2}{\pi}\left[x^2\frac{1}{n}\sin(nx) - \frac{2}{n}\int_{x=0}^{\pi}\sin(nx)xdx\right]_{x=0}^{\pi} \qquad (14.75)$$

where we have taken the factor 2 outside the integral and we need to integrate the second term by parts with:

$$u = x \quad du = 1dx \quad dv = \sin(nx)dx \quad v = \frac{-\cos(nx)}{n} \qquad (14.76)$$

$$\frac{4}{n}\int_{x=0}^{\pi}\sin(nx)xdx = \frac{4}{n}\left[x\cos\frac{(nx)}{n}\right]_{x=0}^{\pi} - \frac{4}{n}\int_{x=0}^{\pi}\cos\frac{(nx)}{n}dx \qquad (14.77)$$

so finally:

$$a_n = \frac{2}{\pi}\left[\frac{x^2}{n}\sin(nx) + \frac{2x}{n^2}\cos(nx) - \frac{1}{n^3}\sin(nx)\right]_{x=0}^{\pi} \qquad (14.78)$$

as $sin(n\pi) = 0$ and the second term is zero if $x = 0$ and using $cosnx = (-1)^n$ we have:

$$a_n = \frac{-2}{\pi}\frac{2\pi}{n^2}\cos(n\pi) \quad = \frac{4\pi}{n^2}(-1)^n \qquad (14.79)$$

And the series this is:

$$f(x) = x^2 = \frac{\pi^2}{3} + 4\left(-\cos(nx) + \frac{1}{4}\cos(2x) - \frac{1}{9}\cos(3x) + \ldots\right) \qquad (14.80)$$

This series decays as $1/n^2$ which is faster than the square wave and so fewer terms are needed for an acceptable approximation. An interesting corollary to Eq. 14.80 is that we can set $x = \pi$ then we have an expansion for π^2

$$\pi^2 = \frac{\pi^2}{3} + 4\Sigma_{n=1}^{\infty}\frac{(-1)^2}{n^2}\cos(n\pi) \qquad (14.81)$$

But returning to our parabola, as a time series Eq. 14.80 is:

$$f(x) = \frac{\pi^2}{3} + 4\left(-\cos(\pi t) + \frac{1}{4}\cos(2\pi t) - \frac{1}{9}\cos(3\pi t) + \ldots\right) \qquad (14.82)$$

Example: Fourier Series of a Parabola

The schematic is ('Parabola.asc') and implements Eq. 14.82 as a function:

$$\textit{func co}(n) = 4/n**2 * \cos{(n * pi * time)}$$

where we remember to insert a minus sign before the odd terms on the schematic. The trace shows that the cusps are not well defined and need higher harmonics to sharpen them. An FFT shows that the first 8 harmonics are significant.

We draw the 'real' parabola in two stages. First the rising voltage of arbitrary source *B3* is converted into integer voltage steps V when time is odd by the **floor** function, that is after t=1,3,5....:

$$\textbf{V} = \textbf{floor}((\textbf{time} + \textbf{1})/\textbf{2})$$

This voltage is named *V(t)* and is used with current source B13 so that **(time-2∗V(t))** resets the current when time is odd and ∗∗2 creates the parabola

$$\textbf{I} = \textbf{10} * (\textbf{time-2} * \textbf{V(t)}) * \textbf{*2}$$

This is trace I(R2).

14.3.5   Full-Wave Rectified Sine Wave

This is very often encountered in electronics where we convert AC to DC. The point is that this shows that we need to be flexible in our approach. The YouTube site quoted below can pull us out of the hole we dig if we try integration by parts.

This function is symmetrical about π and therefore we only need the a_n factors. We take the absolute value of a wave of height *h*:

$$f(x) = h|\sin{(2\pi t)}| \tag{14.83}$$

In terms of angle this is:

$$a_n = \frac{h}{\pi}\int_{x=0}^{2\pi}\sin(x)\cos(nx)dx \tag{14.84}$$

And because of symmetry, we can integrate from *0* to π and double the result.

$$a_n = \frac{2h}{\pi}\int_{x=0}^{\pi}\sin(x)\cos(nx)dx \tag{14.85}$$

Now if we try integrating by parts, we shall soon find ourselves in difficulties. The trick is instead to use the relationship:

$$\sin(a)\cos(b) = \frac{1}{2}[\sin(a+b) + \sin(a-b)] \qquad (14.86)$$

This is well explained at https://www.youtube.com/watch?v=e0F8RKcMoHY which we follow here with slightly different notation. Substituting in Eq. 14.85 (remembering the expansion in Eq. 14.86 divides by 2). we have $sin(x + nx)$ and $sin(x-nx)$ or:

$$a_n = \frac{h}{\pi}\int_{x=0}^{\pi}[\sin([1+n]x) + \sin([1-n]x)]dx \qquad (14.87)$$

and we now have two simple integrations and Eq. 14.87 becomes:

$$a_n = \frac{-h}{\pi}\left[\frac{\cos([1+n]x)}{1+n} + \frac{\cos([1-n]x)}{1-n}\right]_{x=0}^{\pi} \qquad (14.88)$$

Substituting we have:

$$a_n = \frac{-h}{\pi}\left[\cos\frac{([1+n]\pi)}{(1+n)} + \cos\frac{([1-n]\pi)}{1-n} - \cos\frac{([1+n]0)}{1+n} - \cos\frac{([1-n]0)}{1-n}\right]$$
$$(14.89)$$

As $cos([1 + n]0) = 1$ and likewise $cos([1 - n]0) = 1$ we are only left with the awkward case of $n = 1$. But remembering L'Hôpital's Rule, $cos(0)/0$ evaluates to 1 and the second two cosines are always positive. For the first two, if $n = odd$, they will be positive, so the result is zero. If $n = even$, they are negative and we have:

$$a_n = \frac{-h}{\pi}\left[\frac{-2}{1+n} + \frac{-2}{1-n}\right] = \frac{-h}{\pi}\left[\frac{2(1-n) + 2(1+n)}{1-n^2}\right] = \frac{-4h}{\pi}\left[\frac{1}{1-n^2}\right] \quad (14.90)$$

and Eq. 14.90 is 0 if $n = $ odd. Finally $a_0 = 2h/\pi$

Example: Full-Wave Rectified Sine Wave

We use the function *.func co(n) = {-4*h/pi*1/(n**2-1)*cos(n*tpi*time)}* in schematic ('FW Rectified Sine.asc') and find that up to the 12th term there is a very good fit apart from the end of the cycle. We have seen this several times before that sudden changes require high harmonics to follow them properly. The trace of *I(B3)* agrees with the *.four* analysis that it is effectively *10%* of the fundamental and the harmonics are significant up to *Harmonic Number 6* of the *.four* results. An FFT analysis shows the first 6 even harmonics (that is, up to 12 Hz) are significant. The next cluster of harmonics is not until 5 kHz and they are -105 dB down from the fundamental.

Explorations 6
1. Create symmetrical rectangular waves like Fig. 14.7b with different duty cycles, some with no DC, others with.
2. Create rectangular waves like Fig. 14.7c where the waveform is measured from the start of the pulse.
3. Create a hyperbola with different parameters.
4. Explore different amplitudes and frequencies for schematic ('FW Rectified Sine. asc')
5. Add more even harmonics to the previous to resolve the cusps better and note how much they add to the total distortion.

14.4   The Exponential Form

This is a more compact form where we start with Euler's formulae:

$$e^{i\theta} = \cos(\theta) + i\sin(\theta) \quad e^{-i\theta} = \cos(\theta) - i\sin(\theta) \qquad (14.91)$$

This represents a unit length vector (Fig. 14.8). And from Eq. 14.91 we see:

$$\cos(\theta) = \frac{1}{2}\left(e^{i\theta} + e^{-i\theta}\right) \quad \sin(\theta) = \frac{1}{2i}\left(e^{i\theta} - e^{-i\theta}\right) \qquad (14.92)$$

If the angle is a function of time $\theta = \omega t$, the vector rotates anti-clockwise. And if $\theta = -\omega t$, the vector rotates clockwise and we have a negative frequency. And in view of Eq. 14.91 $e^{j\omega t}$ will contain both sine and cosine components. So we have:

$$f(x) = \int_{\substack{-\infty \\ n=-\infty}}^{\infty} c_n e^{jn\omega_0 t} dt \qquad (14.93)$$

where the integration extends from $-\infty$ to ∞ to encompass every possible frequency. This includes $n=0$, the DC component c_0.

As with the trigonometrical form, we need a picking function to uniquely select the frequencies. This is: $c_n = e^{-jm\omega_0 t}$ because if we consider just one harmonic and the periodic time T and insert that in Eq. 14.93:

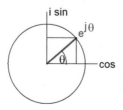

Fig. 14.8 Euler's Formulae

$$f(x) = \int_{t=0}^{T} e^{-jm\omega_0 t} e^{jn\omega_0 t} dt = \int_{t=0}^{T} e^{j\omega_0 t(n-m)} dt \qquad (14.94)$$

If $m = n$ then:

$$f(x) = \int_{t=0}^{T} e^0 dt = T \qquad (14.95)$$

and we have selected one frequency. But we need to check what happens if $m \neq n$. We can write the second Eq. 14.94 using Eq. 14.91 as:

$$f(x) = \int_{t=0}^{T} [\cos(\omega_0 t(n-m)) + j \sin(\omega_0 t(n-m))] dt \qquad (14.96)$$

We evaluate this at $t = 0$ as:

$$f(x) = \cos(0(n-m)) + j\sin(0(n-m)) = n - m \qquad (14.97)$$

As $\omega_0 T = 2\pi$ for $t = T$ we have:

$$f(x) = \cos(2\pi(n-m)) + j\sin(2\pi(n-m)) = n - m \qquad (14.98)$$

and the result is zero so we have successfully picked out just one frequency.

14.4.1   Exponential Coefficients

The two forms must yield identical results. Therefore $c_0 = a_0$. For the rest, we can substitute in Eq. 14.3 which we repeat here:

$$f(x) = \frac{a_0}{2} + \int_{n=1}^{\infty} a_n \cos(nx) dx + \int_{n=1}^{\infty} b_n \sin(nx) dx$$

using the sine and cosine expressions of Eq. 14.92

$$f(x) = \frac{c_0}{2} + \int_{n=1}^{\infty} \left(a_n \frac{1}{2} (e^{jn\omega_0 t} + e^{-jn\omega_0 t}) + b_n \frac{1}{2j} (e^{jn\omega_0 t} - e^{-jn\omega_0 t}) \right) \qquad (14.99)$$

then we group terms with the same exponentials as:

$$f(x) = \frac{c_0}{2} + \int_{n=1}^{\infty} \left[\left(\frac{a_n}{2} + \frac{b_n}{2j} \right) e^{jn\omega_0 t} + \left(\frac{a_n}{2} - \frac{b_n}{2j} \right) e^{-jn\omega_0 t} \right] \qquad (14.100)$$

and we tidy this up to give:

$$f(x) = \frac{c_0}{2} + \int_{n=1}^{\infty} \left(\frac{1}{2}(a_n - jb_n)e^{jn\omega_0 t} + \frac{1}{2}(a_n + jb_n)e^{-jn\omega_0 t} \right) \qquad (14.101)$$

We should note that $\frac{1}{2}(a_n + jb_n)e^{jn\omega_0 t}$ is the complex conjugate of the first term in Eq. 14.101 because if we multiply them together the result is: $\frac{1}{4}(a_n^2 + b_n^2)$ and is a real number.

We define:

$$c_n = \frac{1}{2}(a_n - jb_n) \quad c_n^* = \frac{1}{2}(a_n + jb_n) \qquad (14.102)$$

where c_n^* is the complex conjugate. And because of Eq. 14.91 the coefficients are each a mixture of the cosine and sine terms. As we saw previously, because the integral in Eq. 14.93 starts from $-\infty$ we do not need a separate c_0 term. However, LTspice will not act on an imaginary exponential, so **V=2∗exp(sqrt(−1)∗time)** simply returns a constant 2 in schematic ('sqrt(−1).asc'). Therefore we cannot directly implement the exponential form but must turn it into real sines and cosines as we do now.

14.4.2   *Rectangular Wave Using the Exponential Form*

Using Fig. 14.7b we have a pulse symmetrical about the y-axis so the integration is from $-tp/2$ to $+tp/2$

$$c_n = \frac{1}{t_p} \int_{t=-t_p/2}^{+t_p/2} x(t)e^{-jn\omega_0 t} dt \qquad (14.103)$$

In this case, x is constant and is either h or 0. So we can take it outside the integral and only integrate over the actual pulse:

$$c_n = \frac{h}{t_p} \int_{t=-p_w/2}^{+p_w/2} e^{-jn\omega_0 t} dt \qquad (14.104)$$

then performing the integration:

$$c_n = \frac{-h}{t_p} \frac{1}{jn\omega_0} [e^{-jn\omega_0 t}]_{t=-p_w/2}^{+p_w/2} = \frac{-h}{t_p} \frac{1}{jn\omega_0} \left(e^{-jn\omega_0 p_w/2} - e^{+jn\omega_0 p_w/2} \right) \qquad (14.105)$$

Noting the reversed order of the bracketed terms in Eq. 14.105 we have a minus sine function:

$$e^{-jn\omega_0 p_w/2} - e^{+jn\omega_0 p_w/2} = -j2\sin\left(\omega_0 n\frac{p_w}{2}\right) \tag{14.106}$$

so substituting in Eq. 14.105 and then using $\omega_0 = 2\pi/t_p$ and $p_w = t_p/2$ therefore $\omega_0 = \pi/p_w$

$$c_n = \left[\frac{h}{t_p}\frac{1}{jn\omega_0}\right]\left(-j2\sin\left(\omega_0 n\frac{p_w}{2}\right)\right) = \left[\frac{h}{t_p}\frac{1}{jn}\frac{t_p}{2\pi}\right]\left(-j2\sin\left(\frac{\pi}{p_w}n\frac{p_w}{2}\right)\right) \tag{14.107}$$

$$c_n = \frac{h}{n\pi}\sin\left(\frac{n\pi}{2}\right) \tag{14.108}$$

where $sin(n\pi/2)$ is 0 if $n = even$ and will alternate 1 and -1 if it is odd.

Equation 14.108 is no help in finding c_0. For that we return to Eq. 14.104 with $n = 0$ and find $c_o = h\, p_w/t_p$ then for a square wave $c_0 = h/2$. We can change Eq. 14.108 to suit duty cycles that are not 50%.

$$c_n = \frac{h}{n\pi}\sin\left(n\pi\frac{p_w}{t_p}\right) \tag{14.109}$$

Example: Converting an Exponential Form Square Wave to Sinusoids
We return to the full expression of Eq. 14.93 and substitute for c_n using Eq. 14.108:

$$f(x) = \int_{t=-\infty}^{\infty}\frac{h}{n\pi}\sin\left(\frac{n\pi}{2}\right)e^{jn\omega_0 t}dt \tag{14.110}$$

As an aside, we note that this equation involves negative frequencies, something we might well boggle at. We will find some of the presentations on YouTube follow this path and make use of the relationship that $c_n = c^*_{-n}$ which comes from Eq. 14.102. However, if we remember that the series is symmetrical so every negative n is paired with a positive n we can avoid negative frequencies by taking the terms in pairs, omitting $n = 0$. When n is odd so that $sin(n\pi/2)$ is -1 or 1 the general term is:

$$\frac{h}{n\pi}\left(e^{jn\omega_0 t} + e^{-jn\omega_0 t}\right) \tag{14.111}$$

Now we recall that:

$$\cos(n\omega t) = \frac{e^{jn\omega_0 t} + e^{-jn\omega_0 t}}{2} \tag{14.112}$$

and Eq. 14.111 becomes:

$$\frac{2h}{n\pi}\cos(n\omega t) \tag{14.113}$$

We can take the factor $2h/\pi$ outside the summation and add the term $c_0 = h/2$ to Eq. 14.110. Then starting with $|n| = 1$ we have $sin(\pi/2) = 1$ which gives us the first term. Then when $|n| = 3$ we find $sin(3\pi/2) = -1$ and we subtract the term in cos $(3\omega t)$. When $|n| = 5$ we find $sin(5\pi/2)$ is positive again so we add $cos(5\omega t)$. Thus we build the series.

$$f(x) = \frac{h}{2} + \frac{2h}{\pi} \left[\cos(\omega t) - \frac{\cos(3\omega t)}{3} + \frac{\cos(5\omega t)}{5} - \dots \right] \quad (14.114)$$

And this is the same as Eq. 14.69 except for the multiplying factor of $2h/\pi$ instead of $4h/\pi$ because this time the signal has a DC offset and the height is measured differently.

Explorations 7
1. If we measure the square wave from the start of the pulse (Fig. 14.7c) and not as we have done from the mid-point (Fig. 14.7b) we have only even terms and $c_n = -j2h/(n\pi)$ but as the terms are sines the j will cancel.
2. The web page https://en.wikibooks.org/wiki/Signals_and_Systems/Fourier_ Series has an example of a cubic $f(x) = x^3$ very clearly set out.
3. Create sawtooth and triangular waves using the exponential form.

14.5 Arbitrary Waveforms

These are waveforms which may lack symmetry and which can quickly become quite complicated because they may require both a and b coefficients. In many cases they can be handled by splitting the waveform into sections as we did with the isosceles triangle. A number of examples can be found at https://www.math24.net/ fourier-series-functions-arbitrary-period-page-2/

These come in two types: those where the waveform is only present for part of the period and those where the waveform is constructed from piecewise elements. It can well be argued that the two are essentially the same since we have seen in Fig. 14.7 that we can shift the y-axis by adding a DC component so that the waveform runs from 0 to h or from $h/2$ to $-h/2$ yet this does not affect the harmonic content.

14.5.1 Fractional Waveforms

This is an artificial distinction of those waveforms which are zero for part of the period as opposed to piecewise waveforms. The distinction is artificial because adding or subtracting a DC component can mean that no part of the waveform is zero. Nevertheless the distinction is useful in that it means that we only have to evaluate the one function over part of the period as opposed to piecewise waveforms.

Ramp

This is like a sawtooth only it is zero for some of the time and we need both cosine and sine terms. A basic example is: $f(x) = \frac{x}{\pi} \quad 0 < x \leq \pi \quad f(x) = 0 \quad \pi < x \leq 2\pi$ where the function is an upward slope of $45°$ up to π and 0 thereafter. It is trivial to find that $a_0 = \frac{1}{4}$. We extract the $1/\pi$ then for the other a_n coefficients we have:

$$a_n = \frac{1}{\pi^2} \int_{x=0}^{\pi} x\cos(nx)dx = \frac{1}{\pi^2}\left[x\sin\frac{(nx)}{n} + \cos\frac{(nx)}{n^2}\right]_{x=0}^{\pi} \tag{14.115}$$

The sine term cancels so this evaluates to:

$$a_n = \frac{1}{n^2\pi^2}[(-1)^n - 1] \tag{14.116}$$

and is exactly the same as the Isosceles Triangle only this time there is no symmetry and we need the sine terms:

$$b_n = \frac{1}{\pi^2} \int_{x=0}^{\pi} x\sin(nx)dx \tag{14.117}$$

and after integrating by parts:

$$b_n = \frac{1}{\pi^2}\left[-x\cos\frac{(nx)}{n} + \frac{\sin(nx)}{n^2}\right]_{x=0}^{\pi} \tag{14.118}$$

and again the sine term vanishes

$$b_n = \frac{1}{\pi^2}\left[-x\cos\frac{(nx)}{n}\right]_{x=0}^{\pi} \tag{14.119}$$

we can write Eq. 14.119 as:

$$b_n = \frac{\pi}{n\pi^2}(-1)^{(n+1)} = \frac{1}{n\pi}(-1)^{n+1} \tag{14.120}$$

or as:

$$b_n = \frac{1}{\pi^2}\left[-\pi\cos\frac{(n\pi)}{n}\right] \quad b_n = \frac{-1}{(n\pi)}\cos(n\pi) \tag{14.121}$$

We can extract the terms as:

$$n = odd \ f(x) = \frac{-2}{(n\pi)^2} \cos(n2\pi x) + \frac{1}{n\pi} \sin(n2\pi x) \tag{14.122}$$

$$n = even \ f(x) = \frac{-1}{(n\pi)} \sin(n2\pi x) \tag{14.123}$$

Then using Eqs. 14.122 and 14.123, the first terms are:

$$f(x) = \frac{1}{4} + \frac{1}{\pi}\left[\frac{-2}{\pi}\cos(\pi x) + \sin(\pi x) - \frac{1}{2}\sin(2\pi x) - \frac{2}{9\pi}\cos(3\pi x) + \frac{1}{3}\sin(3\pi x)\right]$$

$$\tag{14.124}$$

and we can compare that to the sawtooth waveform. This shows that the importance of the cosine terms rapidly decreases as they depend upon $1/(n\pi)^2$.

Example: Ramp

The schematic ('Ramp.asc') has a rising ramp for 0.5 s followed by a period of zero voltage until the next ramp. The terms are taken from Eqs. 14.122 and 14.123 and are implemented as:

$$.func \ oddco(n) = \{-2/(n*pi)**2*c(n) + 1/(n*pi)*s(n)\}$$

$$.func \ evenco(n) = \{-1/(n*pi)*s(n)\}$$

where we define the sine and cosine functions as:

$$.func \ s(n) = sin \ (n*tpi*time)$$

$$.func \ c(n) = cos \ (n*tpi*time)$$

to simplify the expressions for the coefficients.

The slope of the ramp for the first 0.4 s is a good fit, but it is more ragged at the drop, and if we cut $h(16)$ and onwards there is a slight improvement.

Example: Ramp 50

We can generalize the ramp so that it runs to some fraction φ of the period to give:

$$a_0 = \frac{h\phi}{4\pi} \tag{14.125}$$

$$a_n = \frac{1}{\pi}\left[x\sin\frac{(nx)}{n} + \cos\frac{(nx)}{n^2}\right]_{x=0}^{\phi} = \frac{1}{\pi}\left[\phi\sin\frac{(n\varphi)}{n} + \cos\frac{(n\phi)}{n^2} - \frac{1}{n^2}\right] \tag{14.126}$$

$$b_n = \frac{1}{\pi}\left[-x\cos\frac{(nx)}{n} + \sin\frac{(nx)}{n^2}\right]_{x=0}^{\phi} = \frac{1}{\pi}\left[-\phi\frac{(n\phi)}{n} + \sin\frac{(n\phi)}{n^2}\right] \tag{14.127}$$

We can extract the factor n to write:

$$\textit{func } a(n) = \{ \, sin \, (n * fi) * fi + \, cos \, (n * fi)/n - \textit{1}/n \}$$
$$\textit{func } b(n) = \{ - \, cos \, (n * fi) * fi + \, sin \, (n * fi)/n \}$$

and reintroduce n outside the brackets to give:

$$x_n = \frac{-h}{n\phi\pi} [a_n \cos(n2\pi time) + b_n \sin(n2\pi time)] \qquad (14.128)$$

and this is realized by the function:

$$\textit{func } co(n) = \{ \textit{-h}/(n * \textit{fi} * \textit{pi}) * (a(n) * \textbf{cos} \, (\textit{tpi} * n * \textit{time}) + b(n) * \textbf{sin} \, (\textit{tpi} * n * \textit{time})) \}$$

This is schematic ('Ramp 50.asc') which runs to 50 terms. What we see is that the period between the ramps becomes flatter as we add more terms and that short pulses need the higher terms to resolve them. The pulse is reasonably well-resolved down to $\varphi = \pi/8$ compared to the 'true' current pulse from current source $I1$ which is:

$$\textbf{\textit{PULSE}}(0 \, \{\textbf{\textit{h}}\} \, 0 \, \{\textbf{\textit{fi}}/(2 * \textbf{\textit{pi}})\} \, \textbf{1}\textit{n} \, \, 0 \, \, 1)$$

This is a good candidate to be turned into a sub-circuit.

Half-Wave Rectified Sine

For the full-wave case, we integrated 0 to π for the cosine and doubled the result. This time we do not. Also a_0 is half the previous because we are lacking half the positive sine loops. But as there is no symmetry, we must also try the sine terms. The expansion is:

$$\sin(x) \sin(nx) = \frac{1}{2} [- \cos([1 + n]x) + \cos([1 - n]x)] \qquad (14.129)$$

and performing the integration:

$$b_n = \frac{h}{2\pi} \left[\frac{-\sin([1 + n]x)}{1 + n} + \frac{\sin([1 - n]x)}{1 - n} \right]_{x=0}^{\pi} \qquad (14.130)$$

The result is zero except when $n = 1$ and then the second terms is $sin(0)/0$. Again using L'Hôpital's Rule, this evaluates as $\pi cos(0)/0 = \pi$ and the result is: $b_1 = h/2$ otherwise $b_n = 0$.

Example: Half-Wave Rectified Sine
The a_n function is Eq. 14.90 divided by 2 in schematic ('HW Rectified Sine.asc') where we define: $\textit{func } co(n) = \{2 * h/pi * 1/(1 - n**2) * cos(n * tpi * time)\}$ and we see that the first 12 terms give a very good fit (admitted, there are only 6 that are not

zero). The term from Eq. 14.130 is handled by: *.func cof(1) = {h/2*sin(tpi*time)}* and the average is $a_0 = h/\pi$.

14.5.2 Piecewise Waveforms

We can analyse these by parts. The function may consist of sequential sections such as a triangle or it may consist of superimposed waveforms – or, indeed a combination of both. We may need to reverse a waveform, for example, the second part of a scalene triangle, or to invert it; both of which we have done before. A number of useful examples explained in detail can be found at: https://www.math24.net/fourier-series-functions-arbitrary-period-page-2/

Manipulating Ramps

We have seen how we can reverse a saw tooth so that it starts at maximum height and falls linearly to zero. We can also invert them about the horizontal axes and add delays. By this means we can construct triangles and rectangular waves.

Reversed Ramp
We can go back to Eq. 14.115 and reverse the slope by replacing $f(x)=x/\pi$ with $f(x)=(1 - x/\pi)$ which we can write as: $f(x)=(\pi - x)/\pi$ and take the $1/\pi$ outside the integration.

$$a_n = \frac{1}{\pi^2} \int_{x=0}^{\pi} (\pi - x)\cos(nx)dx = \frac{1}{\pi^2}\left[(\pi - x)\sin\frac{(nx)}{n} + \cos\frac{(nx)}{n^2}\right]_{x=0}^{\pi} \quad (14.131)$$

and the sine terms cancel and we have Eq. 14.116 again.

$$b_n = \frac{1}{\pi^2} \int_{x=0}^{\pi} (-x)\ \sin(nx)dx = \left[-(-x)\cos\frac{(nx)}{n} + \sin\frac{(nx)}{n^2}\right]_{x=0}^{\pi}$$
$$= \frac{1}{\pi^2}\left[\pi\cos\frac{(n\pi)}{n}\right] \quad (14.132)$$

and finally:

$$b_m = \frac{1}{n\pi} \cos(n\pi) \quad (14.133)$$

which is Eq. 14.121 negated.

Example: Reversed Ramp

The coefficients are identical to the ramp except for the sign of $b(n)$. This also introduces a half-period delay which can be removed by:
$cos(n*tpi*(time+0.5))$ and $.sin(n*tpi*(time+0.5))$ schematic ('Reverse ramp.asc').

Delayed Ramp

We may also need to delay the waveform. As we are working in the time domain we can most easily handle this by an offset to the time: $.func\ c(n)\ =cos(n*tpi*(time-delay))$ Otherwise we can, of course, go back to the function and integrate from *delay* to 2π.

Example: Delayed Ramp

Schematic ('Delayed Ramp.asc') has $delay=0.5$ to delay the ramp by half the period of *1 s* delays more than 1 period roll over so *2.2 s* becomes *0.2 s*.

Inverted Ramp

To achieve this we simply reverse all the signs, schematic ('Delayed Ramp Inverted. asc')

Example: Adding Sawtooths

A trivial example ('Adding Sawtooths.asc') shows that adding a sawtooth shifted by half a period to another sawtooth results in a sawtooth of double the frequency whilst subtracting it gives a square wave.

Example: Isosceles Triangle

The schematic ('Isosceles Triangle A.asc') adds the coefficients of the ramp and the reversed ramp where both ramps have a duration of half the period. What is interesting first is that the sharp drops at the end of each ramp cancel, and second that the outline of the triangle is much smoother than the individual ramps because the raggedness at the apex of the forward ramp is cancelled by the ripple from around 0.3 s to 0.5 s in the reverse ramp.

Example: Scalene Triangle

The previous example is a clue to how to handle piece-wise waveforms: we create each section over the appropriate period and add them to create the full period, schematic ('Scalene Triangle B.asc') Starting with ('Ramp 50.asc') we adjust fi to *0.2 s*. by $fi = 0.2\ x\ (2\pi) = 1.257$. We add a reversed 'Ramp 50' by changing the sign of $b(n)$ and adjusting $fi2 = 5.027$. Note that $fi2$ adjusts the starting point of the reversed ramp. The fit is very good but if we look carefully at the start we find the function starts at some *25 mA* rather than *0 mA*.

Example: Trapezoid Waveform

This example is taken from https://www.math24.net/fourier-series-functions-arbi trary-period-page-2/ using x, which is the time, rather than the angle. The strategy is to take the waveform in three sections and add them. An alternative derivation is to take mid-point, $x = 1.5$ and use symmetry.

The function is defined as: $f(x) = 0 \le x \le 1; 1 < x \le 2; 2 < x \le 3$ and is schematic ('Fourier Trap Example.asc'). The final formula is:

$$f(x) = \frac{2}{3} - \frac{3}{\pi^2 n^2} \sum_{n=1}^{\infty} \left(1 - \cos\left(\frac{2n\pi}{3}\right)\right) \; \cos\left(\frac{2n\pi x}{3}\right) \tag{14.134}$$

and is declared as a function:

func co(n) = {3/(n * pi) * *2 * (1- cos (n * tpi/3)) * cos (n * tpi * time/3)}

and this is plotted up to the 12th harmonic individually. The function is a very close fit, only visibly in error at the start and finish.

Example: Trapezoid Waveform from Ramps
Schematic ('Trapezoid.asc') uses the two ramps in ('Scalene Triangle B.asc') with equal rise and fall times and a gap between which must be filled by a constant current h. Using the symmetry of the figure, we need only an a_n term which we find by:

$$a_n = 1/\pi \int_{x=0}^{\phi} h \; \cos(nx) dx \tag{14.135}$$

and multiplying by 2 to give:

$$a_n = 2\frac{h}{\pi} \sin(n\phi) \tag{14.136}$$

which will give a square pulse centred about half the period. We can cheat and find a_0 by integrating I(R2) over one period. A close examination of the function and the piecewise linear representation by a simple current source at the origin- or by using the *.meas* directives - shows that the current from the synthesis is -34.4 *mA* rather than zero. A slight difference persists throughout the simulation.

Using Parabolas

In the website http://msp.ucsd.edu/techniques/latest/book-html/node187.html Prof. Miller Puckette takes a novel approach by using parabolas to create ramps which can be added to create triangles. This reference is to a far wider opus which is beyond our present scope to explore. It is mentioned here only to show that sometimes there are more ways of solving a problem.

Example: Scalene Triangle
We take two parabolas in schematic ('Scalene Triangle A.asc') where one is inverted and shifted by angle *fi*. The period is *2 s*. It is left to the reader (as they say in all the best books) to calculate the correct pulse for current source *I1* to replicate the waveform and this (of course) requires the DC level to be decided upon – whether the waveform should have no DC as in the schematic or some other value, for example, that the minimum should be 0 V.

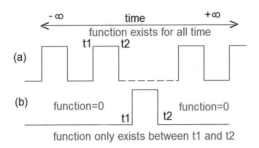

Fig. 14.9 Aperiodic Signal

Explorations 8
1. Reverse the ramp so that it increases from right to left.
2. Explore different ramp periods and amplitudes and note if the DC level is correct.
3. Try different apex times for the scalene triangle (Fig. 14.9c).
4. Note the effect in the scalene triangle if the end and start points do not coincide at the apex. For example, with $f12 = 3$ there is a flat gap between the two ramps.
5. Create an inverted trapezoid that is wider at the top.
6. Can this be extended to a parallelogram?
7. Explore reversed, inverted and delayed ramps and their combination.
8. Derive the coefficient a_o for ('Trapezoidal.asc') from its parts,
9. Have fun creating all sorts of waveforms.

14.6   Aperiodic Signals

The Fourier series strictly applies to an infinitely long train of waves in that the sine and cosine terms repeat and will continue forever, (Fig. 14.10a). But we may be faced with waves that only exist over the interval $t1$ to $t2$; just a single pulse (Fig. 14.10b). This creates a serious problem because, ideally, outside of those limits, from a time of $-\infty$ to $t1$ and from $t2$ to $+\infty$ the waveform must be zero. And we shall sidestep the theological and metaphysical questions of what happened before 4004 BC or the 'Big Bang', and the equally interesting question about what happens after the 'Big Crunch' or the Apocalypse.

We can extend the period between signals. Eventually, if we make the period infinitely long, we are left with only one instance. This means that the duty cycle becomes vanishing small. In section 'Rectangular Waves', we did it the other way round and reduced the signal duration rather than increasing the period and we saw that as the duty cycle became smaller so we needed higher harmonics to resolve the signal to other than a single pulse. The data points become ever closer together until we have a continuum leading to the Continuous Fourier transform described by *Here is a wave – Complex To Real complextoreal.com/wp-content/uploads/2012/12/fft4. pdf* and elsewhere. We shall now explore this.

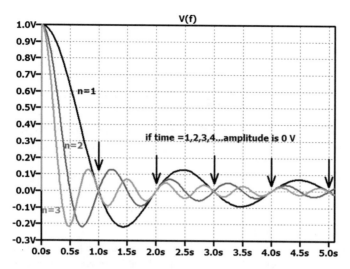

Fig. 14.10 Sinc Amplitude

14.6.1   The Sinc Function

It is important not to confuse this with the *sinh* function. It is a convenient shorthand using the definition:

$$\mathrm{sinc}(x) = \frac{\sin(\pi x)}{\pi x} \qquad (14.137)$$

to somewhat simplify expressions. The nominator oscillates with a period 2π whilst the denominator increases monotonically. If $x = 0$ we have a problem but L'Hôpital's Rule rescues us and the result is 1 and this is the only integer value of x which is not zero. The full function is a mirror image about the y-axis, with negative time, but LTspice only has positive time and Dr. Who was not available, so, sadly, the transient analysis is only for positive time.

We can see it in Fig. 14.10 which is derived from the schematic 'Sinc.asc' and shows the first three harmonics of a fundamental of 1 Hz. The traces are their progression in time. The voltages all fall to zero on multiples of their periods but the maxima and minima are not those of a simple sine at $\pi/4$ because of the denominator πx. If we move the cursor to the first minimum of the fundamental ($n = 1$), it is at 1.43 s, not 1.5 s, and we then find $V(f) = \sin(1.43\pi)/(1.43\pi) = -0.218\ V$. This is discussed in the next section.

The envelope of the positive peaks is $1/(\pi x)$ and in this case it is $1/(\pi*time)$.

Example – Sinc Function
The schematic ('Sinc.asc') traces $V=sin(pi*time)/(pi*time)$ for 10 s. and the envelope $1/\{\pi x\}$ but to see that we need to omit the first *0.25 s* of the run to avoid plotting the very large initial voltage from source *B2*.

Rectangular Wave as Sinc Function

Equation 14.70 is: $a_n = \frac{2h}{n\pi} \sin(\delta n\pi)$ and can be recast as a sinc function:

$$a_n = 2h\delta \frac{\sin(n\pi\delta)}{n\pi\delta} \quad \text{or} \quad a_n = 2h\delta sinc(n\delta) \qquad (14.138)$$

And the multiplier δ cancels the denominator δ to give constant amplitude for all duty cycles. Ignoring any DC, up to the 14th harmonic the function becomes:

$$f(x) = 2h\delta[sinc(\delta) \cos(2\pi time) + sinc(2\delta) \cos(4\pi time)\ldots\ldots sinc(14\delta) \cos(28\pi time)]$$
$$(14.139)$$

With $\delta = 0.5$ we have a square wave.

Figure 14.11 is an FFT with linear axes and $\delta = 0.2$. We are now plotting against frequency so the apexes are the amplitudes of the harmonics. The 5th harmonic is missing because $sin(\delta5\pi) = sin(\pi) = 0$. Likewise the 10th harmonic is missing and the sequence stops at the last harmonic of $n = 14$. We might wonder why the groups of peaks decrease: the answer is the denominator of Eq. 14.137. We can calculate the sinc function from Eq. 14.137 for harmonics 6 to 10, remembering that angles are in degrees:

Table 14.1 shows that the sines change as we should expect, but that when divided by $n\delta\pi$ harmonic 7 is no longer the same as harmonic 8 and harmonic 6 is not identical to harmonic 9. The ratios are normalized in the final row.

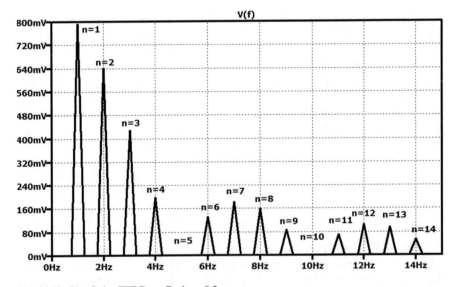

Fig. 14.11 Sinc Pulse FFT Duty Cycle $= 0.2$

Table 14.1 Sinc values for harmonics 6 to 10

n	6	7	8	9	10
sin(nδ × 80)	−0.588	−0.951	−0.951	−0.588	0
nδπ	3.77	4.40	5.02	5.65	6.28
sinc(nδ)	−0.156	−0.216	−0.189	−0.104	0
Normalized to n = 6	1	1.385	1.213	0.667	0

Example: Rectangular Wave as Sinc

LTspice does not have a sinc function so we must create it in the schematic ('Rectangular Sinc.asc'):

$$.func\ sinc(x) = sin\ (pi * x)/(pi * x)$$

then the harmonics are:

$$.func\ co(n) = \{sinc(dc * n) * t(n)\}$$

using

$$.func\ t(n) = cos\ (n * tpi * time)$$

and so the composite current waveform is:

$I = 2 * h * dc * (co(1) + co(2) + co(3) + co(4) + co(5) + co(6) + co(7) + co(8) + co$
$(9) + co(10) + co(11) + co(12) + co(13) + co(14))$

If we step the duty cycle by .*step param dc 0.1 0.7 0.2* we find the same amplitude for all.

An FFT with *dc = 0.5* shows 7 odd harmonics with amplitudes *1.35, 0.45, 0.269, 0.191, 0.147, 0.120, 0.99 V*. Reducing the duty cycle to *0.1* and we find with an FFT with all the harmonics. In both cases the maximum frequency is *14 Hz* created by *co (14)*. There is no amplitude for *10 Hz* and if we set the duty cycle to *dc = 0.2* the *5 Hz* component is also missing, as we found previously. The FFT values are normalized to their RMS values and thus we multiply by $\sqrt{2}$ to find the peak. Taking the fundamental of *800 mV* as an example this becomes *1.122 V* and is the same as the .*four* directive result: the same is true for the other harmonics.

If we normalize the amplitudes against the 6th harmonic, the measured ratios from the FFT are the same as Table 14.1. If we set $\delta = 0.5$, we find only the odd harmonics in the correct ratios.

Duty Cycle and First Zero Frequency of Amplitude

The amplitude falls to zero when $sin(n\delta\pi) = 0$. The first occurrence is when $n\delta = 1$ and therefore as we reduce the duty cycle the first frequency for zero amplitude increases. This is what we have seen above that with a duty cycle of *0.2* the frequency was *5 Hz* and with a duty cycle of *0.1* it is *10 Hz*. We see this in Fig. 14.12 with a cut-off frequency of 200 Hz where we see zeros at frequencies

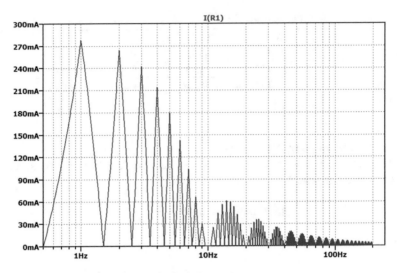

Fig. 14.12 Sinc Pulse FFT with Duty Cycle = 0.1

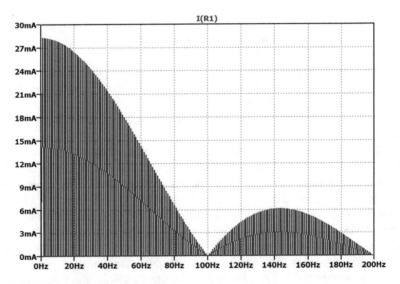

Fig. 14.13 FFT with Duty Cycle = 0.01

which are multiples of 10. With a duty cycles of 0.01 the first zero is not until
100 Hz, Fig. 14.13.

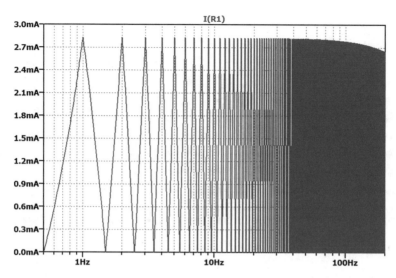

Fig. 14.14 FFT with Duty Cycle $= 0.001$

14.6.2 Aperiodic Pulse

Ideally, we want the period to go to infinity. This is a matter of some practical difficulty using LTspice, not so much because of the computational overhead of a large number of harmonics, but because of the need for short time steps to resolve the pulse.

Duty Cycle and Pulse Height

The classic FFT shape applies only as long as $sin(\delta x)$ does not approach zero. Taking Eq. 14.138 If $n\delta \ll 1$ then

$$a_n = 2h\delta \frac{\sin (n\pi\delta)}{n\pi\delta} \quad \text{if} \quad n\delta \ll 1 \quad a_n \approx 2h\delta \sin \frac{(0)}{0} \approx 2h\delta \qquad (14.140)$$

and the harmonics are all nearly equal height, only falling somewhat at high frequencies. This is clearly seen in Fig. 14.14. This was traced with a duty cycle of 0.001, the amplitudes are almost constant. We also see from the figures that the first pulse height with $\delta = 0.1$ is 300 mA and only 30 mA with $\delta = 0.01$ and slightly less than 3 mA, with a duty cycle of 0.0001, all in agreement with Eq. 14.140.

Example - Duty Cycle Effect

We can run schematic ('Sinc200.asc') to reproduce the figures shown above. This is schematic ('Sinc Pulse.asc') converted to a sub-circuit. With a duty cycle of 0.001 we see that the rectangle is reduced to a sinc pulse and an FFT gives Fig. 14.15.

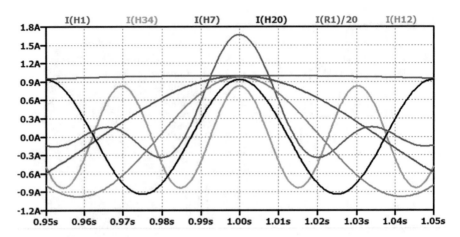

Fig. 14.15 Aperiodic Pulse Harmonics

Example: Sinc Pulse

We can start with 36 harmonics in schematic ('Sinc pulse.asc'), which generates a pulse of variable duty cycle. The amplitudes of the individual harmonics generated by $H1$ to $H34$ agree with an FFT of the current in $R1$. Note in particular that the currents in $H10$, $H20$ and $H30$ are zero (actually, a few femtoamps, but we can ignore that). If we probe the individual current sources with a duty cycle of 0.1, we find that the maximum or minimum of each falls exactly at 1 s intervals. If we reduce the duty cycle to 0.01 we now see a sinc pulse, and if we show the harmonics (Fig. 14.15), we find they are virtually all the same height up to the 20th harmonic or so as we expect because $n\delta \approx 0.2$.

Unlike the schematic ('Rectangular Sinc.asc'), the terms are only the sinc functions and not multiplied by $2h\delta$ and therefore the amplitude is not constant.

If we compare the results with ('Rectangular Sinc.asc'), the pulse with a duty cycle of 0.1 is better resolved. Making the link in schematic ('Sinc Pulse.asc') we have 36 harmonics with zero outputs at multiples of 5 Hz. We double the number of harmonics in schematic ('Sinc Pulse 72.asc') and find that the pulse is better resolved.

Constant Pulse Width

The circuits so far all used a fundamental frequency of 1 Hz so the pulse width was scaled as we changed the duty cycle. Let us instead keep a constant pulse width of 1 s as we change the duty cycle. If the duty cycle is 0.05 that means that the period of the fundamental is $1/0.05 = 20$ s and the fundamental frequency is 50 mHz. A duty cycle of 0.1 means a period of 10 s and a fundamental frequency of 500 mHz and a frequency of 1 Hz for a duty cycle of 0.1.

Example: Constant Pulse Width

We change the parameter tpi to .*param tpi $= 2*pi*dc$ to keep a constant pulse width of 1 s using schematic ('Sinc Pulse 72.asc') for pulse widths of 0.05, 0.1 and 0.5. We also find that the fundamental frequency is 50 mHz, 0.5 Hz and 1 Hz, respectively. Thus the distance between harmonics gets smaller with decreasing duty cycle so the density of harmonics increases. The pulse shape is unchanged but its height is not constant because the multiplier *2∗h∗dc(*. . .. From schematic ('Rectangular sinc.asc') is missing.

Sinc Subcircuit

We thus conclude that for a very short duty cycle we need much higher harmonics n to avoid δn becoming too small. If we take the first zero amplitude frequency as a rough guide, to resolve a 0.001 duty cycle we need the harmonics to extend to at least 1 kHz. It is obviously not possible to draw 1000 current sources on the schematic. We could convert schematic ('Sinc Pulse.asc') into a sub-circuit but it is easier, and computationally more efficient, to add several harmonics to each source rather than just one. If we create 20 sources and assign 10 harmonics to each the sub-circuit is:

```
*sinc pulse of 200 harmonics n = starting harmonic number, h = amplitude, dc = duty cycle
.subckt Sinc200 n001 0 {n} = 1 {h} = 1 {dc} = 0.1
B2 N001 0 I = co(n + 0) + co(n + 1) + co(n + 2) + co(n + 3) + co(n + 4) + co(n + 5)+
        co(n + 6) + co(n + 7) + co(n + 8) + co(n + 9)
B3 N001 0 I = co(n + 10) + co(n + 11) + co(n + 12) + co(n + 13) + co(n + 14)+
        co(n + 15) + co(n + 16) + co(n + 17) + co(n + 18) + co(n + 19)
.........................
B20 N001
I = co(n + 180) + co(n + 181) + co(n + 182) + co(n + 183) + co(n + 184) + co(n + 185)+
        co(n + 186) + co(n + 187) + co(n + 188) + co(n + 189)
B21 N001 0
I = co(n + 190) + co(n + 191) + co(n + 192) + co(n + 193) + co(n + 194) + co(n + 195)+
        co(n + 196) + co(n + 197) + co(n + 198) + co(n + 199)
.func co(n) = {(2 * {h})/(pi * {n}) * sin ({n} * {dc} * pi) * cos ({n} * tpi * time)}
.param tpi = 2 * pi
.ends
```

The amplitude and duty cycle are parameters so that with multiple instances on a schematic, all will change together. The harmonic number n must be stepped by 200 for each instance. It would be possible to have 20 or more harmonics to each source; that is a matter of choice, and does not seem to greatly affect the run time for the same number of harmonics. A simple rectangle assembly is adequate to represent the sub-circuit on the schematic with the output on the centre of the top and ground on the centre of the bottom so instances can easily be connected in parallel.

Example: Aperiodic Pulse

The harmonics of schematic ('Sinc 3000.asc') run from1 Hz to 3 kHz. It is now possible to resolve a duty cycle of only 0.001. A minimum step of 25 μs is needed to see the 'W-shaped' top of the pulse. The FFT of Figure 14.15 shows that the harmonics are almost constant amplitude up to some 100 Hz which is in accord with the duty cycle. But the contributions of these 100 harmonics to the shape of the pulse is negligible in comparison to the next 2900 where the amplitudes follow the pattern of a rectangular pulse and so that is the overall shape of the trace. However, as we might expect, these first 100 harmonics affect the amplitude and without them the peak is some 9 mA less.

We can build up the FFT waveform by probing the individual sub-circuits to show that each contributes the same range of frequencies.

Explorations 9
1. Run ('Sinc Pulse.asc') and compare the measured harmonic amplitudes with those of an FFT and the values predicted by calculation.
2. Run ('Sinc Pulse 72.asc') and note the effect of cutting out some harmonics.
3. Run ('Rectangular Sinc.asc') and note the FFT transform, in particular that the amplitudes of the square wave harmonics are exactly as predicted by calculation.
4. Remove the upper harmonics in schematic ('Sinc 300.asc') and note the changed pulse shape and FFT envelope.
5. Use 'sinc2000.txt' as a base to extend the harmonics to 5 kHz or more.
6. Explore triangular and sawtooth pulses.
7. There are a lot more interesting things that can be gleaned from these schematics: feel free to explore and ponder, and work back to the underlying function.

14.6.3   Impulse Response

If we set a really small product δn, Eq. 14.140 shows that the amplitude of the harmonics will be almost constant. That is what we see up to some 100 Hz in Fig. 14.14 and the function for a rectangular pulse becomes

$$f(x) = \frac{2h\delta}{\pi} \Sigma_{n=1}^{\infty} \cos{(n2\pi t)}dt \tag{14.141}$$

It follows, then, that every harmonic we add to the pulse increases its amplitude by the same amount.

Example: Impulse Response
We set the duty cycle in schematic ('Sinc Pulse.asc') to 0.001 and use the scissors to cut the wires to $H18$ and $H2$ leaving only the fundamental $H1$. We find a current sine wave with a peak of 1 A at 1 s. If we now restore the wire between $H1$ and $H2$ and cut the wire before $H3$ the peak is now 2 A. If we continue in this manner we find every harmonic adds another amp up to 35.97 A. And we may also measure the current from each source. Reduce the duty cycle to 0.00001 and it is exactly 36 A.

A selection of harmonics is shown in Fig. 14.15. All have a maximum at 1 s and all have almost exactly the same amplitude.

Infinite Impulse

If we continue to reduce the duty cycle, we finally have pulses of zero width but all of finite height. Therefore, if we add an infinite number of harmonics, we have a pulse of zero width and infinite height. In effect, this is the Dirac Delta function.

Example: Infinite Pulse

Using schematic ('Sinc3000.asc'), we set a duty cycle of 0.00001 and find that the current contribution from each block is almost the same. However, if we reduce the duty cycle this is no longer true and the lower blocks predominate. Also with a duty cycle of 0.00001 an FFT of each block shows that their shapes are identical but their frequency range and centre frequency increase with the block number.

14.6.4   The Continuous Fourier Transform

Returning to aperiodic pulses in general – and not necessarily rectangular ones – we have seen that as we extend the period towards infinity, the fundamental frequency becomes vanishingly small. The implication is that the spacing between the harmonics becomes ever closer tending to a continuum.

Using δn as an approximate test of the required cut-off frequency, we have seen that $\delta = 0.001$ requires n = 1000 at the least to resolve a rectangular pulse so it does not take much imagination to see that if the duty cycle becomes vanishing small the frequency extends to infinity. And why not *minus infinity*? The idea of a negative frequency may seem daunting but is important in signal processing. So, combining the infinitesimal spacing of the harmonics and the infinite - frequency, we replace the summation by integration. In terms of time this is

$$X(\omega) = \int_{t=\infty}^{\infty} x(t)e^{-j\omega t}dt \tag{14.142}$$

Example - The Rectangular Function 'rect'

This is defined in time as:

$$\text{rect} = 1/2 \quad \text{if} \quad |x| = \frac{\tau}{2}: \quad 0 \quad \text{if} \quad |x| > \frac{\tau}{2}: \quad 1 \quad \text{if} \quad |x| < \frac{\tau}{2}$$

and is show in Fig. 14.17. The integration now is only non-zero between $-\tau/2$ and $\tau/2$. Substituting we have:

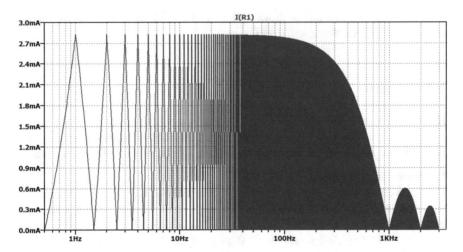

Fig. 14.16 FFT of 'Pulse3000.asc'

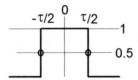

Fig. 14.17 The 'Rect' Function

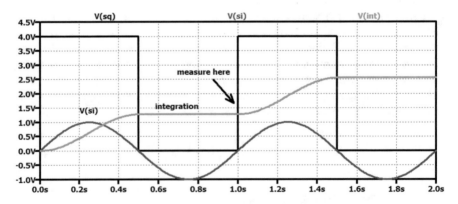

Fig. 14.18 Squarewave Decomposition

$$X(\omega) = \int_{t=-\tau/2}^{\tau/2} e^{-j\omega t} \, dt = \frac{1}{j\omega\tau/2}\left[e^{-j\omega\tau/2} - e^{j\omega}\right] = \tau\left[\frac{2\sin(\frac{\omega t}{2})}{\omega}\right] \quad (14.143)$$

and as $\frac{2}{\omega} = \frac{\tau}{\frac{\omega t}{2}}$ we can write Eq. 14.143 as:

$$\text{rect} = \tau \; sinc\left(\frac{\omega t}{2}\right) \tag{14.144}$$

Example: Squarewave Decomposition
The schematic ('Squarewave Decompostion.asc') is similar to the schematic ('Squarewave Harmonics.asc'). We have a square wave of ± 2 *V* and *1* *V* peak picking sinusoids starting from a fundamental of *1 Hz*. We multiply the picking signal V(si) and the square wave V(sq) and integrate over a cycle: $V = idt(V(si) * V(sq))$. This is not exactly Eq. 14.142 but half of it, remembering that we are only taking positive frequencies. The waveforms are shown in Fig. 14.18. Note that we are integrating the product of the sine wave and the square wave so the result is not a cosine but an increasing waveform during the time the square wave is positive and constant whilst the square wave is zero.

If we step the picking frequency by *.step param f 1 10 1* we shall find that the single sine wave is replaced by 10 and all have zero amplitude at intervals of 0.5 s. If we measure at the end of the period we can plot the stepped 10 values (Fig. 14.19) with the data points marked showing that the even frequencies are zero and the odd function values correctly are half those returned by *.four 1 Vsq* and are exactly in the correct ratios of *1, 1/3, 1/5. 1/7*

The Inverse Fourier Transform

The transform has given us a complete set of the frequencies, amplitudes and phases of all the waveforms in the original signal. To reconstitute the signal we only need to sample them at the same intervals used to create the DFT and add the samples together. Therefore, the exponent is the same but without the minus sign:

$$f(x) = \int_{t=-\infty}^{\infty} \frac{1}{\sqrt{2\pi}} X(\omega)e^{j\omega t} d\omega \tag{14.145}$$

And we switch between time and frequency domains. Note that if we use $2\pi f$ instead of ω there is no $\sqrt{2\pi}$ and the transform and its inverse are symmetrical.

Example: Inverse Fourier Transform
If we create an FFT of schematic ('Squarewave Decomposition.asc') of the integrated 10 harmonics V(int) and then make an FFT of that, we recover the 10 integrated waves.

Explorations 10
1. It is interesting to change the square wave to a ramp in schematic ('Squarewave Decomposition.asc')
2. Run any of the examples and find their FFT. Then right click on the FFT and from the pop-up menu select **View->FFT** and apart from ('Sinc3000.asc') the original signal will be restored.

3. Create a 'rect' wave, explore its composition for different heights and widths in terms of Eq. 14.144 and compare that to the rectangular pulse we have examined before.

Fourier Transform Tables

These can be found at several Internet sites, for example, some at: http://www. thefouriertransform.com/pairs/fourier.php. However, these are mainly of interest for mathematical manipulations. Our concern is with arbitrary signals so we move on to the Discrete Fourier Transform.

14.7   The Discrete Fourier Transform (DFT)

When we are dealing with real-life data, we need a slightly different approach to continuous integration. In fact SPICE in all its forms integrates by measuring the function at short intervals and using numerical integration to move from one data point to the next. We can see these data points by right clicking on the trace panel and selecting **View→Mark Data Points** from the pop-up menu. Taking schematic ('Fourier Test.asc') as an example, if we zoom in we find that these are typically a few *1 ms* apart but closer together where the function is changing more rapidly.

LTspice then does not handle signals continuous but at intervals of time. We must now do something similar by sampling both the signal with the picking function at fixed intervals. And we have two important facts to guide us: the first is that the signal will resolve into a series of harmonics; the second follows that the start and end values are zero, ignoring any DC term. That leaves us three questions to answer:

• How well can we select a harmonic?
• What harmonics can we detect?
• How many data points are needed to find the amplitude of a harmonic?

We address these questions in turn.

14.7.1   Selecting a Harmonic

We may recall that in section 'Derivation of the Coefficients' we found that if we multiply the signal by a cosine and integrate over a cycle, the result is zero except when the picking frequency is identical to the frequency of a component in the

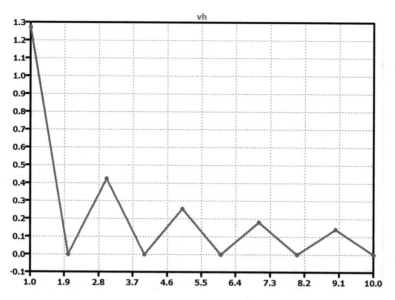

Fig. 14.19 PosNeg

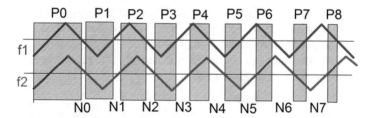

Fig. 14.20 Effect of Frequency Mismatch

signal. We can write the nth component of the signal as a sum of cosine and sine waves where its amplitude An is:

$$A_n = a_n \cos(nx) + b_n \sin(nx)$$

Then when we multiply it by the cosine picking frequency m at every point x we have:

$$A_n = a_n \cos(mx) \cos(nx) + b_n \cos(mx) \sin(nx)$$

And as we discussed at length much earlier, when we integrate we find that the result is zero unless $m = n$. This is what we do now; we multiply the signal by a series of cosine waves of increasing frequency and look for matches.

Selectivity of the Picking Frequency

With the Fourier series we knew the fundamental frequency because we knew the period of the signal. Here, that may not be so clear, especially with real-life signals such as samples of audio data. We must now ask what happens if the picking frequency is not exactly the same as a frequency in the signal, but differs only by a small amount. In other words, we are exploring how selective is the picking function. We can illustrate this with Fig. 14.20 where we have triangular waves rather than sines but that does not affect the argument. The period of $f2$ is about *10%* longer than $f1$.

At the start, the two waves are both positive so their product also is positive during period $P0$ shown hatched. But when $f1$ falls to zero, $f2$ is still slightly positive and therefore there will be a short period $N0$ during which $f1$ is negative but $f2$ is still positive and the product is negative, shown as white. When $f2$ falls to zero and then goes negative, both signals will be negative and their product will again be positive. This is period $P1$ which lasts until $f1$ crosses zero. Because of the increasing mismatched between the waves, this will be shorter than $P0$. At the end of $P1$, $f1$ turns positive, but $f2$ is still negative so the product is again negative, but this time $N1$ is longer than $N0$.

This pattern will continue with positive and negative products alternating as $f1$ or $f2$ crosses zero with the positive periods decreasing whilst the negatives ones increase until around $P5$ $N5$ they are equal. Thereafter the pattern is the reverse of the previous so that over time they are equal and the net result is zero. In the following example we also will test for harmonics. The outcome is that we can safely conclude that only a component with exactly the same frequency as the picking frequency will be measured.

Example - Frequency Mismatch

To test this, we use a BV source with a single frequency $V = 4 * \cos (2 * pi * time)$ in schematic ('Fourier Test.asc') and a test frequency using a voltage source of *SINE* (*0 1 1.1 0 0 90*). This is *1.1 Hz* with a phase angle of *90* to convert it to a cosine wave. A *10 s* run shows that the envelope of the product waveform $V(f)*V(s)$ is a shallow 'V'. It starts at 4 V with the two waves in phase but as time progresses the phase shift causes the product to diminish so that at *2.5 s* it is zero with equal positive and negative excursions. From then until *5 s* the negative portion predominates and the product reaches a minimum of *−4 V* at *5 s*. And if the run stopped here we would find the average was zero. However, continuing to *10 s* and the produce waveform increases to *4 V* at *10 s* which is a left to right inversion of the first part and the average is still zero.

What seems so surprising is the sensitivity of the test. With exactly *1.00 Hz* for both frequencies the average is *1.991 V^2* compared to *128 μV^2* with the mismatch. If we make the frequency difference even smaller, say *1.00 Hz* and *1.01 Hz*, the two can still cancel but it takes a longer run of *100 s* to find *632 uV^2*. We should also note that making the voltage source frequency twice *B1* the product over 10 s is *44 μV^2* and even less for higher harmonics.

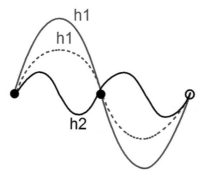

Fig. 14.21 Defining The Harmonics

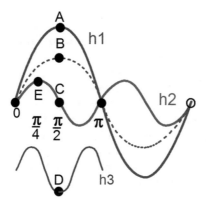

Fig. 14.22 Data Points At 1/4 Cycle

Example - Phase Mismatch

If we add a phase difference between the two waveforms, the amplitudes are not the same. For example, if we change to: *V=4*cos(2*pi*time+pi/2.5)* in schematic ('Fourier Test.asc'), it is only *61.6 mV²*. From this we can conclude that we shall need sine and cosine terms, just as with the Fourier series.

Detecting the Harmonics

We know that the start and finish amplitudes for any harmonic are zero. This is seen in Fig. 14.21 where a cycle of the fundamental *h1* is defined by two solid sample points – one at the start of the cycle and one at the mid-point. The final open point is the start of the next cycle. But we can say nothing about the amplitude since both the solid and the dotted waves pass through these points. Even worse, the second harmonic h2 also will pass through these points as will every other possible harmonic. And to cap it all, the amplitude at these points is zero. Therefore, if we discount the start and finish points, we need more than one intermediate data point.

If we measure at ¼ of the cycle in Fig. 14.22, we find data points A and B are not zero but C is. And adding a third harmonic h3 drawn below and the amplitude at D is non-zero. The fact that these are the peaks of the waves is irrelevant at the moment – we are only concerned to show that we can now detect all the odd harmonics. If we now add measurements at $\pi/4$, the amplitude of the second harmonic at E is no longer zero and similarly for higher even harmonics.

It is usual to add evenly spaced selection points in powers of two. Thus we see that with 8 points every harmonic will have a non-zero value for some points, but we are not able to identify the individual harmonics.

Example - Component Extraction

We can use schematic, ('Fourier Test.asc'). First we set the amplitude of *B1* to *1* V instead of 4 and the frequency of V1 to *1 Hz*. If we try $\mathbf{V(s)}*\mathbf{V(f)}$ to multiply the two *1* V sine waves we find an average of *0.488* V^2. We can then try any waveform we chose such as: $V = 4 * cos\,(2 * pi * time) + cos\,(3 * pi * time) + 2 * cos\,(6 * pi * time)$ to demonstrate how we can pick one waveform out of several. By changing the frequency of the picking wave we find that it returns *1.995* V^2 for the *4 V 1 Hz* component, *497 mV²* for the *1 V 1.5 Hz* and *980 mV²* for the *2 V 3 Hz*. These are close enough to half the peak values. We can run an FFT where we find three clear peaks *2.8 V* at *1 Hz*, *0.72 V* at *1.5 Hz* and *1.4 V* at *3 Hz*. The LTspice FFT window points out that the components are normalized to their RMS values. Hence when we multiply each by √ 2, we get the correct peak values.

We can cut and paste the other functions shown on the schematic as the voltage of source *B1*, or, indeed, invent any function we fancy.

14.7.2   Finding the Amplitude

In the above example, we used the LTspice ability to multiply waves together and find their average over a transient run – a continuous evaluation although actually performed at variable intervals. If we confine ourselves to sine and cosine functions with discrete samples, we expect to use something like:

$$\frac{1}{N}\sum\nolimits_{n=1}^{N} f(t_n)f(pick_n) \qquad (14.146)$$

to find the amplitude of the nth harmonic.

We find the amplitude of the function $f(t_n)$ at point n and multiply it by the value of the picking function $f(pick_n)$ at the same point and add the result to the running total to find the sum over N data points - in effect integrating the value. Finally we divide by the number of samples N to find the average of the summation. We must remember that the waveforms are peak values. Therefore, if we multiply them together, the average will be ½ the peak which gives the correct value. The question now is – How many data points do we need to find the amplitude?

Example: Simple DFT

This is schematic ('FFT Pick.asc') which consists of two cosine waves of *1 Hz 1 V* and *3 Hz 0.5 V* and a transient run of *1 s*. We select N = 8 meaning we measure every *0.125 s* by *.step param x 0 1 0.125*. Then with a *1 Hz* picking wave *I1* for f(pick$_n$) the directive *.meas Vm FIND I(I1)∗I(R1) WHEN time=x* multiplies the total current in *I(R1)* by the picking function at each point. If we run an FFT then right click on the panel and select **File→Export data as text** we can recover the file 'FFT Pick.log.txt' and find:

x	vm
0.000000000000000e + 000	1.500000e + 000
1.250000000000000e − 001	2.503540e − 001
2.500000000000000e − 001	−2.305190e − 014
3.750000000000000e − 001	2.503490e − 001
5.000000000000000e − 001	1.498600e + 000
6.250000000000000e − 001	2.503010e − 001
7.500000000000000e − 001	−2.305190e − 014
8.750000000000000e − 001	2.503490e − 001
1.000000000000000e + 000	1.500000e + 000

We discount the final reading as the start of the next cycle and find a total of 4 then divide by 8 to arrive at 0.5 A – the correct figure. We also find 0.5 A with 10 steps as we might expect since the intervals between the data points is becoming smaller and we are approaching the continuous Fourier transform.

However, with 2 steps the amplitude is 1.5 A, 4 and gives 0.75 A indicating that 8 data points are needed to resolve the amplitude if we have a single cycle of the signal, but if we have several we shall see in section 'Aliasing and the Nyquist-Shannon Sampling Rate' that we only need 2 data points.

Using Sine and Cosine

So far we have glossed over the phase. In the previous examples the two signals both were in phase with zero phase shift. It is not difficult to add a phase-shift to one and see the reduced amplitude of the resultant. In the extreme, if there is a phase difference of 90, there will be no output and that frequency will be missed. Recalling the Fourier series, we remember that, in general, we need both sine and cosine which we can encapsulate in an exponential function $e^{j2\pi}$ which describes movement around a circle (Fig. 14.23), which is drawn using the formula:

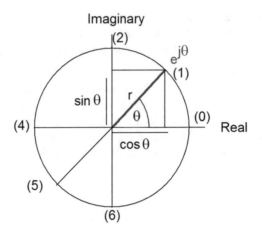

Fig. 14.23 Angles in Terms of $e^{ej\theta}$

Table 14.2 Sine and Cosine Calculated from Equation 14.147

θ	0	$\pi/4$	$\pi/2$	$3\pi/4$	π	$5\pi/4$	$3\pi/2$	$7\pi/4$
$\cos(\theta)$	1	0.707	0	−0.707	−1	−0.707	0	−0/707
$\sin(\theta)$	0	0.707	1	0.707	0	0.707	1	0.707
Point	(0)	(1)	(2)		(4)	(5)	(6)	

$$e^{i\theta} = \cos(\theta) + j\sin(\theta) \qquad (14.147)$$

To demonstrate this we draw the cosine along the x-axis – the 'Real' axis – and the sine along the y-axis – the 'Imaginary' axis. Then taking steps of $\pi/4$, we can calculate the sine and cosine of each angle marked in brackets on the figure 14.23 (Table 14.2):

As sine and cosine are orthogonal, we can calculate the length r from $\sqrt{\cos(\theta)^2 + \sin(\theta)^2}$ but as we know that $\cos(\theta)^2 + \sin(\theta)^2 = 1$ the length r is always 1 for every angle and therefore the locus is a circle. This is seen in the figure showing that as the angle increases the radius rotates anti-clockwise and if $\theta = 2\pi$ it is a complete revolution, or one cycle, corresponding to $e^{j2\pi}$. This leads to the important conclusion that if we use both sine and cosine as the picking function, we can recover the amplitude of the component as the hypotenuse of the right-angled triangle and the phase angle from the tangent.

Example - Sample Data Using Sine and Cosine

We can use schematic ('Fourier Test SC.asc') to independently measure sine and cosine contributions. We set the two voltage sources as **SINE(0 1 1 0 0 90)** which is a *1 V 1 Hz* sinusoid with a *90°* delay and therefore a cosine: the other source is **SINE (0 1 1)** which also is *1 V 1 Hz* but with no delay and thus is the sine source. We add an initial $\pi/4 = 45°$ delay to the *1 Hz* component of the signal by **+pi/4** we have:

Table 14.3 Amplitude Using Sine and Cosine

Delay angle	$\pi/4$	$\pi/3$	$3\pi/4$	π	$4\pi/3$	$3\pi/2$
$V(f)*V(s)$ (V^2)	−1.43	−1.73	−1.41	0.0	1.72	2.00
$V(f)*V(c)$ (V^2)	1.43	1.00	−1.41	−2.00	−1.00	0.0
Amplitude	2.00	2.02	2.00	2.00	2.00	2.00
Angle (°)	−45	−60	−135	−180	150	90

Table 14.4 Sine and Cosine Using $e^{2j\pi}$

Angle	0	$\pi/4$	$\pi/2$	$3\pi/4$	π	$5\pi/4$	$3\pi/2$	$7\pi/4$	2π
Cosine	1	0.707	0	−0.707	−1	−0.707	0	0.707	1
Sine	0	j0.707	j	j0.707	0	−j0.707	−j	−j0.707	0

Table 14.5 Real and Imaginary Parts

Total (V)	7,00	1.80	−0.70	−2.34	−6.00	−0.49	0.707	1.01
Real	7.00	1.27	0	1.65	6	0.35	0	0.71
Imaginary	0	1.27	−0.70	−1.65	0	0.707	−0.707	−0.71

$$V = 4 * cos(2 * pi * time + pi/4) + cos(3 * pi * time) + 2 * cos(6 * pi * time)$$

If we disable the .*step* statement and make a 2 s transient run and take the products *V* (*f*)∗*V(s)* for the sine and *V(f)*∗*V(c)* for the cosine we can hold **Ctrl** and left click the trace name to find the average of each over the whole transient run. Then we use the tangent to calculate the delay angle. The result is tabulated in Table 14.3.

If we take the positive real axis as the origin we see a clockwise rotation with constant amplitude of 2 V. The same picture emerges if we add a delay to the 3 Hz component and find the averages for the sine and cosine elements. However, in effect, this is using a continuous Fourier transform.

If we enable the .*step* statement, we can measure the products at 8 intervals. With a delay of $\pi/4$ and a 1 s run, we find the sine is 1.68 V and the cosine 1.53 V indicating an amplitude of 2.27 V

The Exponential Factor $e^{j2\pi}$

The transform proper uses this exponential as the picking function instead of separate sines and cosine. This is exactly the same as the Fourier series. We can tabulate the value of the exponential at $\pi/4$ intervals, which are close enough for manual calculations.

Table 14.6 Table 14.5 with Phase Shift

Total (V)	4.24	−2.83	−4.50	−3.07	−3.23	4.15	4.51	1.77
Real	4.24	−2.00	0	2.17	3.23	−2.93	0	1.25
Imaginary	0	−2.00	−4.50	−2.17	0	−2.93	−4.51	−1.25

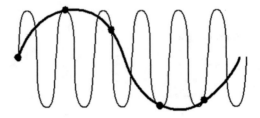

Fig. 14.24 Aliasing

Example: Sample Data Using $e^{j2\pi}$

We shall again use the data from the schematic ('Fourier Test.asc') with a *1 s* run then using Table 14.4, we can calculate the real and imaginary components using the corresponding first eight columns of Table 14.5

We total the sine and cosine components and divide each total by the number of data points *8* to give $-1.79/8 = -0.22$ and $16.98/8 = 2.12$, respectively, then square them to give *0.048* and *4.5* and take the square root of the sum to find the amplitude of *2.1 V*, and use *arctan* to find the angle of *6°*. Given the imprecision of the measurements, these are close enough to the true answers.

As a final demonstration, we add a delay of $\pi/2.5$ and find (Table 14.6):

This returns the correct angle of *72°* but the wrong amplitude, *2.3 V*. In both this case and the previous, if we extend the period the amplitude is closer to the correct *2 V*.

14.7.3   Sampling Rate and Data Points

This leads us to what are the lowest and highest frequencies that can be resolved.

Aliasing and the Nyquist-Shannon Sampling Rate

We have just seen that for a good approximation to the amplitude of a single cycle, we need at least *8* samples. However, the Nyquist-Shannon sampling theorem says that we only need just 2 data points per cycle – but not each at exactly the same time – provided we have a train of waves. This is important - it will not work with just one cycle. Therefore, we need to sample at a frequency at least twice as high as the maximum frequency of interest. If it is not, the reconstructed waveform will be at a lower frequency and the high frequency will be missed (Fig. 14.24), where the high frequency signal is missed and falsely reconstructed as a much lower frequency signal. This is aliasing.

Example: Nyquist-Shannon Sampling Rate

Schematic ('Nyquist-Shannon.asc') consists of a single 1 Hz voltage source. If we step time by .*step param x 0 2 0.04*, this gives us 50 data points. A 10 s transient run where we measure the wave by .*meas V FIND V(s) WHEN time* = *x* and an FFT accurately reproduces the 1 Hz signal. A second FFT shows that the reconstructed signal is almost completely the fundamental frequency with very little higher harmonics.

However, if we reduce the time to 1 s and the increment in x to 0.4, the FFT is completely false.

Run Time

In theory, a single waveform of the lowest frequency is sufficient to meet the Nyquist-Shannon criteria because unless we are investigating a very narrow-band signal the highest harmonic will be at least 10 times the lowest, perhaps from 1 kHz to 10 kHz. So if we take one cycle of the fundamental, the highest harmonic will satisfy the criterion, and lower harmonics, although with fewer cycles, will have many data points per cycle and the fundamental will have 20.

But it makes better resolution if we have more than one cycle. That sets the run time. So if the lowest frequency of interest is known, we try to make the run time an exact multiple of its period. If we set a transient run time that is not an integral of the period, we find low-level spurious even harmonics appearing, schematic ('Non_Integral Cycles.asc'). LTspice uses a special algorithm to mitigate this. If we only have an approximate idea of the lowest frequency, we make the run time long enough to cover several wavelengths so that the inexactitude will tend to even out. Of course, this means that frequencies lower than the minimum of interest will also be measured.

From this we see that if we know the highest frequency of interest and the transient run time, we can set the number of samples.

The DFT Equation

Our tentative Eq. 14.146 can now be fleshed out by substituting the exponential for f *(pick)* and writing X for the result.

$$X() = \frac{1}{N} \sum_{n=0}^{N} f(t_n) e^{\,j2\pi} \qquad (14.148)$$

But there is still a little work to do. First, the DC component, if there is one. This is handled by setting the frequency to zero and we have a constant I as the picking

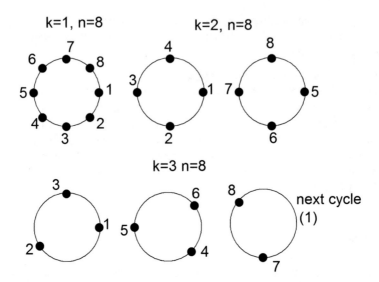

Fig. 14.25 8 Sample Points

function. This will return the average of the waveform which, of course, is the DC. Second, and this is trivial, it is conventional to use a negative exponential. This is optional and we have seen the transform works perfectly well with it positive, but we shall use it now to make it easier to compare these results with the web articles and the angles now rotate clockwise.

The second problem is how to encompass other frequencies. We do it by increasing the angle between sample points, in effect increasing the frequency by multiplying the exponential by an integer k:

$$e^{-j2\pi k\frac{n}{N}}$$

We show this in Fig. 14.25 with 8 sample points. First we see in the left-hand circle that with $k = 1$ there are 8 points spaced evenly round the circle. However, if we make $k = 2$ we only sample at $\pi/4$ intervals, effectively going round twice and omitting sample points 2, 4, 6 and 8. If we make $k = 3$ we now select every third point. And we must not be led astray by thinking that these are points at $120°$ spacing, not so, these are every third one of our original points at $45°$. If we write these as pairs of $k = 3$ points and the corresponding $k = 1$ points as $(k3, k1)$, we have $(1,1)(2,4)(3,7)(4,2)(5,5)(6,8)(7,3)(8,6)$ and we see this is a frequency of exactly three times the fundamental and the next cycle will start again at $(1,1)$ with (1) shown in brackets.

So, finally, we can write the conventional form of the transform equation which is:

$$X(k) = \frac{1}{N}\sum_{n=1}^{N-1}x(n)e^{-j2\pi k\frac{n}{N}} \qquad (14.149)$$

where $X(k)$ is the 'bin' accumulating the value for the kth frequency and $x(n)$ is the quantity of the signal measured at sample n. And we must add all the $X(k)$s for the total transform.

The DFT Matrix

Following the excellent Wikipedia article, we write the matrix for a transform of 7 frequencies and DC. The left-hand column matrix is the 'bin' for each of the frequencies, starting with DC at the top. The central 8x8 matrix are the frequencies calculated from $\omega = e^{-j\frac{2\pi}{N}}$ and the subscripts are the running values of kn which are the exponential terms in Eq. 14.149. The matrix has n increasing left to right and k increasing downwards. Starting from the left, each column increases by k. For the first row, with $k = 0$ all the terms are the same, ω_0. Therefore, the second row with $k = 1$ ω increases by 1 for each column; for the third row where $k = 2$, they are multiplied by 2 and so on.

$$
\begin{bmatrix}
X(0) \\
X(1) \\
X(2) \\
X(3) \\
X(4) \\
X(5) \\
X(6) \\
X(7)
\end{bmatrix}
=
\begin{bmatrix}
\omega_0 & \omega_0 & \omega_0 & \omega_0 & \omega_0 & \omega_0 & \omega_0 & \omega_0 \\
\omega_0 & \omega_1 & \omega_2 & \omega_3 & \omega_4 & \omega_5 & \omega_6 & \omega_7 \\
\omega_0 & \omega_2 & \omega_4 & \omega_6 & \omega_8 & \omega_{10} & \omega_{12} & \omega_{14} \\
\omega_0 & \omega_3 & \omega_6 & \omega_9 & \omega_{12} & \omega_{15} & \omega_{18} & \omega_{21} \\
\omega_0 & \omega_4 & \omega_8 & \omega_{12} & \omega_{16} & \omega_{20} & \omega_{24} & \omega_{28} \\
\omega_0 & \omega_5 & \omega_{10} & \omega_{15} & \omega_{20} & \omega_{25} & \omega_{30} & \omega_{35} \\
\omega_0 & \omega_6 & \omega_{12} & \omega_{18} & \omega_{24} & \omega_{30} & \omega_{36} & \omega_{42} \\
\omega_0 & \omega_7 & \omega_{14} & \omega_{21} & \omega_{28} & \omega_{35} & \omega_{42} & \omega_{49}
\end{bmatrix}
\begin{bmatrix}
x(0) \\
x(1) \\
x(2) \\
x(3) \\
x(4) \\
x(5) \\
x(6) \\
x(7)
\end{bmatrix}
\tag{14.150}
$$

Thus to find $X(1)$ we multiply $x(1)$ by each of the angles in the second row. We can make this clearer by identifying the 8 data points we drew as dots in Fig. 14.25 as the terms of the exponential (Fig. 14.26). Then the angles are shown in the following matrix. So for $k = 1$ we read the second line with $x(1)$, and moving clockwise we find

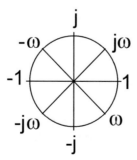

Fig. 14.26 Fourier Data Point Angles

1 for the first point at *0°*; then the second point is *ω* at *45°*, the third is at *−j* and so on, landing on all 8 points. But for *k = 2* we again start at *1* but only measure at *90°* intervals and so the next point is *−j* and the third is *−1*. Likewise for the following values of *x*, we measure every 3rd, 4th, 5th, 6th, and 7th angle as shown in the matrix.

$$
\begin{matrix}
1 & 1 & 1 & 1 & 1 & 1 & 1 & 1 \\
1 & \omega & -j & -j\omega & -1 & -\omega & j & j\omega \\
1 & -j & -1 & j & 1 & -j & -1 & j \\
1 & -j\omega & j & \omega & -1 & j\omega & -j & -\omega \\
1 & -1 & 1 & -1 & 1 & -1 & 1 & -1 \\
1 & -\omega & -j & j\omega & -1 & \omega & j & -j\omega \\
1 & j & -1 & -j & 1 & j & -1 & -j \\
1 & j\omega & j & -\omega & -1 & -j\omega & -j & \omega
\end{matrix}
\qquad (14.151)
$$

And although we have used an 8 × 8 matrix, as we have seen, this is the minimum and we have better results if we extend it to 16 or more.

Errors

There are two potentially important errors.

Aliasing
We have discussed this above. If we want to distinguish high frequencies we must increase the density of the data points. But if we do not want them we should filter them out before making the DFT. We might note in passing that as we add more harmonics, the amplitudes of the previous ones change, albeit slightly. This is seen in the sequence of schematics ('Fourier 1,2,3,5.asc') where the amplitude of the fundamental falls by some 5% as we go from 2 harmonics to 7.

Leakage
If the sample is not taken over exactly one period of the input – and if the sample is not periodic, for example, speech rather than a note played on a musical instrument – then the transform will be corrupted. The solution is to use a *windowing function* to diminish the weighting of the signal at the beginning and end of the sample.

14.8   The Fast Fourier Transform

The solution of the set of Eq. 14.150 is prohibitively time-consuming for all but simple circuits. There are several algorithms in use, the most famous being that reported by Cooley and Tukey in 1965. And LTspice have created their own.

We can gain some insight into the process by noting that the ωs are not all different and relate to the angle of which there are only the values of Fig. 14.25. For instance, ω_2, ω_{10}, ω_{18} and ω_{42} all resolve to $-j$ in Eq. 14.151. And we can see that half of these values are just the negative of the others. This means that direct evaluation results in many redundant calculations. The second point to note is that whilst complex additions are fast, multiplication is not, so the FFT seeks to reduce them.

The starting point is that we can split each summation into odd and even parts using $m = k/2$ if k is even and $m = (k - 1)/2$ if k is odd.

$$X(k) = \sum_{m=0}^{\frac{N}{2}-1} x(2m)e^{-j2\frac{\pi}{N}k2m} + \sum_{m=0}^{\frac{N}{2}-1} x(2m+1)e^{-j2\frac{\pi}{N}k(2m+1)} \qquad (14.152)$$

where each summation is over the appropriate half of the samples; thus the first summation is over $x(0)$, $x(2)$, $x(4)$... and the second over $x(1)$, $x(3)$, $x(5)$...

Turning to the second summation, we note that: $e^{-j2\frac{\pi}{N}k(2m+1)} = e^{j2\frac{\pi}{N}k}e^{-j2\frac{\pi}{N}k2m}$ so we can rewrite the second summation as:

$$\sum_{m=0}^{\frac{N}{2}-1} x(2m+1)e^{-j2\frac{\pi}{N}k(2m+1)} = e^{j2\frac{\pi}{N}k}\sum_{m=0}^{\frac{N}{2}-1} x(2m+1)e^{-j2\frac{\pi}{N}k2m} \qquad (14.153)$$

and we have the same summation as the first. The first part of Eq. 14.152 – the even summation – is conventionally written as $G(k)$ and the odd summation as $H(k)$ and we have:

$$X(k) = G(k) + e^{j2\pi\frac{k}{N}}H(k) \qquad (14.154)$$

Now let us look carefully at the matrix of Eq. 14.150. If we imagine it split vertically in half giving 4 columns each, we find for the even rows starting from 0 that the right-hand 4 columns are identical to the left-hand 4 columns whilst for the odd rows starting from 1 the right-hand 4 columns are the negative of the left-hand 4. For example, taking the second row, the first element is 1 and the fifth -1. Then if we expand the exponentials as sines and cosines and add k, where k is even, 0, 2, 4, 6, the angles are unchanged; but if k is odd, 1, 3, 5, 7..., the sign is reversed. So, for our 8 sample example, $G(k + 4) = G(k)$ and $H(k + 4) = H(4)$, and Eq. 14.154 is:

$$X(k + 4) = G(k) + e^{j2\pi\frac{(k+4)}{N}n}H(k) \qquad (14.155)$$

and the number of multiplications is greatly reduced.

This is the starting point for the FFT. The complete description can be found at: https://www.allaboutcircuits.com/technical-articles/an-introduction-to-the-fast-fourier-transform/ and http://www.robots.ox.ac.uk/~sjrob/Teaching/SP/l7.pdf and also on YouTube.

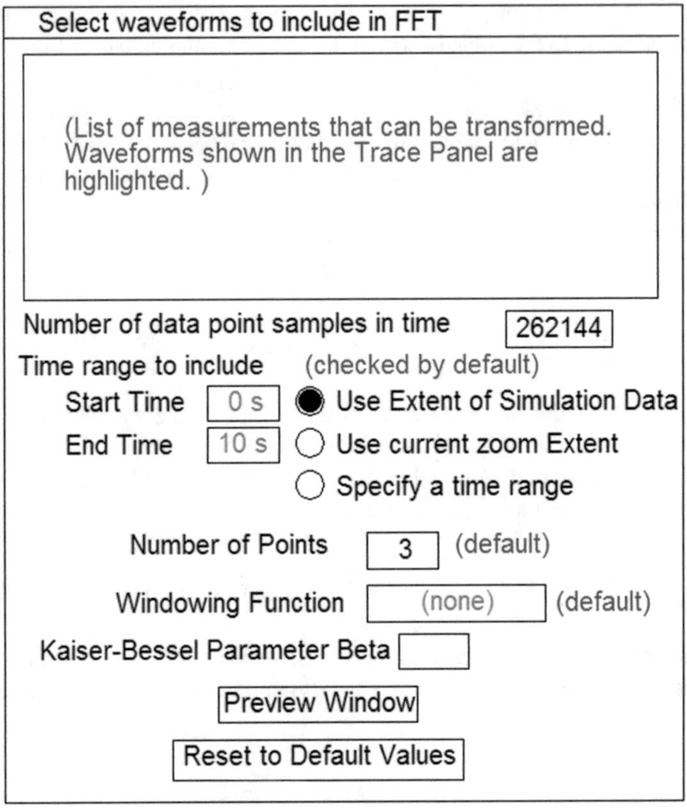

Fig. 14.27 LTspice FFT Dialogue

14.8.1   LTSpice and the FFT

We access this by a right click in the trace panel and **View→FFT.** This will open one
of two dialogues depending on the analysis. If we have made a transient analysis, the
following dialogue will appear, else from a frequency analysis

The FFT Dialogue from 'tran.' Analysis

This will open the dialogue **Select Waveforms to include in the FFT.** The
important parts are sketched in Fig. 14.27. The data is first normalized to the RMS
value of the signal. We shall now discuss the controls starting at the top.

Select Traces to Transform
This shows all the possible values that can be used. It does not matter if they have
been drawn as a trace or not, they can still be transformed.

Number of Data Points in Sample Time

As we have seen, this does not actually set the high frequency limit of the transformation, but the range of frequencies that can be resolved. Going upwards it increases by factors of two but any value can be entered directly; most times the transform runs quickly so 65536 is a good choice to start with.

A quick test is to transform the signal then right click in the transformed signal panel and select **View→FFT** from the pop-up menu to transform it back again and see how close it is to the original.

Example: Selecting Number of Data Points

We want to analyse an audio signal. A good range is from *20 Hz* to *20 kHz*. The low frequency has a period of *50 ms* so a run of *200 ms* would be appropriate. To resolve the amplitude of the high frequency, we need samples at *12.5 μs* intervals. So *16000* samples will suffice. It is usual to select a power of *2* for the data points, although LTspice does not insist on it, and the nearest is *16384*.

If we run schematic ('FFT Audio.asc') with an FFT, we find the *20 Hz* signal well resolved and of the correct height. However, there are many intermediate false frequencies before the *20 kHz* peak – which is too small, being only *385 mA*. Increasing the data points to *131072 Hz* increases it to *686 mA* or a peak of *970 mA* which is still slightly less than the true *1A*.

Example: Low Frequency Limit

The schematic ('FFT Limit.asc') has signals of *1 Hz* and *2 Hz*. Using the previous technique with a run time of less than *1 s*, the *1 Hz* signal is not resolved. It is seen with a *1 s* run and is clearer with *2 s*. However, it is difficult to pick it out with *2.1 s* or *2.2 s*.

Time Range

By default, this is the extent of the simulation shown shaded in the **Start Time** and **End Time** edit boxes which are disabled. But there could be starting transients, so we can zoom the trace and use that or we can set start and finish times by clicking the radio button **Specify a time range**. Note that these are independent of the trace start and finish.

Windowing Function

There is a drop-down list of functions. These all work to reduce spectral leakage by tapering the response, most of them in roughly a bell-shaped curve, with the maximum at the middle of the window. These are well described by the Wikipedia article. https://en.wikipedia.org/wiki/Window_function. Their effects can be seen by clicking the large **Preview Window** button. For example, if we click in the edit box and select the first *Bartlett* and then preview it, we find straight lines from the centre sloping down to each extreme.

Most of the other window functions use a more-or-less bell-shaped function, and in some cases we can modify it, for example the *Kaiser-Bessel Parameter Beta* heightens the centre of the bell-shaped curve and steepens the sides so that with a value of more than *7* the flanks of the curve are reduced to almost zero and the range of significant frequencies continues to narrow as the number is increased. This is easily seen in the preview.

Example: Windowing Function

Examples of the *Bartlett, Bohun, Kaiser-Bessel* and *Hann* functions can be seen by **File→Open** then at the bottom right of the screen selecting **Waveforms(∗.raw;∗.fra)**. We can also open schematic ('Windowing.asc'). The **.four** analysis returns the harmonics in the ratios *1, 1/3, 1/5, 1/7*. If we make a Fourier analysis of the whole run, or change the analysis times to an exact period and using a linear y-axis we find the same ratios between the harmonics. But taking more or less than one period, for example, by 1.7 s to 3.1 s, and FFT is wrong. Using a Windowing Function attenuates the higher frequency components.

Data Density

When we look at an FFT result, the higher frequencies often appear as grass if we use the default logarithmic x-axis. The reason is this: suppose we start with a fundamental of *100 Hz*. Then in the decade to *1 kHz* we will find *200 Hz, 300 Hz...900 Hz*: that is *9* harmonics. Then the next decade from *1 kHz* to *10 kHz* will have the same width on the default logarithmic x-axis. But the harmonics will be at *100 Hz* intervals in every *1 kHz* giving *1.0 kHz, 1.1 kHz...* and so on, then later we have *5.0 kHz, 5.1 kHz* and finally *9.7 kHz, 9.8 k Hz, 9.9 kHz*. In other words, ten times as many data points and thus the data density is ten times greater. And for the next decade it will be ten times more again. This can be resolved by using a linear horizontal scale.

Saving the Data

From the FFT panel, we can right click and then from the pop-up menu **File→Export data as text** will open the dialogue **Select Traces to Export**. Note that the **Format** only offers two choices, and these override the settings of the FFT panel. We used this in 'Example: Simple DFT'.

The Impulse Response

This is accessed from an *.ac* run by right clicking on the trace panel and selecting **View→FFT**. The effect is to give a sharp input pulse and follow the time response of the circuit, very similar to a step input only the pulse is very short. This impulse is theoretically the sum of an infinite series of identical cosines waves so that the central maximum increases with each additional wave but outside of that the waveforms tend to cancel ultimately leading to an infinitely high peak of zero width.

The response of almost all the passive circuits we have developed over the course of this book is just a short burst of high frequency oscillation that decays very rapidly. However, an under-damped tuned circuit does have an interesting response

Example: Tuned Circuit Impulse Response
We open schematic ('Series Resonance Example.asc') in the 'LCR Tuned Circuits' folder and make $R1 = 1\ \Omega$. An **.ac** run shows the expected very sharp resonance. An FFT shows a damped response starting with an alarming initial peak of *629 kV* for the capacitor. However, using the good approximation that this is $V_{in}/\sqrt{(L_1C_1)}$ we have the same answer, given that the input is *1 V*. Also, from the same chapter, the decay envelope is controlled by $e^{-\alpha t}$ where $\alpha = 1/\sqrt{(L_1C_1)}$, so at any time t the voltage is $629 \times e^{-0.987\,t}$, and if we evaluate this at *10 ms* it is *234 V* and we read the same from the FFT.

We can repeat with the current to find an initial current of *200 A* and an identical decay. However, an FFT of the inductor voltage shows a linear fall from *2.0 MV* at time $t = 0$ to zero at time $t = 0.5\ \mu s$. This is because we are looking at the series combination of the inductor and capacitor. If we switch the circuit as schematic ('Impulse.asc') so that the inductor is connected to earth, we find the same decay envelope and the previous initial sharp pulse of *0.5 μs*.

If we perform another FFT on the *V(c)* voltage FFT trace, we find a sharp resonance at *100 kHz* – which is the resonant frequency of the circuit.

We can compare this with exciting the circuit with a brief *1 V* pulse and making a transient run.

Example: Current Impulse
To illustrate the concept of an impulse, we can create a current impulse because we can more easily add currents than voltages, schematic ('Current Pulse.asc'). We can see that as we add more and more frequencies the central peak increases in amplitude and the skirts reduce, although some small ripple remains.

14.9   Summary

The Fourier series and Fourier transform are powerful tools for analysing signals.

- The Fourier series is applicable to a known function and resolves it into a DC term $a_0/2$ and a sum of harmonics. We can use the trigonometrical *form* using sines and cosines, but in many simple cases only one is needed $f(x) = \frac{a_0}{2} + \int_{n=1}^{\infty} a_n \cos{(nx)}dx + \int_{n=1}^{\infty} b_n \sin{(nx)}dx$.
- To select the cosine terms, we use a picking function of *cos(nx)* because *cos (mx)cos(nx) = 0* if $m \neq n$ and *1* if $m = n$. In terms of angle $a_n = \frac{1}{\pi}\int_{x=-\pi}^{\pi} f(x) \cos{(nx)}dx$.
- Similarly we use the picking function *sin(nx)* to find the b terms and: $b_n =$

$\frac{1}{\pi} \int_{x=-\pi}^{+\pi} f(x) \sin(nx) dx.$

- We generally need to integrate the function by parts: $\int u dv = uv + \int v du.$
- If the function is symmetrical, we can integrate over half the period and double the result.
- *Gibb's Phenomenon* means that square waves have an overshoot at start and finish.
- The *exponential form* is more compact $f(x) = \int_{n=-\infty}^{\infty} c_n e^{jn\omega_0 t} dt$ which contains both sine and cosine. The picking function is: $c_n = e^{-jm\omega_0 t}.$
- The relationship to the trigonometrical form is: $c_n = \frac{1}{2}(a_n - jb_n)$ $c_n^* = \frac{1}{2} \times (a_n + jb_n).$
- The series of harmonics often can be written using the function sinc(x) = sin(πx)/ (πx).
- If the duty cycle is very short, the sinc function for low harmonics tends to sin(0)/(0) and all they all have the same amplitude of unity
- To avoid this, if we want to resolve a single pulse where the period tends to infinity and the duty cycle δ to zero, we need high harmonics so that sinc(δn) does not tend to zero.
- The spacing of the harmonics tends to zero, so we now replace summation by integration. This is the continuous time Fourier transform.
- The discrete Fourier transform (DFT) does not rely on a known function but multiplies a given signal by a picking function at discrete intervals of time.
- The resolution depends on the intervals between samples so for a run of t s with n data points the highest frequency whose amplitude can be resolved is $f_{max} = t/2n$
- The fast Fourier transform returns the same result as the DFT but in a vastly shorter time.
- Windowing functions enable us to reduce the weighting of the ends of the spectrum.

Chapter 15
Passive Filters

15.1 Introduction

In past chapters we discussed filters that were made from just a handful or passive components; typically they were configured as a 'T' or 'Π' or two cascaded 'L' sections. The analysis was not too difficult, although in some cases the equations became rather unwieldy. We found that, excluding the special cases of notch filters and band-pass tuned circuits, the slope of the transition from pass band to stop band and vice versa was 60 dB/decade at the most.

Now we shall explore a general approach applicable to both passive and active filters.

15.2 Nyquist Plot

This is a very useful technique for testing the stability of feedback systems. It is more time-consuming to draw by hand than the familiar Bode plot and not so easy to interpret in the case of passive circuits. However, we introduce it here because it is a plotting option offered by LTspice, and it is possible to draw some pretty patterns, so we shall say a little about it. A good starting point is the excellent series of articles by Robert Keim beginning at https://www.allaboutcircuits.com/technical-articles/how-to-use-a-nyquist-plot-for-ac-analysis/.

The difference between it and the Bode plot is that we can show amplitude and phase on the one trace, however we then no longer see the frequency, but LTspice enables us to measure it. In essence it is an Argand Diagram using a real horizontal axis and an imaginary vertical one. So the first task is to separate the real and imaginary parts of the equation.

© Springer Nature Switzerland AG 2020
C. May, *Passive Circuit Analysis with LTspice*®,
https://doi.org/10.1007/978-3-030-38304-6_15

15.2.1   LTspice and the Nyquist Plot

We start with an AC analysis which, by default, creates a Bode plot.

Creating a Nyquist Plot

If we click in the left-hand vertical scale area of the Trace Panel – which by default is in decibels – we open a small dialogue **Left Vertical Axis –> Magnitude**, and in the drop-down list under **Representation**, we select *Nyquist*.

Making Measurements

We can measure the real and imaginary components of the output and the total.

Measuring Real and Imaginary Components
The only way is to move the mouse over the trace when we can read off the real and imaginary parts in the left-hand corner of the status bar at the bottom of the screen. This is subject to the usual inaccuracies of placing the cross-wires which can be mitigated by expanding the trace.

Measuring Frequency, Magnitude and Phase
This is accomplished by setting a cursor on the trace which opens a dialogue. An annotated sketch is Fig. 15.1. The heading [1] is the name of the schematic. Underneath the cursor identification is the name of the quantity being measured [2]. The identification of the horizontal axis [3] may either be **Horiz** or **Freq**, LTspice seems uncertain which to use, but in either case, it is frequency. The magnitude [4] is the resultant of the real and imaginary parts and the phase is [5].

Below are the readings for the second cursor and finally the ratio between the two.

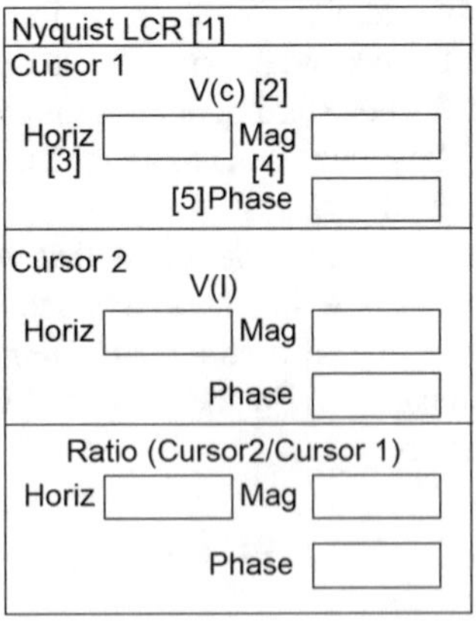

Fig. 15.1 Nyquist dialogue

Moving the Cursor

The Bode plot shows that in some cases, there may be several decades of frequency during which the magnitude and phase change very little. This is exemplified by the schematic ('Nyquist LCR.asc'). The effect is that all these data points will be coincidental at the start or end of the Nyquist plot and it may not be possible to move the cursor with the mouse, only with the arrow keys which must be held down for some considerable time. However, if we first set the cursor position on a Bode plot, this frequency is retained when we switch to a Nyquist plot and this can make things easier. Incidentally, this also demonstrates a very useful feature of the Nyquist plot which is that the resonance region is spread out over the whole width of the plot and can be examined in detail more easily than on a Bode plot.

Also, as there are mainly two imaginary values for each real, to traverse the complete loop, when the left or right extremities are reached, the arrow keys must be used to move to the upper or lower branch of the plot.

15.2.2 Real and Imaginary Parts

The form of the plot is that the x-axis is the real part of the transfer function and the y-axis the imaginary part. Therefore to draw the plot by hand, we must separate the real and imaginary parts of the transfer function although, of course, LTspice does it for us. In the following example, we shall compare the calculated results with those from LTspice.

Example – LR Circuit

The circuit is Fig. 15.2 where the voltage gain and output voltage are:

$$\frac{v_{out}}{v_{in}} = \frac{R}{R + j\omega L} \quad v_{out} = v_{in}\frac{R}{R + j\omega L} \quad (15.1)$$

If we set $v_{in} = 1\ V$ we can cancel it in Eq. 15.1. We now separate the real and imaginary parts by multiplying by the complex conjugate:

$$\frac{R(R - j\omega L)}{R^2 + \omega^2 L^2} \quad \Re = \frac{R^2}{R^2 + \omega^2 L^2} \quad \Im = \frac{-\omega L R}{R^2 + \omega^2 L^2} \quad (15.2)$$

and notice that the imaginary part is negative. The schematic is ('Nyquist LR.asc'). In order to see the calculated points, we set the simulation to:

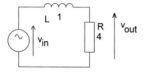

Fig. 15.2 Circuit for Nyquist LR

.ac dec **1** **1m 100**

which is just one point per decade with $R=4$ but omits $\omega = 0$ and creates a very disjointed trace. If we insert the component values we have: $\Re = \frac{16}{16+\omega^2}$ $\Im = \frac{-4\omega}{16+\omega^2}$

We can then tabulate the calculated results at decade intervals in the fourth and fifth rows of the following table. Then in the trace panel, right click and select **View→Mark Data Points** to see them. If we slide the cursor over the trace panel, the measurements on the plot agree. Easier is to set up a dummy variable f and step it with the frequency *.step dec param f 1m 100 1* and measure $V(r1)$ when the *freq = f* - but then we have to convert dBs to numbers.

Frequency (Hz)	0	0.001	0.01	0.1	1	10	100
ω (rad/s)	0	0.00628	0.0628	0.628	6.28	62.8	628
$R^2 + \omega^2 L^2$	16.000	16.000	16.004	16.394	55.44	3960	3.944e5
Calc Real (mV)	1000	1000	1000	976	288.6	4.040	40u
Calc Imag (miV)		−1.57	−15.7	−154	−453	−63.4	−6.4
Magnitude (mV)		999.7	999.6	987	537	63.4	6.37
Phase (degrees)		−0.09	−0.9	−8.9	57.5	83.4	89.6
Point ID	a	b	c	d	e	f	g

We should note that the real part shows a decrease with frequency, whilst the imaginary part increases towards e then decreases. If we increase the number of points to 2 or 3 per decade, we see the data points beginning to crowd at the start and end of the plot. To make accurate measurements, it is essential we increase the number of data points per decade to *1e3* or so; then if we set a cursor on the trace, we will see the frequency and the magnitude and phase in the dialogue at the bottom right of the screen. These are the values in the sixth and seventh rows of the table. The phase is measured between output and input voltages which we can check with schematic ('Nyquist LR Transient.asc'). The final row is identification letters of the points on the Fig. 15.3 where we see frequency increasing clockwise from right to left. We can measure the frequency of the 3 dB point which is *636 mHz* where the phase is *45°*, and this is the maximum imaginary component of *500 miV*, and the real component is *0.5 V*; these agree with calculation.

Example – CR Circuit
This is Fig. 15.4 where the output voltage of the left-hand circuit*(a)* is:

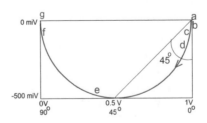

Fig. 15.3 Nyquist LR plot

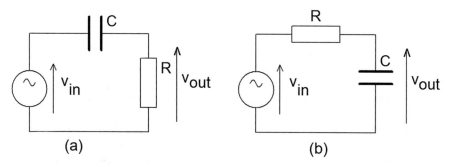

Fig. 15.4 Nyquist CR Circuits

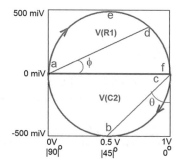

Fig. 15.5 Nyquist CR plot

$$v_{\text{out}} = v_{\text{in}} \frac{R}{R + \frac{1}{j\omega C}} \quad \Rightarrow \quad v_{\text{out}} = v_{\text{in}} \frac{j\omega CR}{1 + j\omega CR} \tag{15.3}$$

being the voltage across the resistor, $V(r1)$. Assuming a $1\ V$ input, we can ignore the input voltage when calculating the voltage gain and separate the real and imaginary parts as:

$$v_{\text{out}} = \frac{j\omega CR(1 - j\omega RC)}{1 + \omega^2 C^2 R^2} \quad \mathfrak{R} = \frac{\omega^2 C^2 R^2}{1 + \omega^2 C^2 R^2} \quad \mathfrak{J} = \frac{j\omega CR}{1 + \omega^2 C^2 R^2} \tag{15.4}$$

For the right-hand circuit(b), the capacitor and resistor are interchanged, and we have:

$$v_{\text{out}} = \frac{\frac{1}{j\omega C}}{\frac{1}{j\omega C} + R} = \frac{1}{1 + j\omega CR} \quad \mathfrak{R} = \frac{1}{1 + \omega^2 C^2 R^2} \quad \mathfrak{J} = \frac{-j\omega CR}{1 + \omega^2 C^2 R^2} \tag{15.5}$$

where the imaginary part has a negative frequency.

If we create the Nyquist plot for both circuits together, we find half an ellipse for each which, in fact, is a distortion of the true circle. If we start with Eq. 15.4 and take a frequency of zero hertz, the real and imaginary parts of $V(r1)$ are both zero at point a on Fig. 15.5, and the phase angle φ is $90°$. At high frequency, we can ignore the

1 in Eq. 15.4 to leave $\Im = \frac{j\omega CR}{\omega^2 C^2 R^2} = \frac{1}{\omega CR}$, and this decreases with frequency. At intermediate frequencies, we note that the form of the equation is similar to 15.1, and we again find an increase up to *500 miV* with a real part also of *500 mV* measured along the x-axis at point *e*. So as the frequency increases, the output moves clockwise and tends to *1* for the real part and *0* for the imaginary at *f*. At any intermediate point *d*, we can measure the phase angle φ and the real and imaginary parts.

Taking the right-hand circuit, we have a negative imaginary part, and at a frequency of zero, the real part is unity and the imaginary part zero so we are at point *c* with a phase of $0°$. As the frequency increases, the output falls towards zero at point *a* with an angle of $-90°$ following a clockwise path. We can again find the 3 dB point at *b* with a phase angle θ of $-45°$ and equal real and imaginary parts.

Using the Laplace 's' Form

If we start with the transfer function in 's' form, we convert it to frequency and expand it and collect the real and imaginary terms, for example:

$$H(s) = \frac{s}{(s+1)(s+20)} = \frac{j\omega}{(j\omega+1)(j\omega+20)} = \frac{j\omega}{(20-\omega^2)+j21\omega} \qquad (15.6)$$

Multiplying by the complex conjugate of the denominator:

$$H(s) = \frac{j\omega[(20-\omega^2)-j21\omega]}{(20-\omega^2)^2+(21\omega)^2} \qquad (15.7)$$

Then we separate the real and imaginary:

$$\Re = \frac{21\omega^2}{\omega^4+441\omega^2+400} \qquad \Im = \frac{j(20\omega-\omega^3)}{\omega^4+441\omega^2+400} \qquad (15.8)$$

And we see that even a fairly simple transfer function is going to take some time to calculate for more than a few points.

Example – 's' Notation
This is schematic ('Nyquist s.asc') where we take a restricted frequency range with only one point per decade *.ac dec 1 1 1e4*.

This, of course, gives a false picture of the true plot but does enable us to see the data points. Solving Eq. 15.8 for *1 Hz*, we find (*46.6 mV, −6.88 miV*) which agrees with measurement. Similarly at *10 Hz*. At higher frequencies, with negligible error, we can keep only the term in ω^4 in the denominator, so the real part is $21/\omega^2$ and the imaginary $-1/\omega$. We tabulate:

Frequency (Hz)	100	1000	10,000
Real (V)	53.2 μ	0.532 μ	5.3 n
Imaginary (V)	1.59 m	159 μ	15.9 μ

If we reduce the start to *1 mHz* by *.ac dec 1 1m 1e4*, we find the start moving to the left and is *(2.10 µV, 315 µiV)*. This time we can approximate the transfer function that the real part is $21\omega^2 400$ and the imaginary is $20\omega 400$, and these agree with the measurements. Or we can set up a dummy parameter *f* as we did for the 'Example - LR Circuit' and get the exact values.

If we increase the number of points per decade to more than *100*, the plot is an ellipse. But note that the data density is highest at the left- and right-hand sides of the trace. This follows, at least qualitatively, from our explorations below.

15.2.3 Second Order Filters

If the filter is simply cascaded RC or LR sections, these do not have complex-conjugate poles, and the Q-factor is always 0.5. This changes dramatically if we have a resonant circuit such as schematic ('Nyquist LCR.asc') which is a repeat of the series resonance circuit in chapter 'Tuned Circuits'. Given a *1 V* input, the output across the capacitor is found by regarding the circuit as a complex potential divider then multiplying top and bottom by *1/jωC*:

$$v_c = \frac{1}{1 - \omega^2 LC + j\omega CR} \tag{15.9}$$

Multiplying by the complex conjugate gives:

$$v_c = \frac{1 - \omega^2 LC - j\omega CR}{\left(1 - \omega^2 LC\right)^2 + \omega^2 C^2 R^2} \tag{15.10}$$

and then:

$$\Re = \frac{1 - \omega^2 LC}{\left(1 - \omega^2 LC\right)^2 + \omega^2 C^2 R^2} \qquad \Im = \frac{-j\omega CR}{\left(1 - \omega^2 LC\right)^2 + \omega^2 C^2 R^2} \tag{15.11}$$

Example – Series Resonance
If we assign values $L = 1\ mH$, $C = 1\ µF$, $R = 0.1\ \Omega$ in Eq. 15.11 at a frequency of *1 kHz* the products $\omega^2 LC \approx 0.04$ and $\omega^2 C^2 R^2 = 3.95 \times 10\text{-}7$ are negligible so the real part is effectively a constant *1.0 V* and the imaginary part is zero. If we make an AC run of *.ac dec 1e5 100m 10k* with a Bode plot, we find a sharp resonant peak at *5.03 kHz* and a magnitude of more than *313 V* and a phase of $-90°$, which agrees

with $Q = \omega_n L/R$. Changing to a Nyquist plot, we find an ellipse with horizontal limits of 156.6 V, $-45°$, 5.025 kHz and -156.1 V, $-135°$, 5.041 kHz both with a vertical value of -157 iV and the cursor reporting magnitudes of 224.4 V and 226 V, respectively, which agree with the resultant of the real and imaginary parts. These are the 3 dB points and can be checked by expanding the Bode plot. The maximum occurs at the very lowest point of the ellipse and is 0 V, 312.6 iV at $90°$ in agreement with the result from the Bode plot.

In the case of $R = 1\ \Omega$ with the values $L = 1$ mH, $C = 1\ \mu F$ then $LC = 10^{-9}$, we can safely approximate the real part to 1 V up to a frequency of 1 kHz. At 10 kHz we have: $\omega^2 = (2\pi f)^2 = 3.95e9$ $\omega^2 LC = 3.95$ $\omega^2 C^2 R^2 = 3.95e{-}3$ we calculate: $\Re = \frac{1-3.95}{(1-3.95)^2+3.95\times10^{-3}} = \frac{-2.95}{8.706} = -0.339V$ and agrees with measurement. Likewise the imaginary part is some $-726\ \mu V$.

We should note here that we are using one possible definition of resonance for a series LC circuit including resistance, and that is the maximum voltage across the resistor. If we increase the resistance to $R = 10\ \Omega$, we can see this more clearly because the resonant frequency is $\omega_0 = \sqrt{\frac{2L-CR^2}{2L^2C}} = \sqrt{\frac{2.10^{-3}-10^{-6}.10^2}{2.10^{-6}.10^{-6}}} = 4.91kHz$ compared to 5.03 kHz in the absence of any resistance. If we measure on the Bode plot, we find the resonance at 4.926 kHz with an amplitude of 10.1091 dB $= 3.202$ V with an angle of $80.7673°$. The maximum voltage on the Nyquist plot is 3.202 V at a frequency of 4.905 kHz; but what is of interest, and easily measured here, is that the phase angle is $80.7°$ not $90°$. This applies also to the 3 dB points where we measure a magnitude of $3.202/\sqrt{2} = 2.26$ V, and we find 4.02 kHz, $-34.7°$ and 5.66 kHz, $-126.6°$ which are the same frequencies as the Bode plot.

We might also note that with a 1 V AC input the Q-factor comes directly from the resonant voltage which we calculate as $Q = \omega_0 \frac{L}{R} = \frac{4905\times2\times\pi\times10^{-3}}{10} = 3.08$. This is in fair agreement with the measurement.

Explorations 1
1. Explore the inductor voltage.
2. We may alternatively define resonance as a phase angle of $90°$. Explore.
3. Open ('Bridged-T.asc') from the chapter 'Second Order RC Filters'. The Nyquist plot is a series of ellipses which all have the point $(1.00$ V, $0.00iV)$ in common. Reduce the number of data points per decade, and explain why the ellipses increase in size with $R1$.
4. This is the transfer function of schematic ('Nyquist RCCR.asc') taken from the chapter 'Second Order RC Filters' section 'Cascaded 2 L-Flters' equation: 4.4.2.6
   $$H(s) = \frac{s}{s^2 C_1 R_1 + s\left(\frac{R}{R_2}+1+\frac{R_1 C_1}{R_2 C_2}\right)+\frac{1}{C_2 R_2}}$$ Construct the Nyquist plot.
5. The filter ('Band-Pass Constant K Pi.asc') creates an interesting Nyquist plot but it is difficult to resolve the points
6. Feel free to draw the Nyquist plots of other circuits, in particular, plot currents as well as voltages.
7. Note the changes in the shape of the plot depending on the start and finish frequencies.

15.3 Pole-Zero (s-Domain) Analysis

This is principally used in the exploration of the stability of feedback systems, but they are also of use here. We have encountered the idea of drawing the real and imaginary parts of impedances. Remembering that $s = \sigma + j\omega$, a passive circuit can never have a positive σ; the plot is entirely confined to the left-hand side of the plane, including the imaginary axis. A very useful website is https://www.maximintegrated.com/en/design/technical-documents/tutorials/7/733.html. Also Chap. 8 of *Analog Devices* handbook https://www.analog.com › media › design-handbooks › Basic-Linear-Design is invaluable.

We should note that LTspice does not create this plot.

15.3.1 First-Order Systems

We can only realize low-pass and high-pass filters with a resistor and either a capacitor or an inductor, but not both. These have been analysed extensively in previous chapters and we only use them here as a convenient lead-in to more complex plots.

Low-Pass Filter

These can be constructed from an RC combination, Fig. 15.6:

$$H(s) = \frac{v_{out}}{v_{in}} = \frac{\frac{1}{sC1}}{R1 + \frac{1}{sC1}} = \frac{1}{sC1R1 + 1} = \frac{\frac{1}{C1R1}}{s + \frac{1}{C1R1}} = \frac{\omega_n}{s + \omega_n} \qquad (15.12)$$

or an RL combination, Fig. 15.7 where we have:

$$H(s) = \frac{R1}{sL1 + R1} = \frac{\frac{R1}{L1}}{s + \frac{R1}{L1}} = \frac{\omega_n}{s + \omega_n} \qquad (15.13)$$

all this is familiar stuff and means that the general transfer function for both circuits is:

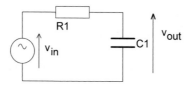

Fig. 15.6 First-order LP RC Filter

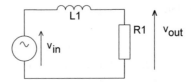

Fig. 15.7 First-order LP LR Filter

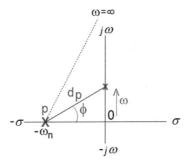

Fig. 15.8 First-order LP pole zero

$$H(s) = A\frac{\omega_n}{s + \omega_n} \tag{15.14}$$

where A is a constant which for all of our passive circuits is unity and will be ignored from now on.

Poles and Zeros

if we replace s by $\sigma + j\omega$ in Eq. 15.14 and set $A = 1$ it becomes:

$$H(s) = \frac{\omega_n}{\sigma + j\omega + \omega_n} \quad \text{normalized } H(s) = \frac{1}{(s + 1)}$$

From this we see that for the transfer function to be infinite, we require $\sigma = -\omega_n$ and $j\omega = 0$. This is what we see in the contours drawn in the first website mentioned above; only there the transfer function is normalized by $R1 = L1 = 1$ and we have $\sigma = -1$. The value $\sigma = -\omega_n$ cancels the real part, and we mark it with an 'X' in red on the $-\sigma$ axis of the pole-zero plot, Fig. 15.8. Then the variable frequency moves along the $j\omega$ axis.

Amplitude and Phase

The voltage gain is found from the modulus of the distance from the pole ω_n to ω which is $d_p = \sqrt{\omega^2 + \omega_n^2}$. This is nothing more than a recasting of the form of previous chapters where, in Eq. 15.13, for example, we multiplied $sL1 + R1$ by its complex conjugate to arrive at $(sL1 + R1)^2$. In this case we have:

$$A_v = \frac{\omega_n}{d_p} = \frac{\omega_n}{\sqrt{\omega^2 + \omega_n^2}} \tag{15.15}$$

If we start at zero frequency, the voltage gain is clearly unity. As the frequency increases, the locus 'x' moves vertically along the axis until at some frequency $\omega = \omega_n$, and then Eq. 15.15 becomes:

$$A_v = \frac{\omega_n}{\sqrt{2\omega_n^2}} = \frac{1}{\sqrt{2}} \tag{15.16}$$

and this is the 3 dB point. We also find the phase angle is $atan(\omega/\omega_n) = 45°$.

As the frequency increases towards infinity, d_p tends to infinity on the $j\omega$ axis and the voltage gain tends asymptotically to zero.

Example – LP Low-Pass
Schematics ('First-Order LP LR.asc') and ('First-Order LP RC.asc') both have breakpoints at *637 mHz*. If we step the components, we find the breakpoint moves according to Eq. 15.16.

High-Pass Filter

We now interchange the capacitor and resistor or the inductor and resistor. The transfer function in the case of a capacitor and resistor, Fig. 15.9, is now:

$$H(s) = \frac{R1}{R1 + \frac{1}{sC1}} = \frac{sC1R1}{+sC1R1 + 1} = \frac{s}{s + \frac{1}{C1R1}} = \frac{s}{s + \omega_n} \tag{15.17}$$

and for an inductance and resistor, Fig. 15.10 we have:

$$H(s) = \frac{sL1}{sL1 + R1} = \frac{s}{s + \frac{R1}{L1}} = \frac{s}{s + \omega_n} \tag{15.18}$$

and in both cases the transfer function or voltage gain reduces to:

$$H(s) = \frac{s}{s + \omega_n} \tag{15.19}$$

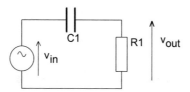

Fig. 15.9 First-order HP CR Filter

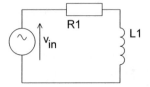

Fig. 15.10 First-order HP RL Filter

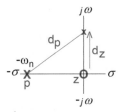

Fig. 15.11 First-order HP pole zero

The difference is s in the numerator rather than a constant. This gives a zero at the origin and we also have a pole at $s = -\omega_n$ as before, Fig. 15.11.

Amplitude and Phase
This is similar to the previous, only slightly complicated by the zero since now we have:

$$A_v = \frac{d_z}{d_p} = \frac{\omega}{\sqrt{\omega^2 + \omega_n^2}} \tag{15.20}$$

The result is that this time if the frequency is zero there is no output voltage but it increases with frequency, and this has a 3 dB point when $\frac{\omega}{\sqrt{\omega^2 + \omega_n^2}} = \frac{1}{\sqrt{2}}$ which is $\omega = \omega_n$ and we again have a phase angle of $45°$. If the frequency continues to increase, we find $d_z \to d_p$ and the voltage gain tends asymptotically to unity.

Example – CR High-Pass
Schematics 'First-Order HP CR,asc' and ('First-Order HP RL.asc') are the high-pass versions of the previous low-pass filters. The only point to note is that if we do not use **Tools->Control Panel→Hacks!** To set the inductor's series resistor to zero, we shall find the low-frequency voltage gain falling.

15.3.2   Cascaded Second-Order Filters

These are nothing more than two cascaded first-order filters. We can realize low-pass and high-pass filters as before, but also band-pass. We should note that in the general case, the amplitude is found from:

$$A_v = \frac{\text{product of distances of zeros to s}}{\text{product of distances of poles from s}} = \frac{(s - z_1)(s - z_2)\cdots}{(s - p_1)(s - p_2)\cdots} \quad (15.21)$$

although in the context of the filters studied here, we shall not have more than two zeros and two poles which is the standard bi-quadratic we have met in previous chapters.

Low-Pass Filter

We again have the choice of LR or RC sections. The problem is to ensure that the second section does not load the first as we found before with the 'Two Tau' circuit. Supposing that we achieve this, the overall response is:

$$H(s) = \frac{\omega_n}{s + \omega_n} \times \frac{\omega_n}{s + \omega_n} = \frac{\omega_n^2}{s^2 + s2\omega_n + \omega_n^2} \quad (15.22)$$

One realization is Fig. 15.12 where:

$$H(s) = \left[\frac{R1}{sL1 + R1}\right]\left[\frac{R2}{sL2 + R2}\right] = \left[\frac{\frac{R1}{L1}}{s + \frac{R1}{L1}}\right]\left[\frac{\frac{R2}{L2}}{s + \frac{R2}{L2}}\right] = \frac{\omega_n^2}{(s + \omega_n)^2} \quad (15.23)$$

And we can easily imagine the corresponding circuit constructed from capacitors and resistors. Equation 15.23 shows that we have two coincidental poles which are shown by two 'X's on top of each other in Fig. 15.13 instead of the single 'X' of the first-order filter. As s does not appear in the numerator, there are no zeros.

Amplitude and Phase

The voltage gain differs mainly in that we need the product of the distance of the two poles from ω and hence:

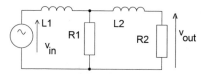

Fig. 15.12 Second-order LR cascaded LP

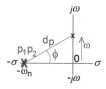

Fig. 15.13 Second-order cascaded LP

$$A_v = \frac{\omega_n^2}{d_p^2} = \frac{\omega_n^2}{\left[\sqrt{\omega_n^2 + \omega^2}\right]^2} \qquad (15.24)$$

At the 3 dB point:

$$A_v = \frac{1}{\sqrt{2}} = \frac{\omega_n^2}{\left[\sqrt{\omega_n^2 + \omega^2}\right]^2} \quad \text{or} \quad A_v = \frac{1}{\sqrt{1.414}} = \frac{\omega_n}{\sqrt{\omega_n^2 + \omega^2}} \qquad (15.25)$$

and then $\omega = 0.643\ \omega_n$. The phase angle is the sum of the phase angles of the individual poles, and as these are identical, we have $2\times\ atan(1/0.643) = 65.5°$. We may also note that from Eq. 15.23, the numerator is squared, which, in terms of decibels, means it is doubled.

Example – Cascaded Low-Pass Filter
This is schematic ('Cascaded LR Filter.asc'). The impedance of the second stage is made much higher than the first to avoid loading it. As a result, the measurements agree very closely with the calculations which are that $\omega_n=10^5$ for each section and that the 3 dB frquency is *10.23 kHz*. The schematic ('Cascaded CR Filter.asc') is the capacitor-resistor implementation and returns the same results.

High-Pass Filter

The analysis follows the same principles as a low-pass filter, for example, using inductors and resistors we have Fig. 15.14:

$$H(s) = \frac{(sL)^2}{(sL)^2 + s2LR + R^2} = \frac{s^2}{s^2 + s2\omega_n + \omega_n^2} = \frac{s^2}{(s + \omega_n)^2} \qquad (15.26)$$

The difference now is that we have two zeros at the origin from the squared s, Fig. 15.15. We can also create a circuit using capacitors and resistors.

Amplitude and Phase
The voltage gain is:

$$A_v = \frac{d_z^2}{d_p^2} = \frac{\omega^2}{\left[\sqrt{\omega_p^2 + \omega^2}\right]^2} \qquad (15.27)$$

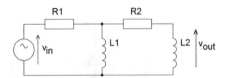

Fig. 15.14 Second-order RL cascaded HP Filter

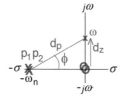

Fig. 15.15 Second-order cascaded HP

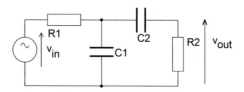

Fig. 15.16 Cascaded BP filter

and this time we see it increasing from zero as ω increases. The 3 dB point is similar to the previous and is:

$$A_v = \frac{1}{\sqrt{1.414}} = \frac{\omega}{\sqrt{\omega^2 + \omega_n^2}} \tag{15.28}$$

This too returns $\omega = 0.643\,\omega_n$ and a phase angle of $65.5°$ and also has a roll-off of *40 dB/decade* the same as the low-pass filter.

Example – Cascaded High-Pass Filter

These are schematics ('Cascaded RL Filter.asc') and ('Cascaded CR Filter.asc') again where the second stage values have been chosen to minimize loading on the first stage and hence the results are in very close agreement with theory.

Band-Pass Filter

Things now become a little more complicated, Fig. 15.16. We now have a high-pass section followed by a low-pass section. Without deriving the equation, it follows from the previous sections that the transfer function is of the form:

$$H(s) = \frac{s}{s + \omega_{n2}} \frac{\omega_{n1}}{s + \omega_{n1}} = \frac{s\omega_{n1}}{s^2 + s(\omega_{n1} + \omega_{n2}) + \omega_{n1}\omega_{n2}} \tag{15.29}$$

The pole-zero plot is Fig. 15.17.

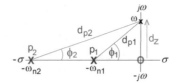

Fig. 15.17 Second-order cascaded BP

Amplitude and Phase

Using the distances in the pole-zero plot we have:

$$A_v = \frac{d_z \omega_{n1}}{d_{p1} d_{p2}} = \frac{\omega \omega_{n1}}{\sqrt{(\omega_{n1}^2 + \omega^2)} \times \sqrt{(\omega_{n2}^2 + \omega^2)}} \tag{15.30}$$

We can divide the frequency response into three stages.

Low Frequency

If $\omega_{n2} \gg \omega$ Eq. 15.30 simplifies to:

$$A_v = \frac{\omega \omega_{n1}}{\sqrt{(\omega_{n1}^2 + \omega^2)}\,\omega_{n2}} = \frac{\omega_{n1}}{\omega_{n2}} \frac{\omega}{\sqrt{(\omega_{n1}^2 + \omega^2)}} \tag{15.31}$$

which is a constant ω_{n1}/ω_{n2} multiplied by a first-order high-pass filter, and therefore starting from a low frequency, we expect a rising output at *20 dB/decade*. As the frequency increases, we reach a point where $\omega = \omega_{n1}$, and as ω is still very much less than ω_{n2}, Eq. 15.31 remains valid, and we expect the downward breakpoint of a first-order filter at ω_{n1} irrespective of ω_{n2} where the phase is *45°* and the voltage gain has fallen by *3 dB*.

Pass Band

As the frequency continues to increase above the point where $\omega = \omega_{n1}$ so that $\omega \gg \omega_{n1}$, we can now ignore ω_{n1} in the square root in Eq. 15.31, and we have a flat response of $A_v = \frac{\omega_{n1}}{\omega_{n2}}$. We can infer at least some attenuation from the potential divider *R1C1*.

High Frequency

At higher frequencies we reach a point where $\omega = \omega_{n2}$, and we can ignore ω_{n1} compared to ω in the first square root of the denominator of Eq. 15.30, and we find the constant ω_{n1}/ω_{n2} multiplied this time by a low pass first-order filter:

$$A_v = \frac{\omega_{n2} \omega_{n1}}{\omega_{n2} \sqrt{\omega_{n2}^2 + \omega^2}} = \frac{\omega_{n1}}{\omega_{n2}} \frac{\omega_{n2}}{\sqrt{\omega_{n2}^2 + \omega^2}} \tag{15.32}$$

with a second downward break-point at ω_{n2} irrespective of ω_{n1}. As the frequency continues to increase asymptotically, we find $d_z = d_{p1} = d_{p2} = \omega$, and thus Eq. 15.30 shows that the voltage gain will be $A_v = \omega_{n1}/\omega = 0$.

Example – Cascaded Band-Pass Filter
There are two versions: schematics ('Cascaded BP Filter.asc') and ('Cascaded BP Filter V2.asc'). In both cases, the breakpoints accord with theory. We also find that $\frac{\omega_{n1}}{\omega_{n2}} = \frac{10^1}{10^4} = 10^{-3} = -60dB$ which is the loss in the pass band. This is inherent in the band-width and will become greater if the band-width is increased.

Explorations 2
1. Explore the cascaded low-pass filters, taking different component values and noting the position of the breakpoints, the phase shifts and the slopes of the traces.
2. Repeat with a cascaded high-pass filter and note the effect of an external load and how it could be incorporated in the filter design.
3. Compare the two versions of the band-pass filter, in particular the low frequency voltage gain. It is interesting to show the *l1r1* and *r1c1* traces and to see if the pass band attenuation can be reduced.

15.3.3   LCR Filters

These can create low-pass, high-pass, band-pass and also band-stop filters. Some of the circuits are identical to, or variations on, those encountered in the chapter 'Tuned Circuits'. The first difference is that we are not necessarily taking the output across the tuned circuit, but across elements of it. The second difference is that it is more convenient here to write the equations in terms of the damping coefficient ζ and the Q-factor where $Q = 1/(2\zeta)$ in order to generalize them so that we do not need to specify the components, and it is left open for us to choose whatever we like to create the transfer function.

Transfer Function and Roots

This is of the general form:

$$H(s) = \frac{z}{s^2 + \frac{\omega}{Q}s + \omega_n^2} \tag{15.33}$$

where z is a constant. For a low-pass filter $z = \omega_n^2$; for a high-pass filter $z = s^2$, for a band-pass filter $z = s\omega_n/Q$; and for a band-stop filter $z = s^2 + \omega_n^2$. The equations all have the same denominator where the roots are the poles. The derivation and validity of Eq. 15.33 will be seen in the following sections on specific types of filter.
From Eq. 15.33 the roots are:

$$p_1, p_2 = \frac{1}{2}\left[\frac{-\omega_n}{Q} \pm \sqrt{\left(\frac{\omega_n}{Q}\right)^2 - 4\omega_n^2} \right] \qquad (15.34)$$

or:

$$p_1, p_1 = \frac{-\omega_n}{2}\left(\frac{1}{Q} \pm \sqrt{\frac{1}{Q^2} - 4} \right) \qquad (15.35)$$

Substituting for Q we may also write Eq. 15.34 as:

$$p1, p2 = \frac{-\omega_n}{2}\left(2\zeta \pm \sqrt{4\zeta^2 - 4} \right) = -\omega_n\left(\zeta \pm \sqrt{\zeta^2 - 1} \right) \qquad (15.36)$$

from which it follows that if $Q < 1/2$, the roots are real and different; if $Q = 1/2$, the roots are real and identical; and if $Q > 1/2$, <1) the roots are complex conjugates. This is re-stating what we found in previous chapters. And by comparison with Eqs. 15.22 and 15.26, the Q-factor for cascaded stages is $Q = 0.5$.

Table 15.1 shows the normalized positions of the two real poles.

Pole Positions

If $Q \leq 0.5$, the poles lie on the negative real axis, and in this case we can write σ for p, and their positions relative to the average are shown in Fig. 15.18 which shows that the higher frequency poles are at approximately twice the natural frequency. The poles in the table above are plotted in Fig. 15.19. This is not to scale, but it does show that for heavy damping $p2$ clusters near the origin. And we also see the two poles move along the σ axis towards ω_n with increasing Q or decreasing ζ.

Table 15.1 LCR Filter Pole Positions

Q	0.01	0.025	0.05	0.25	0.5
ζ	50	20	10	2.0	1.0
p1	99.99	39.975	19.95	3.73	1.0
p2	0.01	0.025	0.05	0.27	1.0

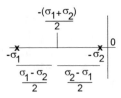

Fig. 15.18 Second-order real poles

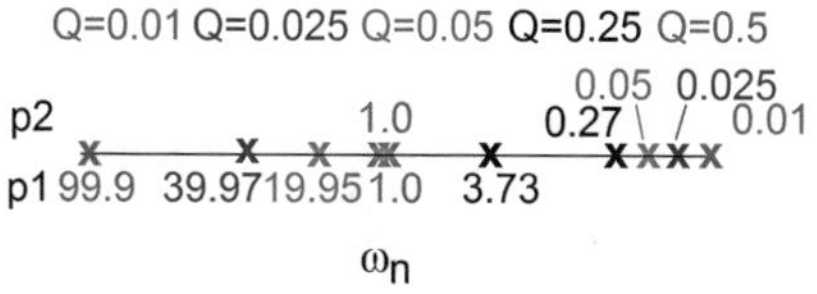

Q=0.01 Q=0.025 Q=0.05 Q=0.25 Q=0.5

p2   1.0   0.05 0.025
                                        0.27 \ / 0.01
x   x   x x   x   x x x x
p1 99.9   39.97 19.95 1.0   3.73

ω_n

Fig. 15.19 LCR Filter Real Pole Positions

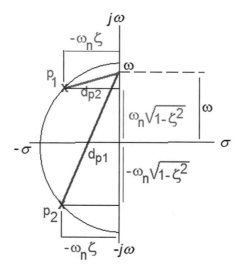

Fig. 15.20 Second-order imaginary pole positions

However, if $\zeta < 1$ we have complex poles, and we can write Eq. 15.36 as:

$$p_1, p_2 = \omega_n \zeta \pm j\omega_n \sqrt{(1 - \zeta^2)} \qquad (15.37)$$

and the modulus is:

$$\sqrt{(\omega_n \zeta)^2 + \omega_n^2 (1 - \zeta^2)} = \omega_n \qquad (15.38)$$

meaning that the locus of the poles is a circle of radius ω_n centred on the origin which we show in Fig. 15.20 where only the negative semi-circle is drawn. Note that this is not the movement of the poles as we change the excitation frequency, but how their positions change with the Q-factor or damping. The real and imaginary pole movements are shown together in Fig. 15.21.

We may also note that:

$$p_1 \times p_2 = \omega_n^2 \qquad (15.39)$$

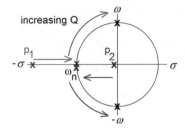

Fig. 15.21   Second-order pole movement

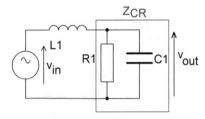

Fig. 15.22   Low-pass LCR Filter

15.3.4   *LCR Filter Types*

We shall consider these in turn in the light of the previous discussion. These all have a denominator of Eq. 15.33.

Low-Pass LCR Filter

We have encountered these in Chapter 'Tuned Circuits' were we used them to select a narrow band of frequencies. However, by choosing where we take the output, we can convert the circuit into a low-pass filter such as Fig. 15.22 where we have a series inductor feeding a parallel combination of a capacitor and a resistor which we replace by:

$$Z_{CR} = \frac{\frac{R1}{sC1}}{R1 + \frac{1}{C1}} = \frac{R1}{sC1R1 + 1} \tag{15.40}$$

Then the transfer function is:

$$H(s) + \frac{Z_{CR}}{sL1 + Z_{CR}} = \frac{\frac{R1}{sC1R1+1}}{sL1 + \frac{R1}{sC1R1+1}} = \frac{R1}{s^2L1C1R1 + sL1 + R1} \tag{15.41}$$

We turn this into the standard form using $Q = R1\sqrt{(C1/L1)}$ and $\omega_n^2 = 1/(C1L1)$, and this is identical to Eq. 15.22 with $z = \omega_n^2$:

$$H(s) = \frac{\frac{1}{L1C1}}{s^2 + \frac{s}{C1R1} + \frac{1}{L1C1}} = \frac{\omega_n^2}{s^2 + s\frac{\omega}{Q} + \omega_n^2} \tag{15.42}$$

Unlike the cascaded sections case, we can change the Q-factor, and if we do it by changing $R1$, it is independent of ω_n.

Real Poles

It is convenient to divide the response into that with real poles and that with complex conjugate poles. With real poles, we can write:

$$H(s) = \frac{\omega_n}{(s + p_1)} \frac{\omega_n}{(s + p_2)} \tag{15.43}$$

and this has the form of two first-order low-pass filters in cascade. We might notice that multiplying out equation 15.43 shows $p_1 p_2 = \omega_n^2$. Even up to $\zeta = 2$, a simulation shows that the poles are separated by more than an order of magnitude so we should reasonably expect to be able to treat each by itself. LTspice can easily measure the first breakpoint, but the second is best found from the phase shift of 135°.

Complex Poles

The website http://www.stades.co.uk/?LMCL=FBHqyY is very useful and perhaps the only place where the derivation is given, although several other websites reproduce the figures. Their treatment is largely followed here. Using Eq. 15.42 if Q is large the term $s\omega_n/Q$ is small and the frequency for maximum output voltage is very close to ω_n. And even at $Q = 5$ simulation shows that the difference is less than 2%.

To accurately determine the position for maximum amplitude, we write Eq. 15.33 in terms of ζ:

$$H(j\omega) = \frac{\omega_n^2}{-\omega^2 + 2\zeta\omega\omega_n + \omega_n^2} \tag{15.44}$$

and find the modulus of the denominator:

$$|H(j\omega)| = \frac{\omega_n^2}{\sqrt{\left(-\omega^2 + \omega_n^2\right)^2 + 4\zeta^2\omega^2\omega_n^2}}$$

$$= \frac{\omega_n^2}{\sqrt{\omega^4 + \omega_n^4 - 2\omega^2\omega_n^2 + 4\zeta^2\omega^2\omega_n^2}} \tag{15.45}$$

We now differentiate the denominator with respect to ω to find the turning point and equate it to zero:

$$0 = 4\omega^3 - 4\omega\omega_n^2 + 8\zeta^2\omega\omega_n^2 \tag{15.46}$$

then:

$$\omega^2 = \omega_n^2 - 2\zeta^2\omega_n^2 \quad \text{finally} \quad \omega = \omega_n\sqrt{1 - 2\zeta^2} \tag{15.47}$$

showing that the resonant frequency tends to the natural resonance as the damping gets smaller.

Pole Positions
These are shown in Fig. 15.20 and lie on a circle of radius ω_n. Figure 15.21 shows that starting with a large damping coefficient, we have two real poles which are widely separated on the negative $-\sigma$ axis with one near the origin with a frequency perhaps of less than 1 Hz and the other at a considerable distance, maybe more than 1 MHz and approximately twice the natural frequency. As the damping is reduced the two poles move towards each other meeting at the natural frequency when $\zeta = 1$. As the damping is further reduced, the poles move in opposite directions around the circle towards the $j\omega$ axis at which point the damping is zero. This is important for the pulse response of the circuit since a low damping implies ringing.

Peak Amplitude
We insert the frequency for maximum output $\omega = \omega_n\sqrt{1 - 2\zeta^2}$ in Eq. 15.44:

$$H(j\omega) = \frac{\omega_n^2}{-\omega_n^2\left(\sqrt{1 - 2\zeta^2}\right)^2 + j2\zeta\omega_n^2\sqrt{1 - 2\zeta^2} + \omega_n^2} \tag{15.48}$$

Divide by ω_n^2 and collect terms:

$$H(j\omega) = \frac{1}{2\zeta^2 + j2\zeta\sqrt{1 - 2\zeta^2}} \tag{15.49}$$

We now find the modulus by squaring and taking the square root:

$$|H(j\omega)| = \frac{1}{\sqrt{4\zeta^2 - 4\zeta^2(1 - 2\zeta^2)}} = \frac{1}{2\zeta\sqrt{1 - \zeta^2}}$$

if $\zeta \ll 1$
$$|H(j\omega)| \simeq \frac{1}{2\zeta} \tag{15.50}$$

and if $\zeta = 0$ H is infinite. It might reasonably be asked what is the point of a filter with a resonant peak and that $Q = 0.5$ (critical damping) is best.

Example – Low-Pass LCR Filter

An AC run of schematic ('Low-Pass LCR.asc') stepping the resistor as *.step param R list 0.3163 0.7906 1.583 3.163 15.83 31.63 63.26 158.3 316.3 3.163k 31.63k* corresponds to Q-factors of 0.01, 0.025, 0.05, 0.1, 0.5 1, 2, 5, 10, 100 and 1000. This shows an asymptotic high frequency fall of -40 *dB/decade* and an asymptotically flat low frequency response of *0 dB*. There are three sets of measurements.

We shall look at the first five readings of each. The first set of readings, *Vmax*, range from highly overdamped with $Q = 0.01$ to critical damping at $Q = 0.5$. All show that the initial value is effectively 0 dB (the largest is only *1.7 mdB*) and is the biggest meaning that there is no resonant peak. The second set of readings of the frequency for maximum output *fmax* confirms this by quoting the frequency for maximum voltage is the start at *1 Hz*. The last set of readings, of *f 3dB*, show a considerable increase in the 3 dB point with increasing Q-factor. If we take the smaller of each pair of poles, we find that the calculated pole frequencies correspond exactly with the measured 3 dB points, for example, taking *step 3* with $Q = 0.05$ using Eq. 15.35 and using frequency rather than rad/s where f_r is the resonant frequency and f_n is the natural frequency we have:

$$f_r = \frac{f_n}{2}(20 - \sqrt{396}) = 2516 \times 0.01 = 251.6 \, Hz$$

which is close enough to the measured *251.941 Hz* considering the rounding errors and are all below the resonance.

Taking the larger of each pair, it is easier to use the cursor to find the point where the phase angle is *135°* and then the agreement with the calculated frequency is good: for example, with $Q = 0.01$ we measure *503 kHz* and calculate *2516 (100-0.02) = 503 kHz*. The next three values are *210 kHz, 105 kHz* and *53 kHz*, and these also agree. The value for $Q = 0.5$ is wrong because both breakpoints coincide.

For the remaining five measurements we have complex conjugate poles. The measurements can be improved by reducing the frequency range from *4 kHz* to *10 kHz* since LTspice has reduced the intended 100,000 points per decade to 15,393 resulting in the 3 dB frequencies being almost the same as the peak. The measured frequencies and amplitudes are in agreement with the previous analysis, for example, the peak amplitude with $Q = 0.1$ and thus $\zeta = 0.2$ is *1/(0.2 x 0.98) = 5.103 = 14.2 dB* - close to the measured value.

A second filter, schematic ('Low-Pass LCR V2.asc'), has the LCR in series and takes the output across the capacitor.

High-Pass LCR Filter

Without analysing the circuit of Fig. 15.23 in detail, the transfer function is:

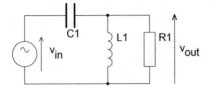

Fig. 15.23 Second-order high pass filter

$$H(s) = \frac{s^2}{s^2 + \frac{s}{C1R1} + \frac{1}{L1C1}} \tag{15.51}$$

which in standard form is:

$$H(s) = \frac{s^2}{s^2 + \frac{\omega}{Q}s + \omega_n^2} \tag{15.52}$$

and now we have two coincidental zeros at $s = 0$. The Bode plot is a left-to-right inversion of the previous.

Real Poles

The transfer function can be written as:

$$H(s) = \frac{s}{(s + p_1)} \frac{s}{(s + p_2)} \tag{15.53}$$

and now we have two cascaded first-order high-pass filters where, once again, if the breakpoints are well separated, we can treat each one separately and we shall find the same results as before.

Complex Poles

Since the denominator of Eq. 15.52 is the same as Eq. 15.42, the previous analysis applies.

Pole and Zero Positions

The poles move in exactly the same way as the low-pass filter so the same discussion applies. The two zeros remain fixed at the origin.

Peak Amplitude

This also is the same as before because the two poles at the origin affect the frequency response only.

Example – High-Pass LCR Filter

This is schematic ('High-Pass LCR.asc') where it can be seen that its performance is identical to the low-pass filter switched left to right.

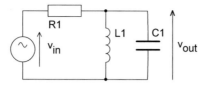

Fig. 15.24 Second order band pass filter

Band-Pass LCR Filter

Using Fig. 15.24 we find that the transfer function is:

$$H(s) = \frac{\frac{s}{R1C1}}{s^2 + \frac{s}{R1C1} + \frac{1}{L1C1}} \tag{15.54}$$

and we have $Q = R1\sqrt{(C1/L1)}$. The poles are located as before but now we have zero in the numerator:

$$H(s) = A\frac{\frac{\omega}{Q}s}{s^2 + \frac{\omega_n}{Q}s + \omega_n^2} \tag{15.55}$$

The circuit is nothing more than a damped parallel tuned circuit with the output taken across the tuned circuit itself. With a small resistor, the pass band is reasonably flat and the sides slope at 20 dB/decade.

Overdamped Low- and High-Frequency Response
We can again factor the transfer function using ζ:

$$H(s) = \frac{s}{(s + p_1)}\frac{2\zeta\omega_n}{(s + p_2)} \tag{15.56}$$

As these are real poles, we can replace them by $\omega_{p1}\ \omega_{p2}$:

$$H(s) = \frac{s2\zeta\omega_n}{\left(s + \omega_{p1}\right)\left(s + \omega_{p2}\right)}$$

If the two poles are well-separated and we have a frequency very much lower than ω_{p2} but comparable to ω_{p1}, we keep the first term in the denominator but ignore s in the second and write:

$$H(s) \approx \frac{s2\zeta\omega_n}{\left(s + \omega_{p1}\right)\omega_{p2}} = \frac{2\zeta\omega_n}{\omega_{p2}}\frac{s}{s + \omega_{p1}} \tag{15.57}$$

and this is a first-order high-pass filter. Similarly if $\omega \gg \omega_{p1}$ but comparable to ω_{p2}, we have:

$$H(s) \approx \frac{s2\zeta\omega_n}{s(s+\omega_{p2})} = 2\zeta\frac{\omega_n}{s+\omega_{p2}} \tag{15.58}$$

and this is a first-order low-pass filter.

Overdamped Pass Band
We can substitute for the poles in Eq. 15.56 using Eq. 15.36:

$$H(s) = \frac{s2\zeta\omega_n}{\left(s+\omega_n\left(\zeta+\sqrt{\zeta^2-1}\right)\right)\left(s+\omega_n\left(\zeta-\sqrt{\zeta^2-1}\right)\right)} \tag{15.59}$$

This expands to:

$$H(s) = \frac{s2\zeta\omega_n}{s^2 + s2\zeta\omega_n + \omega_n^2} \tag{15.60}$$

Then if there is a region where $-s^2 \approx \omega_n{}^2$, Eq. 15.60 reduces to $A_v = 1$, and this is the flat pass band.

Underdamped Peak Frequency and Amplitude
If we substitute for s in Eq. 15.60, the denominator is exactly that of Eq. 15.44 and so we find the same peak frequency of $\omega = \omega_n\sqrt{1-2\zeta^2}$. Inserting that in Eq. 15.60 yields:

$$H(j\omega) = \frac{2\zeta\omega_n^2\sqrt{1-2\zeta^2}}{\omega_n^2\left(2\zeta^2 + j2\zeta\sqrt{1-2\zeta^2}\right)} \tag{15.61}$$

We divide throughout by $\omega_n{}^2$, and the denominator of Eq. 15.61 is that of Eq. 15.49. Thus when we find the modulus we arrive at:

$$A_{vpk} = \frac{2\zeta\sqrt{1-2\zeta^2}}{2\zeta\sqrt{1-2\zeta^2}} = 1 \tag{15.62}$$

and the peak voltage is $0\ dB$.

Example – Second-Order Band-Pass Filter
Using the values of schematic ('Second Order Band Pass.asc') with *step param R list 1 5 10 50,100*, the corresponding values of Q are $Q = 0.0316, 0.158, 0.316, 1.58$, and *3.16*, and the first three are overdamped and yield a quite flat pass region, whilst the last two are more of a notch filter.

Band-Stop Filter

The circuit is Fig. 15.25 which shows that again this is a parallel tuned circuit with the output taken from a series resistor and whose response is essentially a notch that can be broadened by the resistor and is not the inverse of the band-pass filter. So its

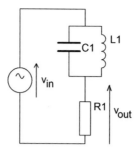

Fig. 15.25 Second-order band stop filter

usefulness can best be described by the limits of the minimum acceptable attenuation.

We have:

$$H(s) = \frac{R1}{\left[\frac{sL1}{s^2 L1C1+1} + R1\right]} \qquad (15.63)$$

which we expand to:

$$H(s) = \frac{s^2 L1C1R1 + R1}{[sL1 + s^2 L1C1R1 + R1]} \qquad (15.64)$$

and divide by $L1C1R1$ to yield:

$$H(s) = \frac{s^2 + \frac{1}{L1C1}}{s^2 + s\frac{1}{C1R1} + \frac{1}{L1C1}} \qquad (15.65)$$

and the denominator is the same as all the previous filters:

$$H(s) = \frac{s^2 + \omega_n^2}{s^2 + 2s\zeta\omega_n + \omega_n^2} \qquad (15.66)$$

Asymptotic Voltage Gain

The behaviour of this filter is different in kind from the others. The numerator shows that there are two zeros, at $+/- \omega_n$. If $\omega = 0$ then Fig. 15.26 shows that the distance of the complex poles p_1 and p_2 from the origin is the same as the zeros and so the voltage gain will be unity. If we have real poles rp_1 and rp_2, the product of their distances from the origin is ω_n and again the voltage gain is unity or 0 dB. If $\omega = \infty$ the four distances from the poles and zeros to ω are the same and we still have 0 dB. Thus at both high and low frequencies, the voltage gain is 0 dB.

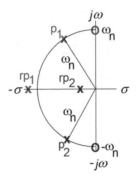

Fig. 15.26 Second-order Band Stop pole and zero positions

Pole and Zero Positions

As it has the same denominator as all the previous circuits, the pole positions are the same. Because the low-frequency response is flat at *0 dB*, the first pole will be a downward break with a slope of −20 dB/decade. The second pole will also create a downward break, but this time it is on the upward slope of 20 dB/decade after the notch, thus causing the asymptotic response again to be constant at 0 dB.

Equation 15.66 shows that as ζ becomes smaller, the poles tend to ω_n and the notch becomes very sharp.

Notch Depth

If $s^2 = -\omega_n^2$ the numerator is zero whatever the denominator of Eq. 15.66, so both real and complex poles will create a notch – there is no region of flat response – and therefore the notch depth is infinite and occurs at ω_n.

Response Shape Near Resonance

So far we have seen how to find the poles and the 3 dB points; we also know how to find the resonant frequency and the height (or depth) of the cusp: but we have not explored the shape of the response in between. For those that are interested in such things (and it can be argued that it is easier and quicker just to simulate the circuit), we need the numerator and denominator. The zeros are at $-j\omega_n$, $+j\omega_n$ and the numerator is:

$$N(j\omega) = \omega_n^2 - \omega^2 \tag{15.67}$$

The denominator for real poles, Fig. 15.27, is found from:

$$D(j\omega) = d_{p1}d_{p2} = \sqrt{(p_1^2 + \omega^2)}\sqrt{(p_2^2 + \omega^2)} \tag{15.68}$$

If we have complex poles, it becomes a little more complicated, Fig. 15.28. The numerator is unchanged; we find the two poles using the hypotenuse:

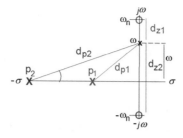

Fig. 15.27 Second Order Band stop real poles

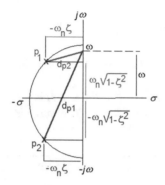

Fig. 15.28 Second order band stop imaginary poles

$$d_{p1} = \sqrt{(\omega_n \zeta)^2 + \left(\omega_n\sqrt{1-\zeta^2} + \omega\right)^2} \qquad (15.69)$$

which expands as:

$$d_{p1} = \sqrt{\omega_n^2 + \omega^2 + 2\omega\omega_n\sqrt{1-\zeta^2}} \qquad (15.70)$$

And:

$$d_{p2} = \sqrt{(\omega_n \zeta)^2 + \left(\omega - \omega_n\sqrt{(1-\zeta^2)}\right)^2} \qquad (15.71)$$

which expands as:

$$d_{p2} = \sqrt{\omega_n^2 + \omega^2 - 2\omega\omega_n\sqrt{1-\zeta^2}} \qquad (15.72)$$

Then:

$$D(j\omega) = \sqrt{\omega_n^4 + \omega^4 - 4\omega^2\omega_n^2(1 - 2\zeta^2)} \qquad (15.73)$$

Example

If we use Table 15.1 the normalized values for $\zeta = 10$ are $p_1 = 0.05$ $p_2 = 19.95$, and if we take $f_n = 5033\,Hz$ and evaluate Eq. 15.68 at $\omega^2 = 4500 \times 2 \times \pi = 7.99 \times 10^8\,Hz^2$, we have:

$$D(j\omega) = \sqrt{(\omega_n^2 p_1^2 + \omega^2)(\omega_n^2 p_2^2 + \omega^2)} = \sqrt{(400.8 \times 10^9)(8 \times 10^8)}$$

$$= 179 \times 10^8 \quad N(j\omega) = 10^9 - 8.10^8$$

Then $A_v = \frac{2.00 \times 10^8}{179 \times 10^8} = 0.0112 = -39.0dB$ and this is in agreement with the $R = 1.583\,\Omega$ trace.

As a second example, we take $\zeta = 0.1$ which is the 8th trace with $R = 158,3\,\Omega$ and measure at a frequency of 5.5 kHz. The numerator now is 1.94×10^8. The denominator is:

$$D(j\omega) = \sqrt{(1.000 + 1.426 - 2.340)10^{18}} = 2.93 \times 10^8$$

hence $A_v = \frac{1.94}{2.93} = 0.992 = -3.58dB$ which agrees with measurement using the cursor.

More Filters

The careful reader will have noticed that, apart from the last example where we used Fig. 15.28 to write Eqs. 15.69, 15.70, 15.71, 15.72, and 15.73, the pole-zero plots have been more of a visual aid than an analytic tool. This is true of many of the examples of passive filters, for example, the very interesting (for audiophiles) website http://www.linkwitzlab.com/crossovers sketches them for a variety of filters.

Higher-Order Filters

We can improve the steepness of the flanks of a filter by adding more stages. This is something we shall touch on later in this chapter. One example is to cascade the previous LCR filters. Schematic ('Second Order 2LCR.asc') shows that where we have real roots the slope is $-80\,dB/decade$ instead of $-40\,dB/decade$.

Prof. P. Cheung www.ee.ic.ac.uk › pcheung › teaching › ee2_signals › Lecture 9 – Poles Ze… draws the pole-zero diagram for some Butterworth filters showing that the poles lie on the circle at equal angles, for example, Fig. 15.29 shows a 5th-order filter with all the poles having the same natural frequency but different damping.

A note in passing: we have only looked at the frequency response here; the temporal response has been dealt with in previous chapters and, apart, from noting

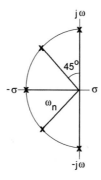

Fig. 15.29 Butterworth pole zero

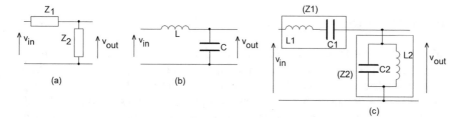

Fig. 15.30 Constant K half section

that as poles approach the $j\omega$ axis, the ringing increases; the pole-zero plot of itself adds nothing to what we already know.

Explorations 3
1. Compare the two schematics for a low-pass filter and try other values. A maximally flat response occurs with $Q = 0.5$, but if we can afford a slight peak of *1.4 dB* then $Q = 1$ has a wider bandwidth. It is worth investigating this and other options for $Q < 1$.
2. The same critique applies to the high-pass filter.
3. The second-order band-pass filter does not suffer from the severe attenuation of the cascaded filter: compare their frequency responses.
4. Take some arbitrary minimum attenuation such as -30 dB and find the bandwidth as a function of ζ.

15.4 LC Filters

We confine ourselves here to simple L, T and Π circuits which are amenable to easy design. These are based on repeated inverted 'L' sections, although, in fact, this is called a half-section because to create the filter we need two which we can connect either to make a 'T' or a 'Π'. The two impedances Z_1 and Z_2 in Fig. 15.30 could be just a capacitor and an inductor as shown at '(b)' or more complex as shown at '(c)'. These sections can be cascaded to form a ladder.

15.4.1   Filter Prototypes

The T and Π circuits are prototypes using generalized impedances. In later sections we shall assign component types to them. Therefore if we do the hard work here, when we later come explore the different filter type, it is mainly a matter of putting numbers in the equations which we shall now derive. But to understand this derivation we must first explain the concept of *image impedance*.

Image Impedance

We introduce this here as a useful concept widely employed in the design of filters. Consider the two-port network shown in Fig. 15.31. If we terminate port *2* with its image impedance Z_{I2} then looking into port *1* the input impedance will be its image impedance Z_{i1}. Similarly if we look in from port *2* with port *1* terminated by its image impedance, we have Z_{i2}. In the particular case of a symmetrical network, these are identical and equal to the characteristic impedance Z_0.

   We shall apply this concept first to find the input impedance for a 'T' and then for a 'Π'.

T- Filter

We start by finding the input impedance.

Input Impedance
The immediate difficulty trying to define or measure the impedances is that to measure Z_{i1}, we need to terminate port *2* by its image impedance but to do that we need the image impedance of port *1*, Fig. 15.32a. We escape from this dilemma by terminating port *2* with a reversed, inverted, 'L', Fig. 15.32b so it too is terminated by Z_{i1}. So we can replace the right-hand Z_2 and Z_1 in series by Z_{i1}, Fig. 15.32b2, and

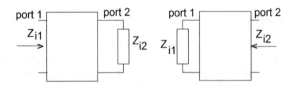

Fig. 15.31 Image impedance

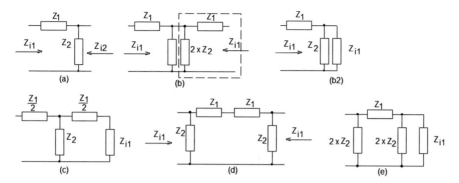

Fig. 15.32 Image impedance derivation

we now have, after expanding the expression for the circuit to the right of Z_1, of Z_2 in parallel with $(Z_1 + Z_{i1})$:

$$Z_{i1} = Z_1 + \frac{Z_1 Z_2 + Z_{i1} Z_2}{Z_2 + 2Z_1 + 2Z_{i1}} \qquad (15.74)$$

then:

$$Z_{i1} Z_2 + 2Z_{i1} Z_1 + 2Z_{i1}^2 = Z_1 Z_2 + 2Z_1^2 + 2Z_1 Z_{i1} + Z_1 Z_2 + Z_{i1} Z_2 \qquad (15.75)$$

After cancelling:

$$Z_{i1}^2 = Z_1^2 + Z_1 Z_2 \qquad (15.76)$$

We define a constant $k^2 = Z_1 Z_2$ which we shall use later. For the present it is enough to note that we can write Eq. 15.76 as $Z_{i1}^2 = Z_1^2 + k^2$.

We draw the circuit as Fig. 15.32c showing the terminating image impedance. By convention we scale the impedances by halving Z_1 and doubling Z_2 to preserve the same k and then from Eq. 15.76:

$$Z_{i1} = \sqrt{\frac{Z_1^2}{4} + \frac{Z_1}{2} 2Z_2} \quad \Rightarrow Z_{i1} = \sqrt{\frac{Z_1^2}{4} + Z_1 Z_2} \qquad (15.77)$$

An alternative way of writing it is:

$$Z_0 = \sqrt{Z_1 Z_2} \sqrt{1 + \frac{Z_1}{4Z_2}} \qquad (15.78)$$

This is the characteristic impedance Z_0.

Π – Filter

We again use two 'L' sections, Fig. 15.32d, with the second one reversed. We define Z_p as the sum of the two impedances Z_1 in series with Z_2 which is in parallel with Z_{i1}:

$$Z_p = 2Z_1 + \frac{Z_2 Z_{i1}}{Z_2 + Z_{i1}} \quad Z_p = \frac{2Z_1 Z_2 + 2Z_1 Z_{i1} + Z_2 Z_{i1}}{Z_2 + Z_{i1}} \tag{15.79}$$

Looking into the circuit $Z_0 = Z_2/Z_p$ and substituting from Eq. 15.79:

$$Z_{i1} = \frac{Z_2 \frac{(2Z_1 Z_2 + 2Z_1 Z_{i1} + Z_2 Z_{i1})}{Z_2 + Z_{i1}}}{Z_2 + \frac{2Z_1 Z_2 + 2Z_1 Z_{i1} + Z_2 Z_{i1}}{Z_2 + Z_{i1}}} \tag{15.80}$$

This expands to:

$$Z_{i1} = \frac{Z_2(2Z_1 Z_2 + 2Z_1 Z_{i1} + Z_2 Z_{i1})}{Z_2^2 + Z_2 Z_{i1} + 2Z_1 Z_1 + 2Z_1 Z_{i1} + Z_2 Z_{i1}} \tag{15.81}$$

Thence:

$$Z_{i1} Z_2^2 + Z_2 Z_{i1}^2 + 2Z_1 Z_2 Z_{i1} + 2Z_1 Z_{i1}^2 + Z_2 Z_{i1}^2$$
$$= 2Z_1 Z_2^2 + 2Z_1 Z_2 Z_{i1} + Z_1^2 Z_{i1} \tag{15.82}$$

Collecting terms we have:

$$2Z_{i1}^2(Z_1 + Z_1) = 2Z_1 Z_2^2 \tag{15.83}$$

We again identify Z_{i1} as Z_0, whence:

$$Z_0 = \sqrt{\left[\frac{Z_1 Z_2^2}{Z_1 + Z_2}\right]} \tag{15.84}$$

As it is conventional to divide Z_1 by 2 and therefore multiply Z_2 by 2 to preserve the same k, Fig. 15.32e, then finally:

$$Z_0 = \sqrt{\left[\frac{4Z_1 Z_2^2}{Z_1 + 4Z_2}\right]} \tag{15.85}$$

By comparison with Eq. 15.78, we can write Eq. 15.85 as:

$$Z_0 = \sqrt{Z_1 Z_2}\sqrt{\frac{4Z_2}{Z_1 + 4Z_2}} \tag{15.86}$$

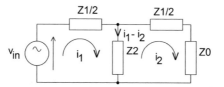

Fig. 15.33 Propagation Coefficient for T-Section

Propagation Coefficient

It is convenient and conventional to use the propagation coefficient γ to describe the attenuation and phase shift from input to output. By definition the propagation coefficient in Fig. 15.33 is:

$$\gamma = \log_e \left[\frac{i_2}{i_1} \right] \tag{15.87}$$

where we have completed the second mesh with the input impedance of a following stage, but the equations relate to the 'T' alone, without Z_0.

To find the ratio between the currents, we only need the equation for the second mesh:

$$(i_1 - i_2)Z_2 = i_2 \left(\frac{Z_1}{2} + Z_0 \right) \tag{15.88}$$

whence:

$$\gamma = \log_e \frac{i_2}{i_1} = \log_e \left[\frac{\frac{Z_1}{2} + Z_2 + Z_0}{Z_2} \right] \tag{15.89}$$

If we substitute for Z_0 we have:

$$e^\gamma = \frac{\frac{Z_1}{2} + Z_2 + \sqrt{\frac{Z_1^2}{4} + Z_1 Z_2}}{Z_2} \tag{15.90}$$

Then, partly following A.L Albert at http://www.vias.org/albert_ecomm/aec05_electric_networks_019.html, we expand Eq. 15.90 as:

$$e^\gamma = \frac{Z_1}{2Z_2} + 1 + \sqrt{\frac{Z_1^2}{4Z_2^2} + \frac{Z_1}{Z_2}} \tag{15.91}$$

and if we define $U = 1 + \frac{Z_1}{2Z_2}$, this reduces to:

$$e^{\gamma} = U + \sqrt{U^2 - 1} \tag{15.92}$$

By definition:

$$\cosh(\gamma) = \frac{1}{2}[e^{\gamma} + e^{-\gamma}] \quad or \quad \cosh(\gamma) = \frac{1}{2}\left[e^{\gamma} + \frac{1}{e^{\gamma}}\right] \tag{15.93}$$

On substituting in Eq. 15.93 for γ:

$$\cosh(\gamma) = \frac{1}{2}\left[\frac{\left(U + \sqrt{U^2 - 1}\right)^2 + 1}{U + \sqrt{U^2 - 1}}\right] \tag{15.94}$$

which expands to:

$$\cosh(\gamma) = \frac{1}{2}\left[\frac{U^2 + U^2 - 1 + 2U\sqrt{U^2 - 1} + 1}{U + \sqrt{U^2 - 1}}\right] \tag{15.95}$$

and reduces to:

$$\cosh(\gamma) = \frac{1}{2}2U\left[\frac{U + \sqrt{U^2 - 1}}{U + \sqrt{U^2 - 1}}\right] \qquad \cosh(\gamma) = U \tag{15.96}$$

Finally:

$$\gamma = \cosh^{-1}\left(1 + \frac{Z_1}{2Z_2}\right) \tag{15.97}$$

It follows from Eqs. 15.93 and 15.96 that:

$$\frac{U - 1}{2} = \frac{1}{4}(e^{\gamma} - 2 + e^{-\gamma}) \tag{15.98}$$

whence:

$$\sqrt{\frac{U - 1}{2}} = \frac{1}{2}\left(e^{\frac{\gamma}{2}} - e^{\frac{-\gamma}{2}}\right) = \sinh\left(\frac{\gamma}{2}\right) \tag{15.99}$$

Substitute for U in Eq. 15.99 and finally:

$$\sinh\left(\frac{\gamma}{2}\right) = \sqrt{\frac{Z_1}{4Z_2}} \tag{15.100}$$

The same applies to a 'Π' network.

Filter Characteristics

The propagation coefficient consists of two parts, identically to the relationship in transmission lines:

$$\gamma = \alpha + j\beta \qquad (15.101)$$

where α is the attenuation in nepers and β the phase shift. The importance of this is in showing the behaviour as a function of frequency, particularly in the transition region from pass to stop. The input and output impedances alone are not really sufficient.

First we rewrite Eq. 15.100 using Eq. 15.101 and expand $sinh((\alpha + \beta)/2)$:

$$\sinh\left(\frac{\alpha + j\beta}{2}\right) = \sinh\left(\frac{\alpha}{2}\right)\cos\left(\frac{\beta}{2}\right) + j\cosh\left(\frac{\alpha}{2}\right)\sin\left(\frac{\beta}{2}\right) = \sqrt{\frac{Z_1}{4Z_2}} \qquad (15.102)$$

In the case of constant-K filters, Z_1 and Z_2 are not the same type of reactance: whichever is inductive, the other is capacitive so:

$$\frac{Z_1}{Z_2} = j\omega C j\omega L = -\omega^2 LC \qquad (15.103)$$

and is always negative. Therefore the square root in Eq. 15.102 is imaginary and hence the real part – the first term – must be zero so that:

$$\cosh\left(\frac{\alpha}{2}\right)\sin\left(\frac{\beta}{2}\right) = \sqrt{\frac{Z_1}{4Z_2}} \qquad (15.104)$$

This leads to two special conditions:

Pass Band
As the filter is entirely reactive consisting only of capacitors and inductors, in the pass band there will be no attenuation and $\alpha = 0$. From Eq. 15.104, this means that $cosh(\alpha/2) = 1$ and hence:

$$\sin\left(\frac{\beta}{2}\right) = \sqrt{\frac{Z_1}{4Z_2}} \qquad (15.105)$$

and the phase is:

$$\beta = 2\sin^{-1}\left[\sqrt{\frac{Z_1}{4Z_2}}\right] \qquad (15.106)$$

Stop Band
The other case is that $|sin(\beta/2)| = 1$ and thus $\beta = n\pi$ where n is an integer. And nowhere is it shown that $n = 1$ which is what is implicitly (and correctly) assumed. Then:

$$\cosh\left(\frac{\alpha}{2}\right) = \sqrt{\frac{Z_1}{4Z_2}} \qquad (15.107)$$

or:

$$\alpha = 2\cosh^{-1}\left[\sqrt{\frac{Z_1}{4Z_2}}\right] \qquad (15.108)$$

and we must remember this is in nepers, where the conversion is *1 dB = 0.115 Np* and that these equations assume no terminating load.

But returning to Eq. 15.104, in the transition region between stop and pass, the product is a constant, so if we pick a frequency, we can at once find the attenuation and vice versa.

15.4.2   Constant-K Filters

These are a direct continuation from the previous sections. A comprehensive coverage of these and m-derived filters is at www.rgcetpdy.ac.in/Notes/ECE/...TLW/Unit%201.pdf' and www.aiktcdspace.org:8080/.../Chapter_on_Filters.pdf, whilst a long, very detailed, typed document is shodhganga.inflibnet.ac.in/jspui/.../07_chapter%203.pdf

We start with the 'L' section of Fig. 15.30a where we have a series impedance Z_1 and a shunt impedance Z_2. We repeat the definition from above that:

$$k^2 = Z_1 Z_2 \qquad (15.109)$$

and we want this to be independent of frequency. A ramification of Eq. 15.109 is that if we double the value of Z_1, we must halve Z_2 to maintain constant *k*.

Low Pass Constant-K Filter

We can build this either as a 'T' or a 'Π'. Both should have the same performance.

Low Pass 'T' Filter
We obviously need a series inductor rather than a series capacitor, and then Eq. 15.109 can be written as:

$$k^2 = Z_1 Z_2 \quad k^2 = \frac{j\omega L}{j\omega C} \quad k^2 = \frac{L}{C} \qquad (15.110)$$

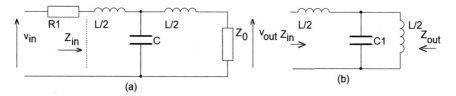

Fig. 15.34 LP constant K T- Filter

and this is constant and independent of the frequency. We build the circuit as Fig. 15.34a and substitute in Eq. 15.78:

$$Z_0 = \sqrt{\frac{L}{C}}\sqrt{1 + \frac{j\omega L}{-4\frac{j}{\omega C}}} \tag{15.111}$$

This reduces to:

$$Z_0 = \sqrt{\frac{L}{C}}\sqrt{\left[\frac{1 - \omega^2 LC}{4}\right]} \tag{15.112}$$

We have, then, that if $\omega = \omega_c = \frac{2}{\sqrt{LC}}$ the impedance will be zero. This is the cut-off frequency f_c where:

$$f_c = \frac{1}{\pi\sqrt{LC}} \tag{15.113}$$

And notice that, because of the 2 in the numerator in the expression for ω, there is no 2 in the denominator of Eq. 15.113. This does not mean that the output voltage is zero, far from it: the voltages across the individual components are not zero, and the voltage across the capacitor feeds the second inductor. Then from Eqs. 15.112 and 15.113, we find that:

$$L = \frac{Z_0}{\pi f_c} \quad C = \frac{1}{Z_0 \pi f_c} \tag{15.114}$$

Input Impedance
Denoting this as Z_{in}, if we substitute ω_c in Eq. 15.112:

$$Z_{in} = Z_0 \sqrt{\left[1 - \left(\frac{\omega}{\omega_c}\right)^2\right]} \tag{15.115}$$

From this we expect the input impedance to be almost constant at Z_0 at low frequencies with a phase of $0°$. We can deduce this from the circuit since the inductors are assumed to be ideal with no resistance. Even when the frequency is $0.1\omega_c$, the impedance has only fallen by 1%.

Above the cut-off frequency from Eqs. 15.115 and 15.116, the impedance is imaginary; the capacitor will tend to a short-circuit, whilst the impedance of L_1 will increase, so we should expect the circuit to approximate to a single inductor. But note that this is only valid if the circuit is terminated by its characteristic impedance.

Effects of Mismatch

If Z_0 is made very small, and ignoring R_1, we have, Fig. 15.34b:

$$Z_{in} = j\omega\frac{L}{2} + \frac{j\omega\frac{L}{2}\frac{1}{j\omega C_1}}{j\omega\frac{L}{2} + \frac{1}{j\omega C_1}} \tag{15.116}$$

$$Z_{in} = j\omega\frac{L}{2} + \frac{j\omega\frac{L}{2}}{1 - \omega^2 C_1\frac{L}{2}} \tag{15.117}$$

and this will result in a maximum impedance when:

$$f = \frac{1}{2\pi\sqrt{C_1\frac{L}{2}}} \tag{15.118}$$

A large increase in Z_0 mainly affects the low frequency input impedance which rises below the cut-off frequency at 20 dB/decade because the circuit becomes a simple LC filter.

With a reduced R_1, which we ignore, the impedances are no longer matched, and we have:

$$Z_{in} = j\omega\frac{L}{2} + \frac{\frac{1}{j\omega C_1}\left(j\omega\frac{L}{2} + Z_0\right)}{\frac{1}{j\omega C_1} + j\omega\frac{L}{2} + Z_0} \tag{15.119}$$

This expands to:

$$Z_{in} = j\omega\frac{L}{2} + \frac{j\omega\frac{L}{2} + Z_0}{1 - \omega^2 C_1\frac{L}{2} + j\omega C_1 Z_0} \tag{15.120}$$

and we have the impedance of the series inductor plus that of the damped tuned circuit whose impedance is a maximum when $\omega = 1/\sqrt{(LC_1/2)}$. The other extreme is if the output impedance is very large, then we divide Eqs. 15.119 and 15.120 throughout by Z_0 and to a good approximation we are left with $1/(j\omega C_1)$ and the input impedance is that of the series resonant circuit of $L/2$ C_1.

Phase Shift and Attenuation

In the pass band, where the frequency is below ω_c and the attenuation is zero, from Eq. 15.106, we have:

$$\beta = 2\sin^{-1}\left[\sqrt{\frac{\omega^2 LC}{4}}\right] \tag{15.121}$$

and substituting using ω_c:

$$\beta = 2\sin^{-1}\left(\frac{\omega}{\omega_c}\right) \tag{15.122}$$

and this is only *11.4°* when $\omega/\omega_c = 0.1$.

In the stop band, as this is a 3-pole network, we expect an asymptotic slope of -60 *dB/decade* and a phase shift of *270°*. From Eq. 15.108 we have:

$$\beta = \pi \quad \alpha = 2\cosh^{-1}\left(\frac{\omega}{\omega_c}\right) \tag{15.123}$$

Example – Low Pass 'T' Filter

Taking $f_c = 200$ *Hz* and $Z_0 = 50$ Ω we have schematic ('LP T Constant-K.asc'). The two inductors each have half the value calculated from Eq. 15.114. From the simulation we find that $f_{3dB} = 200$ *Hz*, and as this is a 3rd-order network the high frequency slope is *60 dB/decade*. The low frequency output voltage is half the input because the two resistors form a potential divider.

To avoid the *.net* output impedance affecting the circuit we make: **.net V(out) V1 Rout = 1e100**. We now find if we plot *Zin(v1)* on a logarithmic y-axis, that the low frequency input impedance correctly is *100* Ω because it includes R_1 and R_2. We also find the impedance rising at 20 dB/decade above the cut-off frequency and the phase shifting to *90°* because the impedance is now only that of L_1.

Making $R_2 = 1$ Ω from Eq. 15.118, we find $f = 140$ *Hz* and this is what we measure. Conversely, if $R_1 = 1$ Ω then we measure a *3 dB* rise in the output voltage at *170 Hz* and a phase angle of some *140°*.

If we make R2 very large – the condition under which Eq. 15.123 applies – we find a low-frequency phase shift of *0°* and an asymptotic high-frequency phase shift of $-180°$. If we pick some points in the transition region, we find agreement.

Of interest is the high-frequency attenuation. The first row are decade ratios of attenuation with the corresponding value in nepers in the second row. But these must be doubled in Eq. 15.123 to give the third row and finally we find the attenuation increasing at *40 dB/decade* in the last row because of the factor 2.

(x_1/x_2)	10	100	1000
$\cosh^{-1}(x_1/x_2)$ Np	3	5.3	7.6
$2 \times$ Np	6	10.6	15.2
dB	52	92	132

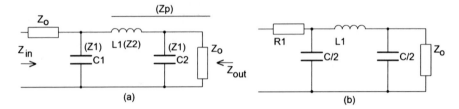

Fig. 15.35 LP constant K Π- Filter

Explorations 4
1. Plot the phase shift and attenuation, and compare the graphs with those shown in the websites above using schematic ('LP T Constant-K.asc').
2. Reducing $R1$ and/or $R2$ and the response is no longer flat but shows a resonance peak.
3. Make $R2 = 1e11\Omega$ and the response is that predicted by Eq. 15.123.
4. Use Eq. 15.104 to calculate attenuation and phase at a few points in the transition region and confirm by measurement.
5. Try cascading two identical stages of schematic ('LP T + P1 Constant-K.asc').

Low Pass 'Π' Filter
Alternatively we may build it as a symmetrical 'Π', Fig. 15.35a. We can analyse the circuit as it stands converting everything after C_1 into a parallel impedance as we did earlier. This will lead us to Eq. 15.86. Like Eq. 15.78, this starts with $\sqrt{(Z_1 Z_2)}$, so if these are identical we have the same k value.

To have the same cut-off frequency, we need the second square root in Eq. 15.86 to have $\omega_c = \frac{2}{\sqrt{LC}}$. By inspection, we see that we need a series inductor again for Z_2, and therefore Z_1 is a capacitor. Inserting this in the second square root of Eq. 15.86 gives:

$$Z_0 = \sqrt{\frac{L}{C}}\sqrt{\left[\frac{4j\omega L}{\frac{1}{j\omega C} + 4j\omega L}\right]} \qquad Z_0 = \sqrt{\frac{L}{C}\left[\frac{4\omega^2 LC}{1 - 4\omega^2 LC}\right]} \qquad Z_0 = \sqrt{\frac{L}{C}\left[\frac{\frac{\omega}{\omega_c}}{1 - \frac{\omega}{\omega_c}}\right]}$$

and this simplifies to:

$$Z_0 = \sqrt{\frac{L}{C}}\sqrt{\frac{1}{\frac{\omega}{\omega} - 1}} \tag{15.124}$$

and we have the same cut-off frequency but the impedance is a maximum, not a minimum, and corresponds, very approximately, to the parallel sum of $C_1 L_1$ and is seen as a sharp resonance if R_2 is made very large.

Phase Shift and Attenuation
At low frequencies the inductor is effectively a short circuit so the output voltage is half the input because of the potential divider formed by the two characteristic

impedances Z_o and the input impedance approximates to $2Z_0$ as the two impedances are in series. Above cut-off, the capacitors will tend to short-circuits so the output voltage will fall at $-60\ dB/decade$ as before and the input impedance will just be Z_0.

Remembering that we have interchanged capacitors and inductors, we restore the inductor to its original value, not half of it, but make the capacitors half the previous size, and k is unchanged.

It is interesting to compare the input and output impedances of the 'T' and 'Π' versions.

Example – Low Pass Constant-K Π- Filter

With the alternative 'Π', schematic ('LP Pi Constant-K.asc'), the frequency response, input current and output current are identical to the 'T' version as evidenced by schematic ('LP T + Pi Constant-K.asc') where the traces coincide. We also know that the previous propagation coefficients of Eq. 15.123 apply.

High-Pass Constant-K Filter

Intuitively, we exchange capacitors for inductors and inductors for capacitors and again we have two configurations.

High-Pass 'T' Filter

This is Fig. 15.36. Substituting in Eq. 15.112:

$$Z_0 = \sqrt{\frac{L}{C}}\sqrt{1 - \frac{\frac{1}{j\omega C}}{4j\omega L}} \quad \Rightarrow Z_0 = \sqrt{\frac{L}{C}}\sqrt{1 - \frac{1}{4}\omega^2 LC} \qquad (15.125)$$

From the second square root we have:

$$\omega_c^2 = \frac{1}{4LC} \ \ or \ \ f_c = \frac{1}{2\pi\sqrt{4LC}} \quad \Rightarrow f_c = \frac{1}{4\pi\sqrt{LC}} \qquad (15.126)$$

From Eq. 15.126

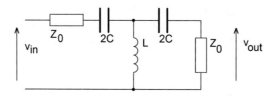

Fig. 15.36 HP constant K T-Filter

$$LC = \frac{1}{16\pi^2 f_c^2} \qquad (15.127)$$

But we also require $\sqrt{L/C} = Z_0$. These conditions are satisfied by:

$$L = \frac{Z_0}{4\pi f_c} \quad C = \frac{1}{4Z_0\pi f_c} \qquad (15.128)$$

We noted above that it is conventional to halve the series impedance, and therefore we must double the capacitor.

Phase Shift and Attenuation

At low frequencies, the impedance of the capacitor in series with the source\(delete index) dominates the input impedance which therefore rises at 20 dB/octave as the frequency falls. We again have a capacitor and inductor in parallel at the breakpoint and another slight rise of the impedance. Above this frequency, the capacitors tend to short circuits and the inductor to an open circuit, so the input impedance is just that of the source and load resistors in series and is $2Z_0$. Similarly, at high frequencies, the output impedance is the two resistors in parallel, whereas at low frequencies, when the impedance of the capacitors is significant, it is just Z_0. So the output impedance doubles in going from pass band to stop band, but the input impedance rises by several orders of magnitude.

Equation 15.123 does not apply. Instead, inserting the component impedances in Eq. 15.106 and substituting ω_c we have:

$$\beta = 2\sin^{-1}\left[\sqrt{\frac{-1}{\omega^2 LC}}\right] = 2\sin^{-1}\left(\frac{\omega_c}{\omega}\right) \qquad (15.129)$$

and this is the inverse of the low-pass filter and we now find the phase shift in the pass band is $0°$ and $180°$ in the stop band. Likewise:

$$\alpha = 2\cosh^{-1}\left(\frac{\omega_c}{\omega}\right) \qquad (15.130)$$

and we find the same slope in the stop band, but in the reverse direction.

Example – High-Pass Constant-K T-Filter

Still with a cut-off frequency of *200 Hz*, we find $C = 7.958 \ \mu F$ $L = 0.0398 \ H$. The capacitors for a 'T' circuit must be twice the value calculated from Eqs. 15.128 and 15.129 so we have $C_1 = C_2 = 15.98 \ \mu F$. The schematic ('HP T Constant-K Filter. asc') shows the correct 3 dB point. The graphs are essentially left-to-right mirror images of the low-pass filter.

High-Pass 'Π' Filter

We may also build a 'Π' following the analysis for a 'Π' low-pass filter, and now the inductors are twice as large and the capacitors halved so that Z_0 is kept the same.

At low frequencies the inductors tend to short circuits, so the input impedance is just the source impedance whilst at high frequencies when the capacitor approximates to a short circuit the input impedance is the two resistors in series. Likewise, the output impedance at high frequencies also is the two resistors in parallel, but at low frequencies, it falls at -20 dB/decade as the impedance of the inductor in parallel with the load falls.

Example High-Pass 'Π' Filter
We can build the 'Π' version, and now the inductor must be doubled instead of the capacitors, schematic ('HP Pi Constant-K Filter.asc'). This, too, shows the correct frequency response.

Explorations 5
1. Build and test the high-pass schematics. Terminate them by high resistance values and confirm that the response is quantified by Eqs. 15.126, 15.129 and 15.130.
2. Note the effects of mismatched components on the cut-off frequency and shape of the response graph.
3. Plot α and β and compare to the graphs shown in the above websites.
4. Try cascading two or three identical stages. In particular, note the asymptotic slope.
5. It is worth noting the change in input impedance with frequency, in particular the differences between the 'T' and 'Π' versions.

Band-Pass Constant-K Filters

Once again we have the choice of a 'T' or 'Π' configuration. If the two frequencies are separated by a factor of *10* or more, we can use the 'Phase shift and Attenuation' results for the low-pass and high-pass sections in turn to find the phase shift and attenuation.

Band-Pass 'T' Filter
One way of deriving the equations can be long and tedious and is presented on YouTube at https://www.youtube.com/watch?v=qJIX07lII_I and runs to 32 min. A neater way is at http://ecoursesonline.iasri.res.in/mod/page/view.php?id=1248 which is followed here.

We start from Fig. 15.30c and using s notation we have:

$$Z_1 = sL_1 + \frac{1}{sC_1} = \frac{1 - s^2 L_1 C_1}{sC_1} \tag{15.131}$$

$$Z_2 = \frac{\frac{sL_2}{sC_2}}{sL_2 + \frac{1}{sC_2}} = \frac{sL_2}{1 - s^2 L_2 C_2} \tag{15.132}$$

we require:

$$k^2 = Z_1 Z_2 = \frac{(1 - s^2 L_1 C_1) s L_2}{(1 - s^2 L_2 C_2) s C_1} \tag{15.133}$$

and this is independent of frequency if:

$$L_1 C_1 = L_2 C_2 = \omega_0^2 \text{ then } k^2 = \frac{L_2}{C_1} = \frac{L_1}{C_2} \tag{15.134}$$

And the resonant frequencies of the series and shunt arms are identical. As this is a constant-k filter, at cut-off:

$$\frac{Z_1}{4Z_2} = -1 \quad \text{or} \quad Z_1 = -4Z_2 \quad \text{then} \quad Z_1^2 = -4Z_1 Z_2 \quad \text{so} \quad Z_1^2 = -4k^2 \tag{15.135}$$

From the last equation of 15.135 we have $Z_1 = j2k$, so there are two solutions with one being the negative of the other. Replacing s by ω_h and ω_l for the two frequencies, we equate the two solutions:

$$\frac{1 - \omega_h^2 L_1 C_1}{j \omega_h C_1} = -\frac{\left(1 - \omega_l^2 L_1 C_1\right)}{j \omega_l C_1} \tag{15.136}$$

Substituting for $L_1 C_1$ and multiplying by $j \omega_h C_1$:

$$1 - \frac{\omega_h^2}{\omega_0^2} = \frac{-\omega_h}{\omega_l} \left[1 - \frac{\omega_l^2}{\omega_0^2} \right] \tag{15.137}$$

Multiplying by $\omega_0^2 \, \omega_l$:

$$\left(\omega_0^2 - \omega_h^2\right) \omega_l = \omega_h \left(\omega_l^2 - \omega_0^2\right) \tag{15.138}$$

Then rearrange:

$$(\omega_h + \omega_l) \omega_0^2 = \omega_h \omega_l (\omega_h + \omega_l) \tag{15.139}$$

and we see that writing frequencies instead of ω we have $f_0 = \sqrt{f_h f_l}$ where f_0 is the geometric mean. To find the component values, we start with $Z_1 = j2k$ and cancel j:

$$\omega_h L_1 - \frac{1}{\omega_h C_1} = -2k \tag{15.140}$$

Then:

$$1 - \omega_h^2 L_1 C_1 = 2k\omega_h C_1 \tag{15.141}$$

But $1/(L_1 C_1) = \omega_0^2 = \omega_h \omega_l$ by Eq. 15.139 so if we substitute in equation 15.141:

$$1 - \frac{\omega_h^2}{\omega_h \omega_l} = 2k\omega_h C_1 \quad \frac{\omega_h - \omega_l}{\omega_l} = 2k\omega_h C_1 \tag{15.142}$$

Then replacing k by Z_0 and using frequencies:

$$C_1 = \frac{f_h - f_l}{4\pi Z_o f_h f_l} \tag{15.143}$$

From Eq. 15.134:

$$L_1 = \frac{1}{C_1 \omega_0^2} \quad L_1 = \frac{4\pi Z_0 f_h f_l}{f_h - f_l} \frac{1}{4\pi^2 f_h f_l} \quad L_1 = \frac{Z_0}{\pi(f_h - f_l)} \tag{15.144}$$

Also from Eq. 15.134:

$$L_2 = C_1 k^2 \quad L_2 = \frac{Z_0(f_h - f_l)}{4\pi f_h f_l} \tag{15.145}$$

$$C_2 = \frac{L_1}{k^2} \quad C_2 = \frac{1}{\pi Z_0(f_h - f_l)} \tag{15.146}$$

The complete circuit is Fig. 15.37a. The simplest way of viewing it is as a low-pass filter overlaid on a high-pass filter. If the two frequencies are spaced far apart, we may approximate the equations by:

$$L_1 = \frac{Z_0}{\pi f_h} \quad L_2 = \frac{Z_0}{4\pi f_l} \quad C_1 = \frac{1}{4\pi f_l Z_0} \quad C_2 = \frac{1}{\pi Z_0 f_h} \tag{15.147}$$

and these are the equations for low- and high-pass filters.

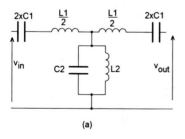
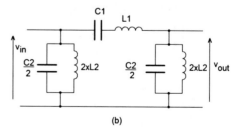

(a)   (b)

Fig. 15.37 Band pass constant K Filter

Example – Constant-K Band-Pass 'T' Filter

We design for a lower frequency $f_l = 50\ Hz$ and $f_h = 1\ kHz$ with a characteristic impedance of $75\ \Omega$. We find $L_1 = 0.0251H$ $C_1 = 21.22\mu F$ $L_2 = 0.113H$ $C_2 = 4.468\mu F$, and we must double the series capacitance and halve the series inductance in the circuit, ('Band-Pass Constant-K T.asc'), to find the 3 dB points are at $49\ Hz$, $998\ Hz$.

Band-Pass 'Π' Filter

The circuit is shown in Fig. 15.37b. We have the same cut-off frequencies for the shunt and series arms; the previous analysis is valid here.

Example – Constant-K Band-Pass 'Π' Filter

We shall design for a lower frequency $f_l = 50\ Hz$ and $f_h = 1\ kHz$ with a characteristic impedance of $75\ \Omega$. We use the values from Eq. 15.147 but we must remember that the capacitor C_2 is halved and the inductor L_2 is doubled. The schematic is ('Band-Pass Constant-K Pi.asc'). The 3 dB points found using the cursor are $48\ Hz$ and $1\ kHz$.

Band-Stop Constant-K Filter

Again we have two choices. The analysis for the 'T' can be found at http://ecoursesonline.iasri.res.in/mod/page/view.php?id=1248 which we follow here. To a first approximation, if the two frequencies are separated by at least a factor of 10, we can again use the 'Phase Shift and Attenuation' of the low-pass and high-pass section estimate the attenuation. It is worth comparing the results with the sketch graphs of the website *ecoursesonline* mentioned above.

Band-Stop 'T' Filter

The schematic is the band-pass filter with the series and parallel arms interchanged Fig. 15.38a. Most of the previous equations apply with the filter designed so that the series and shunt arms resonate at the same frequency.

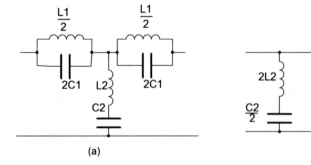

Fig. 15.38 Constant K band stop filter

As we know, $Z_1 = 2jk$, and then from Eq. 15.135 $Z_2 = \frac{jk}{2}$, so from the series arm, we have after cancelling j:

$$Z_2 = \frac{1}{\omega_l C_2} - \omega_l L_2 = \frac{k}{2} \tag{15.148}$$

After multiplying by $\omega_l C_1$:

$$1 - \omega_l^2 L_2 C_2 = \frac{k}{2} \omega_l C_2 \tag{15.149}$$

Then as $L_2 C_2 = \omega_0^2 = \omega_h \omega_l$ and using the same expansion method as before:

$$\frac{\omega_0^2 - \omega_l^2}{\omega_0^2} C_2 = \frac{k}{2} \omega_l C_2 \quad \Rightarrow C_2 = \frac{(f_h - f_l)}{\pi Z_0 f_h f_l} \tag{15.150}$$

Substituting for ω_0^2 in Eq. 15.151 and using frequencies:

$$L_2 = \frac{1}{\omega_0^2 C_2} \quad L_2 = \frac{\pi Z_0 (f_h f_l)}{\omega_0^2 (f_h - f_l)} \quad L_2 = \frac{Z_0}{4\pi(f_h - f_l)} \tag{15.151}$$

$$L_1 = k^2 C_2 \quad L_1 = \frac{Z_0(f_h - f_l)}{\pi f_h f_l} \tag{15.152}$$

$$C_1 = \frac{L_2}{k^2} \quad C_1 = \frac{1}{4\pi Z_0 (f_h - f_l)} \tag{15.153}$$

Example – Constant K Band-Stop 'T' Filter

We shall design for a stop band from *200 Hz* to *1 kHz* with $Z_0 = 600\ \Omega$. The component values are $C_2 = 2.12.\mu F$, $L_2 = 59.68\ mH$, $C_1 = 0.166\ \mu F$ and $L_1 = 0.764\ H$.

But we must remember to halve L_1 and double C_1.

The circuit behaves as a notch filter with broadened 'skirts', schematic 'Band-Stop Constant-K ('T'.asc'), and is certainly not the converse of the band-pass filter. The asymptotic attenuation away from the stop band is the expected −6 dB. The points where the loss falls by −3 dB are exactly as designed.

Band-Stop 'Π' Filter

As with the band-pass versions, this swaps series and parallel elements, Fig. 15.39b. The equations for the 'T' version apply with the scaling shown in the figure.

LC Band-Stop Filter

This is a simple 'L' section, not part of the *k-derived filters* but as a contrast. The circuit is Fig. 15.40. It creates a single sharp notch.

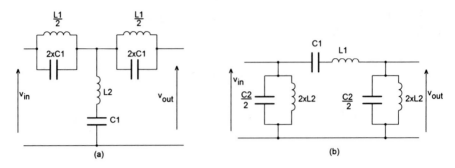

Fig. 15.39 Band stop constant K

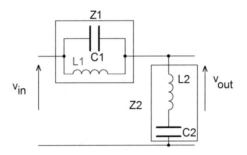

Fig. 15.40 LC band stop filter

Example – LC Band-Stop Filter

A schematic is ('LC Band-Stop Filter.asc'). The values given for an impedance of
50 Ω and a stop band from *100* to *5000 Hz* are $L_1 = 1.62\ mH$, $L_2 = 77.99\ mH$,
$C_1 = 31.19\ uF$ and $C_2 = 649.6\ nH$ using the calculator at the website https://
electronicbase.net/band-stop-filter-calculator/. However, this produced a single
notch at the wrong frequency. Tinkering with the numbers, we can get two notches
at the correct frequencies, but there is no attenuation between them and we cannot
spread the flanks of the notch to give a wider bandwidth.

Explorations 6

1. Compare the performances of the 'T' and 'Π' versions of the band-pass filter.
2. Note the effects of component tolerances.
3. Build a band-stop 'Π' filter.
4. Add .*net* directives and note the changes of attenuation and phase, and compare
   with the graphs at *ecoursesonline* referenced above.
5. Note the change of input and output impedances with frequency and the differences between the 'T' and 'Π' versions.

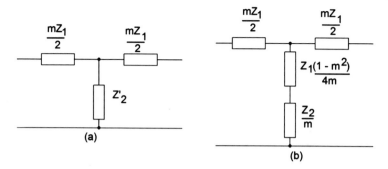

Fig. 15.41 m-Derived T filter

15.4.3   M-Derived Filters

These are derived from the k filter and offer a sharper transition. They can be either 'T' or 'Π' forms.

Derived from T-Prototype
Starting with Eq. 15.77 and Fig. 15.41a, we add a multiplier m to the series Z_1 elements and replace the shunt impedance by Z'_2. As we require the same impedance as the k filter so that we can cascade them, we have:

$$\frac{Z_1^2}{4} + Z_1 Z_2 = \frac{m^2 Z_1^2}{4} + m Z_1 Z_2' \tag{15.154}$$

Then:

$$m Z_1 Z_2' = Z_1^2 (1 - m^2) + Z_1 Z_2 \tag{15.155}$$

And:

$$Z_2' = \frac{Z_1 (1 - m^2)}{4m} + \frac{Z_2}{m} \tag{15.156}$$

and the new Z_2' consists of two elements, Fig. 15.41b, and provided $(1 - m^2)$ is positive, we can realize it physically. This opens the possibility of two resonances.

Derived from Π-Prototype
We again require the same impedance as the k filter, starting with Eq. 15.85. As with the k filter, the shunt arms Z_2 are multiplied 2, Fig. 15.42a, but this time we also divide them by m, and we need to find the new series arm Z_1, Fig. 15.42b:

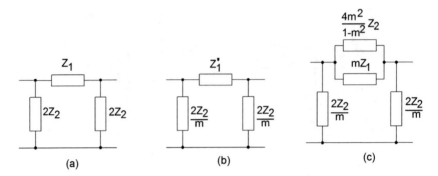

Fig. 15.42 m-derived Π filter

$$\frac{4Z_1 Z_2^2}{Z_1 + 4Z_2} = \frac{4Z_1'\frac{Z_2^2}{m^2}}{Z_1' + 4\frac{Z_2}{m}} \tag{15.157}$$

After cross-multiplying we have:

$$4Z_1 Z_2^2 Z_1' + 16Z_1 \frac{Z_2^3}{m^2} = 4Z_1 Z_1'\frac{Z_2^2}{m^2} + 16Z_1'\frac{Z_2^3}{m^2} \tag{15.158}$$

After dividing by *4* and collecting terms:

$$Z_1' Z_2^2\left(Z_1 - \frac{Z_1}{m^2} - 4\frac{Z_2}{m^2}\right) = -4Z_1 \frac{Z_2^3}{m^2} \tag{15.159}$$

If we again divide by *4* and multiply by m^2:

$$Z_1' = \frac{Z_1 Z_2}{Z_1\frac{(1-m^2)}{4} + Z_2} \tag{15.160}$$

Equation 15.160 is clearly the resultant of two impedances in parallel. If we multiply throughout by $\frac{4m^2}{(1-m^2)}$, we arrive at:

$$Z_1' = \frac{mZ_1 \frac{4m^2}{1-m^2} Z_2}{mZ_1 + \frac{4m^2}{1-m^2} Z_2} \tag{15.161}$$

which is Fig. 15.42c.

Low-Pass m Filter

To create a low-pass filter, we need the series element or elements to be inductors so that they offer a very low impedance at low frequencies, Fig. 15.43. As before, we have 'T' and 'Π' configurations.

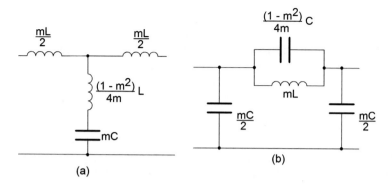

Fig. 15.43 m-derived low pass filter

Low-Pass m Filter 'T'
The resonant frequency for the shunt arm is:

$$\frac{1}{m\omega_r C} = \frac{(1-m^2)}{4m}\omega_r L \tag{15.162}$$

which can be organized as:

$$\omega_r^2 = \frac{4}{(1-m^2)LC} \tag{15.163}$$

$$f_r = \frac{2}{2\pi\sqrt{(1-m^2)LC}} \tag{15.164}$$

And this should be higher than the cut-off frequency which, from before, is:

$$f_c = \frac{1}{\pi\sqrt{LC}} \tag{15.165}$$

Then substituting from Eq. 15.165:

$$f_r = \frac{2\sqrt{LC}}{2\pi\sqrt{(1-m^2)}} \quad f_r = \frac{f_c}{\sqrt{1-m^2}} \tag{15.166}$$

and this is the frequency for infinite attenuation as the impedance of the shunt arm is zero. We may also arrange the second Eq. 15.166 to give:

$$m = \sqrt{\left[1 - \frac{f_c^2}{f_r^2}\right]} \tag{15.167}$$

and from Eqs. 15.166 and 15.167 for a sharp cut-off f_r should be close to f_c, implying a small m.

We also need the correct characteristic impedance $Z_0 = \sqrt{\frac{L}{C}}$ which, with Eq. 15.165, gives:

$$C = \frac{1}{\pi f_c Z_0} \qquad (15.168)$$

Example – Low-Pass 'T' m Filter
We shall design for $f_c = 200\ Hz\ f_r = 300\ Hz$ which gives $m = 0.7454$ and $(1 - m^2)/(4\ m) = 0.149$. If $Z_0 = 50\ \Omega$ from Eq. 15.168 $C = 31.83\ \mu F$ and then from Eq. 15.165 $L = 0.0795\ H$. A check using Eq. 15.165 confirms that $f_c = 200\ Hz$. Then $mL/2 = 29.6\ mH$, $mC = 23.72\ \mu F$ and the shunt inductor is $11.84\ mH$ and the notch is at $300.3\ Hz$. This is schematic ('m-T Low Pass.asc').

The input impedance is correct at $50\ \Omega$. If we reduce the input source resistance to $1\ \Omega$, we find a low frequency resonance at about $177\ Hz$. If we reinstate the source resistance, the pass-band response is flat with just the $-6\ dB$ output loss due to the potential divider of the source and load impedances. The minimum Z_0 and $45°$ phase shift occur at around $185\ Hz$.

Low-Pass m Filter 'Π'
This time the series circuit corresponds to infinite attenuation at its resonant frequency. The impedances have been calculated in Eq. 15.161 and now we assign the series inductor and capacitor so that:

$$m\omega_r L = \frac{1}{\frac{1-m^2}{4m}\omega_r C} \qquad (15.169)$$

and this gives:

$$\omega_r^2 = \frac{4}{LC(1 - m^2)} \qquad (15.170)$$

and this is identical to Eq. 15.163 – as it should be, because we want the same resonant frequency – and we can find m from Eq. 15.167.

Example – Low-Pass 'Π' m Filter
We shall make the notch closer to the cut-off frequency of $200\ Hz$ by making $f_r = 250\ Hz$. Using Fig. 15.43b and schematic ('m-Pi Low Pass.asc') with $C_3 = 9\ \mu F$ and $L_2 = 45\ mH$, we have the correct resonant frequency and characteristic impedance.

Propagation Coefficient
This follows the k definition but with the new impedances. The derivation is set out in full:

$$\alpha = 2\cosh^{-1}\sqrt{\frac{Z_1}{4Z_2}} \tag{15.171}$$

Substituting for $4Z_2$ from Fig. 15.43a:.

$$4Z_2 = \frac{j\omega(1-m^2)L}{m} + \frac{4}{j\omega mC} \tag{15.172}$$

This can be written as:

$$4Z_2 = \frac{-\omega^2(1-m^2)LC + 4}{\omega mC} \tag{15.173}$$

Then:

$$\frac{Z_1}{4Z_2} = \frac{\omega^2 m^2 \frac{LC}{2}}{-\omega^2(1-m^2)LC + 4} \tag{15.174}$$

We can replace $(1-m^2)LC$ by $4/\omega_r^2$ from Eq. 15.170:

$$\frac{Z_1}{4Z_2} = \frac{\frac{\omega}{\omega_c^2}\frac{m^2}{2}}{4 - 4\frac{\omega^2}{\omega_r^2}} \tag{15.175}$$

We then extract the factor 4 and take square roots, finally to give the attenuation in terms of frequency:

$$\alpha = 2\cosh^{-1}\left[\frac{m\frac{f}{f_c}}{\sqrt{1 - \left(\frac{f}{f_r}\right)^2}}\right] \tag{15.176}$$

We have seen before that the numerator means there is virtually no attenuation, even up to $f = 0.1\, f_c$. we also see that at $f = f_r$ the attenuation is infinite and at higher frequencies α is imaginary. But then we can use the inductive potential divider since at high frequencies the capacitor will tend to a short-circuit. This gives us:

$$\alpha = \left[\frac{\frac{1-m^2}{4m}}{\frac{m}{2} + \frac{1-m^2}{4m}}\right] \qquad \alpha = \frac{1}{\frac{4m^2}{2(1-m^2)} + 1} \tag{15.177}$$

and we see that it is impossible to get an attenuation much greater than 20 dB.

We also have $\beta = 2\sin^{-1}\sqrt{\frac{Z_1}{4Z_2}}$ and we can use the factors of Eq. 15.176. At low frequency, the numerator predominates and the phase changes faster than the attenuation, being a sine function, not a cosh. We therefore predict a negative increase from $0°$ at low frequency up to $90°$ at f_r and use the inductive potential divider again to predict an asymptotic phase shift of $0°$.

Example – Attenuation and Phase Shift

If we use $m = 0.7454$ from the Low-Pass T m filter example and insert that in Eq. 15.177, we calculate a high frequency output of 0.285 V which compares well with the measured 0.290 V.

Explorations 7

1. In the 'T' schematic ('m-T Low Pass.asc') the notch frequency is not greatly altered by changes to the series elements, but is very sensitive to the shunt elements.
2. Critically compare the 'T' and 'Π' implementations in terms of input and output impedances, attenuation and phase shift.
3. Redesign for different m values.

High-Pass m Filter

These follow the same logic as k filters with inductors and capacitor interchanged. As before, we have either 'T' or 'Π' configurations shown in Fig. 15.44. Both types are discussed in http://ecoursesonline.iasri.res.in/mod/page/view.php?id=1247.

High-Pass m Filter 'T'

The resonance condition is when the impedance of the shunt arm is zero. That is:

$$\omega_r \frac{L}{m} = \frac{1}{\omega_r \frac{4m}{1-m^2} C} \tag{15.178}$$

which can be written as:

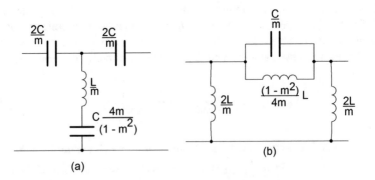

(a)   (b)

Fig. 15.44 m-Derived high pass Filter

$$\omega_r^2 LC \frac{4m}{m(1-m^2)} = 1 \qquad (15.179)$$

and after substituting the frequency this reduces to:

$$f_r = \frac{\sqrt{1-m^2}}{4\pi\sqrt{LC}} \qquad (15.180)$$

And using Eq. 15.126 for the cut-off frequency which is $f_c = \frac{1}{4\pi\sqrt{LC}}$, we have:

$$f_r = f_c\sqrt{1-m^2} \qquad (15.181)$$

and the resonant frequency is lower than the cut-off frequency. Remembering that $Z_0\sqrt{C} = \sqrt{L}$, we can substitute in Eq. 15.126 to find:

$$f_c = \frac{1}{4\pi Z_0 C} \qquad (15.182)$$

High-Pass m Filter 'Π'
We now require the impedance of the series arm to be a maximum:

$$\frac{4m}{1-m^2}\omega_r L = \frac{1}{\omega_r \frac{C}{m}} \qquad (15.183)$$

and this can be written as:

$$\omega_r^2 LC \frac{4m}{m(1-m^2)} = 1 \qquad (15.184)$$

and this is identical to Eq. 15.179 showing that the resonance conditions are identical.

Example – High-Pass m Filter 'T'
We shall design for a resonant frequency of $f_r = 1.0\ MHz$, a cut-off frequency of $f_c = 1.2\ MHz$ and a characteristic impedance of $Z_0 = 75\ \Omega$. We find $m = 0.55$ and from Eq. 15.182 $C = 884\ pF$ and $L = 4.97\ uH$.

We can see that the impedance in the pass band is correct and does not vary so radically in the stop-band as the k filter. But the low-frequency attenuation is very much inferior to the k filter.

Band-Pass and Band-Stop m Filters

The equations are given in www.rgcetpdy.ac.in/Notes/ECE/...TLW/Unit%201.pdf but we shall pass over them.

15.4.4   Combined Filters

We can cascade k low-pass and high-pass filters to achieve a faster roll-off. The m-derived filters have a very steep cut-off but poor attenuation in the stop band. So we could combine one with a k filter to improve the stop band attenuation.

We might in passing note the possibility of creating an m-derived filter response by defining the transfer function using s-parameters, for example, the schematic ('VCVS.asc') where it will be noticed that the response depends critically upon some parameters.

15.5   Tuned-Circuit Filters

We may regard a tuned circuit as a special notch filter. This has been covered in the Chapter 'Tuned Circuits'. Here we are concerned with using two as a band-pass filter whose pass band is determined by the coupling coefficient.

15.5.1   Double-Tuned Band-Pass Filter

This is an extension of mutual inductance. (delete index) The previous section showed us that an inductor and capacitor in parallel create a notch. However we often need a filter with a response like a 'Π', that is, as flat a top as we can get and steep sides. This often arises with communication circuits where the signal extends over a certain band width rather than being a single frequency.

A very common implementation is two tuned circuits, loosely coupled, and tuned to the same frequency, Fig. 15.45. It is not necessary for the capacitors and inductors to be identical, only that the resonant frequencies are the same, but for convenience they are usually identical. We include the circuit here rather than in the Chapter 'Tuned Circuits' because the analysis involves the coupling coefficient K.

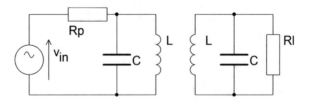

Fig. 15.45 Double tuned Circuit

Analysis

The analysis is surprisingly complicated; see https://slideplayer.com/slide/ 12253218/ and also http://rftoolbox.dtu.dk/book/Ch2.pdf. This is where LTspice comes to our aid if we have no prior knowledge of what coupling coefficient is needed. In the first instance, we run ('Very Wideband double tuned.asc.') and find there are two widely spaced resonant peaks. Reducing the coupling in ('Wideband double tuned.asc') is an improvement, but it is only with ('Narrow band double tuned.asc.') that we home in on the correct value for a maximally flat pass band. And as a bonus we see the effects of over- and under-coupling. Otherwise we should have to wade through some rather abstruse equations to arrive at the same result. But now we know that $K = 0.002$ is the optimum in this case we can proceed with a simplified analysis using Fig. 15.46 where we have removed the load resistor Rl, and the primary resistance Rp. And as K is so small, we can simply keep the primary and secondary inductors at their full value without subtracting M. We have also converted the parallel primary tuned circuit into a series one without, at the moment, changing the component values.

First we have for the primary circuit that:

$$v_{in} = I_p Z_p + j\omega M i_s \qquad (15.185)$$

and for the secondary:

$$0 = j\omega M i_p + i_s Z_s \qquad (15.186)$$

The primary impedance is:

$$Z_{pr} = \frac{v_p}{i_p} = \frac{Z_p i_p + j\omega M i_s}{i_p} \qquad (15.187)$$

and from Eqs. 15.186, we have:

$$i_s = \frac{j\omega M i_p}{Z_s} \qquad (15.188)$$

so on substituting in Eq. 15.187 and dividing by i_p:

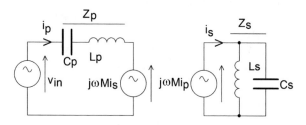

Fig. 15.46 Double tuned equiv Cct

$$Z_{pr} = Z_p - \frac{\omega^2 M^2}{Z_s} \qquad (15.189)$$

Recalling that if $\omega = \omega_r$ then $Z_s = \infty$ we are left with $Z_{pr} = Z_p$, so there is one resonance frequency. And as the inductors are identical, we can write Eq. 15.189 as:

$$Z_{pr} = Z_p + \frac{(\omega KL)^2}{Z_2} \qquad (15.190)$$

and if we substitute for Z_2 we finally arrive at:

$$Z_{pr} = Z_p + \frac{(\omega KM^2)(1 - \omega^2 LC)}{j\omega L} \qquad (15.191)$$

This is 'final' in the sense that we have reached a half-way house. From this equation we can see that the second term modifies Z_p to give two resonant frequencies and their separation will depend upon K. We first calculate the required resonant frequency, and then LTspice enables us quickly to home-in on the coupling coefficient and we can see and measure the effects of over-, critical- and under-coupling and the steepness of the flanks of the curve.

However, to follow this up with the full mathematical treatment, we first need to convert the parallel primary tuned circuit into its series form by choosing components Lx and Cx so that the input impedance is the same, that is:

$$j\omega L_x + \frac{1}{j\omega C_x} = \frac{j\omega L}{1 + \omega^2 LC} \qquad (15.192)$$

and reinstate the resistances and then turn to https://slideplayer.com/slide/12253218/ mentioned above. This will show that resistances do have a considerable impact on the response.

Critical coupling occurs when $K_c = \frac{1}{\sqrt{Q_1 Q_2}}$ if it is over-coupled there are two peaks at $\frac{\pm 1}{2Q} \sqrt{K^2 Q' - 1}$ on either side of the resonant frequency which can be separated by a large distance and only a single peak at under-coupling.

Explorations 8
1. Open ('Narrowband double tuned.asc') and measure the resonant frequency, bandwidth and depth of the 'dip' between the peaks. Note also that the slopes of the flanks off resonance are all the same.
2. Open ('Narrowband double tuned R effect.asc') and compare the resonant conditions and optimum K with the previously.
3. Open ('Narrowband double tuned effect of R inductor.asc') and note that increasing the inductor resistance can result in a single resonance with a low Q-factor.

15.6 Multi-stage Filters

It is not so many decades ago that the design of these filters was a recondite art only practised by the initiated after years of contemplation in a remote Tibetan monastery. However, thanks to the labours of these doughty pioneers, help is at hand to immeasurably lighten the task by using design packages and table of coefficients. We might heretically ask how great a need is there today for the more complex ones of six or more orders given digital signal processing and active filters – especially since it will be seen that for low frequency use, the inductors become impractically large – of the order of several henries. But that still leaves VHF and, in particular, microwaves.

Coilcraft have a downloadable package that designs elliptical filters up to seven poles at https://www.coilcraft.com/appnotes.cfm where ('Coilcraft 7 Elliptic.asc') is an example. And https://www-users.cs.york.ac.uk/~fisher/cgi-bin/lcfilter has an interactive design for low-pass and high-pass filters up to the 10th order. In 'Microwaves & RF' reference is made to commercial design packages, https://www.mwrf.com/software/complete-filter-design-discrete-elements-made-easy. Also of considerable interest is the excellent publication by Analog Devices dealing thoroughly with the theory (https://www.analog.com/media/en/training-seminars/design-handbooks/Basic-Linear-Design/Chapter8.pdf), but there the implementation is largely through operational amplifiers.

15.6.1 Design

The ideal 'brick wall' analogue filter does not exist. It is a filter that offers no attenuation in the pass band, no ripple, and infinite attenuation elsewhere, and an instant transition from one to the other, Fig. 15.47a where a low-pass filter is shown with an abrupt transition at ω_c.

In real life we have to accept inferior performance. The points to be established in the design are:

- The corner or 3 dB frequency ωc.
- The slope above the pass band which determines the *order* of the filter. For this we specify a frequency in the stop band ω_p, Fig. 15.47b, and the required attenuation p.
- The amount of ripple in the pass band.
- The amount of ripple in the stop band.
- The impedance, often 600 Ω, 75 Ω or 50 Ω.

These considerations will determine the form of the filter and the number of elements. We may also choose to have the filter start with a series element or one in parallel, Fig. 15.48, where Z_0 is the characteristic impedance of the filter.

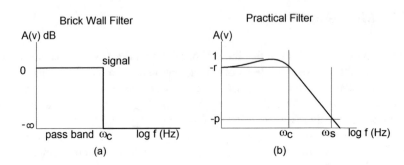

Fig. 15.47 Brick wall filter

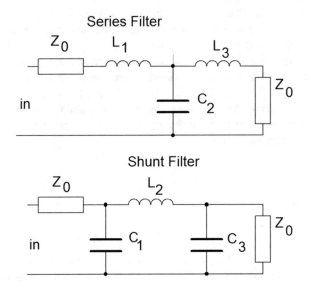

Fig. 15.48 Filter configurations

The more adventurous may turn to http://164.100.133.129:81/econtent/Uploads/ Session%209.pdf which has normalized values for low-pass filters. These can be de-normalized to get component values:

$$C = \frac{C_n}{2\pi f Z_0} \quad L = \frac{L_n Z_0}{2\pi f} \tag{15.193}$$

where C_n and L_n are the normalized values, f is the 3 dB frequency and Z_0 is the impedance. It is heartening to know that these agree with numbers obtained from the previous website.

For a high-pass filter, we design a low pass one, then interchange capacitors and inductors and use transformation formulae given below. Band-pass and band-stop filters take a little more work, also given below in Sect. 6.5.

Additionally we are interested in the *group delay* or the time for the signal to propagate from input to output. This is largely a matter 'how it is' rather than an essential design consideration. It is found from:

$$GD = \frac{d(phase\ angle)}{d(frequency)}$$

which implies a constant group delay if the change of phase is linear with frequency. It will be seen that we can start the filter with a parallel component or one in series. There is no difference in the response of the filter, only the component values.

An extended discussion of these and other matters (such as how to design high-pass and band-pass filters starting from a low-pass design) can be found at https://www.hit.ac.il/.upload/engineering/experiment_5_-_filter_design109.pdf which also has normalized values, whilst for those that want to go the whole hog, http://www.ece.uah.edu/courses/ee426/Chebyshev.pdf expounds everything.

15.6.2 Butterworth Low-Pass Filter

This is a filter of considerable antiquity. The response in the pass band is flat. The roll-off is at the rate of *20N dB/decade* where N is the order of the filter. We determine N by the required attenuation or *insertion loss IL*:

$$IL = 10\log\left(1 + \left(\frac{\omega}{\omega_c}\right)^{2N}\right) \qquad (15.194)$$

and if $\omega = \omega_c$ we have the 3 dB point but note that this is relative to the pass band loss. Then if we specify the insertion loss at some other point ω_p, we can find N.

Example – Low-Pass Butterworth Filter

Suppose we want a low-pass filter with a 3 dB point at 25 Hz and 60 dB attenuation at 125 Hz. Substituting in Eq. 15.194 gives us:

$$60 = 10\log\left(1 + 5^{2N}\right) \approx 10\log\left(5^{2N}\right)$$

so:

$$6 = \log\left(5^{2N}\right)$$

using $\log(a^r) = r\log(a)$ we have:

$$6 = 2N \log(5) \quad 3 = N \log(5)$$

and we find $N = 4.3$ so the next integer is $N = 5$. The schematic is ('25 Butter 5 Series LP.asc') where we correctly find that the loss at 25 Hz is -9 dB and at 125 Hz it is nearly -76 dB rather than -60 dB but that is a result of having to take an integer for N.

Explorations 8
1. Open schematic)'25 Butter 5 Series LP.asc') and measure the 3 dB point and the attenuation. Right click the right-hand vertical axis showing the phase in degrees and select in the exit box **Select Vertical Axis** click **Group Delay**. And note that it is constant where the phase changes linearly up to about 10 Hz and then shows dramatic changes during the transition to a steady slope at about 300 Hz and then is constant since the phase shift with frequency is very small.
2. Note that the voltages at intermediate points are not always flat, but can show a peak.
3. Open ('100 Hz Butter Shunt.asc'). And compare with ('100 Hz Butter Series. asc') They are identical.
4. Explore the effects of an incorrect terminating resistor.
5. Try a pulse input of 10 V 20 ms and note that again the response of both is identical and there is a slight overshoot both on turn on and turn off.

15.6.3   Chebyshev Low-Pass Filter

This achieves steeper attenuation in the transition region at the cost of some ripple in the pass band; greater ripple means faster roll-off. We quote the result that the insertion loss in dBs is:

$$IL = 10 \log \left(1 + a^2 T_N^2 \left(\frac{\omega}{\omega_c} \right) \right) \tag{15.195}$$

where a is a constant soon to be determined and T_N is the *Nth* order Chebyshev polynomial which is:

$$T_N = \cos \left[N \cos^{-1} \left(\frac{\omega}{\omega_c} \right) \right] \qquad \omega \le \omega r \tag{15.196}$$

$$T_N = \cos \left[N \cosh^{-1} \left(\frac{\omega}{\omega_c} \right) \right] \qquad \omega > \omega r \tag{15.197}$$

The first two are $T_0 = 1$ $T_1 = \omega$, and from here on they can be generated by:

$$T_{N+1} = 2\omega T_N - T_{N-1} \tag{15.198}$$

Knowing the insertion loss, the transfer function H which is *input/output* is:

$$H = \frac{1}{\sqrt{\left(1 + a^2 T_N^2 \left(\frac{\omega}{\omega_c}\right)\right)}} \tag{15.199}$$

and hence the attenuation A_v is:

$$A_v = \sqrt{\left(1 + a^2 T_N^2 \left(\frac{\omega}{\omega_c}\right)\right)} \tag{15.200}$$

A special case is $\omega = \omega_c$ when we have $T(1) = 1$ so:

$$H = \frac{1}{\sqrt{(1 + a^2)}} \tag{15.201}$$

And from Fig. 15.47b we have $H = 1 - r$ where r is the ripple, so:

$$1 - r = \frac{1}{\sqrt{(1 + a^2)}} \tag{15.202}$$

and in dBs this is:

$$r = 10 \log \left(1 + a^2\right) \tag{15.203}$$

from which we determine the constant a:

$$a = \sqrt{\left(10^{0.1r} - 1\right)} \tag{15.204}$$

Unfortunately, the order depends not only on the slope but also on the ripple and is found from:

$$N = \frac{\cosh^{-1} \sqrt{\frac{(A_v^2 - 1)}{a^2}}}{\cosh^{-1} \left(\frac{\omega_p}{\omega_c}\right)} \tag{15.205}$$

but in most cases to a very good approximation, we have:

$$N = \frac{\cosh^{-1}\left(\frac{A}{a}\right)}{\cosh^{-1}\left(\frac{\omega_p}{\omega_s}\right)} \tag{15.206}$$

The starting point for the design is the permissible ripple r which use we find a, and from the required attenuation A_v at ω_p, we can find N and take the next higher integer.

There are tables in the above websites for 0.5 dB ripple and 3 dB ripple. However, https://www-users.cs.york.ac.uk/~fisher/lcfilter/ has options of 0.01, 0.1, 0.2 1.0 and 3.0 dB ripple and does the complete design.

Example – Chebyshev Low-Pass Filter
Suppose we want a 3 dB point at 10 MHz, 3 dB ripple and an attenuation of at least 100 dB by 100 MHz. First we find $a = \sqrt{\left(10^{0.3} - 1\right)}$ $a = 0.998$, whence $N = 12.2/3$, and thus we use $N = 5$. This is schematic ('10 MHz Cheb 5 Series 3 dB.asc').

We might note in passing that this is a Chebyshev type I filter – there is also Chebyshev type II which has zeroes as well as poles. But we will pass quickly over that.

Explorations 10
1. Run the various Chebyshev schematics and note the ripple, pass-band loss and group delay .*meas* directives make this easier.
2. Explore the effects of changing the terminating impedance.

15.6.4   Elliptic Filters

These can have ripple in both pass and stop bands and an even steeper roll-off. However, https://www.hit.ac.il/.upload/engineering/experiment_5_-_filter_design109.pdf gives some of the normalized parameters but no equations. If we take the design of a 10 MHz low-pass 3rd-order elliptic filter with a minimum stop band attenuation of 26.53 dB and impedance 50 Ω, they give the parameters $C1 = C3 = 1.0512$ $C2 = 0.2019$ $L2 = 0.9612$, so using Eq. 5.2.1.1, we find the values in schematic ('10 MHz Elliptic LP.asc') where the salient feature is the very sharp notch at 22.7 MHz and a constant group delay of zero apart from the notch.

15.6.5   Transformation of Low Pass to High Pass, Band Pass and Band Stop

This is quite straightforward. We will illustrate it by a Butterworth filter with the same specifications as the Chebyshev above, but, of course, without ripple.

Example – Low-Pass to High-Pass Transformation

From Eq. 15.195 for the low-pass filter, we find $N = 5$. The Session%209.pdf of Dr. Varan gives the normalized parameters as $g1 = 0.618$ $g2 = 1.618$ $g3 = 2.00$ $g4 = 1.618$ and $g5 = 0.618$. As the first element in a series filter is an inductor, we have $L1 = L3 = (0.618 \times 50)/(2 \times \pi \times 10M) = 0.492 \ \mu H$ and then $L2 = (2.000 \times 50)/(2 \times \pi \times 10M = 159 \ \mu H$. And $C1 = C2 = 1.618/(2 \times \pi \times 10M \times 50) = 515 \ pF$. And this is schematic ('10 MHz Butter 5 Series.asc'.

Now we will turn it into a high-pass filter. First we interchange the capacitors and inductors. Next we use the conversion formulae:

$$C_n = \frac{1}{\omega_c g_n Z_o} \quad L_n = \frac{Z_o}{\omega_c g_n} \tag{15.207}$$

where g_n is the coefficient from above and Z_0 is the impedance. Thus we have $C1 = C3 = 1/(2 \times \pi \times 10M \times 0.618 \times 50) = 515 \ pF$ $C2 = 1/(2 \times \pi \times 10M \times 2 \times 50) = 159 \ pF$ $L1 = L2 = 50/(2 \times \pi \times 10M \times 1.618) = 0.492 \ \mu H$

Band-pass filters are a little more tricky: every series inductor becomes an inductor plus a capacitor in series, and every shunt capacitor has an inductor in parallel, Fig .15.49, where g_n and Z_0 are as before and ω_h, ω_l are the higher and lower 3 dB points. The filter thus has twice as many poles as a low pass or high pass.

The series LC combination replacing the simple inductor is:

$$C_s = \frac{(\omega_h - \omega_l)}{\omega_h \omega_l g_n Z_0} \quad L_s = \frac{g_n Z_0}{\omega_h - \omega_l} \tag{15.208}$$

The parallel LC combination replacing the original capacitor is:

$$C_p = \frac{g_n}{(\omega_h - \omega_l)Z_0} \quad L_p = \frac{(\omega_h - \omega_l)Z_0}{\omega_h \omega_l g_n} \tag{15.209}$$

Example – Low-Pass to Band-Pass Transformation

As an example, we shall convert the low-pass Butterworth 10 MHz filter into a band-pass one with a bandwidth of 2 MHz.

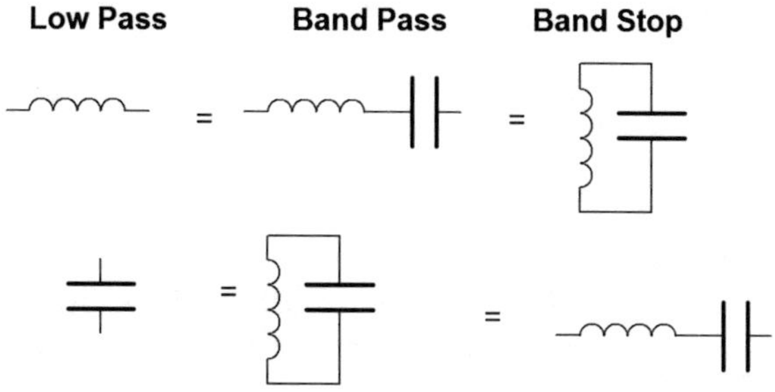

Fig. 15.49 LP to BP transformation

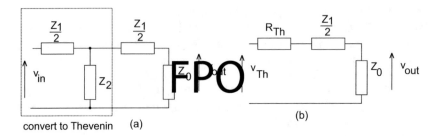

convert to Thevenin   (a)

Fig. 15.50   Constant K T atten

If we take the first inductor $L1$ we use $g1$ and Eq. 15.208 to replace it by an inductor and a capacitor in series where $L1 = (0.618 \times 50)/(2 \times \pi \times 2$ $M) = 2.459\ \mu H$. Again using $g1$ the series capacitor is $C1 = 2M/(2 \times \pi \times 120\ M \times 0.618 \times 50) = 85.84\ pF$.

Taking the first capacitor $C1$ of the low-pass filter, we replace it by a capacitor and inductor in parallel. We use $g2$ with Equation 15.209 to find $C2 = 1.618/(2 \times \pi \times 2\ M \times 50) = 2.575\ nF$. And still with $g2$ we find $L2 = 2 \times 50/(2 \times \pi \times 120\ M \times 1.618) = 81.97\ nH$.

That leaves only $C3$ and $L3$ to find. These use $g3$ so $C3 = 2\ M/(2 \times \pi \times 120\ M \times 2 \times 50) = 26.52\ pF$ (which is a slight difference from the 26.31 pF from York) and $L3 = 2 \times 50/(2 \times \pi \times 2) = 79.6\ \mu H$. This is schematic ('10 MHz BP BW 2 MHz.asc').

A band-stop filter exchanges the band-pass parallel combinations and series combinations, Fig. 15.49. Parameters for 6-pole and 10-pole Butterworth and Chebyshev filters can be found at http://www.wa4dsy.net/cgi-bin/lc_filter4? FilterResponse=Bandstop&poles=10&CF=11&cfunits=MHZ&cutoff=2&funits= MHZ&Z=50. It will be seen that a 10 MHz stop band filter is not a reflection of the pass band but is more of a notch filter with less steep sides (Fig. 15.50).

Following Varun, the series inductor is transformed into a series circuit consisting of an inductor and capacitor in parallel:

$$C_s = \frac{1}{(\omega_h - \omega_l)g_n Z_0} \qquad L_s = \frac{(\omega_h - \omega_l)g_n Z_0}{\omega_h \omega_l} \qquad (15.210)$$

and the shunt capacitor is transformed into a shunt circuit consisting of an inductor and capacitor in series:

$$C_s = \frac{(\omega_h - \omega_l)g_n}{(\omega_h \omega_l)Z_0} \qquad L_s = \frac{Z_0}{(\omega_h - \omega_l)g_n} \qquad (15.211)$$

Example – Pass-Band Filter
An amazing filter with a pass band from 16 kHz to 19.5 kHz can be found at **LTSpiceXVI-→examples->Education** named ('passive.asc'). It is clearly not

Fig. 15.51 First-order LP

Butterworth because there is ripple in the pass band, and the component values are symmetrical about the mid-point, *C6*.

15.6.6 All-Pass Filters

Some of these may only shift the phase of the various frequencies and are used in telecommunications to correct the dispersion caused by the unequal group delay in filters and therefore need only work over a relatively small band of frequencies; see the schematic ('All Pass Bridged-T.asc'), and – like an RC Bridged-T – it has a Q-factor. On the other hand, they can be used to deliberately introduce delays, often in audio work. There are various configurations (see the Wikipedia article) and are mainly built with operational amplifiers to avoid using inductors. They can also shape the frequency response in such as the 'All Pass Crossovers' for audio systems discussed in http://www.linkwitzlab.com/crossovers.htm mentioned earlier (Fig. 15.51).

The schematic ('All pass Bridged-T.asc') is from a paper by Dr. Bose and is intended for use in VHF communications. Referring to the schematic, the component values are:

$$C_1 = C_2 = \frac{Q}{\omega_c Z_0} \quad C_3 = \frac{2Q}{\omega(Q^2 - 1)Z_0}$$

$$L_1 = \frac{QZ_0}{2\omega_c} \quad L_2 = \frac{2Z_0}{\omega_c Q}$$

Two interesting examples have been posted by Dominique Szymik at **LTspiceXVII->examples-→Educational->contrib**. The first is ('gr_del.asc') which explores three bridged-T configurations; the second is ('elip_grd.asc') where the left-hand side uses the same structure as Dr. Bose.

Explorations 11
1. There are various band-pass and band-stop schematics. Test them, note the phase and group delays. Calculate the component values, and they should agree with the schematics, although there might be one or two slight differences.
2. Run ('All Pass Bridged-T.asc') and note the attenuation and phase shift. Recalculate the component values for a centre frequency of 100 MHz and $Q = 5$.

15.6.7   Sensitivity

This is a matter of considerable importance – how the response changes with component values. Components values are never exact, even ignoring temperature effects. A sensitivity analysis tests the effects of changing each component in turn using its maximum and minimum values (which, of course, presumes that they are toleranced) and reports the relative importance of the changes to each component. This is a useful guide for the filter designer on how to assign component tolerances so that all – or an acceptable percentage – will meet specifications. LTspice, however, only provides a Monte Carlo analysis where it needs an external programme to productively process the results.

15.7   Summary

In this concluding chapter, we have extended the analyses of filters begun in previous chapters aided by Nyquist and pole-zero plots. The salient points are:

- A Nyquist plot shows real and imaginary parts of the response, but the LTspice cursor also shows phase and group delay.
- A pole-zero plot visualizes the positions of poles and zeros on a graph with a horizontal axis of σ and a vertical axis of $j\omega$.
- Cascaded CR and LR sections create filters with $Q = 0.5$: low pass and high pass have *40 dB/decade* slopes, band pass only *−20 dB/decade*.
- LCR filters are series or parallel tuned circuits with the output taken from different points giving low-, high- and band-pass filters with resonant peaks if $\zeta < 1$.
- LCR filters also can have an infinitely deep notch response.
- All LCR filters have the same denominator.
- LC filters are based on 'T' or 'Π' prototypes characterized by the *image imped-ance* and the same *propagation constant* as transmission lines.
- *Constant K sections* can be cascaded to make high-, low- and band-pass and band-stop filters from 'T.' or 'Π' sections
- *M-derived filters* are derived from the previous K adding more components to achieve faster transitions.
- Double-tuned circuits have a narrow pass band wider than single-tuned but often with an intervening dip.
- Multi-stage filters Butterworth (no ripple in pass band), Chebyshev (ripple in pass-band, steeper roll-off) and Elliptic (ripple in pass- and stop-bands and even steeper roll-off) are best designed from published tables.
- Low-, high-, and band-pass and band-stop filters can be inter-converted.
- All-pass filters have no attenuation but correct the delay or phase.
- Sensitivity to component tolerances is handled only by a Monte Carlo analysis.

Index

© Springer Nature Switzerland AG 2020
C. May, *Passive Circuit Analysis with LTspice*®,
https://doi.org/10.1007/978-3-030-38304-6

Printed in the United States
by Baker & Taylor Publisher Services